Pocket Companion to
Textbook of
Medical Physiology

GUYTON
HALL

Pocket Companion to
Textbook of Medical Physiology

Arthur C. Guyton, M.D.
Professor Emeritus
Department of Physiology and Biophysics
University of Mississippi Medical Center
Jackson, Mississippi

John E. Hall, Ph.D.
Professor and Chairman
Department of Physiology and Biophysics
University of Mississippi Medical Center
Jackson, Mississippi

W.B. SAUNDERS COMPANY
A Division of Harcourt Brace & Company
Philadelphia   London   Toronto   Montreal   Sydney   Tokyo

W.B. SAUNDERS COMPANY
A Division of Harcourt Brace & Company

The Curtis Center
Independence Square West
Philadelphia, Pennsylvania 19106

Library of Congress Cataloging-in-Publication Data

Guyton, Arthur C.
Pocket companion to textbook of medical physiology / Arthur
C. Guyton, John E. Hall.—1st ed.

p. cm.

ISBN 0–7216–7118–7

1. Human physiology—Handbooks, manuals, etc.
  2. Physiology, Pathological—Handbooks, manuals,
  etc.   I. Hall, John E. (John Edward)   II. Title.
  [DNLM: 1. Physiology.   QT 104 G992t 1998 suppl.]

QP35.G89   1998   612—dc21

DNLM/DLC   97–7746

POCKET COMPANION TO TEXTBOOK
OF MEDICAL PHYSIOLOGY   ISBN 0–7216–7118–7

Printed in the United States of America

Last digit is the print number:
9   8   7   6   5   4   3   2   1

Contributors

Thomas H. Adair, Ph.D.
Professor of Physiology and Biophysics
University of Mississippi Medical Center
Jackson, Mississippi

David J. Dzielak, Ph.D.
Associate Professor of Surgery
University of Mississippi Medical Center
Jackson, Mississippi

Arthur C. Guyton, M.D.
Professor Emeritus of Physiology and
    Biophysics
University of Mississippi Medical Center
Jackson, Mississippi

John E. Hall, Ph.D.
Professsor and Chairman of Physiology and
    Biophysics
University of Mississippi Medical Center
Jackson, Mississippi

Thomas E. Lohmeier, Ph.D.
Professor of Physiology and Biophysics
University of Mississippi Medical Center
Jackson, Mississippi

R. Davis Manning, Ph.D.
Professor of Physiology and Biophysics
University of Mississippi Medical Center
Jackson, Mississippi

Gregory Mihailoff, Ph.D.
Arizona College of Osteopathic Medicine
Glendale, Arizona

David B. Young, Ph.D.
Professor of Physiology and Biophysics
University of Mississippi Medical Center
Jackson, Mississippi

Preface

Human physiology is the discipline that links basic science with clinical medicine. It is integrative and encompasses everything from the study of molecules and subcellular components to the study of organ systems and their interactions, which allow us to function as living beings. Because human physiology is a rapidly expanding discipline and covers a broad scope, the vast amount of information that is potentially applicable for the practice of medicine can be overwhelming. One of our goals in writing a "pocket companion" was to distill this enormous amount of information into a book that would be small enough to be carried in a coat pocket and used frequently, but still contain the basic physiologic principles necessary for the study of medicine.

The pocket companion was designed specifically to accompany Guyton and Hall's *Textbook of Medical Physiology, 9th Edition,* and it cannot serve as a substitute for the parent text. Rather, it is intended to serve as a concise overview of the most important facts and concepts from the parent text, presented in a manner that facilitates rapid comprehension of basic physiologic principles. Some of the most important features of the pocket companion are as follows:

- It has been designed to serve as a guide for students who wish to rapidly and efficiently review a large volume of material from the parent text. The headings of the different sections state succinctly the primary concept in the accompanying paragraphs. Thus, the student can quickly review many of the

main concepts in the textbook by first studying the paragraph headings.

- The table of contents matches that of the
  parent text, and each topic has been cross
  referenced with specific page numbers from
  the parent text.
- The size of the book has been restricted so
  that it will fit conveniently in a coat pocket
  as an immediate source of information when
  needed.

Although the pocket companion contains the
most important facts necessary for studying
physiology, it does not contain the details that
enrich the physiologic concepts or the clinical
examples of abnormal physiology that are contained in the parent book. We therefore recom-
mend that the pocket companion be used in
conjunction with the *Textbook of Medical Physiology, 9th Edition.*

The contributors to the pocket companion
were selected for their knowledge of physiology and their ability to effectively present information to students. I am grateful to each of
the contributors for their careful work on this
pocket companion, especially Dr. Arthur C.
Guyton who was the sole author of the first
eight editions of the *Textbook of Medical Physiology* and very kindly allowed me to help with
the 9th Edition. The pocket companion bor-
rows heavily from the *Textbook of Medical Physiology, 9th Edition,* but I accept sole responsibil-
ity for editing the pocket companion. We have
strived to make this book as accurate as possible and hope that it will be valuable for your
study of physiology. We look forward to your
comments and suggestions.

John E. Hall, Ph.D.

UNIT VI   Blood Cells, Immunity, and Blood Clotting

UNIT VII   Respiration

Introduction to Physiology: The Cell and General Physiology

1

Functional Organization of the Human Body and Control of the "Internal Environment"

The goal of physiology is to understand the *function* of living organisms and their parts. In human physiology, we are concerned with the characteristics of the human body that allow us to sense our environment, move about, think and communicate, reproduce, and perform all of the functions that enable us to survive and thrive as living beings.

Human physiology is a very broad subject that includes the functions of molecules and subcellular components; tissues; organs; and organ systems, such as the cardiovascular system; and the interaction and communication between these different components. A distinguishing feature of physiology is that it seeks to integrate the functions of all of the different parts of the body to understand the function of the entire human body. Life in the human being relies on this total function, which is considerably more complex than the sum of the functions of the individual cells, tissues, and organs.

Cells are the living units of the body. Each organ is an aggregate of many cells held together by intercellular supporting structures. The entire body contains about 75 to 100 trillion cells, each of which is adapted to perform special functions. These individual cell functions are coordinated by multiple regulatory systems operating in cells, tissues, organs, and organ systems.

Although the many cells of the body differ from each other in their special functions, all of them have certain basic characteristics. For example, (1) oxygen combines with breakdown products of fat, carbohydrate, or protein to release energy that is required for the normal function of the cells; (2) most cells have the ability to reproduce, and when-

3

ever cells are destroyed, the remaining cells often regenerate new cells until the appropriate number is restored; and (3) cells are all bathed in extracellular fluid, the constituents of which are very precisely controlled.

"HOMEOSTATIC" MECHANISMS OF THE MAJOR FUNCTIONAL SYSTEMS (p. 4)

Essentially all of the organs and tissues of the body perform functions that help to maintain the constituents of the extracellular fluid relatively constant, a condition called *homeostasis*. Much of our discussion of physiology focuses on the mechanisms by which the cells, tissues, and organs contribute to homeostasis.

Extracellular Fluid Transport System—The Circulatory System

Extracellular fluid is transported throughout the body in two stages. The first stage is movement of blood around the *circulatory system*, and the second stage is movement of fluid between the blood capillaries and cells. The circulatory system keeps the fluids of the internal environment continuously mixed by pumping blood through the vascular system. As blood passes through the capillaries, a large portion of its fluid diffuses back and forth into the interstitial fluid that lies between the cells, allowing continuous exchange of substances between the cells and the interstitial fluid and between the interstitial fluid and the blood.

Origin of Nutrients in the Extracellular Fluid

- The *respiratory system* provides oxygen for the body and removes carbon dioxide.
- The *gastrointestinal system* digests food and absorbs different nutrients, including carbohydrates, fatty acids, and amino acids, into the extracellular fluid.
- The *liver* changes the chemical composition of many of the absorbed substances to more usable forms, and other tissues of the body (e.g., fat cells, kidneys, and endocrine glands) help to modify the absorbed substances or store them until they are needed.

- The *musculoskeletal system* consists of skeletal muscles, bones, tendons, joints, cartilage, and ligaments. Without this system, the body could not move to the appropriate place to obtain the foods required for nutrition. This system also provides protection of internal organs and support of the body.

Removal of Metabolic End Products

- The *kidneys* regulate the extracellular fluid composition by controlling the excretion of salts, water, and waste products of the chemical reactions of the cells. By controlling body fluid volumes and compositions, the kidneys also regulate blood volume and blood pressure.
- The *respiratory system* not only provides oxygen to the extracellular fluid but also removes carbon dioxide, which is produced by the cells, released from the blood into the alveoli, and then released to the external environment.

Regulation of Body Functions

- The *nervous system* directs the activity of the muscular system, thereby providing locomotion. It also controls the function of many internal organs through the autonomic nervous system and allows us to sense our external and internal environment and to be intelligent beings so that we can obtain the most advantageous conditions for survival.
- The *endocrine glands* secrete *hormones* that control many of the metabolic functions of the cells, such as growth, rate of metabolism, and special activities associated with reproduction. Hormones are secreted into the blood stream and carried to all parts of the body to help regulate cell function.
- The *immune system* also acts as a regulatory system by providing the body with a defense mechanism that protects against foreign invaders, such as bacteria and viruses, to which the body is daily exposed.
- The *integumentary system*, which is composed mainly of skin, provides protection against injury and defense against foreign invaders as well as protection of underlying tissue against dehydration. In addition, the skin serves as an important means of regulating the body temperature.

Reproduction

The *reproductive system* provides for formation of new beings like ourselves; even this can be considered a homeostatic function because it generates new bodies in which trillions of additional cells can exist in a well-regulated internal environment.

CONTROL SYSTEMS OF THE BODY (p. 6)

The human body has thousands of control systems that are essential to homeostasis. For example, genetic systems operate in all cells to control intracellular as well as extracellular functions. Other control systems operate within the organs or throughout the entire body to control interaction among the organs.

Regulation of oxygen and carbon dioxide concentrations in the extracellular fluid is a good example of multiple control systems that are operating together. In this instance, the respiratory system operates in association with the nervous system. When the carbon dioxide concentration in the blood increases above normal, the respiratory center is excited, causing the person to breathe rapidly and deeply. This increases the expiration of carbon dioxide and therefore removes it from the blood and the extracellular fluid until the concentration returns to normal.

Normal Ranges of Important Extracellular Fluid Constituents

Table 1–1 shows the more important constituents of extracellular fluid along with their normal values, normal ranges, and maximum limits that can be endured for short periods of time without the occurrence of death. Note the narrowness of the ranges; levels outside these ranges are usually the cause or result of illnesses.

Characteristics of Control Systems

Most control systems of the body operate by *negative feedback*. In the regulation of carbon dioxide concentration as discussed, a high concentration of carbon dioxide in the extracellular fluid increases pulmonary ventilation, which decreases the carbon dioxide concentration toward normal levels. This is an example of *negative feedback*; any stimulus that

TABLE 1-1 SOME IMPORTANT CONSTITUENTS AND PHYSICAL CHARACTERISTICS OF THE EXTRACELLULAR FLUID, NORMAL RANGE OF CONTROL, AND APPROXIMATE NONLETHAL LIMITS FOR SHORT PERIODS

	Units	Normal Value	Normal Range	Approximate Nonlethal Limits
Oxygen	mm Hg	40	35–45	10–1000
Carbon dioxide	mm Hg	40	35–45	5–80
Sodium ion	mmol/L	142	138–146	115–175
Potassium ion	mmol/L	4.2	3.8–5.0	1.5–9.0
Calcium ion	mmol/L	1.2	1.0–1.4	0.5–2.0
Chloride ion	mmol/L	108	103–112	70–130
Bicarbonate ion	mmol/L	28	24–32	8–45
Glucose	mmol/L	85	75–95	20–1500
Body temperature	°F (°C)	98.4 (37.0)	98–98.8 (37.0)	65–110 (18.3–43.3)
Acid-base	pH	7.4	7.3–7.5	6.9–8.0

attempts to cause a change in carbon dioxide concentration is counteracted by a response that is *negative* to the initiating stimulus.

The degree of effectiveness with which a control system maintains constant conditions is determined by the *gain* of the negative feedback. The gain is calculated according to the following formula:

$$Gain = Correction \ / \ Error$$

Some control systems, such as those that regulate body temperature, have feedback gains as high as -33, which simply means that the degree of correction is 33 times greater than the remaining error.

Feed-forward control systems anticipate changes. Because of the many interconnections between different control systems, the total control of a particular body function may be more complex than can be accounted for by simple negative feedback. For example, some movements of the body occur so rapidly that there is not sufficient time for nerve signals to travel from some of the peripheral body parts to the brain and then back to the periphery in time to control the movements. Therefore, the brain uses feed-forward control to cause the required muscle contractions. Sensory nerve signals from the moving parts apprise the brain in retrospect of whether the appropriate movement, as envisaged by the brain, has actually been performed correctly. If it has not, the brain corrects the feed-forward signals that it sends to the muscles the next time the movement is required. This is also called *adaptive control*, which is, in a sense, a delayed negative feedback.

AUTOMATICITY OF THE BODY (p. 9)

The body is a social order of about 75 to 100 trillion cells that are organized into different functional structures, the most important of which are called *organs*. Each functional structure or organ has a role in the maintenance of a constant internal environment. As long as homeostasis is maintained, the cells of the body continue to live and function properly. Thus, each cell benefits from homeostasis, and, in turn, each cell contributes its share toward the maintenance of homeostasis. This reciprocal interplay provides continuous *automaticity* of the body until one or more functional systems lose their ability to contribute their share of function. When this loss happens, all cells of the body suffer. Extreme dysfunction leads to death, whereas moderate dysfunction leads to sickness.

2

The Cell and Its Function

ORGANIZATION OF THE CELL (p. 11)

Figure 2–1 shows a typical cell, including the *nucleus* and *cytoplasm*, which are separated by the *nuclear membrane*. The cytoplasm is separated from *interstitial fluid* that surrounds the cell by a *cell membrane*. The different substances that make up the cell are collectively called *protoplasm*, which is composed mainly of the following:

- *Water* comprises 70 to 85 per cent of most cells.
- *Electrolytes* provide inorganic chemicals for cellular reactions. Some of the most important electrolytes in the cell are *potassium, magnesium, phosphate, sulfate, bicarbonate,* and small quantities of *sodium, chloride,* and *calcium*.
- *Proteins* normally constitute 10 to 20 per cent of the cell mass. These can be divided into two types: *structural proteins* and *globular proteins* (which are mainly *enzymes*).
- *Lipids* constitute about 2 per cent of the total cell mass. Among the most important lipids in the cells are *phospholipids, cholesterol, triglycerides,* and *neutral fats*. In *adipocytes* (fat cells), triglycerides may account for as much as 95 per cent of the cell mass.
- *Carbohydrates* play a major role in nutrition of the cell. Most human cells do not store large amounts of carbohydrates, which usually average about 1 per cent of the total cell mass but may be as high as 3 per cent in muscle cells and 6 per cent in liver cells. The small amount of carbohydrates in the cells is usually stored in the form of *glycogen*, an insoluble polymer of glucose.

PHYSICAL STRUCTURE OF THE CELL (p. 12)

The cell (Fig. 2–2) is not merely a bag of fluid and chemicals; it also contains highly organized physical

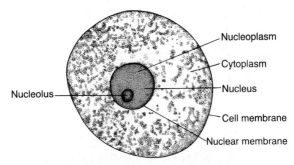

Figure 2-1 Structure of the cell as seen with the light microscope.

structures called *organelles.* Some of the principal organelles of the cell are the *cell membrane, nuclear membrane, endoplasmic reticulum* (ER), *Golgi apparatus, mitochondria, lysosomes,* and *centrioles.*

The cell and its organelles are surrounded by membranes composed of lipids and proteins. These membranes include the *cell membrane, nuclear membrane,* and *membranes of the ER, mitochondria, lysosomes,* and *Golgi apparatus.* They provide barriers that prevent free movement of water and water-soluble substances from one cell compartment to another. Protein molecules in the membrane often penetrate through the membrane providing specialized pathways to allow movement of specific substances through the membranes.

The cell membrane is a lipid bilayer with inserted proteins. The lipid bilayer is composed almost entirely of *phospholipids* and *cholesterol.* Phospholipids have a water-soluble portion *(hydrophilic)* and a portion that is soluble only in fats *(hydrophobic).* The hydrophobic portions of the phospholipids face each other, whereas the hydrophilic parts face the two surfaces of the membrane in contact with the surrounding interstitial fluid.

This lipid bilayer membrane is highly permeable to lipid-soluble substances, such as oxygen, carbon dioxide, and alcohol, but it acts as a major barrier to water-soluble substances, such as ions and glucose. Floating in the lipid bilayer are proteins, most of which are *glycoproteins* (proteins combined with carbohydrates).

There are two types of membrane proteins: the

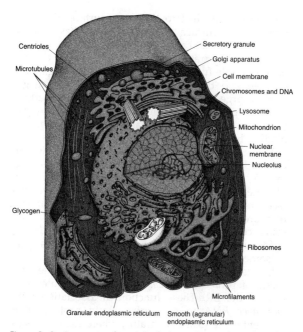

Figure 2–2 Reconstruction of a typical cell, showing the internal organelles in the cytoplasm and in the nucleus.

integral proteins, which protrude through the membrane, and the *peripheral proteins,* which are attached to the inner surface of the membrane and do not penetrate. Many of the integral proteins provide *structural channels* (pores) through which water-soluble substances, especially ions, can diffuse. Other integral proteins act as *carrier proteins* for the transport of substances, sometimes against their natural gradient for diffusion. The peripheral proteins are normally attached to one of the integral proteins and usually function as *enzymes* that catalyze chemical reactions of the cell.

The membrane carbohydrates occur mainly in combination with proteins and lipids in the form of *glycoproteins* and *glycolipids.* The "glyco" portions of these molecules usually protrude to the outside of the cell. Many other carbohydrate compounds, called *proteoglycans,* which are mainly carbohydrate substances bound together by small protein cores, are loosely attached to the outer surface; thus, the entire outer surface of the cell often has a loose carbohydrate coat called the *glycocalyx*.

The carbohydrates on the outer surface of the cell have multiple functions: (1) they are often negatively charged and therefore repel other molecules negatively charged; (2) the glycocalyx of cells may attach to other cells, thus the cells attach to each other; and (3) some of the carbohydrates act as *receptors* for binding hormones.

The *ER* synthesizes multiple substances in the cell. A large network of tubules and vesicles, called the *ER*, penetrates almost all parts of the cytoplasm. The membrane of the ER provides an extensive surface area for the manufacture of many substances that are used inside the cells and released from some cells. These include proteins, carbohydrates, lipids, and other structures such as lysosomes, peroxisomes, and secretory granules.

Lipids are made within the ER wall. For the synthesis of proteins, *ribosomes* attach to the outer surface of the ER. These function in association with *messenger RNA* to synthesize many proteins that then enter the Golgi apparatus, where the molecules are further modified before they are released or used in the cell.

The *Golgi apparatus* functions in association with the ER. The Golgi apparatus has membranes similar to those of the *agranular*, or *smooth*, ER. The Golgi apparatus is prominent in secretory cells and is located on the side of the cell from which the secretory substances are extruded. Small *transport vesicles*, also called *ER* vesicles, continually pinch off from the ER and then fuse with the Golgi apparatus. In this way, substances entrapped in the ER vesicles are transported from the ER to the Golgi apparatus. The substances are then processed in the Golgi apparatus to form lysosomes, secretory vesicles, and other cytoplasmic components.

Lysosomes **provide an intracellular digestive system.** Lysosomes, which are found in great numbers in many cells, are small spherical vesicles surrounded by a membrane that contains digestive enzymes; these enzymes allow lysosomes to break down intracellular substances in structures, especially damaged cell structures; food particles that have been ingested by the cell; and unwanted materials such as bacteria.

Usually, the membranes surrounding the lysosomes prevent the enclosed enzymes from coming in contact with other substances in the cell and therefore prevent their digestive action. When these membranes are damaged, however, the enzymes are released and split the organic substances with which they come in contact into highly diffusible substances such as amino acids and glucose.

Mitochondria **release energy in the cell.** An adequate supply of energy must be available to fuel the chemical reactions of the cell. This is provided mainly by the chemical reaction of oxygen with the three different types of foods: glucose derived from carbohydrates, fatty acids derived from fats, and amino acids derived from proteins. After entering the cell, the foods are split into smaller molecules that, in turn, enter the mitochondria, where other enzymes remove carbon dioxide and hydrogen ions in a process called the *citric acid cycle*. An oxidative enzyme system, which is also in the mitochondria, causes progressive oxidation of the hydrogen atoms. The end products of the reactions of the mitochondria are water and carbon dioxide. The energy liberated is used by the mitochondria to synthesize another substance, *adenosine triphosphate* (ATP), which is a highly reactive chemical that can diffuse through the cell to release its energy whenever it is needed for the performance of cell functions. Mitochondria are also self-replicative, which means that one mitochondrion can form a second one, a third one, and so on, whenever there is a need in the cell for increased amounts of ATP.

There are many other cytoplasmic structures and organelles. There are hundreds of different types of cells in the body, and each has a special structure. Some cells, for example, are rigid and have large numbers of *filamentous* or *tubular structures*, which are composed of *fibrillar proteins*. Some of the tubular structures, called *microtubules*, can transport substances from one area of the cell to another.

One of the important functions of many cells is to secrete special substances, such as digestive enzymes. Almost all of the substances are formed by the ER–Golgi apparatus system and are released into the cytoplasm inside storage vesicles called *secretory vesicles*. After a period of storage in the cell, these are expelled through the cell membrane to be used elsewhere in the body.

The *nucleus* acts as a control center of the cell and contains large amounts of *DNA*, also called *genes*. The genes determine the characteristics of the proteins of the cell, including the enzymes of the cytoplasm. They also control reproduction. They first reproduce themselves through a process of *mitosis* in which two daughter cells are formed, each of which receives one of the two sets of genes.

The *nuclear membrane*, also called the *nuclear envelope*, separates the nucleus from the cytoplasm. This structure is actually composed of two separate membranes; the outer membrane is continuous with the ER, and the space between the two nuclear membranes is also continuous with the compartment inside the ER. Both layers of the membrane are penetrated by several thousand *nuclear pores*, which are almost 100 nanometers in diameter.

The nuclei in most cells contain one or more structures called *nucleoli*, which unlike many of the organelles do not have a surrounding membrane. The nucleoli contain large amounts of RNA and proteins of the type found in ribosomes. A nucleolus becomes enlarged when the cell is actively synthesizing proteins. Ribosomal RNA is stored in the nucleolus and transported through the nuclear membrane pores to the cytoplasm, where it is used to produce mature ribosomes that play an important role in the formation of proteins.

FUNCTIONAL SYSTEMS OF THE CELL (p. 19)

Ingestion by the Cell—Endocytosis

The cell obtains nutrients and other substances from the surrounding fluid through the cell membrane via *diffusion* and *active transport*. Very large particles enter the cell via *endocytosis*, of which the principal forms are *pinocytosis* and *phagocytosis*.

- *Pinocytosis means the ingestion of small vesicles that contain extracellular fluid.* This is the only method by which large molecules, such as proteins, can enter the cells. These molecules usually attach to specialized receptors on the outer surface of the membrane that are concentrated in small pits called *coated pits*. On the inside of the cell membrane beneath these pits is a latticework of a fibrillar protein called *clathrin* and a contractile filament of *actin* and *myosin*. After the protein molecules bind with the receptors, the membrane invaginates and contractile proteins surround the pit, causing its borders to close over the attached proteins and form a *pinocytotic vesicle.*

- *Phagocytosis is the ingestion of large particles, such as bacteria, cells, and portions of degenerating tissue.* This ingestion occurs much in the same way as pinocytosis except that it involves large particles instead of molecules. Only certain cells have the ability to perform phagocytosis, notably the tissue *macrophages* and some of the *white blood cells.* Phagocytosis is initiated when proteins or large polysaccharides under the surface of the particle bind with receptors on the surface of the phagocyte. In the case of bacteria, these usually are attached to specific antibodies, and the antibodies in turn attach to the phagocyte receptors, dragging the bacteria along with them. This intermediation of antibodies is called *opsonization* and is discussed further in Chapters 33 and 34.

Pinocytic and phagocytic foreign substances are digested in the cell by the lysosomes. Almost as soon as a pinocytic or phagocytic vesicle appears inside a cell, lysosomes become attached to the vesicle and empty their digestive enzymes into the vesicle. Thus, a *digestive vesicle* is formed in which the enzymes begin hydrolyzing the proteins, carbohydrates, lipids, and other substances in the vesicle. The products of digestion are small molecules of amino acids, glucose, phosphate, and so on, that can diffuse through the membrane of the vesicle into the cytoplasm. The undigested substances, called the *residual body*, are excreted through the cell membrane via the process of *exocytosis*, which is basically the opposite of endocytosis.

Synthesis of Cellular Structures

The synthesis of most cell structures begins in the ER. Many of the products that are formed in the ER are then passed onto the Golgi apparatus, where they are further processed before release into the cytoplasm. The *granular ER* is characterized by large numbers of ribosomes attached to the outer surface and is the site of protein formation. Ribosomes synthesize the proteins and extrude many of them through the wall of the ER to the interior of the endoplasmic vesicles and tubules, called the *endoplasmic matrix.*

When protein molecules enter the ER, enzymes in the ER wall cause rapid changes, including congregation of carbohydrates to form *glycoproteins*. In addition, the proteins are often cross-linked, folded, and shortened to form more compact molecules.

The ER also synthesizes lipids, especially phospholipid and cholesterol, which are incorporated into the lipid bilayer of the ER. Small ER vesicles, or *transport vesicles*, continually break off from the smooth reticulum. Most of these migrate rapidly to the Golgi apparatus.

The *Golgi apparatus* processes substances formed in the ER. As substances are formed in the ER, especially proteins, they are transported through the reticulum tubules toward the portions of the smooth ER that lie nearest the Golgi apparatus. Small transport vesicles, composed of small envelopes of smooth ER, continually break away and diffuse to the deepest layer of the Golgi apparatus. The transport vesicles instantly fuse with the Golgi apparatus and empty their contents into the vesicular spaces of the Golgi apparatus. Here, more carbohydrates are added to the secretions, and the endoplasmic reticular secretions are compacted. As the secretions pass toward the outermost layers of the Golgi apparatus, the compaction and processing continue; finally, small and large vesicles break away from the Golgi apparatus, carrying with them the compacted secretory substances. These substances can then diffuse throughout the cell.

In a highly secretory cell, the vesicles formed by the Golgi apparatus are mainly *secretory vesicles*, which diffuse to the cell membrane, fuse with it, and eventually empty their substances to the exterior via the mechanism called *exocytosis*. Some of the vesicles made in the Golgi apparatus, however, are destined for intracellular use. For example, specialized portions of the Golgi apparatus form lysosomes.

Extraction of Energy from Nutrients by the Mitochondria

The principal substances from which the cells extract energy are oxygen and one or more of the foodstuffs that react with oxygen—carbohydrates, fats, and proteins. In the human body, almost all carbohydrates are converted into *glucose* by the digestive tract and liver before they reach the cell; similarly, proteins are converted into *amino acids*, and fats are converted into *fatty acids*. Inside the cell, these substances react chemically with oxygen under the influence of enzymes that control the rates of reaction and channel the energy that is released in the proper direction.

Almost all oxidated reactions occur inside the mitochondria, and the energy released is used to form mainly ATP. ATP is a nucleotide composed of the nitrogenous base *adenine*, the pentose sugar *ribose*, and *three phosphate radicals*. The last two phosphate radicals are connected with the remainder of the molecule by *high-energy phosphate bonds*, each of which contains about 12,000 calories of energy per mole of ATP under the usual conditions of the body. The high-energy phosphate bonds are labile so that they can be split instantly on demand whenever energy is required to promote other cellular reactions.

When ATP releases its energy, a phosphoric acid radical is split away, and *adenosine diphosphate (ADP)* is formed. Energy derived from cell nutrients causes the ADP and phosphoric acid to be recombined to form new ATP, with the entire process continuing over and over again.

The major portion of ATP produced in the cell is formed in the mitochondria. After entry into the cells, glucose is subjected to enzymes in the cytoplasm that convert it to *pyruvic acid*, a process called *glycolysis*. Less than 5 per cent of the ATP formed in the cell occurs via glycolysis.

The pyruvic acid that is derived from carbohydrates, the fatty acids that are derived from lipids, and the amino acids that are derived from proteins are all eventually converted into the compound *acetyl-coenzyme A* (acetyl-CoA) in the matrix of the mitochondrion. This substance is then acted on by another series of enzymes in a sequence of chemical reactions called the *citric acid cycle*, or *Krebs' cycle*.

In the citric acid cycle, acetyl-CoA is split into *hydrogen ions* and *carbon dioxide*. The hydrogen ions are very reactive and eventually combine with oxygen that has diffused into the mitochondria. This reaction releases a tremendous amount of energy, which is used to convert large amounts of ADP to ATP. This requires large numbers of protein enzymes that are integral parts of the mitochondria.

The initial event in the formation of ATP is removal of an electron from the hydrogen atom, thus converting it into a hydrogen ion. The terminal event is movement of the hydrogen ion through large globular proteins called *ATP synthetase,* which protrude through the membranes of the mitochondrial *membranous shelves* that protrude into the mitochondrial matrix. ATP synthetase is an enzyme that uses the energy and movement of the hydrogen ions to cause the conversion of ADP to ATP while hydrogen ions are combining with oxygen to form water. The newly formed ATP is transported out of the mitochondria to all parts of the cell cytoplasm and nucleoplasm, where it is used to energize the functions of the cell.

This overall process is called the *chemosmotic mechanism* of ATP formation.

ATP is used for many cellular functions. ATP promotes three types of cell functions: (1) *membrane transport,* as occurs with the sodium-potassium pump, which transports sodium out of the cell and potassium into the cell; (2) *synthesis of chemical compounds throughout the cell*; and (3) *mechanical work,* as occurs with the contraction of muscle fibers or with ciliary and ameboid motion.

Ameboid and Ciliary Movements of Cells

The most important type of cell movement that occurs in the body is that of the specialized muscle cells in skeletal, cardiac, and smooth muscle, which constitute almost 50 per cent of the entire body mass. Two other types of movement also occur in other cells: *ameboid locomotion* and *ciliary movement.*

Ameboid locomotion is the movement of an entire cell in relation to its surroundings. An example of ameboid locomotion is the movement of white blood cells through tissues. Typically, ameboid locomotion begins with protrusion of a pseudopodium from one end of the cell. This results from continual exocytosis, which forms a new cell membrane at the leading edge of the pseudopodium, and continual endocytosis of the membrane in the mid and rear portions of the cell.

Two other effects are also essential to the forward movement of the cell. The first effect is attachment of the pseudopodium to the surrounding tissues so that it becomes fixed in its leading position while the remainder of the cell body is pulled forward, toward the point of attachment. This attachment is effected by receptor proteins that line the insides of the exocytotic vesicles.

The second requirement for locomotion is the presence of the energy needed to pull the cell body in the direction of the pseudopodium. In the cytoplasm of all cells are molecules of the protein *actin*. When these molecules polymerize to form a filamentous network, the network will contract when it binds with another protein, an actin-binding protein such as *myosin*. The entire process is energized by ATP. This takes place in the pseudopodium of a moving cell, in which such a network of actin filaments forms inside the growing pseudopodium.

The most important factor that usually initiates ameboid movement is the process called *chemotaxis*. This results from the appearance of certain chemical substances in the tissue called *chemotactic substances*.

Ciliary movement is a whiplike movement of cilia on the surfaces of cells. Ciliary movement occurs in only two places in the body: on the inside surfaces of the *respiratory airways* and on the inside surfaces of the *uterine tubes* (fallopian tubes of the reproductive tract). In the nasal cavity and lower respiratory airways, the whiplike motion of the cilia causes a layer of mucus to move at a rate of about 1 cm/min toward the pharynx; in this way, these passageways of mucus or particles that become entrapped in the mucus are continually cleared. In the uterine tubes, the cilia cause slow movement of fluid from the ostium of the uterine tube toward the uterine cavity; it is mainly this movement of fluid that transports the ovum from the ovary to the uterus.

The mechanism of the ciliary movement is not fully understood, but there are at least two necessary factors: (1) the presence of ATP and (2) the appropriate ionic conditions, including the appropriate concentrations of magnesium and calcium.

3

Genetic Control of Protein Synthesis, Cell Function, and Cell Reproduction

Cell genes control protein synthesis. The genes control protein synthesis in the cell and in this way control cell function. Proteins play a key role in almost all functions of the cell by serving as enzymes that catalyze the reactions of the cell and as major components of the physical structures of the cell.

Each gene is a double-stranded, helical molecule of *deoxyribonucleic acid (DNA)* that automatically controls the formation of *ribonucleic acid (RNA)*. The RNA, in turn, spreads throughout the cells and controls the formation of a specific protein. Because there are about 100,000 genes, it is possible to form large numbers of different cellular proteins.

Nucleotides are organized to form two strands of DNA that are loosely bound to each other. Genes are attached in an end-on-end manner in long, double-stranded, helical molecules of DNA that are composed of three basic building blocks: (1) *phosphoric acid*, (2) *deoxyribose* (a sugar), and (3) four *nitrogenous bases* (two purines, adenine and guanine, and two pyrimidines, thymine and cytosine).

The first stage in the formation of DNA is the combination of one molecule of phosphoric acid, one molecule of deoxyribose, and one of the four bases to form a *nucleotide*. Four separate nucleotides can therefore be formed, one from each of the four bases. Multiple nucleotides are bound together to form two strands of DNA, and the two strands are loosely bound to each other.

The backbone of each DNA strand is composed of alternating phosphoric acid and deoxyribose molecules. The purine and pyrimidine bases are attached to the side of the deoxyribose molecules, and loose bonds between the purine and pyrimidine bases of the two DNA strands hold them together. *The purine base adenine of one strand always bonds with the pyrimi-*

dine base thymine of the other strand, whereas guanine always bonds with cytosine.

The genetic code consists of triplets of bases. Each group of three successive bases in the DNA strand is called a *code word*, and these code words control the sequence of amino acids in the protein to be formed in the cytoplasm. One code word might be composed of a sequence of adenine, thymine, and guanine, whereas the next code word might have a sequence of cytosine, guanine, and thymine. These two code words have entirely different meanings because their bases are different. The sequence of successive code words of the DNA strand is known as the *genetic code*.

THE DNA CODE IS TRANSFERRED TO AN RNA CODE BY THE PROCESS OF TRANSCRIPTION (p. 28)

Because DNA is located in the nucleus and many of the functions of the cell are carried out in the cytoplasm, there must be some method by which the genes of the nucleus control the chemical reactions of the cytoplasm. This is achieved through RNA, the formation of which is controlled by DNA. In this process, the code of DNA is transferred to RNA, a process called *transcription*. The RNA diffuses from the nucleus to the nuclear pores into the cytoplasm, where it controls protein synthesis.

RNA is synthesized in the nucleus from a DNA template. During the synthesis of RNA, the two strands of the DNA molecule separate, and one of the two strands is used as a template for the synthesis of RNA. The code triplets in the DNA cause the formation of *complementary code triplets* (called *codons*) in the RNA; these codons then control the sequence of amino acids in a protein to be synthesized later in the cytoplasm. Each DNA strand in each chromosome carries the code for an average of about 4000 genes.

The basic building blocks of RNA are almost the same as those of DNA except that in RNA, the sugar *ribose* replaces the sugar deoxyribose and the pyrimidine *uracil* replaces thymine. The basic building blocks of RNA combine to form four different nucleotides, exactly as described for the synthesis of DNA. These nucleotides contain the bases *adenine, guanine, cytosine*, and *uracil*.

The next step in the synthesis of RNA is *activation of the nucleotides*. This occurs through the addition of two phosphate radicals to each nucleotide to form triphosphates. These last two phosphates are combined with the nucleotide by *high-energy phosphate bonds*, which are derived from the ATP of the cell. This activation process makes available large quantities of energy, which is used in promoting the chemical reactions that add each new RNA nucleotide to the end of the RNA chain.

The DNA strand is used as a template to assemble the RNA molecule from activated nucleotides. The assembly of the RNA molecule occurs under the influence of the enzyme *RNA polymerase* as follows:

1. In the DNA strand immediately ahead of the gene that is to be transcribed is a sequence of nucleotides called the *promoter*. An RNA polymerase recognizes this promoter and attaches to it.
2. The polymerase causes unwinding of two turns of the DNA helix and separation of the unwound portions.
3. The polymerase moves along the DNA strand and begins forming the RNA molecules by binding complementary RNA nucleotides to the DNA strand.
4. The successive RNA nucleotides then bind to each other to form an RNA strand.
5. When the RNA polymerase reaches the end of the DNA gene, it encounters a sequence of DNA molecules called the *chain-terminating sequence*; this causes the polymerase to break away from the DNA strand. The RNA strand is then released into the nucleoplasm.

The code present in the DNA strand is transmitted in complementary form to the RNA molecule as follows:

DNA base <····> RNA base
Guanine   <····> Cytosine
Cytosine   <····> Guanine
Adenine   <····> Uracil
Thymine   <····> Adenine

There are three different types of RNA. Each of the three types of RNA plays a different role in protein formation: (1) *messenger RNA (mRNA)* carries the genetic code to the cytoplasm to control the formation of proteins; (2) *ribosomal RNA*, along with other proteins, forms the ribosomes, the structures in which protein molecules are actually assembled; and (3) *transfer RNA (tRNA)* transports activated amino acids to the ribosomes to be used in the assembly of the proteins.

There are 20 separate types of tRNA, each of which combines specifically with one of the 20 different amino acids and carries this amino acid to the ribosomes, where it is combined into the protein molecule. The code in the tRNA that allows it to recognize a specific codon is a triplet of nucleotide bases called an *anticodon*. In the formation of the protein molecule, the three anticodon bases combine loosely by hydrogen bonding with the codon bases of the mRNA. In this way, the different amino acids are lined up along the mRNA chain, thus establishing the proper sequence of amino acids in the protein molecule.

Polypeptides are synthesized on the ribosomes from the genetic code contained in the mRNA through the process of *translation*. To manufacture proteins, one end of the mRNA strand enters the ribosome, and then the entire strand threads its way through the ribosome in just over a minute. As it passes through, the ribosome "reads" the genetic code and causes the proper succession of amino acids to bind together to form chemical bonds called *peptide linkages*. Actually, the mRNA does not recognize the different types of amino acids but instead recognizes the different types of tRNA. Each type of tRNA molecule carries only one specific type of amino acid that will be incorporated into the protein.

Thus, as the strand of mRNA passes through the ribosome, each of its codons attracts to it a specific tRNA that, in turn, delivers a specific amino acid. This amino acid then combines with the preceding amino acids to form a peptide linkage, and this sequence continues to build until an entire protein molecule is formed. At this point, a *chain-terminating codon* appears and indicates the completion of the process, and the protein is released into the cytoplasm or through the membrane of the endoplasmic reticulum to the interior.

CONTROL OF GENETIC FUNCTION AND BIOCHEMICAL ACTIVITY IN CELLS (p. 33)

The genes control the function of each cell by determining the relative proportion of the different types of protein enzymes that are formed as well as the types of structural proteins that are formed.

The *operons* of the chromosomes control biochemical synthesis and are activated by the *promoter.* Formation of the enzymes needed for a specific synthetic process is usually controlled by a sequence of genes that are located on the same chromosomal DNA strand. This area of the DNA strand is called an *operon,* and the genes within the operon that are responsible for forming the respective enzymes are called *structural genes.* Another area on the DNA strand is called the *promoter,* which is a series of nucleotides that has a specific affinity for *RNA polymerase.* The polymerase must bind with this promoter before the polymerase can begin traveling along the DNA strand to synthesize RNA. The promoter is thereby essential to activation of the operon.

The operon is controlled by a *repressor operator.* In the middle of the promoter region is an additional band of nucleotides called a *repressor operator.* A *repressor protein* can bind to the repressor operator and prevent attachment of RNA polymerase to the promoter, thereby blocking transcription of genes in the operon. These repressor proteins can be either stimulated or inhibited by various nonprotein substances in the cell, such as the cell metabolites, allowing feedback control of protein synthesis.

The operon is also controlled by an *activator operator.* Adjacent to the promoter region is another operator called the *activator operator.* When a regulatory protein, called an *activator protein,* binds to this operator, it helps to attract the RNA polymerase to the promoter and in this way activates the operon. The operon can be activated or inhibited through the activator operator in ways that are exactly opposite to control by the repressor operator.

The operon is controlled through negative feedback by the cell product. When the cell produces a critical amount of substance, this causes negative feedback inhibition of the entire operon synthetic process. This inhibition can be accomplished by causing a regulatory repressor protein to bind at the repressor operator or by causing a regulatory activator protein to break this bond with the activator operon. In either case, the operon becomes inhibited.

There are other mechanisms available for control of transcription by the operon, including the following:

1. An operon may be controlled by a *regulatory gene* that is located elsewhere in the genetic complex of the nucleus.
2. In some instances, the same regulatory protein functions as an activator for one operon and as a repressor for another, allowing different operons to be controlled at the same time by the same regulatory protein.
3. The nuclear DNA is packaged in specific structural units, the *chromosomes*. Within each chromosome, the DNA is wound around small proteins called *histones*, which are held together tightly in a compacted state with other proteins. As long as the DNA is in this compacted state, it cannot function to form RNA. Multiple mechanisms exist, however, that can cause selected areas of the chromosomes to become decompacted, allowing RNA transcription.

THE DNA-GENETIC SYSTEM ALSO CONTROLS CELL REPRODUCTION (p. 35)

The genes and their regulatory mechanisms determine not only the growth characteristics of cells but also when and whether these cells will divide to form new cells. In this way, the genetic system controls each stage of the development of the human from the single-cell fertilized ovum to the whole functioning body.

Most cells of the body, with the exception of mature red blood cells, striated muscle cells, and neurons in the nervous system, are capable of reproducing other cells of their own type. Ordinarily, as sufficient nutrients are available, each cell grows larger and larger until it automatically divides via the process of *mitosis* to form two new cells. Different cells of the body have different life cycle periods that vary from as short as 10 hours for highly stimulated bone marrow cells to the entire lifetime of the human body for nerve cells.

Cell reproduction begins with replication of the DNA. Only after all of the DNA in the chromosomes has been replicated can mitosis take place. The DNA is duplicated only once, so that the net result is two exact replicates of all DNA. These replicates then become the DNA of the two daughter cells that will be formed at mitosis. The replication of DNA is similar to the way RNA is transcribed from DNA, except for a few important differences:

1. Both strands of the DNA are replicated, not simply one of them.
2. Both strands of the DNA helix are replicated from end to end rather than small portions of them, as occurs in the transcription of RNA by the genes.
3. The principal enzymes for replication of DNA are a complex of several enzymes called *DNA polymerase*, which is comparable to RNA polymerase.
4. Each newly formed strand of DNA remains attached by loose hydrogen bonding to the original DNA strand that is used as its template. Two DNA helixes are formed, therefore, that are duplicates of each other and are still coiled together.
5. The two new helixes become uncoiled by the action of enzymes that periodically cut each helix along its entire length, rotate each segment sufficiently to cause separation, and then resplice the helix.

DNA strands are "repaired" and "proofread." During the time between the replication of DNA and the beginning of mitosis, there is a period of "proofreading" and "repair" of the DNA strands. Whenever inappropriate DNA nucleotides have been matched up with the nucleotides of the original template strand, special enzymes cut out the defective areas and replace them with the appropriate complementary nucleotides. Because of proofreading and repair, the transcription process rarely makes a mistake. When a mistake is made, however, it is called a *mutation.*

Entire chromosomes are replicated. The DNA helixes of the nucleus are each packaged as a single chromosome. The human cell contains 46 chromosomes arranged in 23 pairs. In addition to the DNA in the chromosome, there is a large amount of protein composed mainly of *histones*, around which small segments of each DNA helix are coiled. During mitosis, the successive coils are packed against each other, allowing the long DNA molecule to be packaged in a coiled and folded arrangement. Replication of the chromosomes in their entirety occurs soon after replication of the DNA helixes. The two newly formed chromosomes remain temporarily attached to each other at a point called the *centromere*, which is located near their center. These duplicated but still-attached chromosomes are called *chromatids.*

Mitosis is the process by which the cell splits into two new daughter cells. Two pairs of *centrioles*, which are small structures that lie close to one pole of the nucleus, begin to move apart from each other. This movement is caused by successive polymerization of protein microtubules growing outward from each pair of centrioles. As the tubules grow, they push one pair of centrioles toward one pole of the cell and the other toward the opposite pole. At the same time, other microtubules grow radially away from each of the centriole pairs, forming a spiny star called the *aster* at each end of the cell. The complex of microtubules extending between the centriole pairs is called the s*pindle,* and the entire set of microtubules plus the pairs of centrioles is called the *mitotic apparatus.* Mitosis then proceeds through several phases:

- *Prophase* is the beginning of mitosis. While the spindle is forming, the chromosomes of the nucleus become condensed into well-defined chromosomes.
- *Prometaphase* is the stage at which the growing microtubular spines of the aster puncture and fragment the nuclear envelope. At the same time, the microtubules from the aster become attached to the chromatids at the centromere, where the paired chromatids are still bound to each other.
- *Metaphase* is the stage at which the two asters of the mitotic apparatus are pushed farther and farther apart by additional growth of the mitotic spindle. Simultaneously, the chromatids are pulled tightly by the attached microtubules to the center of the cell, lining up to form the *equatorial plate* of the mitotic spindle.
- *Anaphase* is the stage at which the two chromatids of each chromosome are pulled apart at the centromere. Thus, all 46 pairs of chromosomes are separated, forming two sets of 46 daughter chromosomes.
- *Telophase* is the stage at which the two sets of daughter chromosomes are pulled completely apart. Then, the mitotic apparatus dissolves, and the new nuclear membrane develops around each set of chromosomes.

Cell differentiation allows different cells of the body to perform different functions. As a human develops from a fertilized ovum, the ovum divides repeatedly until trillions of cells are formed. Gradually, however, the new cells differentiate from each other, with certain cells having different genetic characteristics from other cells. This differentiation process occurs as a result of inactivation of certain genes and activation of others during successive stages of cell division. This process of differentiation leads to the ability of different cells in the body to perform different functions.

UNIT

II

Membrane Physiology, Nerve, and Muscle

4

Transport of Ions and Molecules Through the Cell Membrane

Differences between the composition of intracellular and extracellular fluids are brought about by transport mechanisms of cell membranes. These differences include the following:

- Extracellular fluid has a high sodium concentration and low potassium concentration. The opposite is true of intracellular fluid.
- Extracellular fluid contains a high chloride concentration compared with intracellular fluid.
- The concentrations of phosphates and proteins in intracellular fluid are greater than those in extracellular fluid.

The cell membrane consists of a lipid bilayer with floating protein molecules. The lipid bilayer constitutes a barrier for the movement of most water-soluble substances. However, lipid-soluble substances can pass directly through the lipid bilayer. Protein molecules in the lipid bilayer constitute an alternate transport pathway:

- *Channel proteins* provide a watery pathway for molecules to allow them to move through the membrane.
- *Carrier proteins* bind with specific molecules and then undergo conformational changes that move the molecules through the membrane.

Transport through the cell membrane occurs through diffusion or active transport.

- *Diffusion* means random movement of molecules either through intermolecular spaces in the membrane or in combination with a carrier protein. The energy that causes diffusion is the energy of the normal kinetic motion of matter.
- *Active transport* means movement of substances across the membrane in combination with a car-

rier protein but also against an energy gradient (i.e., from a low concentration state to a high concentration state). This process requires an additional source of energy besides kinetic energy.

DIFFUSION (p. 44)

Diffusion is the continual movement of molecules in liquids or in gases. Diffusion through the cell membrane is divided into the following two subtypes:

- *Simple diffusion* means that molecules move through a membrane without binding with carrier proteins. Simple diffusion can occur via two pathways: (1) through the interstices of the lipid bilayer and (2) through watery channels in transport proteins.
- *Facilitated diffusion* requires a carrier protein. The carrier protein aids in passage of molecules through the membrane, probably by binding chemically with them and shuttling them through the membrane in this form.

The rate of diffusion of a substance through the cell membrane is directly proportional to its lipid solubility. The lipid solubilities of oxygen, nitrogen, carbon dioxide, and alcohols are so high that they can dissolve directly in the lipid bilayer and diffuse through the cell membrane in the same manner as through a watery solution.

Water and other lipid-insoluble molecules diffuse through protein channels in the cell membrane. Water readily penetrates the cell membrane, passing through the protein channels. Other lipid-insoluble molecules can pass through the protein pore channels in the same way as water molecules if they are sufficiently small. As they become larger, their ability to penetrate rapidly diminishes.

Protein channels have selective permeability for transport of one or more specific molecules. This permeability results from the characteristics of the channel itself, such as its diameter, its shape, and the nature of the electrical charges along its inside surfaces.

- *Sodium channels* are specifically selective for the passage of sodium ions. They are calculated to be 0.3×0.5 nm in size, and the inner surfaces are negatively charged. The negative charges are thought to pull sodium ions from their hydrating water molecules. Once in the channel, the sodium ions diffuse according to the usual laws of diffusion.

- *Potassium channels* are specifically selective for the passage of potassium ions. They are calculated to be slightly smaller than the sodium channels, only 0.3×0.3 nm, but they are not negatively charged. Therefore, the ions are not pulled from the water molecules that hydrate them. Because the hydrated form of the potassium ion is considerably smaller than the hydrated form of sodium, the potassium ions can easily pass through this smaller channel, whereas sodium ions are usually rejected.

Gating of protein channels provides a means for controlling their permeability. The gates are thought to be actual gatelike extensions of the transport protein molecule, which can close over the channel opening or be lifted from the opening by a conformational change in the protein molecule itself. The opening and closing of gates are controlled in two principal ways:

- *Voltage gating.* In this instance, the molecular conformation of the gate responds to the electrical potential across the cell membrane. For example, a strong negative charge on the inside of the cell membrane causes the sodium gates to remain tightly closed. When the inside of the membrane loses its negative charge, these gates open suddenly, allowing sodium to pass inward through the sodium pores. The opening of sodium gates is the basic cause of action potentials in nerves.
- *Chemical gating.* Some protein channel gates are opened by the binding of another molecule with the protein; this causes a conformational change in the protein molecule that opens or closes the gate. This is called chemical gating. One of the most important instances of chemical gating is the effect of acetylcholine on the "acetylcholine channel." Acetylcholine opens the gate of this channel, allowing all uncharged molecules as well as positive ions with a smaller diameter to pass through.

Facilitated diffusion is also called carrier-mediated diffusion. A substance transported in this manner usually cannot pass through the membrane without the assistance of a specific carrier protein.

- Facilitated diffusion involves the following two steps: (1) the molecule to be transported enters a blind-ended channel and binds to a specific receptor, and (2) a conformational change occurs in the carrier protein, so that the channel now opens to the opposite side of the membrane.
- Facilitated diffusion differs from simple diffusion through an open channel in the following important way. The rate of simple diffusion increases proportionately with the concentration of the diffusing substance. In facilitated diffusion, the rate of diffusion approaches a maximum as the concentration of the substance increases. This maximum rate is dictated by the rate at which the carrier protein molecule can undergo the conformational change.
- Among the most important substances that cross cell membranes through facilitated diffusion are glucose and most of the amino acids. Insulin can increase the rate of facilitated diffusion of glucose as much as 10- to 20-fold.

Factors That Affect Net Rate of Diffusion (p. 48)

Substances can diffuse in both directions through the cell membrane. Therefore, what is usually important is the net rate of diffusion of a substance in the desired direction. This net rate is determined by the following factors:

- *Permeability.* The permeability of a membrane for a given substance is expressed as the net rate of diffusion of the substance through each unit area of the membrane for a unit concentration difference between the two sides of the membrane (when there are no electrical or pressure differences). Membrane permeability is enhanced by increases in the lipid solubility of the diffusing molecule, the number of channels per unit area of membrane, and the temperature. Membrane permeability is decreased by increases in membrane thickness as well as increases in the molecular weight of the diffusing substance.
- *Concentration difference.* The rate of net diffusion through a cell membrane is proportional to the difference in concentration of the diffusing substance on the two sides of the membrane.
- *Electrical potential.* If an electrical potential is applied across a membrane, the ions will move

through the membrane because of their electrical charges. When large amounts of ions have moved through the membrane, a concentration difference of the same ions will have developed in the direction opposite to the electrical potential difference. When the concentration difference rises to a sufficiently high level, the two effects balance each other. The electrical difference that will balance a given concentration difference can be determined with the Nernst equation.

- *Pressure difference.* At times, a considerable pressure difference develops between the two sides of a membrane. Pressure actually means the sum of all of the forces of the different molecules striking a unit surface area at a given instant. Therefore, when the pressure is higher on one side of a membrane than on the other, increased amounts of energy are available to cause net movement of molecules from the high-pressure side toward the low-pressure side.

Osmosis Across Selectively Permeable Membranes — Net Diffusion of Water (p. 49)

Osmosis is the process of net movement of water caused by a concentration difference of water. Water is the most abundant substance to diffuse through the cell membrane. However, the amount that diffuses in each direction is so precisely balanced under normal conditions that not even the slightest net movement of water occurs. Therefore, the volume of a cell remains constant. A concentration difference for water can develop across a membrane. When this happens, net movement of water does occur across the cell membrane, causing the cell to either swell or shrink, depending on the direction of the net movement. The pressure difference required to stop osmosis is the *osmotic pressure.*

The osmotic pressure exerted by particles in a solution is determined by the number of particles per unit volume of fluid and not by the mass of the particles. The reason for this is that on the average, the kinetic energy of each molecule or ion that strikes a membrane is about the same regardless of its molecular size. Consequently, the factor that determines the osmotic pressure of a solution is the concentration of the solution in terms of number of particles (which is the same as the molar concentration if it is a nondissociated molecule) and not in terms of mass of the solute.

The osmole expresses concentration in terms of number of particles. One osmole is 1 gram molecular weight of undissociated solute. Thus, 180 grams of glucose, which is 1 gram molecular weight of glucose, is equal to 1 osmole of glucose because glucose does not dissociate. A solution that has 1 osmole of solute dissolved in each kilogram of water is said to have an osmolality of 1 osmole per kilogram, and a solution that has 1/1000 osmole dissolved per kilogram has an osmolality of 1 milliosmole per kilogram. The normal osmolality of the extracellular and intracellular fluids is about 300 milliosmoles per kilogram, and the osmotic pressure of these fluids is about 5500 mm Hg.

ACTIVE TRANSPORT (p. 51)

Active transport can move a substance against an electrochemical gradient. An electrochemical gradient is the sum of all the diffusion forces acting at the membrane—the forces caused by concentration difference, electrical difference, and pressure difference. That is, substances cannot diffuse "uphill." When a cell membrane moves molecules or ions uphill against a concentration gradient (or uphill against an electrical or pressure gradient), the process is called active transport. Among the different substances that are actively transported through at least some cell membranes are sodium ions, potassium ions, calcium ions, iron ions, hydrogen ions, chloride ions, iodide ions, urate ions, several different sugars, and most of the amino acids.

Active transport is divided into two types according to the source of the energy used to cause the transport. The two kinds of transport are called primary active transport and secondary active transport. In both instances, transport depends on carrier proteins that penetrate through the membrane, which is also true for facilitated diffusion.

- *Primary active transport.* The energy is derived directly from the breakdown of adenosine triphosphate (ATP) or some other high-energy phosphate compound.
- *Secondary active transport.* The energy is derived secondarily from energy that has been stored in

the form of ionic concentration differences between the two sides of a membrane, originally created by primary active transport.

Primary Active Transport (p. 51)

The sodium-potassium (Na^+-K^+) pump transports sodium ions out of cells and potassium ions into cells. This pump is present in all cells of the body, and it is responsible for maintaining the sodium and potassium concentration differences across the cell membrane as well as for establishing a negative electrical potential inside the cells. The pump operates in the following manner: three sodium ions bind to a carrier protein on the inside of the cell and two potassium ions bind to the carrier protein on the outside of the cell. The carrier protein has ATPase activity, and the binding of ions causes the ATPase function of the protein to become activated. This then cleaves one molecule of ATP, splitting it to form adenosine diphosphate (ADP) and liberating a high-energy phosphate bond of energy. This energy is then believed to cause a conformational change in the protein carrier molecule, extruding the sodium ions to the outside and the potassium ions to the inside.

The Na^+-K^+ pump controls cell volume. The Na^+-K^+ pump transports three molecules of sodium to the outside of the cell for every two molecules of potassium pumped to the inside. This continual net loss of ions initiates an osmotic tendency to move water out of the cell. Furthermore, when the cell begins to swell, this automatically activates the Na^+-K^+ pump, moving to the exterior still more ions that are carrying water with them. Therefore, the Na^+-K^+ pump performs a continual surveillance role in maintaining normal cell volume. Without the function of this pump, most cells of the body would swell until they burst.

Active transport saturates in the same way that facilitated diffusion saturates. When the concentration of the substance to be transported is small, the rate of transport rises approximately in proportion to increases in its concentration. At high concentrations, the rate of transport is limited by the rates at which the chemical reactions of binding, release, and carrier conformational changes can occur.

Co-transport and counter-transport are two forms of secondary active transport. When sodium ions are transported out of cells by primary active transport, a large concentration gradient of sodium can develop. This gradient represents a storehouse of energy because the excess sodium outside the cell membrane is always attempting to diffuse to the interior.

- *Co-transport.* The diffusion energy of sodium can actually pull other substances along with the sodium (in the same direction) through the cell membrane. A carrier protein serves as an attachment point for both the sodium ion and the substance to be co-transported, and a conformational change in the carrier protein causes both the sodium ion and the other substance to be transported together to the interior of the cell.
- *Counter-transport.* The sodium ion binds to a carrier protein on the exterior surface of the membrane, whereas the substance to be counter-transported binds to the interior surface of the carrier protein. Once both have bound, a conformational change occurs again, with the energy of the sodium ion moving to the interior, which causes the other substance to move to the exterior.

Glucose and amino acids can be transported into most cells through sodium co-transport. The transport carrier protein has two binding sites on its exterior side—one for sodium and one for glucose or amino acids. Again, the concentration of sodium ions is very high on the outside and very low on the inside, providing the energy for the transport. A special property of the transport protein is that the conformational change to allow sodium movement to the interior will not occur until a glucose or amino acid molecule also attaches.

- *Sodium co-transport of glucose.* When both sodium and glucose have attached to the carrier protein, the conformational change takes place automatically, and both the sodium and the glucose are transported to the inside of the cell at the same time.
- *Sodium co-transport of amino acids.* This occurs in the same manner as for glucose, except that a different set of transport proteins are used. Five amino acid transport proteins have been identified, each of which is responsible for transporting one subset of amino acids with specific molecular characteristics.

Calcium and hydrogen ions can be transported out of cells through the sodium counter-transport mechanism.

- *Calcium counter-transport* occurs in most cell membranes with sodium ions moving to the interior and calcium ions moving to the exterior, both bound to the same transport protein in a counter-transport mode.
- *Hydrogen counter-transport* occurs especially in the proximal tubules of the kidneys, where sodium ions move from the lumen of the tubule to the interior of the tubular cells, and hydrogen ions are counter-transported into the lumen.

Active transport through cellular sheets occurs at many places in the body. The basic mechanism for transport of a substance through a cellular sheet is (1) to provide active transport through the cell membrane on one side of the cell and then (2) to provide for either simple diffusion or facilitated diffusion through the membrane on the opposite side of the cell. Transport of this type occurs through the intestinal epithelium, the epithelium of the renal tubules, the epithelium of all exocrine glands, the epithelium of the gallbladder, the membrane of the choroid plexus of the brain, and many other membranes.

5

Membrane Potentials and Action Potentials

Electrical potentials exist across the membranes of essentially all cells of the body. In addition, nerve and muscle cells are "excitable" (i.e., capable of self-generation of electrochemical impulses at their membranes). The present discussion is concerned with membrane potentials that are generated both at rest and during action potentials by nerve and muscle cells.

BASIC PHYSICS OF MEMBRANE POTENTIALS (p. 57)

A concentration difference of ions across a selectively permeable membrane can produce a membrane potential.

- *Potassium diffusion potential.* Suppose a cell membrane is permeable to potassium ions but not to any other ions. Potassium ions tend to diffuse outward because of the high potassium concentration inside the cell. Because potassium ions are positively charged, the loss of potassium ions from the cell creates a negative potential inside the cell. Within a few milliseconds, the potential change becomes sufficiently great to block further net diffusion of potassium despite the high potassium ion concentration gradient. In the normal large mammalian nerve fiber, the potential difference required is about 94 millivolts, with negativity inside the fiber membrane.
- *Sodium diffusion potential.* Now suppose a cell membrane is permeable to sodium ions but not to any other ions. Sodium ions tend to diffuse into the cell because of the high sodium concentration outside the cell. Diffusion of sodium ions into the cell creates a positive potential inside the cell. Again, the membrane potential rises sufficiently high within milliseconds to block further net dif-

fusion of sodium ions into the cell; however, this time, for the large mammalian nerve fiber, the potential is about 61 millivolts, with positivity inside the fiber.

The Nernst equation describes the relation of diffusion potential to concentration difference. The membrane potential that prevents net diffusion of an ion in either direction through the membrane is called the *Nernst potential* for that ion. The Nernst equation can be used to calculate the Nernst potential for any univalent ion at normal body temperature. The Nernst potential that is calculated is the potential inside the membrane. The sign of the potential is positive (+) if the ion under consideration is a negative ion and negative (−) if it is a positive ion. The following is the Nernst equation:

$$\text{EMF (millivolts)} = \pm 61 \log \left(\frac{\text{concentration inside}}{\text{concentration outside}} \right)$$

where EMF is the electromotive force.

The Goldman equation is used to calculate the diffusion potential when the membrane is permeable to several different ions. In this case, the diffusion potential that develops depends on three factors: (1) the polarity of the electrical charge of each ion, (2) the permeability of the membrane (*P*) to each ion, and (3) the concentrations (*C*) of the respective ions on the inside (*i*) and outside (*o*) of the membrane. The following is the Goldman equation:

EMF (millivolts) =

$$-61 \log \left(\frac{C_{Na^+_i} P_{Na^+} + C_{K^+_i} P_{K^+} + C_{Cl^-_o} P_{Cl^-}}{C_{Na^+_o} P_{Na^+} + C_{K^+_o} P_{K^+} + C_{Cl^-_i} P_{Cl^-}} \right)$$

Note the following features and implications of the Goldman equation:

- Sodium, potassium, and chloride ions are most importantly involved in the development of membrane potentials in the nerve and muscle fibers as well as in the neuronal cells in the central nervous system.
- The degree of importance of each ion in determining the voltage is proportional to the membrane permeability for that particular ion.
- A positive ion concentration gradient from inside the membrane to the outside causes electronegativity inside the membrane.

The cell membrane functions as an electrical capacitor. When positive ions are pumped out of cells, they line up along the outside of the membrane; on the inside, the negative ions that have been left behind line up. This creates a dipole layer of positive and negative charges between the outside and inside of the membrane, but equal numbers of negative and positive charges are left everywhere else within the fluids. This is the same effect that occurs when the plates of an electrical capacitor become electrically charged. The fact that the nerve membrane functions as a capacitor has one especially important point of significance: only about 1/5,000,000 to 1/100,000,000 of the total positive charges inside the fiber need to be transferred to establish the normal potential of -90 millivolts inside the nerve fiber. Also, an equally small number of positive ions moving from outside to inside of the nerve fiber can reverse the potential from -90 millivolts to as much as $+35$ millivolts within as little as 1/10,000 of a second.

RESTING MEMBRANE POTENTIAL OF NERVES (p. 59)

The resting membrane potential is established by the diffusion potentials, membrane permeability, and electrogenic nature of the Na^+-K^+ pump.

- *Potassium diffusion potential.* A high ratio of potassium ions from inside to outside of the cell, 35 to 1, produces a Nernst potential of -94 millivolts according to the Nernst equation. If potassium ions were the only factor causing the resting potential, the resting potential inside the membrane would be -94 millivolts.
- *Sodium diffusion potential.* The ratio of sodium ions from inside to outside the membrane is 0.1, and this gives a calculated Nernst potential of $+61$ millivolts.
- *Membrane permeability.* The permeability of the nerve fiber membrane to potassium is about 100 times as great as that to sodium, so that diffusion of potassium contributes far more to the membrane potential. The use of this high value of permeability in the Goldman equation gives an internal membrane potential of -86 millivolts, which is near the potassium potential of -94 millivolts.
- *Electrogenic nature of Na^+-K^+ pump.* The Na^+-K^+ pump transports three sodium ions to the outside

for each two potassium ions pumped to the inside, which causes a continual loss of positive charges from inside the membrane. Therefore, the Na^+-K^+ pump is electrogenic because it produces a net deficit of positive ions inside the cell; this causes a negative charge of about -4 millivolts inside the cell membrane.

- *In summary,* the diffusion potentials caused by potassium and sodium diffusion would give a membrane potential of about -86 millivolts, almost all of this being determined by potassium diffusion. Then, an additional -4 millivolts is contributed by the electrogenic Na^+-K^+ pump, giving a net resting membrane potential of -90 millivolts.

NERVE ACTION POTENTIAL (p. 61)

Nerve signals are transmitted by action potentials, which are rapid changes in the membrane potential. Each action potential begins with a sudden change from the normal resting negative potential to a positive membrane potential and then ends with an almost equally rapid change back to the negative potential.

The successive stages of the action potential are as follows.

- *Resting stage.* This is the resting membrane potential before the action potential occurs.
- *Depolarization stage.* At this time, the membrane suddenly becomes permeable to sodium ions, allowing tremendous numbers of positively charged sodium ions to flow to the interior of the axon, and the potential rises rapidly in the positive direction. This is called *depolarization.*
- *Repolarization stage.* Within a few 10,000ths of a second after the membrane becomes highly permeable to sodium ions, the sodium channels begin to close and the potassium channels open more than they normally do. Then, rapid diffusion of potassium ions to the exterior re-establishes the normal negative resting membrane potential. This is called *repolarization.*

Voltage-gated sodium and potassium channels are activated and inactivated during the course of an action potential. The necessary factor in causing both depolarization and repolarization of the nerve membrane during the action potential is the *voltage-gated sodium channel.* The *voltage-gated potassium chan-*

nel also plays an important role in increasing the rapidity of repolarization of the membrane. *These two voltage-gated channels are in addition to the Na⁺-K⁺ pump and the Na⁺-K⁺ leak channels.*

- *Activation of the sodium channel.* This channel has two gates: an activation gate near the outside of the channel and an inactivation gate near the inside. During rest, when the membrane potential is -90 millivolts, the activation gate is closed, preventing entry of sodium ions into the fiber. When the membrane potential rises from -90 millivolts toward zero, it reaches a threshold voltage (between -70 and -50 millivolts), which causes the activation gate to open, increasing sodium permeability as much as 500- to 5000-fold. Sodium ions can then pour inward through the channel.

- *Inactivation of the sodium channel.* The same increase in voltage that opens the activation gate also closes the inactivation gate. The inactivation gate, however, closes a few 10,000ths of a second after the activation gate opens, stopping the passage of sodium ions to the inside of the membrane. At this point, the membrane potential begins to return to the resting membrane state, which is the repolarization process.

- *Activation of the potassium channel.* When the membrane potential rises from -90 millivolts toward zero, this voltage change causes a slow conformational opening of the potassium channel gate, allowing increased potassium diffusion outward through the channel. The decrease in sodium entry to the cell and simultaneous increase in potassium exit from the cell greatly speed the repolarization.

The events summarized that cause the action potential.

- *During the resting state,* before the action potential begins, the conductance for potassium ions is 50 to 100 times as great as the conductance for sodium ions. This is caused by much greater leakage of potassium ions than sodium ions through the leak channels.

- *At the onset of the action potential,* the sodium channels instantaneously become activated and allow up to a 5000-fold increase in sodium conductance. Then, the inactivation process closes the sodium channels within few fractions of a millisecond.

The onset of the action potential also causes voltage gating of the potassium channels, causing them to begin opening more slowly.

- *At the end of the action potential*, the return of the membrane potential to the negative state causes the potassium channels to close back to their original status, but again only after a delay.

A positive-feedback, vicious circle opens the sodium channels. If any event causes the membrane potential to rise from −90 millivolts up toward the zero level, the rising voltage itself will cause many voltage-gated sodium channels to begin opening. This allows rapid inflow of sodium ions, which causes still further rise of the membrane potential, thus opening still more voltage-gated sodium channels. This process is a positive-feedback vicious circle that once the feedback is sufficiently strong will continue until all of the voltage-gated sodium channels have become activated (opened).

An action potential will not occur until the threshold potential has been reached. This happens when the number of sodium ions entering the fiber becomes greater than the number of potassium ions leaving the fiber. A sudden increase in the membrane potential in a large nerve fiber from −90 millivolts up to about −65 millivolts usually causes the explosive development of the action potential. This level of −65 millivolts is said to be the threshold for stimulation.

A new action potential cannot occur as long as the membrane is still depolarized from the preceding action potential. Shortly after the action potential is initiated, the sodium channels become inactivated, and any amount of excitatory signal applied to these channels at this point will not open the inactivation gates. The only condition that will reopen them is when the membrane potential returns either to or almost to the original resting membrane potential level. Then, within another small fraction of a second, the inactivation gates of the channels open, and a new action potential can then be initiated.

- *Absolute refractory period.* An action potential cannot be elicited during the absolute refractory period, even with a strong stimulus. This period for large myelinated nerve fibers is about 1/2500 second, which means that a maximum of about 2500 impulses can be transmitted per second.

- *Relative refractory period*. This period follows the absolute refractory period. During this time, stronger than normal stimuli can excite the fiber, and an action potential can be initiated.

PROPAGATION OF THE ACTION POTENTIAL (p. 65)

An action potential elicited at any one point on a membrane usually excites adjacent portions of the membrane, resulting in propagation of the action potential. Thus, the depolarization process travels along the entire extent of the fiber. The transmission of the depolarization process along a nerve or muscle fiber is called a nerve or muscle impulse.

- *Direction of propagation*. An excitable membrane has no single direction of propagation, but the action potential will travel in both directions away from the stimulus—and even along all branches of a nerve fiber—until the entire membrane has become depolarized.
- *All-or-nothing principle*. Once an action potential has been elicited at any point on the membrane of a normal fiber, the depolarization process will travel over the entire membrane if conditions are right, or it might not travel at all if conditions are not right. This is called the all-or-nothing principle, and it applies to all normal excitable tissues.

RE-ESTABLISHING SODIUM AND POTASSIUM IONIC GRADIENTS AFTER ACTION POTENTIALS—IMPORTANCE OF ENERGY METABOLISM (p. 66)

The transmission of each impulse along the nerve fiber reduces infinitesimally the concentration differences of sodium and potassium between the inside and outside of the membrane. From 100,000 to 50 million impulses can be transmitted by nerve fibers before the concentration differences have decreased to the point that action potential conduction ceases. Even so, with time it becomes necessary to re-establish the sodium and potassium membrane concentration differences. This is achieved by the action of the Na^+-K^+ pump. That is, the sodium ions that have diffused to the interior of the cell during the action potentials and the potassium ions that have diffused to the exterior are returned to their original state by the Na^+-K^+ pump. Because this pump requires energy for operation, the process of "recharging" the nerve fiber is an active metabolic one, using energy derived from the adenosine triphosphate (ATP) energy system of the cell.

SPECIAL ASPECTS OF SIGNAL TRANSMISSION IN NERVE TRUNKS (p. 68)

Large nerve fibers are myelinated, and the small ones are unmyelinated. The central core of the fiber is the axon, and the membrane of the axon is the actual membrane for conducting the action potential. The axon is filled in its center with axoplasm, which is a viscid intracellular fluid. Surrounding the larger axons is a myelin sheath that is often thicker than the axon itself. The myelin sheath, deposited around the axon by Schwann cells, consists of multiple layers of cellular membrane containing the lipid substance sphingomyelin. This substance is an excellent insulator that decreases the ion flow through the membrane by about 5000-fold. At the juncture between two successive Schwann cells, a small noninsulated area only 2 to 3 micrometers in length remains where ions can still flow with ease between the extracellular fluid and the axon. This area is the *node of Ranvier*.

"Saltatory" conduction occurs in myelinated fibers. Even though ions cannot flow significantly through the thick sheaths of myelinated nerves, they can flow with considerable ease through the nodes of Ranvier. Therefore, action potentials can occur only at the nodes, yet the action potentials are conducted from node to node; this is called saltatory conduction. That is, electrical current flows through the surrounding extracellular fluids outside the myelin sheath as well as through the axoplasm from node to node, exciting successive nodes one after another. Thus, the nerve impulse jumps down the fiber, which is the origin of the term "saltatory." Saltatory conduction is of value for two reasons.

- *Increased velocity*. First, by causing the depolarization process to jump long intervals along the axis of the nerve fiber, this mechanism increases the velocity of nerve transmission in myelinated fibers as much as 5- to 50-fold.
- *Energy conservation*. Second, saltatory conduction conserves energy for the axon because only the nodes depolarize, allowing perhaps a hundred times smaller loss of ions than would otherwise

be necessary and therefore requiring little metabolism for re-establishment of the sodium and potassium concentration differences across the membrane after a series of nerve impulses.

Conduction velocity is greatest in large, myelinated nerve fibers. The velocity of conduction in nerve fibers varies from as low as 0.25 m/sec in very small unmyelinated fibers to as high as 100 m/sec in very large myelinated fibers. The velocity increases approximately with the fiber diameter in myelinated nerve fibers and approximately with the square root of fiber diameter in unmyelinated fibers.

6

Contraction of Skeletal Muscle

About 40 per cent of the body is skeletal muscle, and perhaps another 10 per cent is smooth muscle and cardiac muscle. Many of the principles of contraction apply to all of the different types of muscle. In this chapter, the function of skeletal muscle is considered; the functions of smooth muscle are discussed in Chapter 8, and the functions of cardiac muscle are discussed in Chapter 9.

PHYSIOLOGIC ANATOMY OF SKELETAL MUSCLE (p. 73)

Skeletal Muscle Fiber

Figure 6–1 shows the organization of skeletal muscle. In most muscles, the fibers extend the entire length of the muscle; each is innervated by only one nerve ending.

Myofibrils are composed of actin and myosin filaments. Each muscle fiber contains hundreds to thousands of myofibrils; and, in turn, each myofibril (see Fig. 6–1D and E) is composed of, lying side by side, about 1500 myosin filaments and 3000 actin filaments, which are large polymerized protein molecules that are responsible for muscle contraction. In Figure 6–1, the thick filaments are myosin, and the thin filaments are actin; note the following features:

- *Light and dark bands.* The myosin and actin filaments partially interdigitate and thus cause the myofibrils to have alternate light and dark bands. The light bands contain only actin filaments and are called I bands. The dark bands contain the myosin filaments as well as the ends of the actin filaments, where they overlap the myosin, and are called A bands.
- *Cross-bridges.* The small projections from the sides of the myosin filaments are cross-bridges. They protrude from the surfaces of the myosin filament

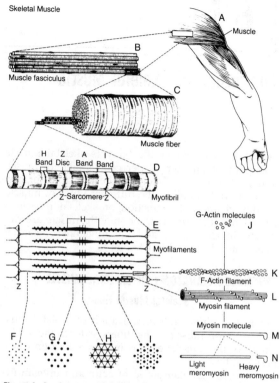

Figure 6–1 Organization of skeletal muscle, from the gross to the molecular level. *F, G, H,* and *I* are cross sections at the levels indicated.

along its entire length except in the center. Interaction between these cross-bridges and the actin filaments causes contraction.

- *Z disc.* The ends of the actin filaments are attached to Z discs (see Fig. 6–1E). The Z disc passes across the myofibril and from one to another, attaching the myofibrils across the muscle fiber. The entire muscle fiber therefore has light and dark bands, giving skeletal and cardiac muscle the striated appearance.
- *Sarcomere.* The portion of a myofibril that lies between two successive Z discs is called a sarcomere. During rest, the actin filaments overlap the myosin filaments and just barely overlap one another.

GENERAL MECHANISM OF MUSCLE CONTRACTION
(p. 74)

The initiation and execution of muscle contraction occur in the following sequential steps.

1. An action potential travels along a motor nerve to its ending on muscle fibers, and the nerve secretes a small amount of the neurotransmitter substance acetylcholine.
2. The acetylcholine acts on a local area of the muscle membrane to open acetylcholine-gated channels, which allow large quantities of sodium ions to flow into the muscle fiber.
3. The action potential travels along the muscle fiber membrane in the same way that the action potential travels along the nerve membrane, causing the sarcoplasmic reticulum to release into the myofibrils large quantities of calcium ions that have been stored in the reticulum.
4. The calcium ions initiate attractive forces between the actin and myosin filaments, causing them to slide together; this is the contractile process.
5. After a fraction of a second, the calcium ions are pumped back into the sarcoplasmic reticulum, where they remain stored until a muscle action potential arrives; this removal of the calcium ions from the myofibrils causes muscle contraction to cease.

MOLECULAR MECHANISM OF MUSCLE CONTRACTION
(p. 76)

Muscle contraction occurs by a sliding filament mechanism. Mechanical forces generated by the interaction of myosin cross-bridges with actin filaments cause the actin filaments to slide inward among the myosin filaments. Under resting conditions, these forces are inhibited, but when an action potential travels over the muscle fiber membrane, the sarcoplasmic reticulum releases large quantities of calcium ions that rapidly penetrate the myofibrils. These calcium ions, in turn, activate the forces between the myosin and actin filaments and contraction begins.

The myosin filament is composed of multiple myosin molecules. The myosin molecule is composed of six polypeptide chains: two heavy chains and four light chains. The two heavy chains wrap around each other to form a double helix. One end of each chain is folded into a myosin head. The four light chains are also parts of the myosin heads. The tails of myosin molecules bundle together to form the body of the filament, whereas the myosin heads and part of each myosin molecule hang outward to the sides of the body, providing an arm that extends the head outward from the body. The protruding arms and heads together are called cross-bridges. An important feature of the myosin head that is essential to muscle contraction is that it functions as an ATPase enzyme. This property allows the head to cleave ATP and to use the energy derived from the high-energy phosphate bond of ATP to energize the contraction process.

The actin filament is composed of actin, tropomyosin, and troponin. Each actin filament is about 1 micrometer long. The bases of the actin filaments are inserted strongly into the Z discs, whereas the other ends protrude in both directions into the adjacent sarcomeres to lie in the spaces between the myosin molecules.

- *F-actin*, a double stranded F-actin protein molecule, is the backbone of the actin filament. Each strand of the double F-actin helix is composed of polymerized G-actin molecules; attached to each G-actin molecule is one molecule of ADP. The ADP molecules are thought to be the active sites where the cross-bridges of the myosin filaments interact to cause muscle contraction.
- *Tropomyosin* molecules are loosely connected to the F-actin strands, being wrapped spirally around the sides of the F-actin helix. In the resting state, the tropomyosin molecules are believed to lie on top of the active sites of the actin strands, so that attraction cannot occur between the actin and myosin filaments. Contraction is thereby prevented.
- *Troponin* is a complex of three protein subunits: troponin I, which has a strong affinity for actin; troponin T, which has a strong affinity for tropomyosin; and troponin C, which has a strong affinity for calcium ions. This complex is believed to attach the tropomyosin to the actin. The strong affinity of troponin C for calcium ions is believed to initiate the contraction process.

Interaction of Myosin, Actin Filaments, and Calcium Ions to Cause Contraction (p. 77)

The actin filament is inhibited by the troponin-tropomyosin complex; activation is stimulated by calcium ions.

- *Inhibition by troponin-tropomyosin complex.* The active sites on the normal actin filament of the relaxed muscle are inhibited or physically covered by the troponin-tropomyosin complex. Consequently, the sites cannot attach to the heads of the myosin filaments to cause contraction until the inhibitory effect of the troponin-tropomyosin complex is inhibited.
- *Activation by calcium ions.* The inhibitory effect of the troponin-tropomyosin complex on the actin filaments is inhibited in the presence of calcium ions. Calcium ions combine with troponin C, causing the troponin complex to undergo a conformational change that is thought to tug on the tropomyosin molecule, causing it to move deeper into the groove between the two actin strands. This "uncovers" the active sites of the actin, thus allowing contraction to proceed.

A "walk along" theory can explain how the activated actin filament and the myosin cross-bridges interact to cause contraction. When a myosin head attaches to an active site, the head tilts automatically toward the arm that is dragging along the actin filament. This tilt of the head is called the power stroke. Then, immediately after tilting, the head automatically breaks away from the active site. Next, the head returns to its normal perpendicular direction. In this position, it combines with a new active site farther along the actin filament. The head then tilts again to cause a new power stroke, and the actin filament moves another step. Thus, the heads of the cross-bridges bend back and forth and step by step walk along the actin filament, pulling the ends of the actin filaments toward the center of the myosin filament.

ATP is the source of energy for contraction—chemical events in the motion of the myosin heads. Large amounts of ATP are cleaved to form ADP during the contraction process. The following is the sequence of events by which this occurs:

1. Before contraction begins, the heads of the cross-bridges bind with ATP. The ATPase activity of the myosin head cleaves the ATP but leaves the cleavage products, ADP plus inorganic phosphate, bound to the head. In this state, the conformation of the head is such that it extends perpendicularly toward the actin filament but is not yet attached to the actin.

2. Next, when the troponin-tropomyosin complex binds with calcium ions, active sites on the actin filament are uncovered, and the myosin heads then bind with these.

3. The bond between the head of the cross-bridge and the active site causes the head to tilt toward the arm of the cross-bridge. This provides the power stroke for pulling of the actin filament. The energy that activates the power stroke is the energy that was stored as a result of the conformational change in the head that occurred earlier when the ATP molecule was cleaved.

4. Once the head of the cross-bridge is tilted, the ADP and inorganic phosphate that were previously attached to the head can be released. At the site of release of the ADP, a new molecule of ATP binds. This binding, in turn, causes detachment of the head from the actin.

5. After the head has detached from the actin, a new molecule of ATP is cleaved to begin the next cycle leading to the power stroke; that is, the energy again "cocks" the head back to its perpendicular condition, at which it is ready to begin the new power stroke cycle.

6. When the cocked head with its stored energy derived from the cleaved ATP binds with a new active site on the actin filament, it becomes uncocked and again provides the power stroke.

Degree of Actin and Myosin Filament Overlap — Effect on Tension Developed by the Contracting Muscle (p. 79)

The strength of contraction is maximal when there is maximal overlap between actin filaments and the cross-bridges of the myosin filaments. A muscle cannot develop tension at very long sarcomere lengths because there is no overlap between actin and myosin filaments. As the sarcomere shortens and actin and myosin filaments begin to overlap, the tension increases progressively until the sarcomere length decreases to about 2.2 micrometers. At this point, the actin filament has already overlapped all of the cross-bridges of the myosin filament. Full tension is maintained until a sarcomere length is about 2.0 micrometers. On further shortening, the ends of the two actin filaments begin to overlap (in addition to overlapping the myosin filaments), causing muscle tension to decrease. When the sarcomere length decreases to about 1.65 micrometers, the two Z discs of the sarcomere abut the ends of the myosin filaments, and the strength of contraction decreases precipitously.

ENERGETICS OF MUSCLE CONTRACTION (p. 80)

Muscle contraction requires ATP to perform three main functions.

- Most of the ATP is used to activate the walk-along mechanism of muscle contraction.
- Calcium is pumped back into the sarcoplasmic reticulum after the contraction ends.
- Sodium and potassium ions are pumped through the muscle fiber membrane to maintain an appropriate ionic environment for the propagation of action potentials.

There are three main sources of energy for muscle contraction. The concentration of ATP in the muscle fiber is sufficient to maintain full contraction for only 1 to 2 seconds. After the ATP is split into ADP, the ADP is rephosphorylated to form a new ATP. There are several sources of the energy for this rephosphorylation.

- *Phosphocreatine* carries a high-energy bond that is similar to that of ATP but has more free energy. The energy released from this bond causes bonding of a new phosphate ion to ADP to reconstitute the ATP. The combined energy of ATP and phosphocreatine is capable of causing maximal muscle contraction for only 5 to 8 seconds.
- The *breakdown of glycogen* to pyruvic acid and lactic acid liberates energy that is used to convert ADP to ATP. The glycolytic reactions can occur in the absence of oxygen. The rate of formation of ATP by the glycolytic process is about two and one-half times as rapid as ATP formation when the cellular food stuffs react with oxygen. Glycoly-

sis alone can sustain maximum muscle contraction for only about 1 minute.

- *Oxidative metabolism* occurs when oxygen is combined with the various cellular food stuffs to liberate ATP. More than 95 per cent of all energy used by the muscles for sustained, long-term contraction is derived from this source. The food stuffs that are consumed are carbohydrates, fats, and proteins.

CHARACTERISTICS OF WHOLE MUSCLE CONTRACTION
(p. 81)

Isometric contractions differ from isotonic contractions.

- *Isometric contraction* occurs when the muscle does not shorten during contraction. True isometric contractions cannot be generated in the intact body because the so-called *series elastic components* will stretch during the contraction, allowing some shortening of the muscle. These elastic elements include the tendons, sarcolemmal ends of muscle fibers, and, perhaps, the hinged arms of the myosin cross-bridges.
- *Isotonic contraction* occurs when the muscle shortens and the tension on the muscle remains constant. The characteristics of the isometric contraction depend on the load against which the muscle contracts as well as on the inertia of the load.

Fast fibers are adapted for powerful muscle contractions, whereas slow fibers are adapted for prolonged muscle activity. Each muscle is composed of a mixture of so-called fast and slow muscle fibers, with still other fibers that are between these two extremes.

- *Slow fibers (red muscle)* (1) are smaller muscle fibers, (2) are innervated by smaller nerve fibers, (3) have high capillarity and large numbers of mitochondria to support high levels of oxidative metabolism, and (4) contain large amounts of myoglobin, which gives the slow muscle a reddish appearance and the name red muscle. The deficit of the red myoglobin in fast muscle provides the name white muscle.
- *Fast fibers (white muscle)* are (1) larger for great strength of contraction, (2) have extensive sarco-

plasmic reticulum for rapid release of calcium ions, (3) have large amounts of glycolytic enzymes for rapid release of energy, and (4) have lower capillarity and fewer mitochondria because oxidative metabolism is of secondary importance.

Mechanics of Skeletal Muscle Contraction (p. 82)

Force summation is the adding together of individual twitch contractions to increase the intensity of overall muscle contraction. Summation occurs in two ways:

- *Multiple motor unit summation.* When the central nervous system sends a weak signal to contract a muscle, the motor units in the muscle that contain the smallest and fewest muscle fibers are stimulated in preference to the larger motor units. Then, as the strength of the signal increases, larger motor units also begin to be excited, with the largest motor units often having up to 50 times as much contractile force as the smallest units.
- *Frequency summation and tetanization.* As the frequency of muscle contraction increases, there comes a point at which each new contraction occurs before the preceding one ends. As a result, the second contraction is added partially to the first, so that the total strength of contraction rises progressively with increasing frequency. When the frequency reaches a critical level, the successive contractions fuse together and appear to be completely smooth; this is called tetanization.

Muscle hypertrophy is an increase in the total mass of a muscle; muscle atrophy is a decrease in the mass.

- *Muscle hypertrophy* results from an increase in the number of actin and myosin filaments in each muscle fiber. When the number of contractile proteins increases sufficiently, the myofibrils split within each muscle fiber to form new myofibrils. Thus, it is mainly this great increase in the number of additional myofibrils that causes muscle fibers to hypertrophy.
- *Muscle atrophy.* When a muscle remains unused for a long period, the rate of decay of the contractile proteins occurs more rapidly than the rate

of replacement; therefore, muscle atrophy occurs. Atrophy begins almost immediately when a muscle loses its nerve supply because it no longer receives the contractile signals that are required to maintain normal muscle size.

7

Excitation of Skeletal Muscle

A. Neuromuscular Transmission and
B. Excitation-Contraction Coupling

TRANSMISSION OF IMPULSES FROM NERVES TO SKELETAL MUSCLE FIBERS: NEUROMUSCULAR JUNCTION
(p. 87)

The skeletal muscle fibers are innervated by large, myelinated nerve fibers that originate in motoneurons of the spinal cord. Each nerve fiber normally stimulates from three to several hundred skeletal muscle fibers. The nerve ending makes a junction, called the *neuromuscular junction*, and the action potential in the fiber travels in both directions toward the muscle fiber ends.

Physiologic Anatomy of the Neuromuscular Junction

- *Motor end plate.* The nerve fiber branches at its end to form a complex of terminals, which invaginate into the muscle fiber but lie outside the muscle fiber plasma membrane. The entire structure is called the motor end plate. It is covered by Schwann cells that insulate it from the surrounding fluids.
- *Synaptic gutter* and *synaptic cleft.* The invagination of muscle fiber membrane is called the synaptic gutter, or synaptic trough, and the space between the terminal and fiber membrane is called the synaptic cleft. At the bottom of the gutter are numerous smaller folds of the muscle membrane called *subneural clefts*, which greatly increase the surface area where the synaptic transmitter can act.
- *Axon terminal.* Acetylcholine, an excitatory transmitter, is synthesized in the terminal but is rapidly absorbed into synaptic vesicles. Attached to the matrix of the basal lamina are large quantities

of the enzyme acetylcholinesterase, which is capable of destroying acetylcholine.

Secretion of Acetylcholine by the Nerve Terminals (p. 87)

When a nerve impulse reaches the neuromuscular junction, vesicles containing acetylcholine are released into the synaptic space. On the inside surface of the neural membrane are linear dense bars. To the side of each dense bar are voltage-gated calcium channels. When the action potential spreads over the terminal, these channels open, allowing calcium to diffuse into the terminal. The calcium ions are believed to exert an attractive influence on the acetylcholine vesicles, drawing them to the neural membrane adjacent to the dense bars. Some of the vesicles fuse with the neural membrane and empty their acetylcholine into the synaptic space via the process of exocytosis.

Acetylcholine opens acetylcholine-gated ion channels on the postsynaptic membrane. Acetylcholine-gated ion channels are located on the muscle membrane immediately below the dense bar areas, where the acetylcholine vesicles empty into the synaptic space. When two acetylcholine molecules attach to the channel receptors, a conformational change opens the channel. The principal effect of opening the acetylcholine-gated channels is to allow large numbers of sodium ions to pour into the inside of the fiber, carrying with them large numbers of positive charges. This effect creates a local potential change at the muscle fiber membrane called the end plate potential. In turn, this end plate potential initiates an action potential at the muscle membrane and thus causes muscle contraction.

Acetylcholine released into the synaptic space is destroyed by acetylcholinesterase or simply diffuses away. The acetylcholine, once released into the synaptic space, continues to activate the acetylcholine receptors as long as it remains in the space. Most of the acetylcholine is destroyed by the enzyme acetylcholinesterase, which is mainly attached to the basal lamina. A small amount diffuses out of the synaptic space and is then no longer available to act on the muscle fiber membrane. The short period during which the acetylcholine remains in the synaptic space—a few milliseconds at most—is almost always sufficient to excite the muscle fiber. Then, the rapid removal of the acetylcholine prevents muscle re-excitation after the fiber has recovered from the first action potential.

Acetylcholine produces an end plate potential that may or may not excite the skeletal muscle fiber. The movement of sodium ions into the muscle fiber causes the internal membrane potential in the local area of the end plate to increase in the positive direction as much as 50 to 75 millivolts, creating a local potential called the *end plate potential*. Because a sudden increase in membrane potential of more than 20 to 30 millivolts is normally sufficient to initiate the positive feedback effect of sodium channel activation, the end plate potential created by the acetylcholine stimulation is normally far greater than that necessary to initiate an action potential in the muscle fiber.

Molecular Biology of Acetylcholine Formation and Release
(p. 90)

Because the neuromuscular junction is sufficiently large to be easily studied, it is one of the few synapses of the nervous system at which most of the details of chemical transmission have been worked out. The formation and release of acetylcholine at this junction occur in the following stages:

- Small vesicles are formed by the Golgi apparatus in the cell body of the motoneuron in the spinal cord. These vesicles are then transported through the core of the axon from the central cell body in the spinal cord to the neuromuscular junction at the tips of the nerve fibers.
- Acetylcholine is synthesized in the nerve terminal and then transported through the membranes of the vesicles to their interior, where it is stored in a highly concentrated form.
- Under resting conditions, an occasional vesicle fuses with the surface membrane of the nerve terminal and releases its acetylcholine into the synaptic gutter. When this occurs, a so-called miniature end plate potential occurs in the local area of the muscle fiber because of the action of this "packet" of 10,000 acetylcholine molecules.
- When an action potential arrives at the nerve terminal, it opens calcium channels in the membrane of the terminal. As a result, the calcium ion concentration in the terminal increases greatly, which in turn increases the rate of fusion of the acetyl-

choline vesicles with the terminal membrane, causing exocytosis of acetylcholine into the synaptic cleft. The acetylcholine is then split by acetylcholinesterase into acetate ion and choline, and the choline is actively reabsorbed into the neural terminal to be reused in the formation of new acetylcholine.

- After exocytosis, the membrane of each vesicle becomes part of the nerve cell membrane. However, the number of vesicles available in the nerve ending is sufficient to allow transmission of only a few thousand nerve impulses. For continued function of the neuromuscular junction, the vesicles must be retrieved from the nerve membrane. Retrieval is achieved through endocytosis.

Drugs That Affect Transmission at the Neuromuscular Junction (p. 90)

Drugs can affect the neuromuscular junction by having acetylcholine-like actions, blocking neuromuscular transmission, and inactivating acetylcholinesterase.

- *Drugs that have acetylcholine-like actions.* Many compounds, including methacholine, carbachol, and nicotine, have the same effect on the muscle fiber as does acetylcholine. The difference between these drugs and acetylcholine is that they are not destroyed by cholinesterase or they are destroyed slowly, so that once applied to the muscle fiber, the action persists for many minutes to several hours.
- *Drugs that block neuromuscular transmission.* A group of drugs known as the curariform drugs can prevent the passage of impulses from the end plate into the muscle. Thus, D-tubocurarine competes with acetylcholine for the acetylcholine receptor sites, so that the acetylcholine generated by the end plate cannot increase the permeability of the muscle membrane acetylcholine channels sufficiently to initiate an action potential, and contraction of the muscle cannot occur.
- *Drugs that inactivate acetylcholinesterase.* Three particularly well known drugs, neostigmine, physostigmine, and diisopropyl fluorophosphate, inactivate acetylcholinesterase. As a result, acetylcholine increases in quantity with successive nerve impulses so that large amounts of acetylcholine can

accumulate and then repetitively stimulate the muscle fiber. Neostigmine and physostigmine last for up to several hours. Diisopropyl fluorophosphate, which has potential military use as a powerful "nerve" gas poison, inactivates acetylcholinesterase for weeks, which makes this a particularly lethal poison.

Myasthenia Gravis Causes Paralysis

Paralysis occurs because of the inability of the neuromuscular junctions to transmit signals from the nerve fibers to the muscle fibers. Pathologically, myasthenia gravis is thought to be an autoimmune disease in which patients have developed antibodies against their own acetylcholine-activated ion channels. Regardless of the cause, the end plate potentials that occur in the muscle fibers are too weak to adequately stimulate the muscle fibers. If the disease is sufficiently intense, the patient dies of paralysis—in particular, paralysis of the respiratory muscles. The disease usually can be ameliorated by the administration of neostigmine or another anticholinesterase drug. This treatment allows acetylcholine to accumulate in the synaptic cleft. Within minutes, some of these paralyzed patients begin to function almost normally.

MUSCLE ACTION POTENTIAL (p. 91)

The conduction of action potentials in nerve fibers is *qualitatively* similar to that of skeletal muscle fibers. Some of the *quantitative* differences and similarities include the following:

- *The resting membrane potential* is about -80 to -90 millivolts in skeletal muscle fibers, which is similar to that of large myelinated nerve fibers.
- *The duration of the action potential* is 1 to 5 milliseconds in skeletal muscle, which is about five times as long as that in large myelinated nerves.
- *The velocity of conduction* is 3 to 5 m/sec in skeletal muscle, which is about $\frac{1}{18}$ the velocity of conduction in the large myelinated nerve fibers that excite skeletal muscle.

The action potential spreads to the interior of the muscle fiber by way of the transverse tubule system. The skeletal muscle fiber is so large that action potentials spreading along its surface membrane cause almost no current flow deep within the fiber. These electrical currents must penetrate to the vicinity of all the separate myofibrils for contraction to occur. This is achieved through transmission of the action potentials along transverse tubules (T tubules) that penetrate through the muscle fiber from one side to the other. The T tubule action potentials in turn cause the sarcoplasmic reticulum to release calcium ions in the immediate vicinity of all the myofibrils, and these calcium ions then initiate contraction. The overall process is called *excitation-contraction coupling*.

EXCITATION-CONTRACTION COUPLING (p. 91)

Transverse tubules are internal extensions of the cell membrane. The transverse tubules (T tubules) run transverse to the myofibrils. They begin at the cell membrane and penetrate from one side of the muscle fiber to the opposite side. Where the T tubules originate from the cell membrane, they are open to the exterior and thus contain extracellular fluid in their lumens. Because the T tubules are internal extensions of the cell membrane, when an action potential spreads over a muscle fiber membrane, it spreads along the T tubules to the interior of the muscle fiber.

The sarcoplasmic reticulum is composed of longitudinal tubules and terminal cisternae. The longitudinal tubules run parallel to the myofibrils and terminate in large chambers called terminal cisternae. The cisternae abut the T tubules. In cardiac muscle, a single T tubule network for each sarcomere is located at the level of the Z disc. In mammalian skeletal muscle, there are two T tubule networks for each sarcomere located near the two ends of the myosin filaments, which are the points at which the actual mechanical forces of muscle contraction are created. Thus, mammalian skeletal muscle is optimally organized for rapid excitation of muscle contraction.

Calcium ions are released from the sarcoplasmic reticulum. Calcium ions located in the vesicular tubules of the sarcoplasmic reticulum are released when an action potential occurs in the adjacent T tubule. The action potential itself is thought to cause rapid opening of calcium channels through the membranes of the cisternae and their attached longitudinal tubules of the sarcoplasmic reticulum. These channels remain open for a few milliseconds; during this time, the calcium ions responsible for muscle contraction are released into the sarcoplasm surrounding the myofibrils.

A calcium pump removes calcium ions from the sarcoplasmic fluid. A continually active calcium pump located in the walls of the sarcoplasmic reticulum pumps calcium ions away from the myofibrils back into the sarcoplasmic tubules. This pump can concentrate the calcium ions about 10,000-fold inside the tubules. In addition, inside the reticulum is a calcium-binding protein called *calsequestrin* that can provide another 40-fold increase in the storage of calcium. This transfer of calcium into the sarcoplasmic reticulum depletes calcium ions in the myofibrillar fluid, thereby terminating the muscle contraction.

8

Contraction and Excitation of Smooth Muscle

CONTRACTION OF SMOOTH MUSCLE

Many of the principles of contraction that apply to skeletal muscle also apply to smooth muscle. Most important, essentially the same attractive forces that occur between myosin and actin filaments cause contraction in smooth muscle, but the internal physical arrangement of smooth muscle fibers is entirely different from that of skeletal muscle.

Types of Smooth Muscle (p. 95)

The smooth muscle of each organ is distinct from that of most other organs in its physical dimensions, organization into bundles or sheets, response to different types of stimuli, characteristics of innervation, and function. In general, smooth muscle can be divided into two major types:

- *Multi-unit smooth muscle.* The most important characteristics of multi-unit smooth muscle fibers are that each fiber can contract independently of the others and that the control is exerted mainly by nerve signals. Examples include the smooth muscle fibers of the ciliary muscle of the eye, the iris of the eye, and the piloerector muscles that cause erection of the hairs when stimulated by the sympathetic nervous system.
- *Single-unit smooth muscle.* This type is also called *unitary smooth muscle, syncytial smooth muscle,* and *visceral smooth muscle.* A mass of hundreds to millions of muscle fibers contract together as a single unit. The cell membranes of the fibers adhere to one another so that force generated in one fiber can be transmitted to the next. In addition, the cell membranes are joined by gap junctions so that action potentials can travel from one fiber to

the next and cause the muscle fibers to contract together. This type of muscle is found in the walls of the gut, bile ducts, ureters, uterus, and blood vessels.

Physical Basis for Smooth Muscle Contraction (p. 96)

Smooth muscle does not have the same striated arrangement of actin and myosin filaments as that found in skeletal muscle.

- *Actin filaments attach to dense bodies.* Some of the dense bodies are dispersed inside the cell and held in place by a scaffold of structural proteins linking one dense body to another. Others are attached to the cell membrane and form bonds with dense bodies of adjacent cells, allowing the force of contraction to be transmitted from one cell to the next.
- *Myosin filaments are interspersed among the actin filaments.* The myosin filaments have a diameter that is more than twice as great as that of the actin filaments.
- *Contractile units.* The individual contractile units consist of actin filaments radiating from two dense bodies; these filaments overlap a single myosin filament that is located midway between the dense bodies. This contractile unit is similar to the contractile unit of skeletal muscle but lacks the regularity of the skeletal muscle structure; in fact, the dense bodies of smooth muscle serve the same role as the Z discs in skeletal muscle.

Comparison of Smooth Muscle Contraction With Skeletal Muscle Contraction (p. 97)

Unlike skeletal muscle contractions, most smooth muscle contractions are prolonged tonic ones that sometimes last hours or even days. Both the physical and chemical characteristics of smooth muscle are different than those of skeletal muscle. The following are some of the differences:

- *Slow cycling of the cross-bridges.* The rapidity of cross-bridge cycling in smooth muscle (i.e., the rate of myosin cross-bridge attachment and release with actin) is much slower in smooth muscle than in skeletal muscle, although the fraction of time during which the cross-bridges remain attached to the actin filaments, which is the major factor that determines the force of contraction, is believed to be greatly increased in smooth muscle.

- *Low energy requirement.* Only $1/10$ to $1/300$ as much energy is required to sustain a contraction in smooth muscle as is required in skeletal muscle. This difference is believed to result from the slow cycling of the cross-bridges and because only one molecule of ATP is required for each cycle, regardless of its duration. This low energy utilization is important because many organs (e.g., intestines, urinary bladder, gallbladder) must maintain tonic muscle contraction almost indefinitely.

- *Slow onset of contraction and relaxation.* A typical smooth muscle tissue begins to contract 50 to 100 milliseconds after it is excited and has a total contraction time of 1 to 3 seconds, which is 30 times as long as that of an average skeletal muscle. The slow onset of contraction and the long duration of contraction are probably caused by the slow cycling of the cross-bridges.

- *Increased maximum force of contraction.* The maximum force of contraction of smooth muscle is often greater than that of skeletal muscle. This increased force of attraction is postulated to result from the prolonged period of attachment of the myosin cross-bridges to the actin filaments.

Smooth muscle can shorten a far greater percentage of its length than can skeletal muscle. Skeletal muscle has a useful distance of contraction of only about one fourth to one third of its stretched length, whereas smooth muscle can often contract more than two thirds of its stretched length. This allows smooth muscle in the gut, bladder, and blood vessels to change their lumen diameters from very large to almost zero. This feature of smooth muscle may be attributed to the following characteristics:

- Some contractile units of smooth muscle are likely to have optimal overlapping of their actin and myosin filaments at one length of the muscle and other units at other lengths, rather than all units being synchronized together, as usually occurs in skeletal muscle. Therefore, a greater distance of contraction can be achieved.

- The actin filaments in smooth muscle are much longer than those in skeletal muscle, so that these filaments can be pulled along the myosin filaments for a much greater distance during smooth muscle contraction than occurs during skeletal muscle contraction.

The "latch mechanism" facilitates prolonged holding contractions. Once smooth muscle has developed full contraction, the degree of activation of the muscle can usually be reduced to far less than the initial level, yet the muscle will maintain its full force of contraction. Furthermore, the energy consumed to maintain contraction is often minuscule, sometimes as little as 1/300 of the energy required for comparable sustained skeletal muscle contraction. This is called the "latch mechanism." The importance of the latch mechanism is that it can maintain prolonged tonic contraction in smooth muscle for hours with little use of energy.

Regulation of Contraction by Calcium Ions (p. 98)

Calcium ions combine with calmodulin to cause activation of myosin kinase and phosphorylation of the myosin head. Smooth muscle does not contain troponin but instead contains *calmodulin,* another regulatory protein. Although this protein reacts with calcium ions, it is different from troponin in the manner in which it initiates the contraction; calmodulin does this by activating the myosin crossbridges. This activation and subsequent contraction occur in the following sequence:

1. The calcium ions bind with calmodulin; the calmodulin-calcium combination then joins with and activates *myosin kinase*, a phosphorylating enzyme.
2. One of the light chains of each myosin head, called the *regulatory chain*, becomes phosphorylated in response to the myosin kinase.
3. When the regulatory chain is phosphorylated, the head has the capability of binding with the actin filament, thus causing muscle contraction. When this chain is not phosphorylated, the attachment-detachment cycling of the head with the actin filament does not occur.

Myosin phosphatase is important for cessation of contraction. When the calcium ion concentration falls below a critical level, the aforementioned processes automatically reverse except for the phosphorylation of the myosin head. Reversal of this requires another enzyme, myosin phosphatase, which splits the phosphate from the regulatory light chain; then, the cycling stops, and the contraction ceases.

NEURAL AND HORMONAL CONTROL OF SMOOTH MUSCLE CONTRACTION (p. 99)

Neuromuscular Junctions of Smooth Muscle

Neuromuscular junctions of the highly structured type found on skeletal muscle fibers do not occur in smooth muscle.

- *Diffuse junctions.* These are the sites of transmitter release. In most instances, the autonomic nerve fibers form so-called diffuse junctions that secrete their transmitter substance into the matrix coating of the smooth muscle; the transmitter substance then diffuses to the cells. Where there are many layers of muscle cells, the nerve fibers often innervate only the outer layer, and the muscle action potential then travels from this outer layer to the inner layer.

- *Varicosities on the axons.* The axons that innervate smooth muscle fibers do not have typical branching end feet of the type in the motor end plate on skeletal muscle fibers. Instead, most of the fine terminal axons have multiple varicosities that are distributed along their axes. The varicosities contain vesicles that are loaded with transmitter substance.

- *Contact junctions.* In the multi-unit type of smooth muscle, the varicosities lie directly on the muscle fiber membrane. These so-called contact junctions have a function similar to that of the skeletal muscle neuromuscular junctions; the latent period of contraction of these smooth muscle fibers is considerably shorter than that of fibers stimulated by the diffuse junctions.

Acetylcholine and norepinephrine can have excitatory or inhibitory effects at the smooth muscle neuromuscular junction. These transmitter substances are secreted by the autonomic nerves innervating smooth muscle, but they are never secreted by the same nerve fibers. Acetylcholine is an excitatory transmitter substance for smooth muscle fibers in some organs but an inhibitory substance for smooth muscle in other organs. When acetylcholine excites a muscle fiber, norepinephrine ordinarily inhibits it, and vice versa. The type of receptor determines whether the smooth muscle is inhibited or excited and which one of the two transmitters, acetylcholine or norepinephrine, is effective in causing the excitation or inhibition.

Membrane Potentials and Action Potentials in Smooth Muscle (p. 99)

The resting membrane potential depends on the type of smooth muscle and the momentary condition of the muscle. It is usually about -50 to -60 millivolts, or about 30 millivolts less negative than in skeletal muscle.

Action potentials occur in single-unit smooth muscle, such as visceral muscle, in the same way that they occur in skeletal muscle. They do not occur in most multi-unit types of smooth muscle. The action potentials of visceral smooth muscle occur in the following two forms:

- *Spike potentials.* Typical spike action potentials occur in most types of single-unit smooth muscle. These can be elicited by electrical stimulation, by stretch, or by the action of hormones and transmitter substances, or as a result of spontaneous generation in the muscle fiber itself.
- *Action potentials with plateaus.* The onset of this type of action potential is similar to that of the typical spike potential. However, the repolarization is delayed for several hundred milliseconds. The plateau accounts for the prolonged periods of contraction that occur in the ureter, the uterus under some conditions, and some types of vascular smooth muscle.

Calcium ions are important for generating the smooth muscle action potential. Sodium participates little in the generation of the action potential in most smooth muscle. Instead, the flow of calcium ions to the interior of the fiber is mainly responsible for the action potential. The calcium channels open much more slowly than do sodium channels, but they also remain open much longer. This accounts in large measure for the slow action potentials of smooth muscle fibers.

Slow wave potentials in single-unit smooth muscle cause the spontaneous generation of action potentials. Slow waves are slow oscillations in membrane potential. The slow wave itself is not an action potential.

- *Cause of slow waves.* Two possible causes of slow waves are (1) oscillations in sodium pump activity, which cause the membrane potential to become more negative when sodium is pumped rapidly and less negative when sodium is pumped slowly, and (2) the conductances of the ion channels, which may increase and decrease rhythmically.
- *Importance of slow waves.* Action potentials can be initiated when the potential of the slow wave rises above threshold (about -35 millivolts). The action potential spreads over the muscle mass, and contraction occurs.

Spontaneous action potentials are often generated when visceral (single-unit) smooth muscle is stretched. Spontaneous action potentials result from a combination of the normal slow wave potentials in addition to a decrease in the negativity of the membrane potential caused by the stretch itself. This response to stretch allows the gut wall, when excessively stretched, to contract automatically and resist the stretch.

Smooth Muscle Contraction Without Action Potentials — Effect of Local Tissue Factors and Hormones (p. 101)

Smooth muscle relaxation in blood vessels occurs in response to local tissue factors. This vasodilatory response is extremely important for the local control of blood flow. Factors that cause vasodilation include a lack of oxygen, excess carbon dioxide, increased hydrogen ion concentration, adenosine, lactic acid, potassium ions, diminished calcium ion concentration, and decreased body temperature.

Most of the circulating hormones in the body affect smooth muscle contraction to some degree. Some of the more important blood-borne hormones are norepinephrine, epinephrine, acetylcholine, angiotensin, vasopressin, oxytocin, serotonin, and histamine. A hormone causes contraction when the muscle cell membrane contains excitatory receptors for the respective hormone. Conversely, the hormone causes inhibition if the membrane contains inhibitory receptors.

Source of Calcium Ions That Cause Contraction (1) Through the Cell Membrane and (2) From the Sarcoplasmic Reticulum (p. 102)

Most of the calcium ions that cause contraction enter the muscle cell from the extracellular fluid. Because the smooth muscle fibers are extremely small (in contrast to the skeletal muscle fibers), these calcium ions can diffuse to all parts of the smooth muscle and elicit the contractile process. Therefore, the force of contraction of smooth muscle is highly dependent on the extracellular fluid calcium ion concentration. The sarcoplasmic reticulum, from which virtually all the calcium ions are derived in skeletal muscle contraction, is only rudimentary in most smooth muscle.

Calcium pumps remove calcium ions from the intracellular fluids and thereby terminate contraction. Calcium is removed by calcium pumps that pump the calcium ions out of the smooth muscle fiber and back into the extracellular fluid or pump the calcium ions into the sarcoplasmic reticulum. These pumps are slow acting in comparison with the fast-acting sarcoplasmic reticulum pump in skeletal muscle. The duration of smooth muscle contraction is often thereby on the order of seconds rather than hundredths to tenths of a second, as occurs for skeletal muscle.

UNIT

III

The Heart

9

Heart Muscle; The Heart as a Pump

The human heart is composed of two pumps: the *right heart*, which receives blood from the peripheral organs and pumps it through the lungs, and the *left heart*, which receives oxygenated blood from the lungs and pumps it back to the peripheral organs. Each pump is composed of an *atrium* and a *ventricle*. The atria function as primer pumps that fill the ventricles with blood. The ventricles contract and impart high pressure to the blood, which is responsible for propelling it through the circulation. The heart has a special conduction system that maintains its own rhythmicity and transmits action potentials throughout the heart muscles.

Distinguishing Features of Cardiac Muscle Compared with Skeletal Muscle (p. 107)

The similarities and differences in cardiac and skeletal muscle include the following:

- *Both cardiac and skeletal muscle are striated and have actin* and *myosin filaments* that interdigitate and slide along each other during contraction.
- *Cardiac muscle has intercalated discs between cardiac muscle cells,* which is different from skeletal muscle. These discs have very low electrical resistance, which allows an action potential to travel freely between cardiac muscle cells.
- The *cardiac muscle is a syncytium* of many heart muscle cells in which the action potential spreads rapidly from cell to cell.
- The *atrioventricular (A-V) bundle conducts impulses from the atria to the ventricles.* This is an exclusive pathway because the atrial syncytium and ventricular syncytium are insulated from one another by fibrous tissue.

Action Potentials in Cardiac Muscle (p. 108)

The resting membrane potential of cardiac muscle is about -85 to -95 millivolts, and the action potential is 105 millivolts. The membranes remain depolarized for 0.2 second in the atria and for 0.3 second in the ventricles.

Slow entry of sodium and calcium ions into the cardiac muscle cells is one of the causes of the action potential plateau. The action potential of skeletal muscle is caused by entry of sodium through *fast sodium channels* that remain open for only a few 10,000ths of a second. In cardiac muscle, the fast sodium channels also open at the initiation of the action potential, but cardiac muscle has unique *slow calcium channels,* or *calcium-sodium channels.* Through the slow channels, calcium and sodium ions flow into the cell after the initial spike of the action potential, and they maintain the plateau. Calcium that enters the cell through these channels also promotes cardiac muscle contraction.

Another cause of the plateau of the action potential is a decrease in the permeability of cardiac muscle cells to potassium ions. This decrease in cardiac potassium permeability also prevents the return of the membrane potential in cardiac muscle; this mechanism is not present in skeletal muscle. When the slow calcium-sodium channels close after 0.2 to 0.3 second, the potassium permeability increases rapidly and thus returns membrane potential to its resting level.

Diffusion of calcium into the myofibrils promotes muscle contraction. The action potential spreads into each cardiac muscle fiber along the *transverse (T) tubules,* and this causes the longitudinal sarcoplasmic tubules to release calcium ions into the sarcoplasmic reticulum. These calcium ions catalyze the chemical reactions that promote the sliding of the actin and myosin filaments along one another to cause muscle contraction. This mechanism is also present in skeletal muscle.

Another means of calcium entry into the sarcoplasm, however, is unique to cardiac muscle. The T tubules of cardiac muscles have 5 times as great a diameter as in skeletal muscle, and the volume of these tubules is 25 times as great. These T tubules contain large amounts of calcium that are released during the action potential. In addition, the T tubules open directly into the extracellular fluid in

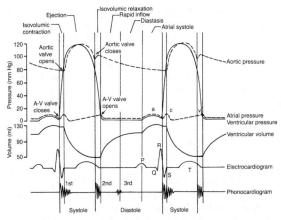

Figure 9–1 Events of the cardiac cycle for left ventricular function showing changes in left atrial pressure, left ventricular pressure, aortic pressure, ventricular volume, the electrocardiogram, and the phonocardiogram.

cardiac muscle, so that their calcium content depends on the extracellular calcium concentration. At the end of the plateau of the action potential, the influx of calcium ions into the muscle fiber abruptly stops, and calcium is pumped back into the sarcoplasmic reticulum and T tubules. Thus, the contraction ends.

THE CARDIAC CYCLE (p. 110)

The events that occur at the beginning of a heartbeat and last until the beginning of the next heartbeat are called the *cardiac cycle*.

- Each beat of the heart begins with a spontaneous action potential that is initiated in the *sinus node* of the right atrium near the opening of the superior vena cava.
- The action potential travels through both atria and the *A-V node and bundle* and into the ventricles.
- A delay of more than $\frac{1}{10}$ of a second occurs in the A-V node and bundle, which allows the atria to contract before the ventricles contract.

Figure 9–1 shows the events of the cardiac cycle. The ventricles fill with blood during *diastole* and contract during *systole*. The top three curves of this diagram show the aortic pressure, left ventricular pressure, and left atrial pressure. The curves below these are the changes in ventricular volume, the electrocardiogram, and the phonocardiogram, a recording of heart sounds.

The spread of the action potential in the heart initiates each heartbeat. The electrocardiogram is a recording of the voltage generated by the heart from the surface of the body during each heartbeat (see Fig. 9–1).

- The *P wave is caused by the spread of depolarization across the atria*, which causes atrial contraction. Atrial pressure increases just after the P wave.
- The *QRS waves appear as a result of ventricular depolarization* about 0.16 second after the onset of the P wave, and this initiates ventricular contraction; then, the ventricular pressure begins to increase.
- The *ventricular T wave is caused by the repolarization of the ventricle.*

The atria function as primer pumps for the ventricles. About 75 per cent of ventricular filling occurs during diastole before the contraction of the atria, which causes the remaining 25 per cent of ventricular filling. When the atria fail to function properly, such as in atrial fibrillation, little difficulty is encountered unless a person exercises, and then shortness of breath and other symptoms of heart failure occur. The atrial pressure waves (see Fig. 9–1) include

- The *a wave, which is caused by atrial contraction*
- The *c wave, which occurs during ventricular contraction* due to a slight backflow of blood and bulging of the A-V valves toward the atria
- The *v wave, which is caused by in-filling of the atria* from the venous return.

The ventricles fill with blood during diastole. The following events occur just before and during diastole:

- During *systole, the A-V valves are closed*, and the atria fill with blood.
- At the beginning of *diastole*, when ventricular pressure decreases below that of the atria, *the A-V valves open.*
- The higher pressure in the atria pushes blood into the ventricles during diastole.

- The *period of rapid filling of the ventricles* occurs during the first third diastole and provides most of the ventricular filling.
- *Atrial contraction* occurs during the last third of diastole and contributes about 25 per cent of the filling of the ventricle.

Outflow of blood from the ventricles occurs during systole. The following events occur during systole:

- *At the beginning of systole, ventricular contraction occurs,* the A-V valves close, and pressure begins to build up in the ventricle. No outflow of blood occurs during the first 0.2 to 0.3 second of ventricular contraction *(period of isovolumic contraction).*
- When left ventricular pressure exceeds the aortic pressure of about 80 mm Hg, and the right ventricular pressure exceeds the pulmonary artery pressure of 8 mm Hg, the aortic and pulmonary valves open. Ventricular outflow occurs, and this is called the *period of ejection.*
- Most ejection occurs during the first part of this period *(period of rapid ejection).*
- This is followed by the *period of slow ejection;* during this period, aortic pressure may exceed ventricular pressure because the momentum of the blood leaving the ventricle is converted to pressure in the aorta, which slightly increases its pressure.
- The last period of systole is called the period of *isovolumic relaxation* and is caused by ventricular relaxation, which, in turn, causes the ventricular pressure to fall below the aortic and pulmonary artery pressures. Thus, the semilunar valves close at this time.

The fraction of the end-diastolic volume that is ejected is called the *ejection fraction.*

- At the end of diastole, the volume of each ventricle is 110 to 120 milliliters; this volume is called the *end-diastolic volume.*
- The *stroke volume,* which has a value of about 70 milliliters, is the amount of blood that is ejected with each beat.
- The *end-systolic volume* is the remaining volume in the ventricle at the end of systole and has a value of about 40 to 50 milliliters.

- The *ejection fraction* is calculated by dividing the stroke volume by the end-diastolic volume; it has a value of about 60 per cent. By both increasing the end-diastolic volume and decreasing the end-diastolic volume, the stroke volume of the heart can be doubled.

Ventricular ejection increases pressure in the aorta to 120 mm Hg (systolic pressure). When ventricular pressure exceeds the diastolic pressure in the aorta, the aortic valve opens and blood is ejected into the aorta. Pressure in the aorta increases to about 120 mm Hg and distends the elastic aorta and other arteries.

When the aortic valve closes at the end of ventricular ejection, there is a slight backflow of blood followed by a sudden cessation of flow, and this causes an *incisura,* or a slight increase in aortic pressure. During diastole, blood continues to flow into the peripheral circulation, and arterial pressure decreases to 80 mm Hg (diastolic pressure).

The heart valves prevent backflow of blood. The A-V valves (the *tricuspid* and *mitral* valves) prevent backflow of blood from the ventricles to the atria during systole. In a similar fashion, the *semilunar valves* (the *aortic* and *pulmonary* valves) prevent backflow of blood from the aorta and pulmonary artery into the ventricle during diastole. The A-V valves have papillary muscles attached to them by the *chordae tendineae.* During systole, the papillary muscles contract to help prevent the valves from bulging back too far into the atria. The aorta and pulmonary valves are thicker than the A-V valves and do not have any papillary muscles attached.

Work Output of the Heart (p. 113)

The *stroke work output* of the ventricles is the output of energy by the heart during each heart beat. The heart performs two types of work:

- *The volume-pressure work of the heart* is the work to be done to increase the pressure of the blood; in the left heart, this equals stroke volume multiplied by the difference between left ventricular mean ejection pressure and left ventricular mean input pressure. The volume-pressure work of the right ventricle is only about one sixth that of the left ventricle, because the ejection pressure of the right ventricle is much lower.

- *The work to be done to supply kinetic energy to the blood* equals $MV^2/2$, in which M is the mass of blood ejected and V is the velocity.

Usually, only about 1 per cent of the work of the heart creates kinetic energy. However, during a condition such as aortic stenosis, the opening of the aortic valve is very small, and the velocity of blood is very high. Kinetic energy,therefore, can be as high as 50 per cent of the total work output of the heart.

The volume-pressure diagram of the left ventricle describes the cardiac work output. The cardiac cycle can be depicted in a *volume-pressure diagram*, which plots intraventricular pressure as a function of left ventricular volume. The different phases of the cardiac cycle include the following:

- *Phase I*: *Period of filling* during which the left ventricular volume increases from the *end-systolic* volume to the *end-diastolic* volume, or from 45 milliliters to 115 milliliters, an increase of 70 milliliters.
- *Phase II*: *Period of isovolumic contraction* during which the volume of the ventricle remains at the end-diastolic volume but the intraventricular pressure increases to aortic diastolic pressure, or 80 mm Hg.
- *Phase III*: *Period of ejection* during which the systolic pressure increases further because of additional ventricular contraction, and the ventricular volume decreases by 70 milliliters, which is the *stroke volume*.
- *Phase IV*: *Period of isovolumic relaxation during which ventricular volume remains at a value of 45 milliliters but intraventricular pressure decreases to its diastolic pressure level.*

The area inside of the volume-pressure diagram represents the volume-pressure work (or external work output) of the ventricle during each cardiac cycle. This diagram and cardiac work are affected by *preload* and *afterload* on the heart. Preload is usually considered to be the end-diastolic pressure, and the afterload is considered to be pressure in the artery exiting the ventricle (aorta or pulmonary artery).

Oxygen consumption by the heart depends on cardiac work. Cardiac oxygen consumption mainly depends on the volume-pressure type of work. This oxygen consumption has also been found to be proportional to the tension of the heart multiplied by the time the tension is maintained. Wall tension in the heart is proportional to the pressure times the diameter of the ventricle. Ventricular wall tension, therefore, increases at high systolic pressures or when the heart is dilated.

REGULATION OF HEART PUMPING (p. 115)

The Frank-Starling mechanism intrinsically regulates cardiac pumping ability. When *venous return* of blood increases, the heart muscle stretches more, which makes it pump with a greater force of contraction. The *Frank-Starling* mechanism of the heart can be stated in another way: *Within physiological limits, the heart pumps all the blood that comes to it without allowing excess damming of blood in the veins.* The extra stretch of the cardiac muscle during increased venous return, within limits, causes the actin and myosin filaments to interdigitate at a more optimal length for force generation. In addition, more stretch of the right atrial wall causes a reflex increase in heart rate of 10 to 20 per cent, which helps the heart to pump more blood.

The ability of the heart to pump blood can be illustrated graphically in several ways. First, stroke work output can be plotted for each ventricle as a function of its corresponding atrial pressure. Ventricular output (or *cardiac output*) can also be plotted as a function of atrial pressure (refer to Fig. 20–1).

The autonomic nervous system affects cardiac pumping. Under strong sympathetic stimulation, the heart rate of an adult increases from a resting value of 72 up to 180 to 200 beats per minute, and the force of contraction of the heart muscles increases dramatically. Sympathetic stimulation, therefore, can increase cardiac output twofold to threefold. The heart has a resting sympathetic tone; therefore, inhibition of the sympathetics decreases heart rate and the force of contraction of the heart, and thus cardiac output decreases. This is explained further in Chapter 20.

Parasympathetic stimulation mainly affects the atria and can decrease the heart rate dramatically and the force of contraction of the ventricles slightly. The combined effect decreases cardiac output by 50 per cent or more.

Cardiac contractility is affected by several factors. Among the factors that affect cardiac contractility are *extracellular electrolyte concentrations*. Excess potassium in the extracellular fluid causes the heart to become very flaccid and reduces heart rate, thus causing a large decrease in contractility. Excess calcium in the extracellular fluid causes the heart to go into spastic contraction. In contrast, a decrease in calcium ions causes the heart to be very flaccid.

Assessment of cardiac contractility has proved to be difficult. The *rate of change of ventricular pressure*, or dP/dt, has been used as an index of contractility, especially the peak dP/dt. This index, however, is affected by both preload and afterload; another index that is more reliable is (dP/dt)/P.

10

Rhythmical Excitation of the Heart

The heart has a special system for self-excitation of rhythmical impulses to cause repetitive contraction of the heart. This system conducts impulses throughout the heart and causes the atria to contract one-sixth second before the ventricles contract, which allows extra filling of the ventricles with blood before contraction.

SPECIALIZED EXCITATORY AND CONDUCTIVE SYSTEM OF THE HEART (p. 121)

The parts of the rhythmical conduction system and their function are

- The *sinus node* (or the *sinoatrial node*), which initiates the cardiac impulse
- The *internodal pathway*, which conducts impulses from the sinus node to the atrioventricular (A-V) node
- The *A-V node*, which delays impulses from the atria to the ventricle
- The *A-V bundle*, which conducts impulses from the A-V node to the ventricles
- The right and left bundles of *Purkinje fibers*, which conduct impulses to all parts of the ventricles.

The sinus node controls the rate of beat of the entire heart. The membrane potential of a sinus nodal fiber is -55 to -60 millivolts compared with -85 to -90 millivolts in a ventricular muscle fiber.

The action potential in the sinus node is caused by the following:

- The *fast sodium channels* are inactivated at the normal resting membrane potential, but there is a slow leakage of sodium into the fiber.
- Between action potentials, the resting potential gradually increases because of this *slow leakage of sodium* until the potential reaches -40 millivolts.

TABLE 10–1 TIME OF ARRIVAL OF IMPULSE	
Sinus node	0.00 sec
A-V node	0.03 sec
A-V bundle	0.12 sec
Ventricular septum	0.16 sec

- At this time, the *calcium-sodium channels* become activated, allowing rapid entry of sodium and calcium and thus causing the action potential.
- The *potassium channels open*, allowing potassium to escape from the cells, within about 100 to 150 milliseconds after the calcium-sodium channels open. This returns membrane potential to its resting potential, and the self-excitation cycle starts again with sodium leaking slowly into the sinus nodal fibers.

The internodal and interatrial pathways transmit impulses in the atrium. The parts of the *internodal pathway* are the *anterior internodal pathway, middle internodal pathway,* and *posterior internodal pathway,* and these pathways carry impulses from the sinoatrial node to the A-V node. Small bundles of atrial muscle fibers transmit impulses more rapidly than the normal atrial muscle, and one of these, the *anterior interatrial band,* conducts impulses from the right atrium to the anterior part of the left atrium.

The A-V node delays impulses from the atria to the ventricles. This delay time allows the atria to empty their contents into the ventricles before ventricular contraction occurs. Table 10–1 shows the time of the arrival of impulses at the parts of the conduction system from an impulse initiated at the sinus node.

Note that a delay of 0.09 second occurs between the A-V node and the A-V bundle. The velocity of conduction of this system is only 0.02 to 0.05 m/sec, or one fourth that of normal cardiac muscle. The reason for this slow conduction in the A-V node and bundle is that (1) membrane potential is much less negative in the A-V node and bundle than in normal cardiac muscle and (2) few gap junctions exist between the cells in the A-V node and bundle, so that the resistance to ion flow is very great.

The transmission of impulses through the Purkinje system and cardiac muscle is rapid. The *Pur-*

kinje fibers lead from the A-V node through the A-V bundle and into the ventricles. The A-V bundle divides into the *left and right bundles* and lies just under the endocardium, an area that receives the cardiac impulses first. The following are characteristics of the Purkinje system:

- The action potentials travel at a velocity of 1.5 to 4.0 m/sec, which is 6 times the velocity in cardiac muscle.
- The high permeability of the gap junctions at the intercalated discs between the Purkinje fiber cells likely causes the high velocity of transmission.
- The transmission time of only 0.03 second occurs from the bundle branches in the ventricular septum to the termination of the Purkinje fibers.

The atrial and ventricular syncytia are separate and insulated from one another. The methods of this separation are the following:

- *The action potentials do not travel backward through the A-V bundle.* This prevents ventricular impulses from re-entering the atria.
- *The atria and ventricles are separated by a fibrous barrier that acts as an insulator.* This forces the atrial impulses to enter the ventricles through the A-V bundle.

The transmission of impulses through cardiac muscles travels at a velocity of 0.3 to 0.5 m/sec. Because the Purkinje fibers lie just under the endocardium, the action potential spreads into the rest of the ventricular muscle from this area. Then, the cardiac impulses travel up the spirals of the cardiac muscle and finally reach the epicardial surface. The endocardium-to-epicardium transit time is 0.03 second. The transmission time from the initial bundle branches to the epicardial surface of the last part of the heart to be stimulated, therefore, is 0.06 second.

CONTROL OF EXCITATION AND CONDUCTION IN THE HEART (p. 125)

The sinus node is the normal pacemaker of the heart. The intrinsic rhythmical rates of the different areas of the heart are shown in Table 10–2.

The reason the sinus node is the normal pacemaker is that it discharges faster than the other tissues in the cardiac conduction system. When the

TABLE 10–2 INTRINSIC DISCHARGE RATE	
	Times/Min
Sinus node	70–80
A-V node	40–60
Purkinje system	15–40

sinus node discharges, it sends impulses to the A-V node and Purkinje fibers and thereby discharges them before they can discharge intrinsically. The tissues and sinus node then repolarize at the same time, but the sinus node loses its hyperpolarization faster and discharges again—before the A-V node and Purkinje fibers can undergo self-excitation. Occasionally, some cardiac tissue develops a rhythmical rate faster than that of the sinus node; this is called an *ectopic pacemaker*. The most common location of this new pacemaker is the A-V node or the penetrating portion of the A-V bundle.

A-V block occurs when impulses fail to pass from the atria to the ventricles. During *A-V block*, the atria continue to beat normally, but the ventricular pacemaker lies in the Purkinje system that discharges at a normal rate of 15 to 40 beats per minute. After a sudden block, the Purkinje system does not emit its rhythmical impulses for 5 to 30 seconds because it has been overdriven by the sinus rhythm. During this time, therefore, the ventricles fail to contract, and the person may faint because of the lack of cerebral blood flow. This condition is called the *Stokes-Adams syndrome.*

Control of Heart Rhythmicity and Conduction by the Cardiac Nerves: Sympathetic and Parasympathetic Nerves
(p. 126)

Parasympathetic (vagal) stimulation slows the cardiac rhythm. Stimulation of parasympathetic nerves to the heart releases the neurotransmitter *acetylcholine* from the vagal nerve endings. Acetylcholine causes the following effects:

• The rate of sinus node discharge decreases.
• The excitability of the fibers between the atrial muscle and the A-V node decreases.

The heart rate decreases to one-half normal under mild or moderate vagal stimulation, but strong stimulation can temporarily stop the heartbeat. This results in a lack of impulses traversing the ventricles, but under these conditions, the Purkinje fibers develop their own rhythm at 15 to 40 beats per minute. This phenomenon is called *ventricular escape*.

The mechanisms of vagal effects on heart rate are as follows:

1. Acetylcholine increases the permeability of the sinus node and A-V fibers to potassium; this causes *hyperpolarization* of these tissues and makes them less excitable.
2. The membrane potential of the sinus nodal fibers decreases from -55 to -60 millivolts to -65 to -75 millivolts.

The normal upward drift in membrane potential that is caused by sodium leakage in these tissues requires a much longer time to reach self-excitation.

Sympathetic stimulation increases the cardiac rhythm. Stimulation of the sympathetic nerves to the heart has the following three basic effects:

- *The rate of sinus node discharge increases.*
- *The cardiac impulse conduction rate increases* in all parts of the heart.
- *The force of contraction increases* in both atrial and ventricular muscle.

Sympathetic stimulation releases *norepinephrine* at the sympathetic nerve endings. The mechanisms of the norepinephrine effects on the heart are not clear, but they are believed to involve two basic effects. First, norepinephrine is believed to increase the permeability of cardiac muscle fibers to sodium and calcium, which increases the resting membrane potential and makes the heart more excitable; therefore, heart rate increases. Second, the greater calcium permeability increases the force of contraction of cardiac muscle.

11

The Normal Electrocardiogram

As the depolarization wave passes through the heart, electrical currents pass into surrounding tissue, and a small part of the current reaches the surface of the body. The electrical potential generated by these currents can be recorded from electrodes placed on the skin on the opposite sides of the heart; this recording is called an *electrocardiogram*.

The normal electrocardiogram (see Fig. 9–1) is composed of

- A *P wave* caused by the electrical potential generated from depolarization of the atria before their contraction
- A *QRS complex* caused by the electrical potential generated from the ventricles before their contraction
- A *T wave* caused by the potential generated from repolarization of the ventricles.

Atrial and ventricular contractions are related to the electrocardiogram waves. In Figure 9–1, the relationships between the electrocardiogram and atrial and ventricular contractions can be seen and indicate that

- *The P wave immediately precedes atrial contraction.*
- *The QRS complex immediately precedes ventricular contraction.*
- *The ventricles remain contracted until a few milliseconds after the end of the T repolarization wave.*
- *The atria remain contracted until the atria are repolarized,* but an atrial repolarization wave cannot be seen on the electrocardiogram because it is obscured by the QRS wave.
- *The P-Q or P-R interval on the electrocardiogram has a normal value of 0.16 second* and is the duration of time between the beginning of the P wave and the beginning of the QRS wave; this represents the

time between the beginning of atrial contraction and the beginning of ventricular contraction.

- *The Q-T interval has a normal value of 0.35 second* and is the duration of time from the beginning of the Q wave to the end of the T wave; this approximates the time of ventricular contraction.
- *The heart rate can be determined with the reciprocal of the time interval between each heartbeat.*

During the depolarization process, the average electrical current flows from the base of the heart toward the apex. The heart is suspended in a highly conductive medium, so that when one area of the heart depolarizes, current flows from this area toward a polarized area. The first area that depolarizes is the ventricular septum, and current flows quickly from this area to the other endocardial surfaces of the ventricle. Then, current flows from the electronegative inner surfaces of the heart to the electropositive outer surfaces, with the average current flowing from the base of the heart to the apex in an elliptical type of pattern. An electrode placed near the base of the heart will be electronegative, and one placed near the apex will be electropositive.

ELECTROCARDIOGRAPHIC LEADS (p. 132)

Bipolar limb leads involve an electrocardiogram recorded from electrodes on two different limbs. There are three bipolar limb leads:

- To record from *lead I*, the *negative terminal of the electrocardiogram is connected to the right arm, and the positive terminal is connected to the left arm.* During the depolarization cycle, the point at which the right arm connects to the chest is electronegative compared with the point at which the left arm connects, so that the electrocardiogram records positively when this lead is used.
- To record from *lead II*, the *negative terminal of the electrocardiogram is connected to the right arm, and the positive terminal is connected to the left leg.* During most of the depolarization cycle, the left leg will be electropositive compared with the right arm, so that the electrocardiogram records positively when this lead is used.
- To record from *lead III*, the *negative terminal is connected to the left arm, and the positive terminal is connected to the left leg.* During most of the depolarization cycle, the left leg will be electropositive

compared with the left arm, so that the electrocardiogram records positively when this lead is used.

Einthoven's law states that the electrical potential of any limb lead equals the sum of the potentials of the other two limb leads. The positive and negative signs of the different leads must be observed when using this law. The following example illustrates *Einthoven's law*. We first assume that the right arm is 0.2 millivolt negative with respect to the average potential in the body, the left arm is 0.3 millivolt positive, and the left leg is 1.0 millivolt positive. Therefore, lead I will have a potential of 0.5 millivolt, because this is the difference between -0.2 millivolt in the right arm and 0.3 millivolt in the left arm. Similarly, lead II will have a potential of 1.2 millivolts, and lead III will have a potential of 0.7 millivolt.

Chest leads (precordial leads) can be used to detect minor electrical abnormalities in the ventricles. Chest leads, known as leads V_1, V_2, V_3, V_5, and V_6, are connected to the positive terminal of the electrocardiograph, and the *indifferent electrode*, or the negative electrode, is simultaneously connected to the left arm, left leg, and right arm. The QRS recording from the V_1 or V_2 leads, which are placed over the heart near the base, usually read negatively, and the QRS recording from leads V_4, V_5, and V_6, which are closer to the apex, usually read positively. Because these leads can record the electrical potential immediately beneath the electrode, small changes in electrical potential of the cardiac musculature can be detected, such as that generated by a small myocardial infarction.

12

Electrocardiographic Interpretation of Cardiac Muscle and Coronary Abnormalities

Vectorial Analysis

Any change in the transmission of impulses through the heart alters the electrical potentials around the heart, which causes changes in the electrocardiogram waves. Therefore, most abnormalities in the cardiac muscle can be detected through analysis of the electrocardiogram.

PRINCIPLES OF VECTORIAL ANALYSIS OF ELECTROCARDIOGRAMS (p. 135)

Vectors can be used to represent electrical potentials. Several principles are employed in the vectorial analysis of electrical potentials:

- The current in the heart flows from the area of depolarization to the polarized areas, and the electrical potential generated can be represented by a vector, with the *arrowhead pointing in the positive direction.*
- The length of the vector is *proportional to the voltage of the potential.*
- The generated potential at any instance can be represented by an *instantaneous mean vector.*
- When a vector is horizontal and points toward the subject's left side, the axis is said to be zero degrees.
- The scale of the vectors rotates clockwise from the zero-degree reference point.
- If a vector points directly downward, it is said to have direction of +90 degrees.
- If a vector points horizontally to the subject's right side, it is said to have a direction of +180 degrees.

- If a vector points directly upward, it is said to have a direction of −90 or +270 degrees.
- *The axis of lead I is zero degrees* because the electrodes lie in the horizontal direction on each of the arms.
- *The axis of lead II is +60 degrees* because the right arm connects to the torso in the top right corner, and the left leg connects to the torso in the bottom left corner.
- *The axis of lead III is 120 degrees.*
- When the vector representing the mean direct current flow in the heart is perpendicular to the axis of one of the bipolar limb leads, the voltage recorded in the electrocardiogram in this lead is very low.
- When the vector has approximately the same direction as the axis of one of the bipolar limb leads, nearly the entire voltage will be recorded in this lead.

The normal electrocardiogram represents the vectors that occur during electrical potential changes in the cardiac cycle.

- *The QRS complex represents ventricular depolarization* that begins at the ventricular septum and proceeds toward the apex of the heart with an average direction of 59 degrees.
- *The ventricular T wave represents repolarization of the ventricle* that begins near the apex of the heart and proceeds toward the base. Because the cardiac muscle near the apex becomes electropositive after it repolarizes and the muscle near the base is still electronegative, the T wave vector is positive with a similar direction as the QRS wave.
- *The atrial P wave represents depolarization of the atria* that begins at the sinus node and spreads in all directions, but the average vector points toward the atrioventricular node. Thus, the average vector during the P wave is in a positive direction.

Several factors shift the mean electrical axis of the ventricles to the left (counterclockwise), including the following:

- *Changes in the position of the heart,* such as occur during expiration, when a person is *recumbent* and the abdominal contents press upward against the diaphragm
- *Accumulation of abdominal fat,* which also presses upward on the heart

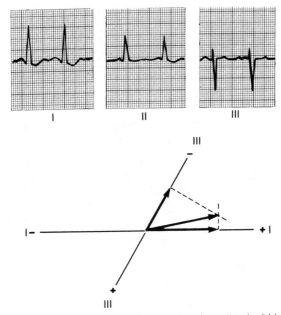

Figure 12–1 Left axis deviation in hypertensive heart disease. Note the slightly prolonged QRS complex.

- *Left bundle branch block,* which is when the cardiac impulse spreads through the right ventricle 2 to 3 times as fast as in the left ventricle. Consequently, the left ventricle remains polarized much longer than the right, and a strong electrical vector points from the right ventricle to the left
- *Hypertrophy of the left ventricle,* which is caused by hypertension, aortic valvular stenosis, or aortic valvular regurgitation.

An example of left axis deviation caused by hypertension and the resulting effects on left ventricular hypertrophy of the electrocardiogram is shown in Figure 12–1. *Note that the lead I and III vectors are plotted in this figure and that a vertical dotted line is extended from the ends of these vectors. The resultant vector is drawn from the origin to the intersection of the two dotted lines and represents the mean electrical axis in this condition.*

Several factors shift the mean electrical axis of the ventricles to the right (clockwise), including the following:

- *Inspiration*
- *Standing up*
- *Lack of abdominal fat,* which allows the heart to rotate clockwise compared with the normal individual
- *Right bundle branch block*
- *Right ventricular hypertrophy.*

CONDITIONS THAT CAUSE ABNORMAL VOLTAGES OF THE QRS COMPLEX (p. 142)

Hypertrophy of the heart increases the voltage of the QRS complex. When the sum of the voltages of the QRS waves from the three standard leads is greater than 4 millivolts, a high-voltage electrocardiogram is considered to exist. The most common cause of high-voltage QRS complexes is right or left ventricular hypertrophy.

The following conditions decrease voltage of the QRS complex:

- *Hearts with old myocardial infarctions* and resultant decreased cardiac muscle mass—This condition also slows the conduction wave through the heart and decreases the amount of muscle that is depolarized at one time. Therefore, decreased QRS voltage and prolongation of the QRS complex result.
- *Conditions surrounding the heart that effectively "short-circuit" the cardiac electrical potential. Fluid in the pericardium* and *pleural effusion* both conduct currents from around the heart and prevent much of the voltage from reaching the surface of the body. *Pulmonary emphysema* also decreases conduction of the cardiac potentials because the excess volume of air in the lungs insulates the heart.

The following conditions cause a prolonged QRS complex:

- The most common cause of an extended QRS complex is *prolonged conduction* through the ventricles. This occurs in both hypertrophied and dilated hearts and increases the duration of the QRS waves by about 0.02 to 0.05 second. A prolonged QRS wave that is caused by left ventricular hypertrophy is shown in Figure 12–1.
- *Blockade of the impulses in the Purkinje system* prolongs the QRS complex because the duration of

ventricular depolarization increases in one or both ventricles.

CURRENT OF INJURY (p. 144)

Several abnormalities cause a portion of the heart to remain *depolarized all of the time*, and the current that flows from the depolarized area to the polarized areas of the heart is called the *current of injury*. Some of the abnormalities that can cause a current of injury are as follows:

- Mechanical trauma
- Infectious processes that damage the cardiac muscle membrane
- Coronary ischemia.

The axis of the current of injury can be determined with the electrocardiogram. When a portion of the heart is injured and emits a current of injury, the only time when the heart returns to zero potential is at the end of the QRS wave because all of the heart is depolarized at this time (see Fig. 9–1). The axis of the current of injury is determined in the following way:

1. First, *determine the J point*, which is the point of zero potential at the end of QRS wave.
2. *Determine the level of the T-P segment* with respect to the J point on the three standard leads.
3. *Plot the voltages on the coordinates of the three leads* to determine the axis of the current of injury, and note that the negative end of the vector points toward the *injured* area of the ventricles.

Acute anterior and posterior wall infarctions can be diagnosed with the electrocardiogram. The current of injury is also useful in determining whether an infarction is on the anterior or posterior part of the heart. A negative injury potential found on one of the precordial leads indicates that this electrode is in an area of strong negative potential and that the current of injury originates in the anterior wall of the ventricles. In contrast, a positive T-P segment with respect to the J point indicates the existence of a posterior ventricular wall infarction.

ABNORMALITIES IN THE T WAVE (p. 147)

Normally, the apex of the ventricle repolarizes before the base, and the resultant T wave has a mean electrical axis similar to that of the QRS wave. Several conditions alter the electrical axis of the T wave, including the following:

- *During bundle branch block, one of the ventricles depolarizes before the other.* The first ventricle to depolarize is also the first to repolarize, and this causes an axis deviation in the T wave. Therefore, a left bundle branch block causes a rightward axis deviation of the T wave.
- During prolongation in depolarization of the apex of the heart, the base of the heart repolarizes before the apex, and this *inverts the T wave.* The most common cause of prolonged depolarization is *mild ischemia* of cardiac muscle in the apex of the ventricles.

13

Cardiac Arrhythmias and Their Electrocardiographic Interpretation

Often, the heart malfunctions not because of abnormal heart muscle but instead because of an abnormal rhythm of the heart. The causes of cardiac arrhythmias include (1) abnormal rhythmicity of the sinus node, (2) shift of the pacemaker function from the sinus node to other parts of the heart, (3) block of impulse transmission in the heart, (4) abnormal pathway of transmission in the heart, and (5) spontaneous generation of abnormal impulses from any part of the heart.

ABNORMAL SINUS RHYTHMS (p. 149)

Stimulation of the pacemaker of the heart causes tachycardia. An increase in heart rate, called *tachycardia*, is usually defined as a heart rate greater than 100 beats per minute. The causes of sinus-initiated tachycardia include the following:

- *Increased body temperature*
- *Sympathetic stimulation of the heart*, which occurs, for example, after a blood loss that decreases arterial pressure and increases sympathetic stimulation through baroreceptor mechanisms. In this instance, heart rate may increase up to 150 to 180 beats per minute.
- *Toxic conditions of the heart* (e.g., digitalis intoxication).

Vagal stimulation of the heart causes a decrease in heart rate. A slow heart rate, usually less than 60 beats per minute, is called *bradycardia*. Stimulation of the vagus nerve decreases heart rate because of the release of the parasympathetic transmitter agent acetylcholine. In *carotid sinus syndrome*, an atherosclerotic process causes excess sensitivity of the baroreceptors in the arterial wall. As a result, increased external pressure on the neck causes the athero-sclerotic plaque in the carotid sinus to stimulate the baroreceptors, which then stimulate the vagus nerve and cause bradycardia.

ABNORMAL CARDIAC RHYTHMS THAT RESULT FROM IMPULSE CONDUCTION BLOCK (p. 150)

Rarely, the impulse from the sinoatrial node is blocked before it enters the atrial muscle. In this condition, the atrial P wave is absent from the electrocardiogram, and the ventricles pick up a rhythm that usually originates from the atrioventricular (A-V) node.

A-V block inhibits or completely blocks impulses originating in the sinoatrial node. The conditions that cause A-V block include the following:

- *Ischemia of the A-V node or A-V bundle*, which occurs during coronary ischemia if the region of ischemia includes the A-V node or bundle
- *Compression of the A-V bundle*, which is caused by scar tissue or calcified portions of the heart
- *Inflammation of the A-V node or bundle*, which can occur during myocarditis, diphtheria, or rheumatic fever
- *Strong vagal stimulation of the heart*.

The types of A-V block include:

- *First degree block*—In this condition, the *P-R (or P-Q) interval increases* from a normal value of 0.16 second to about 0.20 second in a heart beating at a normal rate.
- *Second degree block*—When conduction through the A-V junction slows sufficiently for the *P-R interval to increase to 0.25 to 0.45 second*, only a portion of the impulses passes through to the ventricle. Therefore, the atria beat faster than the ventricles, and "dropped beats" of the ventricles occur.
- *Third degree block*—This is *complete A-V junction block*, and a complete dissociation of the P waves and QRS waves occurs. Therefore, the ventricles "escape" from the influence of the sinoatrial pacemaker. A condition in which A-V block comes and goes is called *Stokes-Adams syndrome*.

PREMATURE CONTRACTIONS (p. 152)

Most premature contractions *(extrasystoles)* result from *ectopic foci* that generate abnormal cardiac impulses. The causes of ectopic foci include the following:

- Local ischemia
- Irritation of cardiac muscle as a result of pressure from calcified plaque
- Toxic irritation of the A-V node, Purkinje system, or myocardium by drugs, nicotine, or caffeine.

Ectopic foci can cause premature contractions that originate in the atria, A-V junction, or ventricle. The consequences of premature contractions are as follows:

- *Premature atrial contraction*–The P-R interval decreases in this condition, with the amount dependent on how far the origin of the ectopic foci is from the A-V junction. Premature atrial contraction causes premature ventricular beats that may have a *pulse deficit* if the ventricles do not have sufficient time to fill with blood.
- *A-V nodal or A-V bundle premature contractions*–The P wave is often missing from the electrocardiogram because it is superimposed on the QRS wave.
- *Premature ventricular contractions (PVCs)*–The ectopic foci originate in the ventricle, and the QRS complex is often prolonged because the impulses must pass through muscle, which conducts at a much lower rate than the Purkinje system. The QRS voltage increases because one side of the heart depolarizes ahead of the other, causing a large electrical potential between the depolarized and polarized muscle.

PAROXYSMAL TACHYCARDIA (p. 153)

The cause of these foci is believed to be re-entrant pathways that set up local repeated self-re-excitation. The rapid rhythm of the area causes it to become the new pacemaker of the heart. *Paroxysmal tachycardia* means that the heart rate increases in rapid bursts and, after a few seconds, minutes, or hours, returns to normal. Treatment is the administration of quinidine or lidocaine because they each increase sodium permeability of the cardiac muscle and thus block the rhythmical discharge of the irritable area.

Two basic types of paroxysmal tachycardia occur:

- *Atrial paroxysmal tachycardia*—When the origin of the tachycardia is in the atrium but is not close to the sinoatrial node, an inverted P wave will occur, which is caused by atrial depolarization in the opposite direction from normal. When the abnormal rhythm originates in the A-V node, P waves are missed or obscured; this condition is called *supraventricular tachycardia*.
- *Ventricular paroxysmal tachycardia*—This type of tachycardia usually does not occur unless significant ischemic damage is present in the ventricles, and this abnormality frequently initiates lethal fibrillation.

VENTRICULAR FIBRILLATION (p. 154)

Ventricular fibrillation is the most serious of all cardiac arrhythmias. It occurs when an impulse stimulates first one portion of the ventricular muscles and then another and finally stimulates itself. This stimulation causes many portions of the ventricles to contract at the same time while other portions relax. Therefore, impulses travel around the heart muscle; the phenomenon is also referred to as *circus movements*.

Circus movements are the basis for ventricular fibrillation. When an impulse travels through the extent of a normal ventricle, it dies because all of the ventricular muscle is in a refractory state. However, three conditions allow the impulse to continue around the heart and start circus movements:

- *Increased pathway around the ventricle.* By the time the impulses return to an originally stimulated muscle, it is no longer in a refractory state, and the impulse will continue to travel around the heart. This is especially likely to occur in hearts that are dilated or have valvular disease or other conditions that have a long pathway of conduction.
- *Decreased velocity of conduction.* By the time the slower impulse travels around the heart, the muscle is no longer refractory to a new impulse and is stimulated again. This frequently occurs in the Purkinje system during ischemia of the cardiac muscle or during high blood potassium concentration.
- *Shortened refractory period of the muscles.* This allows repeated stimulation as the impulse travels

around the heart and occurs after epinephrine administration or repetitive electrical stimulation.

Defibrillation of the heart causes essentially all parts of the ventricles to become refractory. Clinically, the heart is defibrillated by the application of a high-voltage direct current through the chest with electrodes placed on either side of the heart.

ATRIAL FIBRILLATION (p. 156)

Because the atria and ventricles are insulated from one another, ventricular fibrillation can occur without atrial fibrillation, and atrial fibrillation can occur without ventricular fibrillation. The causes of atrial fibrillation are identical to those of ventricular fibrillation, and a frequent cause of atrial fibrillation is an enlarged atrium that is caused by heart valve lesions. The atria do not pump if they are fibrillating, and the efficiency of ventricular pumping decreases 20 per cent to 30 per cent. A person can live for years with atrial fibrillation, although there will be some cardiac debility.

Atrial flutter is different from atrial fibrillation in that a single large wave front travels around and around the atria. Thus, the atria contract at 250 to 300 times per minute; because one side of the atria contracts while the other relaxes, the amount of blood that is pumped is small.

The Circulation

14

Overview of the Circulation; Medical Physics of Pressure, Flow, and Resistance

The main function of the circulation is to serve the needs of the tissues by transporting nutrients to the tissues, transporting away waste products, carrying hormones from one part of the body to another, and in general maintaining homeostatic conditions in the tissue fluids for optimal survival and function of the cells.

The circulation is divided into the *pulmonary circulation*, which supplies the lungs, and the *systemic circulation*, which supplies the tissues of the remainder of the body. The functional parts of the circulation are

- The *arteries*, which transport blood under high pressure to the tissues and have strong vascular walls and rapid blood flow
- The *arterioles*, which are the last branches of the arterial system and act as control valves through which blood is released into the capillaries; these vessels have strong muscular walls that can be constricted or dilated, giving them the capability of markedly altering blood flow to the capillaries in response to the needs of the tissues.
- The *capillaries*, which exchange fluids, nutrients, and other substances between the blood and the interstitial fluid; they have thin walls and are highly permeable to small molecules
- The *venules*, which collect blood from the capillaries and gradually coalesce into progressively larger veins.

The circulation is a complete circuit. Contraction of the left heart propels blood into the systemic circulation through the aorta, which empties into smaller arteries, arterioles, and eventually capillaries.

Because the blood vessels are distensible, each contraction of the heart distends the vessels; during relaxation of the heart, the vessels recoil, thereby continuing flow to the tissues, even between heartbeats. Blood leaving the tissues enters the venules and then flows into increasingly larger veins, which carry the blood to the right heart.

The right heart then pumps the blood through the pulmonary artery, small arteries, arterioles, and capillaries, where oxygen and carbon dioxide are exchanged between the blood and the tissues. From the pulmonary capillaries, blood flows into venules and large veins and empties into the left atrium and left ventricle before it is again pumped into the systemic circulation.

Because blood flows around and around the same vessels, any change in flow in a single part of the circuit alters flow in other parts. For example, strong constriction of the arteries in the systemic circulation can reduce the total cardiac output, in which case blood flow to the lungs will decrease equally as much as flow through the systemic circulation.

Another feature of the circulation is that sudden constriction of a blood vessel must always be accompanied by opposite dilation of another part of the circulation, because blood volume cannot change rapidly and blood itself is not compressible. For instance, strong constriction of the veins in the systemic circulation displaces blood into the heart, dilating the heart and causing it to pump with increased force. This is one of the mechanisms by which cardiac output is regulated. With prolonged constriction of a portion of the circulation, changes in total blood volume can occur through exchange with the interstitial fluid or because of changes in fluid excretion by the kidneys.

Most of the blood volume is distributed in the veins of the systemic circulation. About 84 per cent of the total blood volume is in the systemic circulation, with 64 per cent in the veins, 13 per cent in the arteries, and 7 per cent in the systemic arterioles and capillaries. The heart contains about 7 per cent and the pulmonary vessels contain 9 per cent of the blood volume.

Velocity of blood flow is inversely proportional to vascular cross-sectional area. Because approximately the same volume of blood flows through each segment of the circulation each minute, vessels with a large cross-sectional area, such as the capillaries, have a slower blood flow velocity. The approximate total cross-sectional areas of the systemic vessels of each type are as follows:

	cm^2
Aorta	2.5
Small arteries	20
Arterioles	40
Capillaries	2500
Venules	250
Small veins	80
Venae cavae	8

Thus, under resting conditions, velocity in blood flow in the capillaries is only about $1/1000$ the velocity of flow in the aorta.

Pressures vary in the different parts of the circulation. Because the pumping action of the heart is pulsatile, the aortic arterial pressure rises to its highest point, the *systolic pressure*, during systole and falls to its lowest point, the *diastolic pressure*, at the end of diastole. In the normal adult, systolic pressure is approximately 120 mm Hg and diastolic pressure is 80 mm Hg. This is usually written as 120/80 mm Hg. The difference between systolic and diastolic pressure is called the *pulse pressure* (120 − 80 = 40 mm Hg). As the blood flows through the systemic circulation, its pressure falls progressively to approximately 0 mm Hg by the time it reaches the termination of the venae cavae in the right atrium of the heart.

Pressure in the systemic capillaries varies from as high as 35 mm Hg near the arteriolar ends to as low as 10 mm Hg near the venous ends, but the average functional capillary pressure is about 17 mm Hg.

Pressures in the pulmonary circulation are much lower than those in the systemic circulation. Pressure in the pulmonary arteries is also pulsatile, but systolic arterial pressure is about 25 mm Hg and diastolic pressure is 8 mm Hg, with the mean pulmonary artery pressure of only 16 mm Hg. Pulmonary capillary pressure averages only 8 mm Hg, yet the total blood flow through the lungs is the same as that in the systemic circulation because of a lower vascular resistance of the pulmonary blood vessels.

BASIC THEORY OF CIRCULATORY FUNCTION (p. 162)

The details of circulatory function are complex and are described later, but there are three basic principles that underlie the major functions of the circulatory system, as follows:

- *The blood flow to each tissue of the body is precisely controlled relative to the tissue needs.* When tissues are active, they need more blood flow than at rest, occasionally as much as 20 times more. The microvessels of each tissue continuously monitor the tissue needs and control the blood flow at the level required for the tissue activity. In addition, nervous and hormonal mechanisms provide additional control of tissue blood flow.
- *The cardiac output is controlled mainly by the sum of all the local tissue flows.* After blood flows through a tissue, it immediately returns by way of the veins to the heart. The heart responds to the inflow of blood by pumping almost all of it immediately back into the arteries. In this sense, the heart responds to the demands of the tissues, although it sometimes needs help in the form of nervous stimulation to make it pump the required amounts of blood flow.
- *In most instances, the arterial pressure is controlled independently of local blood flow or cardiac output control.* The circulatory system is provided with a very extensive system for controlling arterial pressure. If arterial pressure falls below a normal level, a barrage of nervous reflexes elicits a series of circulatory changes that raise the pressure back toward normal, including increased force of heart pumping, contraction of large venous reservoirs to provide more blood to the heart, and constriction of most of the arterioles throughout the body. Over more prolonged periods of time, the kidneys play additional roles by secreting pressure-controlling hormones and regulating blood volume.

INTERRELATIONSHIPS AMONG PRESSURE, FLOW, AND RESISTANCE (p. 163)

Blood flow through a vessel is determined by the pressure gradient and vascular resistance. The flow of blood through a vessel can be calculated by the formula $Q = \Delta P/R$, where Q is blood flow, ΔP is the pressure difference between the two ends of the vessel, and R is the vascular resistance. Note that it is the *difference in pressure* between the two ends of the vessel that provides the driving force for flow, not the absolute pressure in the vessel. For example, if the pressure at both ends of the vessel were 100 mm Hg, there would be no flow despite the presence of high pressure.

Because of the extreme importance of the relationship among pressure, flow, and resistance, the reader should become familiar with the other two algebraic forms of this relationship: $\Delta P = Q \times R$ and $R = \Delta P/Q$. Blood pressure is usually expressed in millimeters of mercury (mm Hg), and blood flow is expressed in milliliters per minute (ml/min); vascular resistance is expressed as mm Hg/ml per minute. In the pulmonary circulation, the pressure gradient is much lower than that in the systemic circulation, whereas the blood flow is the same as that in the systemic circulation; *therefore, the total pulmonary vascular resistance is much lower than the systemic vascular resistance.*

Vessel diameter has a very large effect on resistance to blood flow—Poiseuille's law. According to the *theory of Poiseuille,* vascular resistance is directly proportional to the viscosity of the blood and the length of the blood vessel and inversely proportional to the radius of the vessel raised to the fourth power:

$$\text{Resistance } \alpha = \frac{(\text{Constant} \times \text{Viscosity} \times \text{Length})}{\text{Radius}^4}$$

Decreased radius of the blood vessels markedly increases vascular resistance. Because vascular resistance is inversely related to the fourth power of the radius, even small changes in radius can cause very large changes in resistance. For example, if the radius of a blood vessel increases from one to two (two times normal), resistance will decrease to $1/16$ of normal ($1/2^4$) and flow will increase to 16 times normal if the pressure gradient remains unchanged. Small vessels in the circulation have the greatest amount of resistance, whereas large vessels have very little resistance to blood flow.

For a parallel arrangement of blood vessels, as occurs in the systemic circulation in which different organs are each supplied by an artery that branches into multiple vessels, the total resistance can be expressed as follows:

$$\frac{1}{R_{total}} = \frac{1}{R_1} + \frac{1}{R_2} + \cdots \frac{1}{R_n}$$

where R_1, R_2, and R_n are the resistances of each of the various vascular beds in the circulation. The total resistance is less than the resistance of any of the individual vascular beds.

For a series arrangement of blood vessels, as occurs within a tissue in which blood flows through arteries, arterioles, capillaries, and veins, the total resistance is the sum of the individual resistances, as follows:

$$R_{total} = R_1 + R_2 + \cdots R_n$$

where R_1, R_2, and R_n are the resistances of the different blood vessels in series within the tissues.

Conductance is a measure of the ease of which blood can flow through a vessel and is the reciprocal of resistance: Conductance = 1/Resistance.

Increased blood hematocrit and increased viscosity raise vascular resistance and decrease blood flow. One important factor in Poiseuille's law is the *viscosity* of the blood. The greater the viscosity, the less is the flow of blood in a vessel if all other factors remain constant. The normal viscosity of blood is about three times as great as the viscosity of water. The main factor that makes blood so viscous is that it has large numbers of suspended red blood cells, each of which exerts frictional drag against adjacent cells and against the wall of the blood vessel.

The percentage of blood that is cells, called the *hematocrit*, is normally about 40; this indicates that about 40 per cent of the blood is cells and the remainder is plasma. The greater the percentage of cells in the blood—that is, the greater the hematocrit— the greater is the viscosity of blood and therefore the greater is the resistance to blood flow.

Increased pressure can distend blood vessels and decrease vascular resistance. Based on the previous discussion, one might expect an increase in arterial pressure to cause a proportional increase in blood flow throughout the various tissues of the body; however, the effect of pressure on blood flow is sometimes greater than might be expected. Increased pressure not only increases the force that tends to push blood flow toward the vessels but also distends the vessels and decreases their resistance. As discussed in Chapter 17, most tissues have local autoregulatory mechanisms that prevent excess blood flow in the tissues by causing the arterioles to constrict when arterial pressure is increased and blood vessels are overstretched.

15

Vascular Distensibility, and Functions of the Arterial and Venous Systems

VASCULAR DISTENSIBILITY (p. 171)

The distensibility of the arteries allows them to accommodate the pulsatile output of the heart and average out the pressure pulsations; this ability provides a continuous flow of blood through the very small blood vessels of the tissue. The veins are even more distensible than the arteries, allowing them to store large quantities of blood that can be called into use when needed. On the average, veins are about eight times as distensible as arteries in the systemic circulation. In the pulmonary circulation, the distensibility of veins is similar to that of the systemic circulation. The arteries are more distensible than those of the systemic circulation.

Vascular distensibility is normally expressed as follows:

$$\text{Vascular Distensibility} = \frac{\text{Increase in Volume}}{\text{Increase in Pressure} \times \text{Original Volume}}$$

Vascular compliance (capacitance) is the total quantity of blood that can be stored in a given part of the circulation for each mm Hg of pressure and is calculated as follows:

$$\text{Vascular Compliance} = \frac{\text{Increase in Volume}}{\text{Increase in Pressure}}$$

The greater the compliance of the vessel, the more easily it can be distended by pressure. Compliance is related to distensibility as follows:

$$\text{Compliance} = \text{Distensibility} \times \text{Volume}$$

The compliance of a vein in the systemic circulation is about 24 times as great as its corresponding artery because it is about eight times as distensible and has a volume that is three times as great ($8 \times 3 = 24$).

Sympathetic stimulation decreases vascular capacitance. Sympathetic stimulation increases vascular smooth muscle tone in veins and arteries, causing a shift of blood to the heart, which is an important method that the body uses to increase heart pumping. For example, during hemorrhage, enhanced sympathetic tone of the vessels, especially of the veins, reduces vessel size so that the circulation can continue to operate almost normally even when as much as 25 per cent of the total blood volume has been lost.

Vessels exposed to increased volume will at first exhibit a large increase in pressure, but delayed stretch of the vessel wall allows pressure to return toward normal. This phenomenon is often referred to as *"delayed compliance"* or *"stress relaxation."* Delayed compliance is a valuable mechanism by which the circulation can accommodate extra amounts of blood when necessary, such as after a transfusion that was too large. Delayed compliance in the reverse direction permits the circulation to readjust itself over a period of minutes or hours to diminished blood volume after serious hemorrhage.

ARTERIAL PRESSURE PULSATIONS (p. 173)

With each heartbeat, there is a new surge of blood in the arteries. Were it not for the distensibility of the arterial system, blood flow through the tissues would occur only during cardiac systole, and no blood would flow during diastole. The combination of distensibility of the arteries and their resistance to flow reduces the pressure pulsations almost to no pulsations by the time the blood reaches the capillaries, allowing a continuous rather than pulsatile flow through the tissues.

In the normal young adult, the pressure at the height of each pulse, the *systolic pressure*, is about 120 mm Hg, and pressure at its lowest point, the *diastolic pressure*, is about 80 mm Hg. The difference between these two pressures, about 40 mm Hg, is called the *pulse pressure*.

The two most important factors that can *increase* pulse pressure are (1) *increased stroke volume* (the amount of blood pumped into the aorta with each heartbeat) and (2) *decreased arterial compliance*. Decreased arterial compliance can result when the arteries "harden" with aging or *arteriosclerosis*. Several other pathophysiological conditions of the circulation can cause abnormal contours of the pulse pressure.

- In *aortic stenosis*, the pulse pressure is greatly decreased because of diminished blood flow through the stenotic aortic valve.
- In *patent ductus arteriosus,* some of the blood pumped into the aorta flows immediately through the open ductus arteriosus into the pulmonary artery, allowing the diastolic pressure to fall very low before the next heartbeat and increasing pulse pressure.
- In *aortic regurgitation,* the aortic valve is absent or functions poorly. After each heartbeat, the blood that flows into the aorta flows immediately back into the left ventricle, causing the aortic pressure to fall to a very low level between heartbeats, increasing the pulse pressure.

The pressure pulses are damped in the smaller vessels. Pressure pulsations in the aorta are progressively diminished (damped) by (1) the resistance to blood movement in the vessels and (2) the compliance of the vessels. The resistance damps the pulsations because a small amount of blood must flow forward to distend the next segment of the vessel; the greater the resistance, the more difficult it is for this to occur. The compliance damps the pulsation because the more compliant a vessel, the greater quantity of blood is required to cause a rise in pressure. The degree of damping of arterial pulsations is directly proportional to the product of resistance and compliance.

Blood pressure can be measured indirectly by the auscultatory method. With this method, a stethoscope is placed over a vessel, such as the antecubital artery, and a blood pressure cuff is inflated around the upper arm proximal to the vessel. As long as the cuff inflation is not sufficiently great to collapse the vessel, no sounds are heard with the stethoscope despite the fact that blood within the artery is pulsing. When the cuff pressure is sufficiently great to close the artery during part of the arterial pressure cycle, a sound is heard with each pulsation; these sounds are called *Korotkoff sounds*.

In determining blood pressure by the auscultatory method, pressure in the cuff is first inflated well above arterial systolic pressure. As long as the pressure is higher than systolic pressure, the brachial artery remains collapsed and no blood jets into the lower artery during the cardiac cycle; therefore, no Korotkoff sounds are heard in the lower artery. As soon as the pressure in the cuff falls below systolic pressure, blood slips through the artery beneath the cuff during the peak of systolic pressure, and one begins to hear *tapping sounds* in the antecubital artery in synchrony with the heartbeat. As soon as these sounds are heard, the pressure level indicated by the manometer connected to the cuff is about equal to the systolic pressure.

As pressure in the cuff is further lowered, the Korotkoff sounds change in quality, having a rhythmical, harsher sound. Finally, when the pressure in the cuff falls to equal diastolic pressure, the artery no longer closes during diastole; the sounds suddenly change to a muffled quality and then usually disappear entirely after another 5- to 10-millimeter drop in cuff pressure. When the Korotkoff sounds change to the muffled quality, the manometer pressure is about equal to the diastolic pressure.

The mean arterial pressure can be estimated from the systolic and diastolic pressures measured by the auscultatory method as follows: Mean Arterial Pressure = $\frac{2}{3}$ Diastolic Pressure + $\frac{1}{3}$ Systolic Pressure. For the normal person, mean arterial pressure is about ($\frac{2}{3} \times 80$ mm Hg) + ($\frac{1}{3} \times 120$ mm Hg), or 93.3 mm Hg.

VEINS AND THEIR FUNCTION (p. 176)

The veins, as discussed previously, are capable of constricting and enlarging and thereby storing either small or large quantities of blood, making this blood available when needed by the remainder of the circulation. Veins can also propel blood forward by means of a "venous pump," and they help to regulate cardiac output.

Venous pressures: as related to the right atrial pressure (central venous pressure) and peripheral venous pressures. Because blood from systemic veins flows into the right atrium, anything that affects the right atrial pressure usually affects venous pressure everywhere in the body. *Right atrial pressure is regulated by a balance between the ability of the*

heart to pump blood out of the right atrium and a tendency of blood to flow from the peripheral vessels back to the right atrium.

The normal right atrial pressure is about 0 mm Hg, but it can rise to as high as 20 to 30 mm Hg under abnormal conditions, such as serious heart failure or massive transfusion.

Increased venous resistance can raise peripheral venous pressure. When large veins are distended, they offer little resistance to blood flow. Many of the large veins entering the thorax are compressed, however, by the surrounding tissues, so they are at least partially collapsed or collapsed to an ovoid state. For these reasons, large veins usually offer a significant resistance to blood flow, and because of this, the pressure in the peripheral veins is usually 4 to 7 mm Hg greater than the right atrial pressure. Partial obstruction of a large vein markedly increases the peripheral venous pressure distal to the obstruction.

Increased right atrial pressure raises peripheral venous pressure. When the right atrial pressure rises above its normal state of 0 mm Hg, blood begins to back up in large veins and open them up. Pressures in the peripheral veins do not rise until the collapsed points between the peripheral veins and the large central veins have opened up, which usually occurs at a right atrial pressure of about 4 to 6 mm Hg. As right atrial pressure rises still further, as occurs in severe heart failure, this causes a corresponding rise in peripheral venous pressure.

"Hydrostatic" pressure affects venous pressure. The pressure at the surface of a body of water is equal to atmospheric pressure, but the pressure rises 1 mm Hg for each 13.6 mm Hg distance below the surface. This pressure results from the weight of the water and therefore is called *hydrostatic pressure*.

Hydrostatic pressure also occurs in the vascular system because of the weight of the blood in the vessels. In an adult who is standing absolutely still, the pressure in the veins of the feet is approximately +90 mm Hg because of the hydrostatic weight of the blood in the veins between the heart and feet.

The venous valves and "venous pump" influence venous pressure. Were it not for the valves of the veins, the hydrostatic pressure effect would cause venous pressure in the feet to always be about +90 mm Hg in a standing adult. Each time one tightens the muscles and moves the legs, however, this com-

presses the veins either in the muscles or adjacent to them and squeezes the blood out of the veins.

The valves in the veins are arranged so that the direction of blood flow can only be toward the heart. Consequently, each time a person moves the legs or tenses the muscles, a certain amount of blood is propelled toward the heart, and the pressure in the veins is lowered. This pumping system is known as the "venous pump" or "muscle pump," and it keeps the venous pressure in the feet of a walking adult near 25 mm Hg.

If a person stands perfectly still, however, the venous pump does not work, and venous pressure quickly rises to the full hydrostatic value of 90 mm Hg. If the valves of the venous system become incompetent or even destroyed, this also decreases the effectiveness of the venous pump. When valve incompetence develops, greater pressure in the veins of the legs may further increase the size of the veins and finally destroy the function of the valves entirely. When this occurs, the person develops *varicose veins*, and the venous and capillary pressures become very high, causing leakage of fluid out of the capillaries and edema in the legs when standing.

The veins function as blood reservoirs. More than 60 per cent of the blood in the circulatory system is usually in the veins. For this reason and because the veins are so compliant, the venous system can serve as a blood reservoir for the circulation. For example, when blood is lost from the body, activation of the sympathetic nervous system causes the veins to constrict, and this takes up much of the "slack" of the circulatory system caused by the lost blood.

Certain portions of the circulatory system are so compliant that they are especially important as blood reservoirs. These include (1) the *spleen*, which can sometimes decrease in size to release as much as 100 milliliters of blood into the reservoir of the circulation; (2) the *liver*, the sinuses of which can release several hundred milliliters of blood into the rest of the circulation; (3) the *large abdominal veins*, which can contribute as much as 300 millimeters; and (4) the *venous plexus beneath the skin*, which can contribute several hundred milliliters.

16

The Microcirculation and the Lymphatic System: Capillary Fluid Exchange, Interstitial Fluid, and Lymph Flow

The most important function of the circulation—to transport nutrients to the tissues and remove waste products—occurs in the capillaries. The capillaries have only a single layer of highly permeable endothelial cells, permitting rapid interchange of nutrients and cellular waste products between the tissues and circulating blood. About 10 billion capillaries, which probably have a total surface area of the size of a football field, provide this function for the body.

Structure of the Microcirculation. Blood enters the capillaries through an *arteriole* and leaves through a venule. Blood from the arteriole passes into a series of *metarterioles,* which have structures midway between those of arterioles and capillaries. Arterioles are highly muscular and play a major role in controlling blood flow to the tissues. The metarterioles do not have a continuous smooth muscle coat, but smooth muscle fibers encircle the vessel at intermittent points called *precapillary sphincters.* Contraction of the muscle in these sphincters can open and close the entrance to the capillary.

This arrangement of the microcirculation is not found in all parts of the body, but similar arrangements serve the same purposes. Both the metarterioles and arterioles are in close contact with the tissues that they serve, and local conditions, such as changes in the concentration of nutrients of waste products of metabolism, can have direct effects on these vessels in controlling the local blood flow.

The capillary wall is very thin, consisting of a single layer of endothelial cells. They are also very

porous, with several million slits, or *pores*, between the cells that make up their walls (the width of the pores are about 8 nanometers) to each square centimeter of capillary surface. Because of the high permeability of the capillaries for most solutes and the high surface area, as blood flows through the capillaries, large amounts of dissolved substances diffuse in both directions through these pores. In this way, almost all dissolved substances in the plasma, except the plasma proteins, continually mix with the interstitial fluid.

Blood flows intermittently through capillaries, a phenomenon called "vasomotion." In many tissues, blood flow through capillaries is not continuous but instead turns on and off every few seconds. The cause of this intermittence is contraction of the metarterioles and precapillary sphincters, which are influenced mainly by *oxygen* and *waste products of tissue metabolism*. When oxygen concentrations of the tissue are reduced (e.g., due to increased oxygen utilization), the periods of blood flow occur more often and last longer, thereby allowing the blood to carry increased quantities of oxygen and other nutrients to the tissues.

EXCHANGE OF NUTRIENTS AND OTHER SUBSTANCES BETWEEN THE BLOOD AND INTERSTITIAL FLUID (p. 185)

Diffusion is the most important means for transfer of substances between the plasma and interstitial fluid. As blood traverses the capillary, tremendous numbers of water molecules and dissolved substances diffuse back and forth through the capillary wall, providing continual mixture of the interstitial fluid and plasma. Lipid-soluble substances, such as oxygen and carbon dioxide, can diffuse directly through the cell membranes without having to go through the pores. Water-soluble substances, such as glucose and electrolytes, diffuse only through intercellular pores in the capillary membrane. The rate of diffusion for most solutes is so great that cells as far as 50 micrometers away from the capillaries can receive adequate quantities of nutrients.

The three primary factors that affect the rate of diffusion across the capillary walls are as follows:

1. *The pore size in the capillary*; in most capillaries, the pore size is 6 to 7 nanometers. The pores of some capillary membranes, such as the liver cap-

illary sinusoids, are much larger and are therefore much more highly permeable to substances dissolved in plasma.

2. *The molecular size of the diffusing substance;* water and most electrolytes, such as sodium and chloride, have a molecular size that is smaller than the pore size, allowing rapid diffusion across the capillary wall. Plasma proteins, however, have a molecular size that is slightly greater than the width of the pores, restricting their diffusion.

3. *The concentration difference of the substance between the two sides of the membrane;* the greater the difference between the concentrations of a substance on the two sides of the capillary membrane, the greater the rate of diffusion in one direction though the membrane. The concentration of oxygen in the blood is normally greater than in the interstitial fluid, allowing large quantities of oxygen to move from the blood toward the tissues. Conversely, the concentrations of the waste products of metabolism are greater in the tissues than in the blood, allowing them to move into the blood and to be carried away from the tissues.

THE INTERSTITIUM AND INTERSTITIAL FLUID (p. 186)

About one sixth of the body consists of spaces between cells, which collectively are called the *interstitium.* The fluid in these spaces is the *interstitial fluid.* The interstitium has two major types of solid structures: (1) *collagen fiber bundles* and (2) *proteoglycan filaments.* The collagen provides most of the tensional strength of the tissues, whereas the proteoglycan filaments, composed mainly of *hyaluronic acid,* are very thin and form a filler of fine reticular filaments, often described as a "brush pile."

"Gel" in the interstitium consists of proteoglycan filaments and entrapped fluid. Fluid in the interstitium is derived by filtration and diffusion from the capillaries and contains almost the same constituency as plasma except with lower concentrations of protein. The interstitial fluid is mainly entrapped in the minute spaces among the proteoglycan filaments and has the characteristics of a *gel.*

Because of the large number of proteoglycan filaments, fluid does not flow easily through the tissue gel. Instead, it mainly *diffuses* through the gel. This diffusion occurs about 95 to 99 per cent as rapidly as it does through free fluid.

The amount of "free" fluid in the interstitium in normal tissues is less than 1 per cent. Although almost all the fluid in the interstitium is entrapped within the tissue gel, small amounts of *"free" fluid* are also present. When the tissues develop *edema*, these small pockets of free fluid can expand tremendously.

THE PROTEINS AND HYDROSTATIC PRESSURES OF THE PLASMA AND INTERSTITIAL FLUID MAINLY DETERMINE THE DISTRIBUTION OF FLUID BETWEEN THE PLASMA AND INTERSTITIUM (p. 187)

Although the exchange of nutrients, oxygen, and metabolic waste products across the capillaries occurs almost entirely by diffusion, the distribution of fluid across the capillaries is determined by another process—the *bulk flow* or *ultrafiltration* of protein-free plasma. As discussed previously, capillary walls are highly permeable to water and most plasma solutes, except the plasma proteins; therefore, hydrostatic pressure differences across the capillary wall push protein-free plasma (ultrafiltrate) through the capillary wall into the interstitium. In contrast, osmotic pressure caused by the plasma proteins (called *colloid osmotic pressure*) tends to produce fluid movement by osmosis from the interstitial spaces into the blood. Interstitial fluid hydrostatic and colloid osmotic pressures also influence fluid filtration across the capillary wall.

The rate at which ultrafiltration occurs across the capillary depends on the difference in hydrostatic and colloid osmotic pressures of the capillary and interstitial fluid. These forces are often called *Starling forces* in honor of Ernest Starling, the physiologist who described their functional significance more than a century ago.

Four forces determine fluid filtration through the capillary membrane. The four primary forces that determine the fluid movement across the capillaries are shown in Figure 16–1; these forces are as follows:

- The *capillary hydrostatic pressure* (Pc), which forces fluid outward through the capillary membrane
- The *interstitial fluid hydrostatic pressure* (Pif), which forces fluid inward through the capillary membrane when Pif is positive but outward into the interstitium when Pif is negative

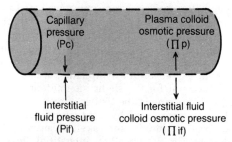

Figure 16–1 Forces operative at the capillary membrane tending to move fluid either outward or inward through the membrane pores.

- The *plasma colloid osmotic pressure* (Πp), which tends to cause osmosis of the fluid inward through the capillary membrane
- The *interstitial fluid colloid osmotic pressure* (Πif), which tends to cause osmosis of fluid outward through the capillary membrane.

The net rate of filtration out of the capillary is determined by the balance of these forces as well as by the *capillary filtration coefficient* (K_f) as follows:

$$\text{Filtration} = K_f \times (Pc - Pif - \Pi p + \Pi if)$$

Functional capillary hydrostatic pressure averages about 17 mm Hg. When blood is flowing through many capillaries, the pressure averages 30 to 40 mm Hg on the arterial ends and 10 to 15 mm Hg on the venous ends, or about 25 mm Hg in the middle. When the capillaries are closed, the pressure in the capillaries beyond the closure is about equal to the pressure at the venous ends of the capillaries (10 mm Hg). When averaged over a period of time, including the periods of opening and closure of the capillaries, the *functional mean capillary pressure* is closer to the pressure in the venous ends of the capillaries than to the pressure in the arteriole ends and averages about 17 mm Hg.

Interstitial fluid hydrostatic pressure is subatmospheric (negative pressure) in loose subcutaneous tissue and positive in tightly encased tissues. Measurements of interstitial fluid hydrostatic pressure have yielded an average value of about -3 mm Hg in loose subcutaneous tissue. One of the basic reasons for this negative pressure is the lymphatic pumping system (discussed later). When fluid enters

the lymphatic capillaries, any movement of the tissue propels the fluid forward through the lymphatic system and eventually back into the circulation. In this way, free fluid that accumulates in the tissue is pumped away as a consequence of tissue movement. This pumping action of lymphatic capillaries appears to account for the slight intermittent negative pressure that occurs in the tissues at rest.

In tissues surrounded by tight encasements, such as the brain, kidney, and skeletal muscle (surrounded by fibrous sheaths), interstitial fluid hydrostatic pressures are usually positive. For instance, the brain interstitial fluid hydrostatic pressure averages about +4 to +6 mm Hg. In the kidneys, interstitial fluid hydrostatic pressure averages about +6 mm Hg.

Plasma colloid osmotic pressure averages about 28 mm Hg. The proteins are the only dissolved substances in the plasma that do not readily pass through the capillary membrane. These substances exert an osmotic pressure that is referred to as *colloid osmotic pressure*. The normal concentration of plasma protein averages about 7.3 gm/dl. About 19 mm Hg of the colloid osmotic pressure is due to the dissolved protein, but an additional 9 mm Hg is due to the positively charged cations, mainly sodium ions, that bind to the negatively charged plasma proteins. This is called the *Donnan equilibrium effect*, which causes the colloid osmotic pressure in the plasma to be about 50 per cent greater than that produced by the proteins alone.

The plasma proteins are mainly a mixture of albumin, globulin, and fibrinogen. About 80 per cent of the total colloid osmotic pressure of the plasma results from the albumin fraction, 20 per cent from the globulin, and only a very tiny amount from the fibrinogen.

Interstitial fluid colloid osmotic pressure averages about 8 mm Hg. Although the size of the usual capillary pore is smaller than the molecular size of the plasma protein, this is not true of all pores; therefore, small amounts of plasma protein leak through the pores into the interstitial spaces. The average protein concentration of the interstitial fluid is around 40 per cent of that in the plasma, or about 3 gm/dl, giving a colloid osmotic pressure of about 8 mm Hg. In some tissues, such as the liver, the interstitial fluid colloid osmotic pressure is much

TABLE 16–1 PRINCIPLES OF EQUILIBRIUM OF FORCES ACROSS CAPILLARIES	
	mm Hg
Mean forces tending to move fluid outward:	
Mean capillary hydrostatic pressure	17.3
Negative interstitial free fluid pressure	3.0
Interstitial fluid colloid osmotic pressure	8.0
Total outward force	28.3
Mean force tending to move fluid inward:	
Plasma colloid osmotic pressure	28.0
Total inward force	28.0
Summation of mean forces:	
Outward	28.3
Inward	−28.0
Net outward force	0.3

greater because the capillaries are much more permeable to the plasma proteins.

Summary of Fluid Volume Exchange Through the Capillary Membrane. The average capillary pressure at the arteriolar ends of the capillaries is 15 to 25 mm Hg greater than at the venular ends. Because of this difference, fluid filters out of the capillaries at the arteriolar ends, and at their venular ends, fluid is reabsorbed back into the capillaries. A small amount of fluid actually flows through the tissues from the arteriolar ends of the capillaries to the venular ends.

Under normal conditions, however, a state of near-equilibrium exists between the amount of fluid filtering outward at the arteriolar ends of the capillaries and the amount of fluid that is returned to the circulation by absorption at the venular ends of the capillaries. There is a slight disequilibrium that occurs, and a small amount of fluid is filtered in excess of that reabsorbed. This fluid is eventually returned to the circulation by way of the lymphatics.

Table 16–1 shows the *average* forces that exist across the entire capillaries and illustrates the principles of this equilibrium. For this table, the pressures in the arterial and venous capillaries are averaged to calculate the mean functional capillary pressure, which is about 17.3 mm Hg.

The small imbalance of forces, 0.3 mm Hg, causes slightly more filtration of fluid into the interstitial spaces than reabsorption.

The rate of filtration in the capillaries is determined not only by the net filtration force, but also by the *capillary filtration coefficient* **(K_f).** The filtration coefficient in the average tissue is about 0.01 milliliter of fluid per minute per millimeter of mercury per 100 grams of tissue. For the entire body, the capillary filtration coefficient is about 6.67 milliliters of fluid per minute per millimeter of mercury. Thus, the net rate of capillary filtration for the entire body is

$$\text{Net Filtration} = K_f \times \text{Net Force}$$
$$= 6.67 \times 0.3$$
$$= 2 \text{ ml/min}$$

Because of the extreme differences in the permeabilities of the capillary systems in different tissues, the capillary filtration coefficient may vary more than 100-fold among the different tissues. For example, the capillary filtration coefficient in the kidneys is about 4.2 milliliters of fluid per minute per millimeter of mercury per 100 grams of kidney weight, a value almost 400 times as great as the K_f of many other tissues. This obviously causes a much greater rate of filtration in the glomerular capillaries of the kidney.

An abnormal imbalance of forces at the capillary can cause edema. *If the mean capillary pressure rises* above the normal value of 17 mm Hg, the net force causing filtration of fluid into the tissue spaces also rises. A rise in mean capillary pressure of 20 mm Hg causes an increase in the net filtration pressure from 0.3 to 20.3 mm Hg, which results in 68 times as much net filtration of fluid into the interstitial spaces as normally occurs. Prevention of accumulation of excess fluid in the spaces would require 68 times the normal flow of fluid into the lymphatic system, an amount that is too great for the lymphatics to carry away. As a result, large increases in capillary pressure can cause accumulation of fluid in the interstitial spaces, a condition referred to as *edema*.

Similarly, a *decrease in plasma colloid osmotic pressure* will also increase the net filtration force and therefore the net filtration rate of fluid into the tissues.

THE LYMPHATIC SYSTEM (p. 193)

The lymphatic system carries fluid from the tissue spaces into the blood. Importantly, the lymphatics also carry away proteins and large particulate matter from the tissue spaces, neither of which can be removed through absorption directly into the blood capillary.

Almost all tissues of the body have lymphatic channels. Most of the lymph from the lower part of the body flows up the *thoracic duct* and empties into the venous system at the juncture of the left interior jugular vein and subclavian vein. Lymph from the left side of the head, left arm, and parts of the chest region also enters the thoracic duct before it empties into the veins. Lymph from the right side of the neck and head, right arm, and parts of the thorax enter the *right lymph duct*, which then empties into the venous system at the juncture of the right subclavian vein and internal jugular vein.

Lymph is derived from interstitial fluid. As lymph first flows from the tissue, it has almost the same composition as the interstitial fluid. In many tissues, the protein concentration averages about 2 gm/dl, but in other tissues such as the liver, the protein concentration may be as high as 6 gm/dl. As fluid flows along lymphatics, passing through lymph nodes, some reabsorption of fluid occurs, causing the lymph in large lymphatic vessels to have a slightly higher protein concentration than that in the terminal lymphatic capillaries.

In addition to carrying fluid and protein from the interstitial spaces to the circulation, the lymphatic system is one of the major routes for absorption of nutrients from the gastrointestinal tract, as discussed in Chapter 65. After a fatty meal, for instance, thoracic duct lymph sometimes contains as much as 1 to 2 per cent fat.

The rate of lymph flow is determined by interstitial fluid hydrostatic pressure and the lymphatic pump. The total rate of lymph flow is approximately 120 ml/hr, or 2 to 3 liters per day. This rate of formation can change dramatically, however, in certain pathological conditions associated with excessive fluid filtration from the capillaries into the interstitium.

- *Increased interstitial fluid hydrostatic pressure increases lymph flow rate.* At the normal interstitial fluid hydrostatic pressures in the subatmospheric range, lymph flow is very low. As the pressure rises to values slightly greater than 0 mm Hg, the lymph flow increases by more than 20-fold. When interstitial pressure reaches +1 or +2 mm Hg, lymph flow fails to rise further. This probably results from the fact that rising tissue pressure not only increases the entry of fluid into the lymphatic capillaries but also compresses the larger lymphatics, thereby impeding lymph flow.
- *The lymphatic pump increases lymph flow.* Valves exist in all lymph channels. In addition, each segment of the lymphatic vessel functions as a separate automatic pump; that is, filling of a segment causes it to contract, and the fluid is pumped through the valve into the next lymphatic segment. This fills the lymphatic segment, and within a few seconds, it too contracts, with the process continuing along the lymph vessel until the fluid is finally emptied. This pumping action propels the lymph forward toward the circulation. In addition to pumping caused by intrinsic contraction of the vessels, external factors that compress lymph vessels can cause pumping. For example, contraction of muscles surrounding lymph vessels or movement of body parts may increase lymphatic pumping. In some conditions, such as during exercise, the lymphatic pump may increase lymph flow by as much as 10- to 30-fold.

The lymphatic system plays an important role as an "overflow mechanism" that returns to the circulation excess proteins and fluid volume that enter the tissue spaces. In conditions in which the lymphatic system fails, such as occurs with blockade of a major lymphatic vessel, proteins and fluid accumulate in the interstitium, causing edema. The accumulation of protein in the interstitium is especially important in causing edema because the lymphatics provide the only mechanism by which protein that leaks out of the capillaries can re-enter the circulation in significant quantities. With protein accumulation in the interstitial spaces, due to lymphatic failure, there is an increase in colloid osmotic pressure of the interstitial fluid that tends to allow more fluid filtration into the interstitium. As a result, complete blockade of the lymphatic vessels results in severe edema.

Bacteria and debris from the tissues are removed by the lymphatic system at lymph nodes. Because of the very high degree of permeability of the lymphatic capillaries, bacteria and other small particulate matter in the tissues can pass into the lymph. The lymph passes through a series of nodes on its way out to the blood. In these nodes, bacteria and other debris are filtered out, phagocytized by macrophages in the nodes, and finally digested into amino acids, glucose, fatty acids, and other small-molecular-weight substances before being released into the blood.

17

Local Control of Blood Flow by the Tissues, and Humoral Regulation

The local tissues autoregulate blood flow in response to tissue needs. In most tissues, blood flow is autoregulated, which means that the tissue regulates its own blood flow. This is beneficial to the tissue because it allows the delivery of oxygen and nutrients and removal of waste products to parallel the rate of tissue activity. Autoregulation permits blood flow from one tissue to be regulated independently of flow to another tissue.

In certain organs, blood flow serves purposes other than supplying nutrients and removing waste products. For instance, blood flow to the skin determines heat loss from the body and in this way helps control the body temperature. Delivery of adequate quantities of blood to the kidneys allows them to rapidly excrete the waste products of the body.

The ability of the tissues to regulate their own flow permits them to maintain adequate nutrition and perform necessary functions in maintaining homeostasis. In general, the greater the rate of metabolism in an organ, the greater its blood flow; in Table 17–1, for example, there is very great blood flow in glandular organs such as the thyroid and adrenal glands, which have a very high metabolic rate. In contrast, blood flow in resting skeletal muscles of the body is very low because in the resting state, metabolic activity of the muscle is also very low; however, during heavy exercise, muscle metabolic activity can increase by more than 60-fold and the blood flow can increase by as much as 20-fold.

MECHANISMS OF LOCAL BLOOD FLOW CONTROL
(p. 200)

Local tissue blood flow control can be divided into two phases: (1) acute control and (2) long-term con-

TABLE 17–1 BLOOD FLOW TO DIFFERENT ORGANS AND TISSUES UNDER BASAL CONDITIONS

	Flow		
	Per Cent of Cardiac Output	ml/min	ml/min/100 gm of tissue
Brain	14	700	50
Heart	4	200	70
Bronchi	2	100	25
Kidneys	22	1100	360
Liver	27	1350	95
Portal	(21)	(1050)	
Arterial	(6)	(300)	
Muscle (inactive state)	15	750	4
Bone	5	250	3
Skin (cool weather)	6	300	3
Thyroid gland	1	150	160
Adrenal glands	0.5	25	300
Other tissues	3.5	175	1.3
Total	100.0	5000	

trol. Acute control occurs within seconds to minutes through constriction or dilation of arterioles, metarterioles, and precapillary sphincters. Long-term control, however, occurs over a period of days, weeks, or even months and, in general, provides better control of flow in proportion to the needs of the tissues. Long-term control occurs mainly as a result of increases or decreases in the physical sizes and numbers of blood vessels supplying the tissues.

Acute Control of Local Blood Flow

Increased tissue metabolic rate raises local blood flow. In many tissues, such as skeletal muscle, increases in metabolism up to eight times normal raise blood flow acutely about fourfold. Initially, the rise in flow is less than that in metabolism, but once the metabolism increases sufficiently high to remove most of the nutrients from the blood, a further rise in metabolism can occur only with a concomitant rise in blood flow to supply the required nutrients.

Decreased oxygen availability increases local blood flow. One of the required nutrients for tissue metabolism is oxygen. Whenever the availability of oxygen in the tissues decreases, such as at high altitude, in pneumonia, or with carbon monoxide poisoning (which inhibits the ability of hemoglobin to transport oxygen), the tissue blood flow increases markedly. Cyanide poisoning, for instance, which reduces the ability of the tissues to utilize oxygen, can increase tissue blood flow by as much as sevenfold.

Increased demand for oxygen and nutrients increases local blood flow. In the absence of an adequate supply of oxygen and nutrients as a result of either increased tissue metabolism or decreased delivery of these substances to the tissues, the arterioles, metarterioles, and precapillary sphincters relax, thereby decreasing vascular resistance and allowing more flow to the tissues. The relaxation of precapillary sphincters allows flow to occur more frequently in capillaries that are closed due to periodic contraction of precapillary sphincters *(vasomotion).*

Accumulation of vasodilator metabolites increases local blood flow. The greater the rate of metabolism in the tissue, the greater the rate of production of tissue metabolites, such as *adenosine, adenosine phosphate compounds, carbon dioxide, lactic acid, potassium ions,* and hydrogen ions. Each of these substances has been suggested to act as a *vasodilator* that contributes to increased blood flow associated with stimulation of tissue metabolism.

A lack of other nutrients may also cause vasodilation. For example, a deficiency of glucose, amino acids, or fatty acids may contribute to local vasodilation, although this has not been proven. Vasodilation occurs in the vitamin deficiency disease *beriberi,* in which the patient usually has a deficiency of the vitamin B substances thiamine, niacin, and riboflavin. Because these vitamins are all involved in the oxidative phosphorylation mechanism for generating ATP, a deficiency of these vitamins may lead to diminished ability of the smooth muscle to contract, thereby causing local vasodilation.

Special Examples of Local Blood Flow Control (p. 202)

"Reactive hyperemia" occurs after the blood supply to a tissue is blocked for a short time. If blood flow is blocked for a few seconds to several hours and then unblocked, flow to the tissue usually increases to four to seven times normal; the increased flow will continue for a few seconds or for much longer if the flow has been stopped for 1 hour or longer. This phenomenon is called *reactive hyperemia* and appears to be a manifestation of local "metabolic" blood flow regulation mechanisms. After vascular occlusion, there is an accumulation of tissue vasodilator metabolites and the development of an oxygen deficiency in the tissues. The extra blood flow during reactive hyperemia lasts long enough to almost exactly repay the tissue oxygen deficiency and to wash out the accumulated vasodilator metabolites.

"Active hyperemia" occurs when the tissue metabolic rate increases. When a tissue becomes highly active, such as muscle during exercise or even the brain during rapid mental activity, the blood flow to the tissue increases. Again, this appears to be related to increases in local tissue metabolism that cause accumulation of vasodilator substances and possibly a slight oxygen deficit. The dilation of local blood vessels helps the tissue to receive the additional nutrients required to sustain its new level of function.

Tissue blood flow is "autoregulated" during changes in arterial pressure. In any tissue of the body, acute increases in arterial pressure will cause an immediate increase in blood flow. Within less than 1 minute, however, the blood flow in many tissues returns back toward the normal level, which is called "autoregulation of blood flow."

- *The metabolic theory of autoregulation* suggests that when arterial pressure becomes too great, the excess provides too much oxygen and too many nutrients to the tissues, causing the blood vessels to constrict and the flow to return toward normal despite the increased arterial pressure.

- *The myogenic theory of autoregulation* suggests that sudden stretch of small blood vessels will cause the smooth muscles in the vessel walls to contract. This is an intrinsic property of smooth muscles that allows them to resist excessive stretching. Conversely, at low pressures, the degree of stretch of the vessel is less, and the smooth muscle relaxes, decreasing vascular resistance and allowing flow to be maintained relatively constant despite the lower blood pressure.

The relative importance of these two mechanisms for autoregulation of blood flow is still debated by physiologists. It seems likely that both mechanisms contribute to the maintenance of a relatively stable blood flow during variations in arterial pressure.

There are additional mechanisms for blood flow control in specific tissues. Although the general mechanisms for local blood flow control discussed thus far are present in most tissues of the body, there are special mechanisms that control blood flow in special areas. These are discussed in relation to specific organs, but the following are two notable mechanisms.

- *In the kidneys, blood flow control is vested, in part, in a mechanism called tubuloglomerular feedback,* in which the composition of fluid in the early distal tubule is detected by the *macula densa*, which is located where the tubule abuts the afferent arteriole at the *juxtaglomerular apparatus*. When too much fluid filters from the blood through the glomerulus into the tubular system, feedback signals from the macula densa cause constriction of the *afferent arterioles*, thereby reducing renal blood flow and glomerular filtration rate back toward normal.

- *In the brain, the concentrations of carbon dioxide and hydrogen play prominent roles in local blood flow control.* An increase in either dilates the cerebral blood vessels, which allows rapid washout of the excess carbon dioxide and hydrogen ions.

Endothelium-derived relaxing factor (nitric oxide) dilates large upstream arteries when microvascular blood flow increases. The local mechanisms for controlling tissue blood flow act mainly on the very small microvessels located in the immediate tissue because local feedback by vasodilator substances or oxygen deficiency can reach only these vessels, not the larger arteries upstream. When blood flow through the microvascular portion of the circulation increases, however, the endothelial cells lining the larger vessels release a vasodilator substance called *endothelium-derived relaxing factor*, which appears to be mainly *nitric oxide*. This release of nitric oxide is caused, in part, by increased *shear stress* on the endothelial walls that occurs as blood flows more rapidly through the larger vessels. The release of nitric oxide then relaxes the larger vessels, causing them to dilate.

Without the dilation of larger vessels, the effectiveness of local blood flow would be compromised, because a significant part of the resistance in blood flow is in the upstream arterioles and small arteries.

Long-Term Blood Flow Regulation (p. 204)

Most of the mechanisms that have been discussed thus far act within a few seconds to a few minutes after the local tissue conditions have changed. Even with full function of these acute mechanisms, blood flow usually is adjusted only about three fourths of the way back to the exact requirements of the tissues. Over a period of hours, days, and weeks, a long-term local blood flow regulation develops that helps to precisely adjust the blood flow to match the metabolic needs of the tissues.

The mechanism for long-term regulation of blood flow is a change in tissue vascularity. If the metabolism of a tissue is increased for prolonged periods of time, the physical size of the vessels in a tissue increases; under some conditions, the number of blood vessels also increases. One of the major factors that stimulate this increased vascularity is a *low oxygen concentration* in the tissues. Animals that live at high altitudes, for instance, have increased vascularity. Likewise, fetal chicks hatched at low oxygen levels have up to two times as much vascular conductivity as normally seen. This growth of new vessels is called *angiogenesis*.

Angiogenesis occurs mainly in response to the presence of angiogenic factors released from (1) ischemic tissues, (2) tissues that are growing rapidly, and (3) tissues that have excessively high metabolic rates.

Many angiogenic factors are small peptides. Three of the best characterized angiogenic factors are *vascular endothelial growth factor (VEGF), fibroblast growth factor (FGF),* and *angiogenin,* each of which has been isolated from tumors or from other tissues that are rapidly growing or have inadequate blood supply.

Essentially all of the angiogenic factors promote new vessel growth by causing the vessels to sprout from small venules or, occasionally, capillaries. The basement membrane of the endothelial cells is dissolved, followed by the rapid production of new endothelial cells that stream out of the vessel in extended cords directed toward the source of the angiogenic factor. The cells continue to divide and eventually fold over into a tube. The tube then connects with another tube budding from another donor vessel and forms a capillary loop through which blood begins to flow. If the flow is sufficiently great, smooth muscle cells eventually invade the wall so that some of these vessels grow to be small arterioles or perhaps even larger arteries.

Collateral blood vessels develop when an artery or a vein is blocked. New vascular channels usually develop around a blocked artery or vein and allow the affected tissue to be at least partially resupplied with blood. An important example is the development of collateral blood vessels after a thrombosis of one of the coronary arteries. In many people over the age of 60, there is a blockage of at least one of the smaller coronary vessels, yet most people do not know that this has happened because collateral blood vessels have gradually developed as the vessels have begun to close, thereby providing sufficient blood flow to the tissue to prevent myocardial damage. It is in those instances in which thrombosis occurs too rapidly for the development of collaterals that serious heart attacks occur.

HUMORAL REGULATION OF THE CIRCULATION (p. 205)

Several hormones are secreted into the circulation and transported into the blood throughout the entire body. Some of the most important of these that can influence circulatory function include the following:

- *Norepinephrine* and *epinephrine* released by the adrenal medullae can act as vasoconstrictors in many tissues by stimulating alpha-adrenergic receptors; however, epinephrine is much less potent as a vasoconstrictor and may even cause mild vasodilation through stimulation of beta-adrenergic receptors in some tissues such as skeletal muscle.
- *Angiotensin II* is a powerful vasoconstrictor substance that is usually formed in response to volume depletion or decreased blood pressure.
- *Vasopressin*, or *antidiuretic hormone*, is one of the most powerful vasoconstrictors in the body. It is formed in the hypothalamus and transported to the posterior pituitary, where it is released in response to decreased blood volume, as occurs in hemorrhage, or increased plasma osmolarity, as occurs in dehydration.
- *Endothelin* is a powerful vasoconstrictor released by endothelial cells in response to damage to these cells. For example, after severe blood vessel

damage, endothelin release and subsequent vasoconstriction may help to prevent excessive bleeding from the vasculature.

- *Prostaglandins* are formed in almost every tissue in the body. These substances have important intracellular effects, but some of them are released in the circulation, especially *prostacyclin* and *prostaglandins of the E series*, which are *vasodilators*. Some prostaglandins, such as *thromboxane A_2* and *prostaglandins of the F series*, are *vasoconstrictors*.

- *Bradykinin*, which is formed in the blood and in tissue fluids, is a powerful vasodilator that also increases capillary permeability. For this reason, increased levels of bradykinin may cause marked edema as well as increased blood flow in some tissues.

- *Histamine* is a powerful vasodilator that is released into the tissues when they become damaged or inflamed. Most of the histamine is released from *mast cells* and damaged tissues or from *basophils* in the blood. Histamine, like bradykinin, increases capillary permeability and causes tissue edema as well as greater blood flow.

Ions and other chemical factors can also alter local blood flow. Many different ions and chemical factors can either dilate or constrict local blood vessels. Their specific effects are as follows:

- *Increased calcium ion concentration* causes vasoconstriction.

- *Increased potassium ion concentration* causes vasodilation.

- *Increased magnesium ion concentration* causes vasodilation.

- *Increased sodium ion concentration* causes vasodilation.

- *Increased osmolarity of the blood*, caused by increased quantities of glucose or other nonvasoactive substances, causes vasodilation.

- *Increased hydrogen ion concentration* (decreased pH) causes vasodilation.

- *Increased carbon dioxide concentration* causes vasodilation in most tissues and marked vasodilation in the brain.

18

Nervous Regulation of the Circulation, and Rapid Control of Arterial Pressure

Except for certain tissues, such as skin, nervous control normally has little to do with adjustments of blood flow to the tissues. Nervous control mainly affects more global functions, such as redistributing blood flow to different parts of the body, increasing the pumping activity of the heart, and providing rapid control of arterial pressure. This control of the circulation by the nervous system is exerted almost entirely through the *autonomic nervous system*.

AUTONOMIC NERVOUS SYSTEM (p. 209)

The two components of the autonomic nervous system are the *sympathetic nervous system*, which is most important for controlling the circulation, and the *parasympathetic nervous system*, which contributes to regulation of heart function.

Sympathetic nerves innervate the blood vessels and heart. Sympathetic vasomotor fibers leave the spinal cord through all of the thoracic and the first one or two lumbar spinal nerves. They pass into the sympathetic chain and then go via two routes to the circulation: (1) through specific *sympathetic nerves* that innervate mainly the vasculature of the internal viscera and heart and (2) through *spinal nerves* that innervate mainly the vasculature of the peripheral areas. Almost all of the blood vessels, except the capillaries, are innervated by sympathetic nerve fibers. Sympathetic stimulation of the small arteries and arterioles increases the vascular resistance and decreases the rate of blood flow through the tissues. Innervation of the large vessels, especially the veins, makes it possible for sympathetic stimulation to decrease the volume of the vessels.

Sympathetic fibers also go to the heart and stimulate the activity of the heart, increasing both the rate and strength of pumping.

Parasympathetic stimulation decreases heart rate and heart muscle contractility. Although the parasympathetic system plays an important role in controlling many other autonomic functions of the body, its main role in controlling the circulation is to markedly decrease heart rate and slightly decrease heart muscle contractility.

Control of the Sympathetic Vasoconstrictor System by the Central Nervous System (p. 210)

The sympathetic nerves carry large numbers of vasoconstrictor fibers and only a few vasodilator fibers. The vasoconstrictor fibers are distributed to almost all segments of the circulation. Their distribution is greater in some tissues, such as skin, gut, and spleen.

Vasomotor centers of the brain control the sympathetic vasoconstrictor system. Located bilaterally in the reticular substance of the medulla and lower third of the pons is an area called the *vasomotor center* that transmits parasympathetic impulses through the vagus nerves to the heart and sympathetic impulses through the cord and peripheral sympathetic nerves to almost all the blood vessels of the body.

Although the organization of the vasomotor centers is not completely understood, certain areas appear to be especially important.

- *A vasoconstrictor area*, called *area C-1*, is located bilaterally in the anterolateral portions of the upper medulla. The neurons in this area secrete *norepinephrine*, and their fibers are distributed throughout the cord, where they excite vasoconstrictor neurons of the sympathetic nervous system.
- *A vasodilator area*, called *area A-1*, is located bilaterally in the anterolateral portions of the lower half of the medulla. The fibers from these neurons inhibit vasoconstrictor activity of the C-1 area, causing vasodilation.
- *A sensory area*, called *area A-2*, is located bilaterally in the *tractus solitarius* in the posterolateral portions of the medulla and lower pons. The neurons of this area receive sensory nerve signals from the

vagus and glossopharyngeal nerves, and the output signals from this sensory area help to control the activities of both the vasoconstrictor and vasodilator areas, providing "reflex" control of many circulatory functions. An example is the baroreceptor reflex for controlling arterial pressure (discussed later).

Continuous sympathetic vasoconstrictor tone causes partial constriction of most blood vessels. Normally, the vasoconstrictor area of the vasomotor center transmits signals continuously to the sympathetic vasoconstrictor nerve fibers over the entire body, causing slow firing of these fibers at a rate of about 1 impulse per second. This *sympathetic vasoconstrictor tone* maintains a partial state of contraction of the blood vessels. When this tone is blocked, for example, by spinal anesthesia, the blood vessels throughout the body dilate and arterial pressure may fall to as low as 50 mm Hg.

The vasomotor system is influenced by higher nervous centers. Large numbers of areas throughout the *reticular substance* of the *pons, mesencephalon,* and *diencephalon* can either excite or inhibit the vasomotor center.

The *hypothalamus* plays a special role in control of the vasoconstrictor system and can exert powerful excitatory or inhibitory effects on the *vasomotor center.*

Many different parts of the *cerebral cortex* can also excite or inhibit the vasomotor center; for example, stimulation of the *motor cortex* excites the vasomotor center. Many areas of the brain can have profound effects on cardiovascular function.

Norepinephrine is the transmitter substance of the sympathetic vasoconstriction system. Norepinephrine, which is secreted at the endings of the vasoconstrictor nerves, acts directly on alpha adrenergic receptors of vascular smooth muscle to cause vasoconstriction.

The adrenal medullae release norepinephrine and epinephrine during sympathetic stimulation. Sympathetic impulses are usually transmitted to the adrenal medullae at the same time that they are transmitted to the blood vessels, stimulating release of epinephrine and norepinephrine into the circulating blood. These two hormones are carried into the blood stream to all parts of the body, where they act directly on the blood vessels to cause vasoconstriction through stimulation of alpha adrenergic receptors. Epinephrine, however, also has potent beta adrenergic effects, which cause vasodilation in certain tissues, such as skeletal muscle.

ROLE OF THE NERVOUS SYSTEM FOR RAPID CONTROL OF ARTERIAL PRESSURE (p. 212)

One of the most important functions of the sympathetic nervous system is to provide rapid control of arterial pressure by causing vasoconstriction and stimulation of the heart. At the same time that sympathetic activity is increased, there often is a reciprocal inhibition of parasympathetic vagal signals to the heart that also contribute to a greater heart rate. As a consequence, there are three major changes that take place to increase arterial pressure through stimulation of the autonomic nervous system.

- *Arterioles throughout the body are constricted*, causing increased total peripheral resistance and raising blood pressure.
- *The veins and larger vessels of the circulation are constricted*, displacing blood from the peripheral vessels toward the heart and causing the heart to pump with greater force, which also helps to raise the arterial pressure.
- *The heart is directly stimulated by the autonomic nervous system, further enhancing cardiac pumping.* Much of this is caused by increased heart rate, sometimes to as great as three times normal. In addition, sympathetic stimulation directly increases the contractile force of the heart muscle, raising its capability to pump larger volumes of blood.

An important characteristic of nervous control is that it is very rapid, beginning within seconds. Conversely, sudden inhibition of nervous stimulation can decrease the arterial pressure within seconds.

The autonomic nervous system contributes to increased arterial pressure during muscle exercise. During heavy exercise, the muscles require greatly increased blood flow. Part of this increase results from local vasodilation, but additional increase in flow results from simultaneous elevation of arterial pressure during exercise. During very heavy exercise, arterial pressure may rise as much as 30 per cent to 40 per cent.

The rise in arterial pressure during exercise is believed to result mainly from the following effect: At the same time that the motor areas of the nervous system become activated to cause exercise, most of the reticular activating system in the brain is also activated, which greatly increases stimulation of the vasoconstrictor and cardioaccelerator areas of the vasomotor center. These raise the arterial pressure instantly to keep pace with increased muscle activity. The vasodilation of the muscle, however, is maintained despite increased sympathetic activity because of the overriding effect of local control mechanisms in the muscle.

The autonomic nervous system increases arterial pressure during the "alarm reaction." For instance, during extreme fright, the arterial pressure often rises to as high as 200 mm Hg within a few seconds. This *alarm reaction* provides the necessary increase in arterial pressure that can immediately supply blood to any of the muscles of the body that might need to respond instantly to cause flight from danger.

Reflex Mechanisms Help Maintain Normal Arterial Pressure (p. 213)

Aside from special circumstances such as stress and exercise, the autonomic nervous system operates to maintain the arterial pressure at or near its normal level through *negative feedback reflex mechanisms.*

The Arterial Baroreceptor Reflex Control System. This reflex is initiated by stretch receptors, called *baroreceptors,* that are located in the walls of large systemic arteries, particularly in the walls of the *carotid sinus* and the *aortic arch.* Signals from the carotid sinus receptors are transmitted through *Herring's nerve* to the *glossopharyngeal nerve* and then to the *tractus solitarius* in the medullary area of the brain stem. Signals from the aortic arch are transmitted through the *vagus nerves* to the same area of the medulla. The baroreceptors function to control arterial pressure as follows:

- Increased pressure in the baroreceptors causes increased impulse firing.
- Baroreceptor signals enter the tractus solitarius, inhibit the vasoconstrictor center of the medulla, and excite the vagal center.

- The net effects are stimulation of sympathetic activity and inhibition of parasympathetic activity, which cause (1) *vasodilation of veins and arterioles* and (2) *decreased heart rate* and *strength of heart contraction.*
- This causes the arterial pressure to decrease because of a decline in peripheral vascular resistance and cardiac output.

The baroreceptor functions as a "buffer" to maintain arterial pressure relatively constant during changes in body posture and other daily activities. When a person stands up after lying down, the arterial pressure in the head and upper parts of the body tends to fall. The reduction in pressure decreases the signals sent from the baroreceptors to the vasomotor centers, eliciting a strong sympathetic discharge that minimizes the reduction in arterial pressure. In the absence of functional baroreceptors, marked reductions in arterial pressure can decrease cerebral blood flow to the extent of loss of consciousness.

Daily activities that tend to increase blood pressure, such as eating, excitement, defecation, and so forth, can cause extreme increases in blood pressure in the absence of normal baroreceptor reflexes. A primary purpose of the arterial baroreceptor system is to reduce the daily variation in arterial pressure to about one half to one third of the pressure that would occur if the baroreceptor system were not present.

The baroreceptors are relatively unimportant for long-term regulation of arterial pressure. One of the reasons that the baroreceptors play only a minor role in long-term regulation of arterial pressure is that the baroreceptors reset within 1 to 2 days to the pressure level to which they are exposed. In experimental animals in which the baroreceptors have been denervated, blood pressure is extremely variable, swinging to very high and very low levels, but the average arterial pressure measured during a 24-hour period is relatively constant. Prolonged regulation of blood pressure requires other control systems, especially the renal–body fluid pressure control system (see Chapter 19).

Control of Arterial Pressure by the Carotid and Aortic Chemoreceptors—Effect of Oxygen Lack on Arterial Pressure. Closely associated with the baroreceptor control system is a chemoreceptor reflex that operates in much the same way as the baroreceptor reflex, except that *chemoreceptors*, instead of stretch receptors, initiate the response.

The chemoreceptors are sensitive to *oxygen lack, carbon dioxide excess*, or *hydrogen ion excess*. They are located in two *carotid bodies*, one of which lies in the bifurcation of each common carotid artery, and in several *aortic bodies* adjacent to the aorta. Whenever the arterial pressure falls below a critical level, the chemoreceptors become stimulated because of diminished blood flow to the bodies and the resulting diminished availability of oxygen and excess buildup of carbon dioxide and hydrogen ions that are not removed by the slow blood flow. Signals transmitted from the chemoreceptors into the vasomotor center *excite* the vasomotor center, and this elevates the arterial pressure.

Cardiopulmonary reflexes help to regulate arterial pressure. Both the atrial and pulmonary arteries have stretch receptors, called *cardiopulmonary receptors* or *low pressure receptors*, in their walls that are similar to the baroreceptor stretch receptors of the systemic arteries. These low-pressure receptors play an important role in minimizing arterial pressure changes in response to blood volume changes. Although the low-pressure receptors do not directly detect systemic arterial pressure, they detect increases in pressure in the heart and pulmonary circulation caused by changes in volume, and they elicit reflexes parallel to the baroreceptor reflexes to make the total reflex system more potent for control of arterial pressure.

Increased stretch of the atria causes reflex decreases in sympathetic activity to the kidney, which causes vasodilation of the afferent arterioles and increases in glomerular filtration rate, as well as decreases in tubular reabsorption of sodium. These cause the kidney to excrete more sodium and water, thereby ridding the body of excess volume.

The Central Nervous System Ischemic Response Raises Arterial Pressure in Response to Diminished Blood Flow in the Vasomotor Center of the Brain (p. 217)

When blood flow to the vasomotor center in the lower brain stem becomes sufficiently decreased to cause *cerebral ischemia* (i.e., nutritional deficiency), the neurons of the vasomotor center become strongly excited. When this occurs, systemic arterial pressure often rises to a level as high as the heart can pump. This effect may be caused by failure of the slowly flowing blood to carry carbon dioxide away from the vasomotor center, since increased carbon dioxide concentration is a potent agent for stimulating the sympathetic nervous control areas of the medulla of the brain. Other factors, such as buildup of lactic acid, may also contribute to marked stimulation of the vasomotor center and increased arterial pressure.

This arterial pressure elevation in response to cerebral ischemia is known as the *central nervous system ischemic response*. This response is an emergency arterial pressure control system that acts rapidly and powerfully to prevent further decline in arterial pressure when blood flow to the brain decreases dangerously; it is sometimes called the "last ditch" mechanism for blood pressure control.

The *Cushing reaction* is a special type of central nervous system ischemic response that results from increased pressure in the cranial vault. For instance, when cerebrospinal fluid pressure rises to equal the arterial pressure, a central nervous system ischemic response is initiated that can raise arterial pressure to as high as 250 mm Hg. This response helps to protect the vital centers of the brain from loss of nutrition, which could occur if pressure in the cranial vault exceeds the normal arterial pressure and compresses the blood vessels supplying the brain.

If cerebral ischemia becomes so severe that a maximal increase in arterial pressure still cannot relieve the ischemia, the neuronal cells begin to suffer metabolically, and within 3 to 10 minutes they become inactive. This causes the arterial pressure to decrease.

19

Dominant Role of the Kidneys in Long-Term Regulation of Arterial Pressure and in Hypertension: The Integrated System for Pressure Control

RENAL–BODY FLUID SYSTEM FOR ARTERIAL PRESSURE CONTROL (p. 221)

Although the reflex mechanisms play a very important role in the short-term control of arterial pressure, when the arterial pressure changes slowly over many hours or days, the mechanisms gradually lose most of their ability to oppose the changes. The most important mechanism for the long-term control of blood pressure is linked to control of circulatory volume by the kidneys, a mechanism known as the *renal–body fluid feedback system*. When arterial pressure rises too high, the kidneys excrete increased quantities of sodium and water because of *pressure natriuresis* and *pressure diuresis,* respectively. As a result of the increased renal excretion, the extracellular fluid volume and blood volume both decrease until blood pressure returns to normal and the kidneys excrete normal amounts of sodium and water.

Conversely, when arterial pressure falls too low, the kidneys reduce their rate of sodium and water excretion, and over a period of hours to days, if the person drinks enough water and eats enough salt to increase blood volume, arterial pressure will return to its previous level. This mechanism for blood pressure control is very slow to act, sometimes requiring several days, 1 week, or longer to come to equilibrium; therefore, it is not of major importance in the acute control of arterial pressure. In contrast, it is by far the most potent of all long-term arterial pressure controllers.

159

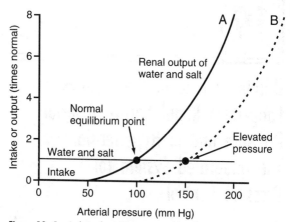

Figure 19-1 Analysis of arterial pressure regulation by equating the renal output curve with the salt and water intake curve. The equilibrium point describes the level to which the arterial pressure will be regulated. Curve A *(solid line)* shows the normal renal output curve. Curve B *(dashed line)* shows the renal output curve in hypertension.

Renal output of salt and water must be balanced with the intake of salt and water. Figure 19-1 shows the effect of different arterial pressures on urine volume output, demonstrating marked increases in the output of volume (pressure diuresis) and sodium (pressure natriuresis) as arterial pressure rises. Note that as long as arterial pressure is above the normal equilibrium point at which water and salt intakes are balanced with renal output, renal output will exceed the intake of salt and water, resulting in a progressive decline in extracellular fluid volume. Conversely, if blood pressure falls below the equilibrium point, the renal output of water and salt will be lower than the intake, resulting in a progressive increase in extracellular fluid volume. The only point on the curve at which a balance between renal output and intake of salt and water can occur is at the normal arterial pressure (the equilibrium point).

The renal–body fluid feedback mechanism demonstrates an "infinite gain" in long-term blood pressure control. To understand why this mechanism demonstrates "infinite gain" in controlling blood pressure, assume that the arterial pressure rises to 150 mm Hg. At this level, renal output of water and salt is about three times greater than intake. The body loses fluid, blood volume decreases, and arterial pressure decreases. *Furthermore, this loss of fluid will not cease until the arterial pressure declines all the way to the exact equilibrium point (see Fig. 19–1A).* Conversely, if blood pressure falls below the equilibrium point, the kidneys will decrease salt and water excretion to a level below intake, causing accumulation of fluid and greater blood volume until arterial pressure returns to the exact equilibrium point. Because there is no remaining error in arterial pressure after full correction, this feedback system has *infinite gain.*

There are two primary determinants of the long-term arterial pressure. From the curve shown in Figure 19–1, one can see that the two factors that determine long-term arterial pressure are (1) *the renal output curve for salt and water* and (2) *the level of salt and water intake.* As long as these two factors remain constant, arterial pressure will also remain exactly at the normal level of 100 mm Hg. For arterial pressure to deviate from the normal level for long periods of time, one of these two factors must be altered.

In Figure 19–1B, an abnormality of the kidney has caused the renal output curve to shift 50 mm Hg toward higher blood pressure. This results in a new equilibrium point, and arterial pressure will follow to this new pressure level within a few days. Although greater salt and water intake can theoretically elevate arterial pressure (discussed later), the body has multiple neurohumoral mechanisms that protect against large increases in arterial pressure when salt and water intake is elevated. This is accomplished mainly by decreasing the formation of angiotensin II and aldosterone, which increases the ability of the kidneys to excrete salt and water and results in a very steep renal output curve. In most persons, therefore, very large increases in salt and water output can be accomplished with minimal increases in arterial pressure.

Increased total peripheral resistance cannot elevate the long-term arterial pressure if fluid intake and renal function do not change. When total peripheral resistance is acutely increased, the arterial pressure increases almost immediately. If the vascular resistance of the kidneys is not increased and they continue to function normally, the acute rise in arterial pressure is not maintained. The reason for this is that increasing resistance everywhere else in the body does not change the equilibrium point for blood pressure as dictated by the renal output curve. With increased peripheral resistance and arterial pressure, the kidneys undergo pressure diuresis and pressure natriuresis, causing loss of salt and water from the body. This loss continues until arterial pressure returns to the normal equilibrium point (see Fig. 19–1*A*).

In many cases, when total peripheral resistance increases, renal vascular resistance also increases; this causes hypertension by shifting the renal function curve to higher blood pressures. When this shift occurs, it is the increase in renal resistance, not the increase in total peripheral resistance, that causes the long-term increase in arterial pressure.

Increased fluid volume elevates arterial pressure. The sequential events that link increased extracellular fluid volume and increased arterial pressure are (in order of occurrence)

1. Increased extracellular fluid volume and increased blood volume
2. Increased mean circulatory filling pressure
3. Increased venous return of blood to the heart
4. Increased cardiac output
5. Increased arterial pressure.

The increased cardiac output, by itself, tends to elevate the arterial pressure; however, the increased cardiac output also causes excess blood flow in many of the tissues of the body that respond by vasoconstriction, which tends to return the blood flow back toward normal. This phenomenon is called *autoregulation* and tends to raise the total peripheral vascular resistance. With greater extracellular fluid volume, there is an initial rise in cardiac output and rise in tissue blood flow, but after several days, the total peripheral resistance begins to increase because of autoregulation, so that cardiac output usually returns toward normal.

HYPERTENSION (HIGH BLOOD PRESSURE) (p. 225)

The normal systolic/diastolic arterial pressures are about 120/80 mm Hg, with a mean arterial pressure of 93 mm Hg under resting conditions. Hypertension is said to occur when the diastolic pressure is greater than 90 mm Hg and the systolic pressure is greater than 135 or 140 mm Hg.

Even moderate elevation of the arterial pressure leads to shortened life expectancy through at least three ways.

1. Excessive work load on the heart and high arterial pressure lead to coronary artery disease, congestive heart disease, or both, often causing death as a result of a *heart attack*.
2. High blood pressure frequently leads to rupture of a major blood vessel in the brain or hypertrophy and eventual obstruction of a cerebral blood vessel. In either case, this leads to cerebral ischemia and death of a major portion of the brain, a condition called *stroke*.
3. High blood pressure almost always causes damage to the kidneys, eventually leading to *kidney failure*.

There are multiple ways by which hypertension can occur. In all types of hypertension that have been studied so far, however, there has been a shift of the renal output curve toward higher blood pressures. Lessons learned from one type of hypertension called *volume loading hypertension* have been crucial in understanding the role of the renal–body fluid feedback mechanism for arterial pressure regulation.

Sequential changes that occur in circulatory function during the development of volume-loading hypertension. In experimental animals in which the kidney mass has been surgically reduced to about 30 per cent of normal, an increase in the salt and water intake causes marked hypertension. Although reduction of the functional kidney mass, by itself, does not cause significant hypertension, it reduces the ability of the kidney to effectively excrete a large load of salt and water. When salt and water intake are increased, the following sequence of events occurs:

- Extracellular fluid volume and blood volume are expanded.
- Increased blood volume raises mean circulatory filling pressure, venous return, and cardiac output.
- Increased cardiac output raises arterial pressure.
- During the first day after increased salt and water intake, there is a *decrease* in total peripheral resistance, caused mainly by the baroreceptor reflex

mechanism, which attempts to prevent the rise in pressure.

- After several days, there is a gradual return of cardiac output toward normal due to long-term blood flow autoregulation, which simultaneously causes a secondary increase in total peripheral resistance.
- As arterial pressure increases, the kidneys excrete the excess volume of fluid through pressure diuresis and pressure natriuresis, and a balance between intake and renal output of salt and water is re-established.

This sequence illustrates how an initial abnormality of kidney function and excess salt and water intake can cause hypertension and how the volume-loading aspects of hypertension may not be apparent after the kidneys have had sufficient time to reestablish sodium and water balance and after the autoregulatory mechanisms have caused an increase in total peripheral resistance. The following are two clinical examples of volume-loading hypertension:

- *Volume-loading hypertension can occur in patients who have no kidneys and are being maintained on an artificial kidney.* If the body fluid volume of a patient maintained on an artificial kidney is not regulated at the normal level and is allowed to increase, hypertension will develop in almost exactly the same way as previously discussed.
- *Excessive secretion of aldosterone causes volume-loading hypertension.* Occasionally, a tumor of the adrenal glands causes excessive secretion of aldosterone, which increases reabsorption of salt and water by the tubules of the kidneys (see Chapter 29). This reduces urine output, causing an increase in extracellular fluid volume and initiating the same sequence described previously for volume-loading hypertension.

RENIN-ANGIOTENSIN SYSTEM: ITS ROLE IN PRESSURE CONTROL AND HYPERTENSION (p. 227)

In addition to its capability of controlling arterial pressure through changes in extracellular fluid volume, the kidneys control pressure through the *renin-angiotensin system*. When the arterial pressure falls too low, the kidneys release a small protein enzyme, *renin*, that activates the renin-angiotensin system and helps to raise arterial pressure in several ways, thus helping to correct for the initial fall of pressure.

The components of the renin-angiotensin system play an important role in the regulation of arterial pressure. The renin-angiotensin system acts in the following manner for acute blood pressure control:

- A decrease in arterial pressure stimulates the secretion of *renin* from the *juxtaglomerular cells* of the kidney into the blood.
- Renin catalyzes the conversion of *renin substrate (angiotensinogen)* to release a 10–amino acid peptide, *angiotensin I.*
- Angiotensin I is converted into *angiotensin II* by the action of *converting enzyme* present in the endothelium of vessels throughout the body, especially in the lung.
- Angiotensin II, the primary active component of this system, is a potent vasoconstrictor and helps to raise arterial pressure.
- Angiotensin II persists in the blood until it is rapidly inactivated by multiple blood and tissue enzymes collectively called *angiotensinases.*

Angiotensin II has two principal effects that can elevate the arterial pressure:

1. *Angiotensin II constricts arterioles and veins throughout the body,* thereby raising total peripheral resistance and decreasing vascular capacity, which promotes increased venous return to the heart. These effects are important in preventing excessive reductions in blood pressure during acute circumstances such as hemorrhage.
2. *Angiotensin II acts on the kidneys to decrease excretion of salt and water.* This action slowly increases extracellular fluid volume, which increases arterial pressure over a period of hours and days.

The effect of angiotensin II that causes renal retention of salt and water is especially important for the long-term control of arterial pressure. Angiotensin II causes salt and water retention by the kidneys in two ways:

- *Angiotensin acts directly on the kidneys to cause salt and water retention.* Angiotensin II constricts the efferent arterioles, which diminishes blood flow through the peritubular capillaries, allowing rapid osmotic reabsorption from the tubules. In addition, angiotensin II directly stimulates the epithelial cells of the renal tubules to increase reabsorption of sodium and water.

- *Angiotensin II stimulates the adrenal glands to secrete aldosterone, and aldosterone increases salt and water reabsorption by the epithelial cells of the renal tubule.*

The renin-angiotensin system helps maintain normal arterial pressure despite wide variations in salt intake. One of the most important functions of the renin-angiotensin system is to allow a person to ingest either a very small or very large amount of salt without causing great changes in either extracellular fluid volume or arterial pressure. For example, when salt intake is increased, there is a tendency for extracellular fluid volume and arterial pressure to increase. This greater arterial pressure also decreases renin secretion and angiotensin II formation, which in turn decreases renal tubular salt and water reabsorption. The reduced tubular reabsorption allows the person to excrete the extra amounts of salt and water with minimal increases in extracellular fluid volume and arterial pressure.

When salt intake is decreased below normal levels, the opposite effects take place. As long as the renin-angiotensin is fully operative, salt intake can be as low as $\frac{1}{10}$ or as high as 10 times normal with only a few millimeters of mercury change in arterial pressure. On the other hand, when the renin-angiotensin system is blocked, the same changes in salt intake will cause large variations in blood pressure, often as much as 50 mm Hg.

Excessive angiotensin II formation causes hypertension. Occasionally, a tumor of the juxtaglomerular cells occurs that secretes tremendous quantities of renin, which in turn cause excessive formation of angiotensin II. This almost invariably leads to severe hypertension through the potent actions of angiotensin II to elevate arterial pressure.

The effect of angiotensin II to increase total peripheral resistance is the primary cause of the rapid rise in blood pressure that occurs when angiotensin II levels are suddenly elevated. The long-term increase in blood pressure associated with excessive angiotensin II formation is due mainly to the various actions of angiotensin II that cause renal salt and water retention.

Impairment of the Renal Circulation Causes Hypertension
(p. 231)

Any condition that seriously reduces the ability of the kidneys to excrete salt and water will cause hypertension. One type of renal dysfunction that can cause severe hypertension is renal vascular damage, such as occurs with stenosis of the renal arteries, constriction of the afferent arterioles, or increased resistance to fluid filtration through the glomerular membrane. Each of these factors reduces the ability of the kidney to form glomerular filtrate, which in turn causes salt and water retention as well as increased blood volume and increased arterial pressure. The rise in arterial pressure then helps to return glomerular filtration rate toward normal and reduces tubular reabsorption, permitting the kidneys to excrete normal amounts of salt and water despite the vascular disorders.

Constriction of the renal arteries causes hypertension. When one kidney is removed and a constrictor is placed on the renal artery of the remaining kidney, the immediate effect is greatly reduced pressure in the renal artery beyond the constriction. Within a few minutes, systemic arterial pressure begins to rise, and it continues to rise for several days, until the renal arterial pressure beyond the constriction has returned almost to normal levels. The hypertension produced in this way is called *one-kidney Goldblatt hypertension*, in honor of Harry Goldblatt, who first described the features of hypertension caused by this method in experimental animals.

The rapid rise in arterial pressure in Goldblatt hypertension is caused by activation of the renin-angiotensin vasoconstrictor mechanism. Because of poor blood flow through the kidney after reduction of renal artery pressure, large quantities of renin are secreted, causing increased angiotensin II formation and a rapid rise in blood pressure. The more delayed rise in blood pressure, occurring over a period of several days, is caused by fluid retention. The fluid retention and expansion of extracellular fluid volume continue until the arterial pressure has risen sufficiently to return the renal perfusion pressure to almost normal levels.

Hypertension also occurs when the artery of one kidney is constricted and the artery of the other kidney is normal; this is often called *two-kidney Goldblatt hypertension*. The constricted kidney retains salt and water because of decreased arterial pressure in this kidney. The "normal" kidney retains salt and water because of the renin produced in the ischemic kidney and the increase in circulating angiotensin II, which causes the opposite kidney to retain salt and water. Both kidneys become salt and water retainers, and hypertension develops.

Coarctation of the aorta above the renal arteries also causes hypertension, with characteristics very similar to those described for one-kidney Goldblatt hypertension. Aortic coarctation results in decreased perfusion pressure to both kidneys, stimulating the release of renin and angiotensin II formation as well as salt and water retention by the kidneys. These changes increase the arterial pressure in the upper part of the body above the coarctation, thereby helping to return the perfusion pressure of the kidneys toward normal.

Patchy ischemia of one or both kidneys can also cause hypertension. When this occurs, the characteristics of the hypertension are almost identical to those of two-kidney Goldblatt hypertension; the patchy ischemic kidney tissue secretes renin, and this in turn stimulates formation of angiotensin II, causing the remaining kidney mass to retain salt and water. This type of hypertension is much more common than hypertension caused by constriction of the main renal arteries or aortic coarctation, especially in older patients with atherosclerosis.

Toxemia of pregnancy is also associated with hypertension. Although the precise cause of hypertension of this condition is not completely understood, many physiologists believe that this is due to *thickening of the glomerular membranes* (perhaps caused by an autoimmune process), which reduces the glomerular capillary filtration coefficient and rate of fluid filtration from the glomeruli into the renal tubules.

The Causes of Human Essential Hypertension Are Unknown

Approximately 25 per cent to 30 per cent of adults in industrialized societies have high blood pressure, although the incidence of hypertension is higher in elderly people. The precise cause of hypertension in about 90 per cent of these people is unknown; this type of hypertension is called *essential hypertension.*

Although the exact causes of essential hypertension are not fully understood, most patients who develop essential hypertension slowly over many years have significant changes in kidney function. Most important, the kidneys cannot excrete adequate quantities of salt and water at normal arterial pressures; instead, they require a high arterial pressure to maintain a normal balance between the intake and output of salt and water unless they are treated with drugs that enhance their ability to excrete salt and water at lower blood pressures.

Abnormal renal excretory capability could be caused by renal vascular disorders that reduce glomerular filtration or tubular disorders that increase reabsorption of salt and water. Because patients with essential hypertension are very heterogeneous with respect to the characteristics of the hypertension, it seems likely that both types of disorders contribute to increased blood pressure.

SUMMARY OF THE INTEGRATED, MULTIFACETED SYSTEM FOR ARTERIAL PRESSURE REGULATION (p. 234)

By now, it is clear that arterial pressure is regulated by several systems, each of which performs a specific function. Some systems are most important for acute regulation of blood pressure and react rapidly, within seconds or minutes. Others respond over a period of minutes or hours. Some provide long-term arterial pressure regulation over days, months, and years.

Rapidly acting pressure control mechanisms are the nervous system reflexes. The three nervous reflexes that act very rapidly (within seconds) are (1) the *baroreceptor feedback mechanism*, (2) the *central nervous ischemic mechanism*, and (3) the *chemoreceptor mechanism*. These mechanisms not only act within seconds but also are powerful in preventing acute decreases in blood pressure (e.g., during severe hemorrhage). They also operate to prevent excessive increases in blood pressure, as might occur in response to excessive blood transfusion.

There are intermediate time period pressure control mechanisms. Three mechanisms that are important in blood pressure control after several minutes of acute pressure change are the (1) *renin-angiotensin vasoconstrictor mechanism*, (2) *stress relaxation of the vasculature*, and (3) *shift of fluid through the capillary walls* in and out of the circulation to readjust the blood volume as needed.

The role of the renin-angiotensin vasoconstrictor mechanism has been described. The *stress relaxation mechanism* is demonstrated by the following example: When pressure in the blood vessels becomes too high, the vessels become stretched and continue to stretch for minutes or hours. As a result, the pressure in the vessels tends to fall back toward normal.

The *capillary fluid shift mechanism* means simply that any time the capillary pressure falls too low, fluid is absorbed via osmosis from the tissue into the circulation, thus increasing the blood volume and helping to return the blood pressure toward normal. Conversely, when capillary pressure rises too high, fluid is lost out of the circulation, thus reducing blood volume and arterial pressure.

The long-term mechanism for arterial pressure regulation involves the renal–body fluid feedback. The main goal of this chapter is to explain the role of the kidneys in long-term control of arterial pressure. The renal–body fluid feedback control mechanism takes several hours to show any significant response, but it operates very powerfully to control arterial pressure over days, weeks, and months. As long as kidney function is unaltered, disturbances that tend to alter arterial pressure, such as increased total peripheral resistance, have a minimal effect on blood pressure over long periods of time. Factors that alter the ability of the kidneys to excrete salt and water can cause major long-term changes in arterial pressure. This mechanism, if given sufficient time, controls arterial pressure at the level that provides normal output of salt and water by the kidneys.

Many different factors can affect the renal–body fluid feedback mechanism and, therefore, long-term blood pressure control. One of the most important of these is the renin-angiotensin system, which allows a person to have very low or very high salt intake with minimal changes in arterial pressure. Thus, arterial pressure control begins with lifesaving measures of the nervous reflexes, continues with the sustaining characteristics of the intermediate pressure controls, and finally is stabilized at the long-term pressure level by the renal–body fluid feedback mechanism.

20

Cardiac Output, Venous Return, and Their Regulation

Cardiac output is the amount of blood that is pumped into the aorta each minute by the heart. It also represents the quantity of blood that flows to the peripheral circulation; the cardiac output transports substances to and from the tissues. The cardiac output of an average adult is approximately 5 liters/min or 3 liters/min/m² of body surface area.

Venous return is the amount of blood that flows from the veins back to the right atrium each minute.

CONTROL OF CARDIAC OUTPUT BY VENOUS RETURN— ROLE OF THE FRANK-STARLING MECHANISM OF THE HEART (p. 239)

In the absence of changes in cardiac strength, cardiac output is controlled by factors that affect venous return. One of the most important regulators of venous return is metabolism of the tissues. An increase in the tissue metabolic rate will result in local vasodilation, which will cause a decrease in total peripheral resistance and thus an increase in venous return. This greater venous return causes an increase in diastolic filling pressure in the ventricles, which in turn results in a greater force of contraction by the ventricles. This mechanism for increasing cardiac pumping ability is called the *Frank-Starling law of the heart*. This law states that within limits, an increase in the volume of blood returning to the heart will stretch the cardiac muscle a greater amount, and the heart will contract with a greater force and pump out all the excess venous return.

An important concept that can be learned from the Frank-Starling law is that except for very short-term changes, cardiac output equals venous return. Therefore, factors that control venous return also control cardiac output. If this were not so—for example, if cardiac output were greater than venous return—the lungs would quickly be emptied of blood. In contrast, if cardiac output were less than venous return, the lungs would rapidly fill with blood.

During increases in venous return, right atrial stretch elicits two reflexes that help to raise cardiac output. First, the stretch of the sinus node causes a direct effect on the rhythmicity of the node, which causes a 10 to 15 per cent increase in heart rate. This increase in heart rate helps to pump the extra blood that is returning to the heart. Second, the extra stretch in the right atrium elicits a *Bainbridge reflex*, with impulses going first to the vasomotor center and then back to the heart by way of sympathetic nerves and the vagi. This reflex causes an increase in heart rate, which also helps to pump out the excess venous return.

Cardiac output regulation is the sum of all tissue blood flow regulation. Because venous return is the sum of all local blood flows, anything that affects local blood flow also affects the venous return and cardiac output.

One of the main ways by which local blood flow can be changed is through local metabolism. For example, if the biceps muscle of the right arm is used repetitively to lift a weight, the metabolic rate of that muscle will increase. This causes a local vasodilation. Blood flow to the biceps muscle will increase because of the extra oxygen use. The extra blood flow to the biceps will also cause an increase in venous return and an increase in cardiac output. Remarkably, the increased cardiac output goes primarily to the area of increased metabolism, the biceps, because of its vasodilation.

Changes in cardiac output can be predicted with the use of Ohm's law. Ohm's law, as applied to the circulation, can be stated as the following relationship:

$$\text{Cardiac Output} = \frac{(\text{Arterial Pressure} - \text{Right Atrial Pressure})}{\text{Total Peripheral Resistance}}$$

If right atrial pressure is equal to its normal value of 0 mm Hg, the relationship can be simplified to

$$\text{Cardiac Output} = \frac{\text{Arterial Pressure}}{\text{Total Peripheral Resistance}}$$

If arterial pressure is constant, this formula can be accurately used to predict changes in flow that occur due to changes in total peripheral resistance. If we return to the example of an increase in metabolic rate in a peripheral tissue, the increase in oxygen use that also occurs elicits local vasodilation and decreases total peripheral resistance, which causes an increase in oxygen delivery to the local tissue, an increase in venous return, and an increase in cardiac output. Thus, if arterial pressure is constant, the long-term cardiac output will vary in a reciprocal manner with total peripheral resistance. Therefore, a decrease in total peripheral resistance will increase cardiac output, and an increase in total peripheral resistance will decrease cardiac output.

Maximum Cardiac Output Achieved by the Heart Is Limited by the Plateau of the Cardiac Output Curve
(p. 241)

The *cardiac output curve*, in which cardiac output is plotted as a function of right atrial pressure, can be affected by several factors, and their net effect is a change in the plateau level of this curve. Some of these factors are

- Increased sympathetic stimulation, which increases the plateau
- Decreased parasympathetic stimulation, which increases the plateau
- Cardiac hypertrophy, which increases the plateau
- Myocardial infarction, which decreases the plateau
- Cardiac valvular disease, such as stenotic or insufficient valves, which decreases the plateau
- Abnormal cardiac rhythm, which can cause a decrease in the plateau.

PATHOLOGICALLY HIGH AND PATHOLOGICALLY LOW CARDIAC OUTPUTS (p. 242)

High cardiac output is almost always caused by reduced total peripheral resistance. A distinguishing feature of many conditions with high cardiac output is that they result from a chronic decrease in total peripheral resistance. Among these conditions are

- *Beriberi*—This disease is caused by a lack of thiamine, and the associated diminished ability to use cellular nutrients results in marked vasodilation, decreased total peripheral resistance, and increased cardiac output.
- *Arteriovenous fistula (shunt)*—This condition is caused by a direct opening between an artery and a vein, which causes a decrease in total peripheral resistance and thus an increase in cardiac output.
- *Hyperthyroidism*—This condition causes an increase in oxygen use, which causes a release of vasodilatory products, total peripheral resistance decreases, and cardiac output increases.
- *Anemia*—The decrease in total peripheral resistance in this condition is caused by (1) a lack of oxygen delivery to the tissue, which causes vasodilation, and (2) a decrease in viscosity of the blood due to the lack of red blood cells. Cardiac output rises.

Low cardiac output can be caused by cardiac or peripheral factors. Severe myocardial infarction, severe valvular disease, myocarditis, cardiac tamponade, and certain metabolic derangements can decrease cardiac output by lowering the plateau of the cardiac output curve (see Chapter 22).

The peripheral factors that acutely reduce cardiac output also reduce venous return; they include

- Decreased blood volume
- Acute venous dilation
- Obstruction of the large veins.

A MORE QUANTITATIVE ANALYSIS OF CARDIAC OUTPUT REGULATION (p. 244)

The cardiac output curve is used to describe the ability of the heart to increase its output when right atrial pressure rises. Figure 20–1 shows the intersection of the cardiac output curve with two separate venous return curves; the cardiac output curve plateaus at 13 liters/min of cardiac output. This is a normal cardiac output curve, and sympathetic stimulation elevates the plateau of this curve, whereas sympathetic inhibition or depressed cardiac function lowers the plateau of the curve.

The normal cardiac output curve (see Fig. 20–1) is plotted for an intrapleural pressure of −4 mm Hg (the normal external pressure on the outside of the heart). As the intrapleural pressure increases, the

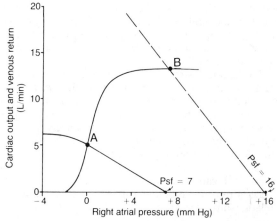

Figure 20–1 Two solid curves demonstrate an analysis of cardiac output and right atrial pressure when the cardiac output and venous return curves are normal. Transfusion of blood equal to 20 per cent of the blood volume causes the venous return curve to become the *dashed curve;* as a result, the cardiac output and right atrial pressure shift from point A to point B. (Psf = mean systemic filling pressure.)

heart tends to collapse, particularly the atria. For example, if intrapleural pressure increases from −4 to −1 mm Hg, the volume of the right atrium will decrease. To return the right atrial size to normal, an additional 3 mm Hg of right atrial pressure is required to overcome the extra 3 mm Hg of intrapleural pressure. Therefore, the cardiac output curve will shift to the right by exactly 3 mm Hg. The cardiac output curve can be shifted to the right or left by several factors, including

- *Normal inspiration,* which shifts the curve leftward
- *Normal expiration,* which shifts the curve rightward
- *Negative pressure breathing,* which shifts the curve to the left
- *Positive pressure breathing,* which shifts the curve to the right
- *Surgical opening of the thoracic cage,* which shifts the curve to the right
- *Cardiac tamponade,* which shifts the curve to the right and rotates it downward.

The venous return curve describes the relationship between venous return and right atrial pressure. The normal *venous return curve* (see Fig. 20–1, solid line) intersects the normal cardiac output curve at point A, a right atrial pressure of 0 mm Hg; this is the normal right atrial pressure. The mean systemic filling pressure (P_{sf}) is located where the venous return curve intersects the abscissa; this pressure has a value of 7 mm Hg.

The mean systemic filling pressure is a measure of the *tightness* with which the circulatory system is filled with blood. This pressure is proportional to the amount of blood volume that exceeds the unstressed vascular volume and is inversely proportional to the total vascular compliance. The slope of the linear portion of the venous return curve is equal to 1 divided by the resistance to venous return. If mean systemic filling pressure is known, venous return can be determined with the following relationship:

$$\text{Venous Return} = \frac{\left(\begin{array}{cc}\text{Mean Systemic} & \text{Right Atrial} \\ \text{Filling Pressure} & \text{Pressure}\end{array}\right)}{\text{Resistance to Venous Return}}$$

The numerator of this formula equals the pressure gradient for venous return, which is the average pressure from the peripheral vessels to the heart. Therefore, if the pressure gradient for venous return increases, venous return increases.

In Figure 20–1, the dashed venous return curve represents a condition of excess blood volume. Therefore, this hypervolemia increased the mean systemic filling pressure to 16 mm Hg and decreased the resistance to venous return, because the excess blood volume distended the blood vessels and decreased their resistance.

The resistance to venous return is the average resistance between the peripheral vessels and the heart. Most of the resistance to venous return occurs in the veins, although some occurs in the arterioles and arteries. The venous resistance is an important determinant of the resistance to venous return because if venous resistance increases, blood dams up in the highly distensible veins, and venous pressure increases by a little. The venous return would, therefore, decrease dramatically.

The venous return curve is shifted upward and to the right during sympathetic stimulation and is shifted downward and to the left during sympathetic inhibition or decreased blood volume. The cardiac output curve is elevated dramatically during sympathetic stimulation; when combined with this upward and rightward shifted venous return curve, the cardiac output increases markedly. Sympathetic stimulation also increases the venous resistance, which by itself increases the resistance to venous return; however, the mean systemic filling pressure increases even more, and therefore, venous return increases.

METHODS FOR MEASURING CARDIAC OUTPUT (p. 250)

Cardiac output can be measured utilizing several methods, including the following:

- An electromagnetic flowmeter
- An ultrasonic flowmeter
- The indicator dilution method
- The oxygen Fick method.

The Fick procedure can be used to calculate cardiac output with the following relationship:

$$\frac{\text{Cardiac Output}}{\text{(liters/min)}} = \frac{\text{(Oxygen Absorbed in the Lungs [ml/min])}}{\text{(Arteriovenous Oxygen Difference [ml/liter of blood])}}$$

With this technique, the venous blood sample is removed from the pulmonary artery, and the arterial blood sample is taken from any artery in the body.

21

Muscle Blood Flow and Cardiac Output During Exercise; The Coronary Circulation and Ischemic Heart Disease

BLOOD FLOW IN SKELETAL MUSCLE INCREASES MARKEDLY DURING EXERCISE (p. 253)

Resting blood flows through skeletal muscle at an average rate of 3 to 4 ml/min/100 gm of muscle. During exercise, this rate can increase by 15- to 25-fold, and cardiac output may increase up to 6 or 7 times normal. This rise in blood flow is necessary to deliver extra nutrients to the exercising muscle and to carry away the by-products of muscular contraction. During skeletal muscle contraction, the muscle blood flow drops markedly (because of mechanical compression of the vessels), but it rises rapidly between contractions.

Vasodilator factors increase skeletal muscle blood flow during exercise. Muscle contraction increases the metabolic rate of the tissue, which in turn reduces *oxygen* concentration in the tissues; the decreased oxygen concentration vasodilates the blood vessels. In addition, the exercising skeletal muscle releases vasodilator factors, including the following:

- Adenosine
- Potassium ions
- Hydrogen ions
- Lactic acid
- Carbon dioxide.

Sympathetic activation reduces skeletal muscle blood flow. During massive sympathetic stimulation, such as occurs in circulatory shock, blood flow to skeletal muscle can decrease to as little as one-fourth normal. This effect is due to the direct effects of sympathetic nerve stimulation and the adrenal release of norepinephrine and epinephrine. The sympathetic nerve stimulation and the norepinephrine release from the adrenals predominantly stimulate alpha adrenergic receptors, and the epinephrine release from the adrenal predominantly stimulates beta adrenergic receptors. Stimulation of alpha receptors causes vasoconstriction, whereas stimulation of peripheral beta receptors causes vasodilation.

Cardiovascular changes during exercise deliver more nutrients and remove greater amounts of metabolic by-products from exercising muscle. The cardiovascular changes that occur during exercise include the following:

- *Massive sympathetic discharge,* which increases heart rate and heart strength and causes arteriolar constriction and venoconstriction in all vasculature except exercising muscle, brain, and coronary bed
- *Decreased parasympathetic impulses,* which also increases heart rate
- *Local vasodilation in exercising muscle,* which decreases resistance to venous return
- *Increased mean systemic filling pressure,* which is due mainly to venoconstriction but also to arteriolar constriction
- *Increased venous return and cardiac output,* which is due to increased mean systemic filling pressure and decreased resistance to venous return
- *Increased mean arterial pressure,* which is an important result of the increased sympathetic activity in exercise. The cause of this elevated pressure is (1) arteriolar and small artery constriction, (2) increased cardiac contractility, and (3) increased mean systemic filling pressure.

The increase in arterial pressure can range from 20 to 80 mm Hg depending on the type of exercise being performed. When exercise is performed under tense conditions, such as isometrics, in which many of the muscles are contracted for significant periods of time, a large increase in arterial pressure will occur. When a more isotonic exercise is performed, such as swimming or running, the arterial pressure increase is much less.

If arterial pressure is prevented from increasing during exercise, such as in a patient with a congenitally impaired sympathetic nervous system, cardiac output can rise to only about one third of what it does normally. When arterial pressure is allowed to increase normally, blood flow through skeletal muscle increases normally from about 1 liter/min during rest to 20 liters/min during exercise. If arterial pressure is prevented from increasing during exercise, skeletal muscle blood flow seldom increases more than about eight-fold.

The rise in arterial pressure helps to increase blood flow by (1) pushing the blood through the arterial system and back toward the heart and (2) dilating the arterioles, which reduces total peripheral resistance and allows more blood to flow through the skeletal muscle and back to the heart.

CORONARY CIRCULATION (p. 256)

The resting coronary blood flow is about 225 ml/min and can increase by three- to four-fold during exercise. The coronary flow is delivered to the cardiac muscle primarily through the *left coronary artery*, which supplies most of the left ventricle, and the *right coronary artery*, which supplies the right ventricle and part of the posterior part of the left ventricle. Like skeletal muscle, the flow into the cardiac muscle decreases during muscle contraction, which in the heart coincides with systole. Flow particularly decreases a large amount in the subendocardial vessels because they lie in the midportion of the heart muscle. The surface vessels, the *epicardial vessels*, experience a much smaller decrease in flow during systole.

Local Metabolism Is a More Important Controller of Coronary Flow than Is Nervous Control

Several vasodilator factors are released during decreases in the cardiac muscle oxygen concentration, including the following:

- Adenosine
- Adenosine phosphate compounds
- Potassium ions
- Hydrogen ions
- Carbon dioxide
- Bradykinin
- Prostaglandins.

The release of these vasodilator factors occurs in response to changes in local metabolism and is an important regulator of coronary flow; most of these factors contribute to vasodilation in exercising skeletal muscle. One of the most important regulators of coronary flow is adenosine. There also are some sympathetic effects on coronary flow. Compared with the vasodilator factors, the sympathetic effects on coronary flow are usually modest. The epicardial vessels have a preponderance of alpha receptors and therefore are constricted during sympathetic stimulation. In contrast, the subendocardial arteries have more beta receptors and are vasodilated during sympathetic stimulation. The overall effect of sympathetic stimulation is usually a small decrease in coronary flow.

The control of coronary flow is very important because a constant of delivery of oxygen is necessary for normal cardiac metabolism. Fat metabolism, which requires oxygen, normally supplies 70 per cent of the energy for the heart. Under moderate ischemic conditions, anaerobic glycolysis can supply energy for cardiac metabolism.

Ischemic Heart Disease Is Responsible for About 35 Per Cent of Deaths in the United States Each Year

Atherosclerosis is the primary cause of ischemic heart disease. People who eat excessive quantities of fat or cholesterol and are overweight have a high risk of developing atherosclerosis. The stages of development of atherosclerosis and its effects on the heart are as follows:

1. First, large quantities of cholesterol are deposited beneath the endothelium in arteries throughout the body, including the coronary arteries.
2. Later, these areas are invaded by fibrous tissue.
3. This is followed by a necrotic stage.
4. Finally, a stage of calcification occurs.
5. The final result is the development of atherosclerotic plaques, which can protrude into the lumen of the vessel and because of their rough surface initiate blood clots.
6. The blood clot is called a *thrombus* and can either partially or fully occlude the coronary vessels.
7. Sometimes, the clot breaks away and flows downstream; this is an *embolism*.
8. A thrombus or an embolism can totally block blood flow to an area of the heart, which causes death (infarction) of myocardial tissue.
9. The final result is a myocardial infarction.

When atherosclerosis slowly constricts coronary vessels over many years, collateral vessels can develop and maintain coronary flow at nearly a normal level. This development of vessels can prevent or even postpone a myocardial infarction for many years.

Coronary spasm can also cause a myocardial infarction. Coronary spasm can cause a temporary occlusion in the coronary vessels and thus cause a myocardial infarction. The etiology of the spasm can be irritation of a vessel by roughened atherosclerotic plaque or the result of nervous reflexes or circulating factors. Coronary spasm can also occur in vessels that have no atherosclerotic damage.

Death may occur after a myocardial infarction. The causes of death after myocardial infarction include the following:

- Decreased cardiac output
- Pulmonary edema
- Ventricular fibrillation
- Rupture of the heart.

Decreased cardiac output occurs after myocardial infarction because the mass of cardiac tissue that contracts normally is decreased. Further weakening of the heart may occur as some of the ischemic muscle actually bulges outward during the high intraventricular pressures of systole; this is called *systolic stretch*. If a large mass of the heart is damaged, cardiac output may decline to very low levels, which can cause a reduction in arterial pressure. The decreased pressure will, in turn, reduce coronary flow and further weaken the heart. This vicious cycle of events is called *cardiogenic shock*.

If the left side of the heart is damaged severely, blood will back up into the pulmonary system and cause *pulmonary edema*. Pulmonary capillary pressure increases in this condition, which will cause leakage of fluid into the pulmonary interstitium. This edema prevents proper oxygenation of blood and can also lead to death.

Ventricular fibrillation, the uncoordinated contraction of the ventricle, usually occurs within 10 minutes of a myocardial infarction. The factors that increase the tendency of the heart to fibrillate are

- *Increased extracellular potassium concentration* due to loss of potassium from ischemic muscles

- *Current of injury* from the infarct area
- *Increased irritability of cardiac muscle* due to sympathetic reflexes after a myocardial infarction
- *Circus movements*, which occur because the dilation of the heart after a myocardial infarction causes an increased pathway length for impulse conduction in the heart.

Cardiac rupture is another cause of death after a myocardial infarction. If systolic stretch is very severe after an infarction, rupture of the area sometimes occurs. Rapid blood loss occurs into the pericardial area, which causes cardiac tamponade and marked decreases in cardiac output because of the inability of the heart to fill properly during diastole.

Proper treatment of a patient with myocardial infarction often leads to a recovery of much of the myocardial function. If a patient lives past the critical early periods after a myocardial infarction, proper medical treatment can enhance the probability of recovery. After an infarct occurs, the necrotic tissue in the center of the damaged area of the myocardium is gradually replaced by fibrous tissue. During the early phases of recovery from a myocardial infarction, tissues on the margin of the infarct usually have just the minimal amount of blood flow necessary to prevent tissue death. Any increase in activity of the heart may cause the normal cardiac tissue to rob the marginal tissue of its blood flow and cause the *coronary steal syndrome*. This condition can cause ischemia of the tissue on the margins of the infarct and may actually cause death of these marginal tissues. Therefore, it is critical that patients maintain complete bed rest after experiencing a myocardial infarction. In addition, patients are usually administered oxygen to breathe during recovery, which may help deliver a little more oxygen to the heart and can help to improve cardiac function. Over weeks and months, some of the normal cardiac tissue hypertrophies and thereby helps return cardiac function to normal.

Occasionally, after recovery from an extensive myocardial infarction, cardiac function returns nearly to normal. In most cases, however, cardiac function remains below that of a normal heart. Cardiac reserve is significantly decreased below the normal level of 300 per cent in these patients, which means that the heart can normally pump 300 per cent more blood per minute than is needed during rest. Although the resting cardiac output may be

normal after partial recovery from a myocardial infarction, the amount of strenuous activity that can
be performed is limited.

**Angina pectoris is pain that originates in the
heart.** In many cases, patients with partially recovered hearts and patients without myocardial infarction but ischemic heart disease will experience heart
pain, called *angina pectoris.* This occurs when the
heart is overloaded in relation to the amount of coronary blood flow supplied. Cardiac ischemia occurs.
The pain associated with this ischemia is felt beneath the sternum but can be referred to the surface
areas of the body, such as the left arm, left shoulder,
neck, face, and, sometimes, right arm and shoulder.

This angina pain is caused by a lack of oxygen
supply to the heart. Anaerobic glycolysis occurs,
which produces lactic acid or other pain-producing
compounds. Several treatments for angina pain and
cardiac ischemia can be very helpful, including the
following:

- *Nitrovasodilators,* such as nitroglycerin
- *Beta blockers,* which decrease the need of the heart
  for oxygen during stressful conditions
- *Coronary angioplasty,* in which a balloon is inflated
  in a coronary artery that has atherosclerotic narrowing in an attempt to increase the lumen diameter
- *Coronary bypass surgery,* in which vascular grafts
  are attached from the aorta to a point on the
  coronary artery distal to the constricted area.

22

Cardiac Failure

The term cardiac failure means that the heart is unable to pump sufficient blood to satisfy the needs of the body. The cause usually is decreased myocardial contractility resulting from diminished coronary blood flow. However, failure can result from heart valve damage, external pressure around the heart, vitamin B deficiency, or primary cardiac muscle disease.

DYNAMICS OF THE CIRCULATION IN CARDIAC FAILURE
(p. 265)

Rapid compensations for heart failure occur primarily by the sympathetic nervous system. Immediately after the heart becomes damaged in patients with heart failure, myocardial contractility decreases dramatically. This results in a lower plateau of the cardiac output curve. Within a few seconds, the sympathetic reflexes are activated, and the parasympathetic reflexes are reciprocally inhibited at the same time. Sympathetic stimulation has two major effects on the circulation:

- The heart is strongly stimulated.
- The peripheral vasculature is constricted.

Under the influence of increased sympathetic impulses, the heart becomes a much stronger pump, increasing the plateau of the cardiac output curve. This increased contractility helps to restore the cardiac output.

Sympathetic stimulation during heart failure also increases the vascular tone of the peripheral blood vessels, especially the veins, which aids in restoration of cardiac output. Mean systemic filling pressure increases to a value of 12 to 14 mm Hg, and this increases the tendency of the blood to flow back to the heart.

Chronic responses to heart failure involve renal sodium and water retention and recovery of the damaged heart. The depressed cardiac output that occurs during heart failure reduces arterial pressure and urinary output. This results in sodium and water retention and an increase in blood volume. The resulting hypervolemia increases the mean systemic filling pressure and the pressure gradient for venous return, which in turn increases venous return. The hypervolemia distends the veins and thus decreases venous resistance, further adding to the increase in venous return.

Recovery of the heart also helps to restore the cardiac output during heart failure. The cardiac recovery process depends on the factors that initiated cardiac failure. If the initiating factor was, for example, a myocardial infarction, collateral blood supply rapidly begins to develop after the initial cardiac damage. The undamaged myocardium hypertrophies, which offsets much of the cardiac damage and helps to increase cardiac output. Recovery of cardiac output to normal levels for sustained periods of time is referred to as *compensated failure*. The features of this compensated failure are:

- Relatively normal cardiac output as long as the person remains at rest and places no additional demands on the heart
- Increased right atrial pressure, which causes engorgement of the jugular veins
- Decreased cardiac reserve
- Increased heart rate
- Pale or clammy skin (this normalizes after recovery)
- Sweating and nausea, which also normalize after recovery
- Air hunger (dyspnea)
- Weight gain as a result of fluid retention

One of the key diagnostic features of a patient in compensated heart failure is the increased right atrial pressure and the resultant distended neck veins. The increase in right atrial pressure in compensated failure occurs because (1) blood from the damaged heart backs up into the right atrium, (2) venous return increases because of sympathetic stimulation, and (3) the kidney retains sodium and water and thus increases blood and venous return.

Sodium and water retention occur in heart failure because of sympathetic reflexes, decreased arte-

rial pressure, and stimulation of the renin-angio-
tensin-aldosterone system. Retention of sodium and
water by the kidneys during heart failure is a criti-
cally important factor in the compensatory increases
in blood volume and mean systemic filling pressure.
The causes of sodium and water retention are as
follows:

- *Decreased arterial pressure,* which decreases the
  glomerular filtration rate
- *Sympathetic constriction of the afferent arterioles,*
  which also decreases glomerular filtration rate
- *Increased angiotensin II formation,* which occurs in
  the kidney because of an increase in renin release—
  Decreases in arterial pressure and renal blood
  flow, as well as an increase in sympathetic output,
  contribute to the increase in renin release. The
  increased angiotensin II blood concentration constricts the efferent arterioles in the kidney, which
  decreases peritubular capillary pressure and thus
  promotes sodium and water retention.
- *Increased aldosterone release,* which occurs because
  of stimulation by the increased angiotensin II
  blood and plasma potassium concentrations that
  occur in heart failure. This increased aldosterone
  concentration causes renal sodium retention in the
  distal parts of the nephron.
- *Increased antidiuretic hormone release,* which occurs
  because of renal sodium retention during heart
  failure; this hormone promotes water retention in
  the kidney.

In decompensated heart failure, compensatory
responses cannot maintain an adequate cardiac output. In some patients, the hearts are too weak to
restore cardiac output to a level adequate to maintain the nutritional needs of the body and to make
the kidneys excrete the necessary daily amounts of
fluid. Therefore, the kidneys continue to retain fluid
and the heart muscle continues to be stretched until
the interdigitation of the actin and myosin filaments
is past optimum levels. Thus, cardiac contractility
decreases further, and a vicious cycle ensues. The
causes of decompensated heart failure are believed
to be the following:

- Longitudinal tubules of the sarcoplasmic reticulum fail to accumulate sufficient calcium, which is
  one of the basic causes of myocardial weakness.

- Myocardial weakness causes excess fluid retention, which in turn causes overstretched sarcomeres and further decreases cardiac contractility.
- Excess fluid retention also causes edema of the heart muscle, which results in a stiffened ventricular wall of the heart and in turn inhibits diastolic filling.
- Norepinephrine content of the sympathetic nerve endings of the heart decreases to very low levels, which further decreases cardiac contractility.

There are several treatments for decompensated heart failure, including the following:

- Use of a *cardiotonic drug* such as digitalis—This drug is believed to block the calcium pump, which normally pumps calcium out of the myocardial cells. Thus, more calcium accumulates in the cell, which increases cardiac contractility.
- *Use of diuretics* such as furosemide
- *Decreased sodium and water intake*—When combined with the use of diuretics, decreased sodium and water intake reduces the excess fluid in the body, which improves cardiac function and allows a balance between fluid intake and output despite a low cardiac output.

UNILATERAL LEFT HEART FAILURE (p. 268)

In unilateral left heart failure, the blood backs up into the lungs, which increases pulmonary capillary pressure and the tendency for pulmonary edema. The features of left-sided heart failure are:

- Increased left atrial pressure
- Pulmonary congestion
- Pulmonary edema if pulmonary capillary pressure exceeds approximately 28 mm Hg
- Arterial pressure and cardiac output maintained near normal as long as the patient remains at rest
- Intolerance to exercise, which if attempted may worsen pulmonary edema

In contrast, unilateral right-sided heart failure is accompanied by increased right atrial pressure and peripheral edema, and elevated left atrial pressure and pulmonary edema are not present.

"HIGH-OUTPUT CARDIAC FAILURE"—THIS CAN OCCUR EVEN IN A NORMAL HEART THAT IS OVERLOADED
(p. 268)

In many types of high output failure, the pumping ability of the heart is not diminished but is overloaded by excess venous return. Most often, this is caused by a circulatory abnormality that drastically decreases total peripheral resistance, such as the following:

- *Arteriovenous fistulae*
- *Beriberi*—The lack of B vitamins in this condition, especially thiamine, markedly decreases peripheral resistance, which increases the venous return. Also, the cardiac output curve is depressed, reflecting a decrease in cardiac contractility. However, cardiac output remains elevated because of the increased venous return.
- *Thyrotoxicosis*—The increased metabolic rate due to the increase in thyroid hormone causes an autoregulatory decrease in total peripheral resistance and an increase in venous return. The cardiac output curve is often depressed, but the cardiac output still increases.

LOW-OUTPUT CARDIAC FAILURE—CARDIOGENIC SHOCK
(p. 269)

Cardiogenic shock can occur in a number of conditions associated with depressed myocardial function, but the most common occurrence is after a myocardial infarction, when cardiac output and arterial pressure often decrease rapidly. The decreased pressure results in a decrease in coronary flow that can weaken the heart and further decrease cardiac output and arterial pressure. To break this vicious cycle, the following treatments are used:

- *Digitalis* is used to increase cardiac strength
- *A vasopressor* drug is given to increase arterial pressure
- *Whole blood or plasma* is given to increase arterial pressure. This increase in pressure helps to increase coronary flow.
- *Tissue plasminogen activator* can be infused to dissolve the coronary thrombosis if treatment is started during or soon after the clot forms

Acute progressive pulmonary edema sometimes occurs in patients with long-standing heart failure. If a patient already has some degree of pulmonary edema and an acute event further depresses left ventricular function, more pulmonary edema fluid can quickly form. This increase in edema fluid reduces the oxygenation of blood, which in the peripheral tissues causes vasodilation. An increase in venous return thus results from the vasodilation, and the resulting increase in pulmonary capillary pressure can cause more pulmonary edema fluid to form and further reduce blood oxygenation. The treatment of this cycle of pulmonary edema in many ways requires heroic measures and in some ways is opposite of that for cardiogenic shock:

- Application of tourniquets to both arms and legs, which will sequester blood in these limbs and thus reduce the pulmonary blood volume, and thus the amount of pulmonary edema will decrease
- Bleeding the patient
- Administering a rapidly acting diuretic such as furosemide
- Administering oxygen for the patient to breathe
- Administering digitalis to increase heart strength

Although volume-expanding agents are sometimes given in cardiogenic shock to increase arterial pressure, volume-reducing measures are used to decrease the edema fluid in the lungs in acute progressive pulmonary edema.

Cardiac reserve decreases in all types of heart failure. Cardiac reserve is the percentage increase in cardiac output that can be achieved during maximum exertion; it can be calculated with the following relationship:

Cardiac Reserve

$$= \frac{\left[\left(\begin{array}{c}\text{Maximum} \\ \text{Cardiac Output}\end{array} - \begin{array}{c}\text{Normal} \\ \text{Cardiac Output}\end{array}\right) \times 100\right]}{\text{Normal Cardiac Output}}$$

If a patient with a decrease in cardiac reserve undergoes an exercise test, the following often occur:

- Dyspnea (shortness of breath and air hunger)
- Extreme muscle fatigue
- Excessive increase in heart rate

23

Heart Sounds; Dynamics of Valvular and Congenital Heart Defects

HEART SOUNDS (p. 275)

Listening to the sounds of the heart is one of the oldest methods of examining a patient. Heart sounds are associated with *closure of the heart valves*; no sounds occur when the valves are open.

When one listens to the heart with a stethoscope, the sounds are described as *lub, dub, lub, dub*. The lub is associated with closure of the atrioventricular (A-V) valves at the beginning of systole, and the dub occurs at the end of systole and is caused by closure of the aortic and pulmonary valves.

The *first heart sound* is associated with closure of the A-V valves. Vibration of the valves and surrounding blood, ventricular wall, and major vessels around the heart causes the first heart sound. The closure of these valves at the beginning of systole is caused by the effects of ventricular contraction, which increases intraventricular pressure and results in a backflow of blood against the A-V valves. After these valves close, the back-and-forth vibration of the elastic valve leaflets and chordae tendineae causes reverberation of the surrounding blood and ventricular walls.

The *second heart sound* is associated with closure of the aortic and pulmonary valves. The second heart sound occurs at the end of systole, when the total energy of the blood in the ventricles is less than that in the aorta and pulmonary artery. This causes the semilunar valves (aortic and pulmonary) to close and again starts a vibration in the valve leaflets and the surrounding blood, ventricular wall, and blood vessels. When the vibration of these structures contacts the chest wall, the sound, with proper amplification, can be heard from the outside of the body.

Comparison of the first and second heart sounds shows that the first sound, or the *lub*, is louder because of the high rate of change of pressure across the A-V valves. In addition, the first heart sound has a lower pitch than that of the second heart sound because of the low elastic modulus of the valves and greater amount of blood vibrating in the ventricles than in the aorta and pulmonary artery. This effect is analogous to the lower pitch made by the thick strings of a piano or guitar after being struck.

The *third heart sound* occurs at the beginning of the middle third of diastole. The cause of the sound is thought to be an in-rushing of blood into the ventricles. Not much sound occurs at the beginning of diastole because insufficient blood has entered the ventricles to create much elastic tension in the walls, which is necessary for reverberation. This sound cannot be normally heard with a stethoscope, but it can be recorded with a phonocardiogram.

The *fourth heart sound* is associated with atrial contraction. An atrial heart sound is also very difficult to hear with a stethoscope, but it sometimes can be recorded with a phonocardiogram. The sound is associated with atrial contraction and the associated inflow of blood into the ventricles. It occurs during the last third of diastole.

The Greatest Number of Cardiac Valvular Lesions Result From Rheumatic Fever (p. 277)

Rheumatic fever is an autoimmune disease in which a patient's immune system damages or destroys the heart valves. In this disease, the patients contract a group A hemolytic streptococcal infection, and the *M antigen* is released by the streptococci. Antibodies form against the M antigen, and the antigen-antibody complex has a propensity for attachment to the heart valves. The immune system then attacks the M antigen-antibody-heart valve complex and causes damage, including hemorrhagic, fibrinous, bulbous lesions.

Two types of heart valve lesions occur in rheumatic fever:

- *Stenotic valves* occur if damage to the valves causes the leaflets to adhere to one another.
- *Insufficient* or *regurgitant valves* result if the valves are partially destroyed or cannot properly close; both cause a backleak of blood.

Heart Murmurs Are Abnormal Heart Sounds Caused by Valvular Lesions (p. 277)

Aortic stenosis **causes a harsh-sounding systolic murmur.** Because of the small opening in the aortic valve in this condition, intraventricular pressure must increase to as much as 300 to 400 mm Hg to eject the ventricular blood through the small opening. The jet-like ejection of blood intensely vibrates the aortic wall. The resultant sound is very harsh and sometimes can be heard from several feet away. The vibration can be felt on the upper chest. The following are common features of aortic stenosis:

- *Left ventricular hypertrophy* occurs due to increased ventricular work load.
- *Chronic increase in blood volume* occurs as a renal compensation to an initial decrease in arterial pressure; red cell mass also increases because of mild hypoxia.
- *Chronic increase in left atrial pressure*, secondary to hypervolemia, which increases venous return to the heart. The greater venous return also increases both ventricular end-diastolic volume and end-diastolic pressure, which are necessary for the heart to forcefully contract sufficiently to overcome the outflow resistance.
- *Angina pectoris pain* occurs in severe stenosis.

Aortic regurgitation **causes a "blowing" type of diastolic murmur**. Because of the lack of ability to completely close the aortic valve, blood leaks backward through this valve and into the left ventricle during diastole. The murmur has a relatively high pitch that is caused by the jetting of blood back into the ventricle. The associated vibration is best heard over the left ventricle. The following are features of aortic regurgitation:

- *Stroke volume increases* to as high as 300 milliliters, with 70 milliliters to the periphery and 230 milliliters leaking back into the heart.
- *Left ventricular hypertrophy* is caused by the increased stroke volume required by the heart.
- *Diastolic pressure decreases rapidly* and causes back-leaking of blood into the left ventricle.
- *Blood volume chronically increases.*

Coronary ischemia often occurs during aortic valvular lesions. The amount of left ventricular hypertrophy is particularly large during both aortic stenosis and regurgitation and often is associated with coronary ischemia. During aortic stenosis, the ventricular muscle must develop a very high tension to create the high intraventricular pressure needed to force blood through the stenosed aortic valve. The oxygen consumption of the ventricle increases, necessitating a rise in coronary flow to deliver this oxygen. The high wall tension of the ventricle causes marked decreases in coronary flow during systole, particularly in the subendocardial vessels.

Intraventricular diastolic pressure is increased in this condition, which may cause compression of the inner layers of the heart muscle and result in reduced coronary flow. Coronary ischemia is likely to occur in severe aortic stenosis. In aortic regurgitation, the intraventricular diastolic pressure also increases, which compresses the inner layer of the heart muscle and decreases coronary flow. Aortic diastolic pressure falls during aortic regurgitation, which can cause a direct decrease in coronary flow. Both of these mechanisms can lead to a decrease in coronary flow and result in coronary ischemia.

Mitral stenosis **is a weak-sounding diastolic murmur that is heard best during mid to late diastole.** In mitral stenosis, blood passes with difficulty from the left atrium to the left ventricle. The left atrium is unable to develop a pressure of much more than 30 mm Hg; therefore, the velocity of blood flow through the mitral valve never increases dramatically. Sufficient velocity does develop to create a low-frequency, weak murmur that is best detected using a phonocardiogram. The following are features of mitral stenosis:

- *Cardiac output and mean arterial pressure decrease* but not as much as in aortic stenosis.
- *Atrial volume increases*, which may lead to atrial fibrillation.
- *Left atrial pressure increases*, which can cause pulmonary edema.
- *Right ventricular failure* occurs in severe stenosis because the right ventricle must pump much harder owing to an increase in pulmonary artery pressure.

ABNORMAL CIRCULATORY DYNAMICS IN CONGENITAL CARDIAC DEFECTS (p. 280)

Occasionally, the heart and associated blood vessels are malformed during fetal life. The three major congenital abnormalities are

- *Stenosis* of a channel of blood flow in the heart or one of the surrounding blood vessels
- A *left-to-right shunt*, an abnormality in which blood flows from the left side of the heart or aorta to the right side of the heart or pulmonary artery
- A *right-to-left shunt (tetralogy of Fallot)*, an abnormality in which blood bypasses the lungs and goes directly to the left side of the heart

One of the most common causes of congenital heart defects is viral infection, such as German measles, during the first trimester of pregnancy. At this time, the fetal heart is being formed and is susceptible to damage.

Patent ductus arteriosus is a left-to-right shunt. Because the lungs are collapsed during fetal life, most blood flow bypasses the lungs and enters the aorta through the ductus arteriosus, which connects the pulmonary artery and aorta. After birth, the high oxygen concentration in the aortic blood that passes through the ductus causes closure of the ductus in the majority of newborns. Occasionally the ductus does not close, and the condition is called *patent ductus arteriosus*.

In patent ductus arteriosus, the high pressure in the aorta forces blood through the open ductus and into the pulmonary artery, and blood recirculates several times through the lungs. Arterial blood oxygen saturation is, therefore, actually greater than normal unless heart failure has occurred. The following are features of patent ductus arteriosus:

- *Blood volume increases* to compensate for the decrease in cardiac output.
- *This murmur is heard throughout systole and diastole.*
- *Cardiac reserve decreases.*
- *Left ventricular hypertrophy* occurs because of the extra blood the left ventricle must pump.
- *Right ventricular hypertrophy* occurs because of high pulmonary artery pressure.
- *Pulmonary edema* can occur if the left heart is too overloaded.

Other left-to-right shunts that can occur include the *interventricular septal defect* and *interatrial septal defect*.

Tetralogy of Fallot is a right-to-left shunt. In *tetralogy of Fallot*, four abnormalities of the heart occur simultaneously:

1. The aorta is displaced over the ventricular septum and originates from the right ventricle.

2. A ventricular septal defect is also present, causing the right ventricle to pump both left and right ventricular blood through the aorta.
3. Pulmonary artery or pulmonary valve stenosis is also present, and because of the high pulmonary arterial resistance, much of the right ventricular blood shunts around the lungs and enters the aorta.
4. Right ventricular hypertrophy occurs because the right side of the heart must pump large quantities of blood against the high pressure in the aorta.

Surgical treatment of this condition is very helpful.

24

Circulatory Shock and Physiology of Its Treatment

Circulatory shock can occur when blood flow is inadequate to meet tissue demands, causing widespread tissue damage throughout the body. Damage to the tissues of the cardiovascular system, including the heart, blood vessels, and sympathetic nervous system, causes the shock to become progressively worse.

Because shock results from inadequate cardiac output, factors that decreases cardiac output can lead to shock, including the following:

- *Cardiac abnormalities* that decrease the pumping ability of the heart, including myocardial infarction, toxic states of the heart, dysfunction of the heart and heart valves, and cardiac arrhythmias
- *Factors that reduce venous return*, including decreased blood volume, decreased vascular tone (especially the tone of the veins), and obstruction to blood flow

Cardiac output does not always decrease during shock. Inadequate cardiac output can result from excessive increases in metabolic rate or from abnormal perfusion patterns that route blood through vessels other than those that supply the local tissues with nutrition. In these cases, a normal cardiac output is not sufficient to meet the needs of the tissues.

SHOCK CAUSED BY HYPOVOLEMIA—HEMORRHAGIC SHOCK (p. 286)

Nonprogressive (Compensated) Shock

One of the most common causes of shock is rapid loss of blood. If the sympathetic reflexes and other factors compensate sufficiently to prevent further deterioration of the circulation, this type of reversible shock is called *compensated shock*. The mechanisms

199

that compensate for the blood loss and its cardiovascular effects are

- *The sympathetic nervous system,* which is the first reflex mechanism that increases arterial pressure back toward normal; the baroreceptors are the main activators of the sympathetic nervous system during moderate hypotension. The decrease in blood volume in compensated shock causes a decrease in mean systemic filling pressure, cardiac output, and arterial pressure. The arterial pressure decrease stimulates the sympathetic nervous system through the baroreceptors, and this causes several cardiovascular effects, including constriction of the arterioles, which increases the total peripheral vascular resistance; constriction of the veins, which increases mean systemic filling pressure and venous return; and higher heart rate. Without these reflexes, a person would die after a loss of only 15 to 20 per cent of the blood volume over a period of 30 minutes; this is in contrast to the 30 to 40 per cent loss in blood volume that a person with normal sympathetic reflexes can sustain.
- *The central nervous system ischemic response,* which occurs during more severe hypotension, when arterial pressure decreases below 50 mm Hg
- *Reverse stress-relaxation,* which causes the blood vessels to contract down around the diminished volume and so helps to prevent the decrease in arterial pressure and cardiac output
- *Increased angiotensin II formation,* which constricts peripheral arterioles and causes sodium and water retention by the kidneys
- *Increased vasopressin release,* which constricts the peripheral blood vessels and causes water retention by the kidneys
- *Other mechanisms that increase blood volume back toward normal,* including absorption of fluid from the intestines and interstitial spaces, decreased urinary volume output, increased thirst, and increased appetite for sodium

Progressive Shock Is Caused by a Vicious Circle of Cardiovascular Deterioration (p. 287)

When shock becomes sufficiently severe, various circulatory system structures begin to deteriorate and cause a progressive vicious circle of decreasing cardiac output.

Cardiac deterioration in progressive shock is due to poor coronary flow. With severe decreases in arterial pressure, particularly diastolic pressure, the coronary blood flow also decreases, and coronary ischemia occurs. This weakens the myocardium and further decreases cardiac output. A positive feedback cycle can develop and cause progressive cardiac deterioration.

Peripheral circulatory failure can also occur during progressive hemorrhagic shock. During decreases in cardiac output, the flow to the brain and heart is usually preserved. When the arterial pressure falls sufficiently low, the cerebral blood flow begins to decrease, and flow to the vasomotor center also decreases. If flow decreases sufficiently, the sympathetic discharge of the vasomotor center dramatically falls, which can result in further reductions in arterial pressure and progressive peripheral circulatory failure.

Blood clotting in minute vessels also occurs during progressive hemorrhagic shock. Because of the low blood flow in shock, tissue metabolites, including large amounts of carbonic and lactic acid, are not carried away from the tissues properly, allowing local concentrations to build up. The increased concentration of hydrogen ions and other ischemic products that result cause local agglutination of blood and formation of blood clots. The thickened blood in these minute blood vessels is called *sludged blood*.

Increased capillary permeability causes a further decrease in blood volume during progressive hemorrhagic shock. Because of capillary hypoxia and lack of other nutrients during shock, the capillary permeability increases, allowing fluid and protein to transude into the tissues. This loss of fluid into the interstitium causes a decrease in blood volume that can make the shock progressively worse.

Release of toxins may cause cardiac depression in progressive hemorrhagic shock. Dead gram-negative bacteria in the intestines release a toxin called *endotoxin*. This toxin causes an increase in cellular metabolism that can be harmful in shock because the cells that are alive have barely adequate nutrition. Endotoxin specifically depresses the heart. Both of these factors can lead to progressive cellular damage and shock.

Widespread cellular deterioration occurs during progressive hemorrhagic shock. During shock, generalized cellular damage usually occurs first in highly metabolic tissues, such as the liver. Among the damaging cellular effects are the following:

- Decreases occur in active transport of sodium and potassium through cell membranes; sodium accumulates in the cells and potassium is lost, and the cells begin to swell.
- Mitochondrial activity is decreased.
- Lysosomes begin to split the widespread tissues, which release hydrolases that cause widespread intracellular damage.
- The cellular metabolism of glucose decreases.

Irreversible Shock (p. 290)

During *irreversible shock*, even though a transfusion of blood may temporarily increase cardiac output and arterial pressure to normal, the cardiac output begins to fall, and death soon ensues. The temporary increase in cardiac output does not prevent the widespread tissue damage caused by acidosis, release of hydrolases, blood clots, and other destructive factors. Therefore, a stage is reached after which even rigorous therapy is of no avail.

One of the main causes of irreversible shock is the *depletion of high-energy phosphate compounds*. Once adenosine triphosphate has been degraded in the cell to adenosine diphosphate, adenosine monophosphate, and, finally, adenosine, the adenosine diffuses out of the cell and is converted to uric acid, which cannot re-enter the cell. New adenosine can be synthesized at a rate of only 2 per cent of the total cellular amount per hour. The high-energy phosphate compounds are, therefore, difficult to regenerate during shock, and this contributes to the final stage of irreversible shock.

PHYSIOLOGY OF TREATMENT IN SHOCK (p. 292)

Replacement Therapy

Because blood loss is the cause of hemorrhagic shock, the appropriate therapy is replacement of the blood. Intravenous infusion of whole blood has proved to be very helpful in the treatment of hemorrhagic shock. Other therapies, such as norepinephrine infusion, have been of little benefit. Under battlefield conditions, whole blood often is not available, and plasma has been substituted. Plasma maintains the colloid osmotic pressure of the blood, but the hematocrit decreases with this therapy, placing an extra load on the heart, because cardiac output must increase to maintain oxygen delivery to the tissues. Whole blood administration is, therefore, the better therapy for hemorrhagic shock.

If plasma is not available for a patient in hemorrhagic shock, a plasma substitute may be used. The substitute must have a high colloid osmotic pressure so that the fluid will not rapidly transude through the capillary pores and into the interstitium. *Dextran* and other high-molecular-weight polysaccharide polymers have been developed and have proved to remain in the blood compartment after intravenous infusion.

Because plasma loss is the cause of hypovolemic shock in patients with intestinal destruction or burns, plasma infusion is the appropriate therapy. During intestinal obstruction, blockage and distention of the intestines partly impede the venous blood flow and thus increase capillary pressure and leakage of highly proteinaceous fluid into the intestinal lumen. In severe intestinal blockage, shock can ensue; however, if an intravenous plasma infusion is started soon, hemodynamic conditions are rapidly restored to normal. In patients with severe burns, plasma transudes through the damaged areas of the skin, causing a marked decrease in plasma volume. The appropriate therapy for the shock that might occur in a burn patient, therefore, is intravenous infusion of plasma.

Because water and electrolyte loss is the cause of hypovolemic shock in patients with dehydration, intravenous infusion of a balanced electrolyte solution is the appropriate therapy. A number of conditions can result in dehydration, including vomiting, diarrhea, excess perspiration, diabetes mellitus, diabetes insipidus, excessive use of diuretics, destruction of the adrenal cortices with loss of aldosterone, and loss of fluid by nephrotic kidneys. If the dehydration is very severe, shock can occur. If a balanced electrolyte solution such as lactated Ringer's solution is soon infused intravenously, the problem is corrected.

Traumatic shock can be caused by hypovolemia and pain. Often, a patient with trauma due to severe contusion of the body also experiences hypovolemia. Whole blood administration can correct this hypovolemia, but the pain associated with trauma is an additional aggravating factor. This pain sometimes inhibits the vasomotor center, resulting in a decrease in sympathetic output. This can reduce arterial pressure and venous return of blood to the heart. The administration of a proper analgesic can help alleviate the pain and its effects on the sympathetic nervous system.

Neurogenic shock is caused by increased vascular capacity; therefore, therapy must decrease this capacity toward normal. *Neurogenic shock* results from a sudden loss of vasomotor tone throughout the body, thus increasing total vascular capacity. The normal blood volume is inadequate to fill the circulatory system properly; therefore, a decrease in mean systemic filling pressure results. Some causes of neurogenic shock include the following:

- *Deep general anesthesia*, which depresses the vasomotor center
- *Spinal anesthesia*, especially when the anesthetic migrates all the way up the spinal cord, blocking sympathetic outflow
- *Brain damage*, such as a brain concussion or contusion in the basal areas of the brain near the vasomotor center, that dramatically decreases sympathetic outflow from the vasomotor center.

The therapy of choice for neurogenic shock is intravenous infusion of a sympathomimetic drug, such as norepinephrine or epinephrine, that replaces the lost neurogenic vascular tone.

Anaphylactic shock is caused by an allergic reaction. When an antigen enters the bloodstream in a person who is highly allergic, an antigen-antibody reaction takes place. One of the main effects is the release of histamine or histamine-like substances from basophils and mast cells. The histamine has several effects, including

- An increase in vascular capacity because of venodilation
- Arteriolar dilation, which decreases arterial pressure
- Increased capillary permeability, causing a loss of fluid from the vascular compartment.

These effects of histamine can decrease arterial pressure and venous return, which leads to anaphylactic shock. A person may die minutes after anaphylactic shock symptoms appear. The rapid administration of a sympathomimetic drug, which decreases vascular capacity and constricts the arterioles, is often life saving.

Septic shock is caused by widespread dissemination of bacteria in the body. There are many causes of *septic shock*; all start with a bacterial infection. When sufficient bacteria spread throughout the body, there are many effects, including the following:

- High fever
- High metabolic rate
- Marked vasodilation throughout the body
- High cardiac output, caused by peripheral vasodilation, in perhaps one half of patients
- Sludging of blood due to red cell agglutination
- Disseminated intravascular coagulation.

A special case of septic shock occurs when the colon bacteria, containing a toxin called *endotoxin*, are released during strangulation of the gut.

Therapy for shock other than that previously mentioned includes

- Treatment with the head-down position, which will promote venous return
- Oxygen
- Glucocorticoids, which stabilize the lysosomes (this has been shown to be helpful during anaphylactic shock).

STILL OTHER EFFECTS OF SHOCK ON THE BODY
(p. 292)

During shock, especially hypovolemic shock, the decrease in cardiac output reduces the delivery of oxygen and other nutrients to the tissues. Widespread cellular damage may occur, including the ability of the mitochondria to synthesize adenosine triphosphate and a depressed sodium-potassium cellular membrane pump. Other effects include

- Muscle weakness
- Decreased body temperature because of decreased metabolism
- Depressed mental function
- Decreased renal function and renal deterioration.

25

The Body Fluid Compartments: Extracellular and Intracellular Fluids; Interstitial Fluid and Edema

The total amount and composition of the body fluids are maintained relatively constant under most physiological conditions, as required for homeostasis. Some of the most important problems in clinical medicine, however, arise because of abnormalities in the control systems that maintain this constancy. In this section, we discuss the overall regulation of body fluid volume, control of the constituents of the extracellular fluid, regulation of the fluid exchange between the extracellular and intracellular compartments, and regulation of the acid-base balance.

FLUID INTAKE AND OUTPUT ARE BALANCED DURING STEADY-STATE CONDITIONS (p. 297)

The total intakes of water and electrolytes must be carefully matched by equal outputs from the body to prevent fluid volumes and electrolyte concentrations from increasing or decreasing. Table 25–1 shows the routes of daily water intake and output from the body. Under most conditions, the primary means of regulating output is by altering renal excretion. Urine volume can be as low as 0.5 liter/day in a dehydrated person or as high as 20 liters/day in a person who has been drinking large amounts of fluids. This capability of the kidneys to adjust the output to such an extreme to match intake also occurs for the electrolytes of the body (sodium, chloride, and potassium).

TOTAL BODY FLUID IS DISTRIBUTED BETWEEN THE EXTRACELLULAR FLUID AND THE INTRACELLULAR FLUID

The total amount of body water averages about 60 per cent of the body weight, or about 42 liters in a

TABLE 25–1 DAILY INTAKE AND OUTPUT OF WATER (in ml/day)

	Normal	**Prolonged Heavy Exercise**
Intake		
Fluids ingested	2100	?
From metabolism	200	200
Total intake	2300	?
Output		
Insensible skin	350	350
Insensible lungs	350	650
Sweat	100	5000
Feces	100	100
Urine	1400	500
Total output	2300	6600

70-kilogram adult. This fluid is distributed into two main compartments: (1) the intracellular fluid, which is about 40 per cent of body weight, or 28 liters, and (2) the extracellular fluid, which is about 20 per cent of body weight, or 14 liters in a 70-kilogram person. The two main compartments of the extracellular fluid are the *interstitial fluid,* which makes up about three fourths of the extracellular fluid, and the *plasma,* which makes up about one fourth of the extracellular fluid, or about 3 liters. The plasma is the noncellular portion of the blood that mixes continuously with interstitial fluid through the pores of the capillary membranes.

Blood contains extracellular and intracellular fluids. Blood is considered to be a separate fluid compartment because it is contained in its own chamber, the circulatory system. The average blood volume in a normal adult is 8 per cent of the body weight, or about 5 liters. About 60 per cent of the blood is plasma, and about 40 per cent is red blood cells. The *hematocrit,* the fraction of blood that is composed of red blood cells, is normally about 0.42 in men and about 0.38 in women. In severe *anemia,* the hematocrit may fall to as low as 0.10, a value that is barely sufficient to sustain life. When there is excessive production of red blood cells, resulting in *polycythemia,* the hematocrit can rise to as high as 0.65.

The constituents of extracellular and intracellular fluids differ. Table 25–2 is a comparison of the

TABLE 25–2 CHEMICAL COMPOSITIONS OF EXTRACELLULAR AND INTRACELLULAR FLUIDS

	Intracellular	Extracellular
Na^+ (mmol/liter)	10	10
K^+ (mmol/liter)	140	4
Cl^- (mmol/liter)	4	108
HCO_3^- (mmol/liter)	10	24
Ca^{++} (mmol/liter)	0.0001	2.4
Mg^{++} (mmol/liter)	58	1.2
$SO_4^=$ (mmol/liter)	2	1
Phosphates (mmol/liter)	75	4
Glucose (mg/dl)	0–20	90
Amino acids (mg/dl)	200?	30
Protein (mg/dl)	16	2

compositions of the intracellular and extracellular fluids.

Because plasma and interstitial fluid of the extracellular compartment are separated by highly permeable capillary membranes, their ionic compositions are similar. The most important difference between these two compartments is that plasma has a higher protein concentration. The capillaries have a low permeability to proteins and therefore leak only small amounts of protein into the interstitial spaces in most tissues.

The intracellular fluid is separated from the extracellular fluid by a highly selective cell membrane that is permeable to water but not to most electrolytes in the body. For this reason, the concentration of water and the osmolarity of intracellular and extracellular fluids are approximately equal under steady-state conditions, although the concentrations of various solutes are markedly different in these fluid compartments.

THE INDICATOR-DILUTION PRINCIPLE CAN BE USED TO MEASURE VOLUMES OF THE DIFFERENT BODY FLUID COMPARTMENTS (p. 300)

The volume of a fluid compartment in the body can be estimated by injecting a substance into the compartment, allowing it to disperse evenly throughout the compartment, and then analyzing the extent to which the substance has become diluted. This method is based on the assumption that the total amount of substance remaining in the fluid compartment after dispersion is the same as the total amount that was injected into the compartment. Thus, when a small amount of substance contained in syringe A is injected into compartment B and the substance is allowed to disperse throughout the compartment until it becomes mixed in equal concentrations in all areas, the following relationship can be expressed:

$$\text{Volume B} = \frac{\text{Volume A} \times \text{Concentration A}}{\text{Concentration B}}$$

This method can be used to measure the volume of virtually any compartment in the body if (1) the amount of indicator injected into the compartment (the numerator of the equation) is known, (2) the concentration of the indicator in the compartment is known, (3) the indicator disperses evenly throughout the compartment, and (4) the indicator disperses only in the compartment that is being measured.

Table 25–3 shows some of the indicators that can be used to measure the fluid volumes of the different body fluid compartments. The volumes of two of the compartments, the intracellular and extracellular interstitial fluids, cannot be measured directly but instead are calculated from the values of other body fluid volumes.

FLUID DISTRIBUTION BETWEEN INTRACELLULAR AND EXTRACELLULAR COMPARTMENTS IS DETERMINED MAINLY BY THE OSMOTIC EFFECT OF ELECTROLYTES ACTING ACROSS THE CELL MEMBRANE (p. 302)

Because the cell membrane is highly permeable to water but relatively impermeable to even small ions, such as sodium and chloride, the distribution of fluid between the intracellular and extracellular compartments is determined mainly by the osmotic effects of these ions. The basic principles of osmosis and osmotic pressure are presented in Chapter 4. Only the most important principles as they apply to volume regulation are therefore discussed in this section.

Osmosis **is the net diffusion of water from a region of high water concentration to one of lower water concentration.** The addition of a solute to

TABLE 25–3 MEASUREMENT OF BODY FLUID VOLUME	
Volume	Indicators
Total body water	^3H_2O, ^2H_2O, antipyrine
Extracellular fluid	^{22}Na, ^{125}I-iothalamate, inulin
Intracellular fluid	Calculated as: Total body water − Extracellular fluid volume
Plasma volume	^{125}I-albumin, Evans blue dye (T-1824)
Blood volume	^{51}Cr-labeled red blood cells, calculated as: Blood volume = Plasma volume/(1 − Hematocrit)
Interstitial fluid	Calculated as: Extracellular fluid volume − Plasma volume

pure water dilutes the water concentration and causes water to move toward the region of high solute concentration. The concentration term used to measure the total number of solute particles in solution is the *osmole*. One osmole is equal to 1 mole (6.02×10^{23}) solute particles. For biological solutions, the term *milliosmole* (mOsm), which equals $\frac{1}{1000}$ osmole, is commonly employed.

The osmolar concentration of a solution is called *osmolarity* when the concentration is expressed as *osmoles per kilogram of water*, and *osmolarity* when it is expressed as *osmoles per liter of solution*. The precise amount of pressure required to prevent osmosis of water through a semipermeable membrane is called the *osmotic pressure*. Expressed mathematically, the osmotic pressure (π) is directly proportional to the concentration of osmotically active particles in that solution:

$$\pi = CRT$$

where C is the concentration of solutes in osmoles per liter, R is the ideal gas constant, and T is the absolute temperature in degrees kelvin. If π is expressed in millimeters of mercury (the unit of pressure commonly used for biological fluids), π calculates to be about 19.3 mm Hg for a solution with an osmolarity of 1 mOsm/liter. Thus, for each mOsm concentration gradient across the cell membrane, 19.3 mm Hg of force is required to prevent water diffusion across the membrane. Very small differences in solute concentration across the cell membrane can therefore cause rapid osmosis of water.

A solution can be isotonic, hypotonic, or hypertonic. A solution is said to be *isotonic* if no osmotic force develops across the cell membrane when a normal cell is placed into the solution. An isotonic solution has the same osmolarity as the cell, and the cells will not shrink or swell if placed into the solution. Examples of isotonic solutions include a 0.9 per cent sodium chloride solution and a 5 per cent glucose solution.

A solution is said to be *hypertonic* when it contains a higher concentration of osmotic substances than does the cell. In this case, an osmotic force develops that causes water to flow out of the cell into the solution, thereby reducing intracellular fluid volume and increasing intracellular fluid concentration.

A solution is said to be *hypotonic* if the osmotic concentration of substances in the solution is less than the concentration of the cell. The osmotic force develops immediately when the cell is exposed to the solution, causing water to flow by osmosis into the cell until the intracellular fluid has about the same concentration as the extracellular fluid or until the cell bursts as a result of excessive swelling.

VOLUMES AND OSMOLARITIES OF EXTRACELLULAR AND INTRACELLULAR FLUIDS IN ABNORMAL STATES (p. 305)

Abnormalities of the compositions and volumes of the body fluids are among the most common and important problems in clinical medicine. Some of the different factors that can cause extracellular and intracellular volumes to change markedly are ingestion of large amounts of water, dehydration, intravenous infusion of different types of solutions, loss of large amounts of fluid from the gastrointestinal tract, and loss of abnormal amounts of fluid by sweating or from the kidneys.

One can approximate the changes in intracellular and extracellular fluid volumes and the types of therapy that must be instituted if the following basic principles are kept in mind:

- *Water moves rapidly across cell membranes;* therefore, the osmolarities of intracellular and extracellular fluids remain almost exactly equal to each other except for a few minutes after a change in one of the compartments.

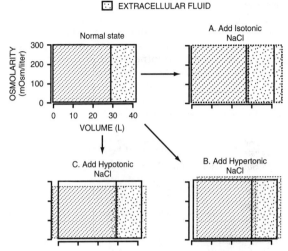

Figure 25–1 Effect of adding isotonic, hypertonic, and hypotonic solutions to the extracellular fluid after osmotic equilibrium. The normal state is indicated by the solid lines and the shifts from normal are shown by the dashed lines. The volumes of intracellular and extracellular fluid compartments are shown in the abscissa of each diagram, and the osmolarities of these compartments are shown on the ordinates.

- *The cell membrane is almost completely impermeable to most solutes;* therefore, the number of osmoles in the extracellular and intracellular fluids remains relatively constant unless solutes are added to or lost from the extracellular compartment.

The addition of saline solution to the extracellular fluid has an effect. If an *isotonic* solution is added to the extracellular fluid compartment, the osmolarity of the extracellular fluid does not change and there is no osmosis through the cell membranes. The only effect is an increase in extracellular fluid volume (Fig. 25–1). Sodium and chloride in large part remain in the extracellular fluid because the cell membrane behaves as though it were virtually impermeable to sodium chloride.

If a *hypertonic* solution is added to the extracellular fluid, the extracellular fluid osmolarity increases and causes osmosis of water out of the cells into the extracellular compartment. The net effect is an increase in extracellular volume (greater than the volume of fluid that was added), a decrease in intracellular fluid volume, and an increase in the osmolarity of both compartments.

If a *hypotonic* solution is added to the extracellular fluid, the osmolarity of the extracellular fluid decreases, and some of the extracellular water diffuses into the cells until the intracellular and extracellular compartments have the same osmolarity. Both the intracellular and extracellular volumes are increased by the addition of hypotonic fluid, although the intracellular volume is increased to a greater extent.

EDEMA: EXCESS FLUID IN THE TISSUES (p. 308)

Intracellular Edema: Increased Intracellular Fluid

Two conditions especially likely to cause intracellular swelling are as follows: (1) depression of the metabolic systems of the tissues and (2) lack of adequate nutrition to the cells. When these conditions occur, sodium ions that normally leak into the interior of the cells can no longer be pumped out of the cells, and the excess sodium ions cause osmosis of water into the cells.

Intracellular edema can also occur in inflamed tissues. Inflammation usually has a direct effect on the cell membranes to increase their permeability, allowing sodium and other ions to diffuse into the interior of the cells with subsequent osmosis of water into the cells.

Extracellular Edema: Increased Fluid in the Interstitial Spaces

The two general causes of extracellular edema are (1) an abnormal leakage of fluid from the plasma to the interstitial spaces across the capillaries and (2) a failure of the lymphatics to return fluid from the interstitium to the blood.

Factors can increase capillary filtration and cause interstitial fluid edema. To understand the causes of excessive capillary filtration, it is useful to review the determinants of capillary filtration discussed in Chapter 13, as shown in the following relationship:

$$\text{Filtration} = K_f \times (P_c - P_{if} - \pi_c + \pi_{if})$$

where K_f is the capillary filtration coefficient (the product of permeability and surface area of the capillaries), P_{if} is the interstitial fluid colloid osmotic pressure, π_c is the capillary plasma colloid osmotic pressure, and π_{if} is the interstitial fluid colloid osmotic pressure. From this equation, it can be seen that any of the following changes can increase the capillary filtration rate:

1. *Increased capillary filtration coefficient*, which allows leakage of fluids and plasma proteins through the capillary membranes; this can occur as a result of allergic reactions, bacterial infections, and toxic substances that injure the capillary membranes and increase their permeability to plasma proteins

2. *Increased capillary hydrostatic pressure*, which can result from obstruction of veins, excessive flow of blood from the arteries into the capillaries, or failure of the heart to pump blood rapidly out of the veins (heart failure)

3. *Decreased plasma colloid osmotic pressure*, which can occur as a result of a failure of the liver to produce sufficient quantities of plasma proteins (cirrhosis), loss of large amounts of proteins in the urine in certain kidney diseases (nephrotic syndrome), or loss of large quantities of proteins through burned areas of the skin or other denuding lesions

4. *Increased interstitial fluid colloid osmotic pressure*, which draws fluid out of the plasma into the tissues spaces; this occurs most frequently as a result of lymphatic blockage, which prevents the return of proteins from the interstitial spaces to the blood (discussed in the following sections).

Lymphatic blockage causes edema. When lymphatic blockage occurs, edema can become especially severe because plasma proteins that leak into the interstitium have no other way to be returned to the plasma. The rise in protein concentration increases the colloid osmotic pressure of the interstitial fluid, which draws even more fluid out of the capillaries.

Blockage of lymph flow can be especially severe with infections of the lymph nodes, as occurs with infection by *filaria nematodes*. Blockage of the lymph vessels can also occur in certain types of cancer or after surgery in which the lymph vessels are removed or obstructed.

Safety Factors That Normally Prevent Edema

Although many abnormalities can cause fluid accumulation in interstitial spaces, the disturbances must be severe before clinically significant edema develops. Three major safety factors prevent fluid accumulation in the interstitial spaces:

1. *The compliance of the tissues is low as long as interstitial fluid hydrostatic pressure is in the negative range.* A low compliance (defined as the change in volume per millimeter of mercury pressure change) means that small increases in interstitial fluid volume are associated with relatively large increases in interstitial fluid hydrostatic pressure. When interstitial fluid volume increases, interstitial fluid hydrostatic pressure increases markedly, which opposes further excess capillary filtration. The safety factor that protects against edema for this effect is about 3 mm Hg in many tissues such as the skin.
2. *Lymph flow can increase as much as 10- to 50-fold.* Lymph vessels carry away large amounts of fluid and proteins in response to increased capillary filtration. The safety factor for this effect has been calculated to be about 7 mm Hg.
3. *There is a "washdown" of interstitial fluid protein as lymph flow increases.* As increased amounts of fluid are filtered into the interstitium, the interstitial fluid pressure increases, causing greater lymph flow. This decreases protein concentration of the interstitium because more protein is carried away than can be filtered by the capillaries. A decrease in tissue fluid protein concentration lowers the net filtration force across the capillaries and tends to prevent further fluid accumulation. The safety factor for this effect has been calculated to be about 7 mm Hg in most tissues.

Combining all of the safety factors, the total safety factor that protects against edema is about 17 mm Hg. Capillary pressure in peripheral tissues could, therefore, theoretically rise 17 mm Hg before significant interstitial edema would occur.

26

Urine Formation by the Kidneys:

I. Glomerular Filtration, Renal Blood Flow, and Their Control

The multiple functions of the kidney in the maintenance of homeostasis include:

- Regulation of water and electrolyte balances
- Regulation of body fluid osmolarity and electrolyte concentrations
- Excretion of metabolic waste products and foreign chemicals
- Regulation of arterial pressure through excretion of varying amounts of sodium and water and secretion of substances, such as renin, that lead to formation of vasoactive products such as angiotensin II
- Regulation of acid-base balance through excretion of acids and regulation of body fluid buffer stores
- Regulation of erythrocyte production through secretion of erythropoietin, which stimulates red blood cell production
- Regulation of 1,25-dihydroxy vitamin D_3 production
- Synthesis of glucose from amino acids (gluconeogenesis) during prolonged fasting.

URINE FORMATION RESULTS FROM GLOMERULAR FILTRATION, TUBULAR REABSORPTION, AND TUBULAR SECRETION (p. 319)

One of the primary functions of the kidney is to "clear" unneeded substances from the blood and excrete them in the urine while returning needed substances to the blood. The first step in the performance of this function is the filtration of fluid from the glomerular capillaries into the renal tubules, a process called *glomerular filtration*. As glomerular filtrate flows through the tubules, the volume of fil-

Basic Mechanisms of Renal Excretion

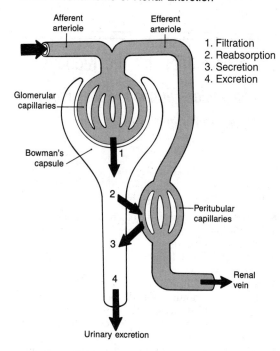

1. Filtration
2. Reabsorption
3. Secretion
4. Excretion

Excretion = Filtration – Reabsorption + Secretion

Figure 26–1 Basic kidney processes that determine the composition of the urine. Urinary excretion rate of a substance is equal to the rate at which the substance is filtered minus its reabsorption rate plus the rate at which it is secreted from the peritubular capillary blood into the tubules.

trate is reduced, and its composition is altered by *tubular reabsorption* (the return of water and solutes from the tubules back into the blood) and by *tubular secretion* (the net movement of water and solutes into the tubules), each of which is highly variable depending on the body's needs. Thus, the excretion of each substance in the urine involves a specific combination of filtration, reabsorption, and secretion (Fig. 26–1), as expressed by the following relationship:

Urinary excretion rate = Filtration rate
  – Reabsorption rate + Secretion rate

Each of these processes is physiologically controlled, and changes in excretion rate can obviously occur via changes in glomerular filtration, tubular reabsorption, or tubular secretion.

The nephron is the structural and functional unit of the kidney. Each kidney has about 1 million nephrons, each of which is capable of forming urine. Each nephron is composed of a tuft of *glomerular capillaries,* called the *glomerulus,* in which large amounts of fluid are filtered from the blood; a capsule around the glomerulus called *Bowman's capsule;* and a long *tubule* in which the filtered fluid is converted into urine on its way to the *renal pelvis,* which receives urine from all of the nephrons.

The renal tubule is subdivided into the following major sections, each of which has different structural and functional characteristics: (1) the *proximal tubule,* which lies in the outer portion of the kidney (cortex); (2) the *loop of Henle,* which includes descending and ascending limbs that dip into the inner part of the kidney (medulla); (3) the *distal tubule,* which lies in the renal cortex; and (4) the *connecting tubule,* the *cortical collecting tubule,* and the *cortical collecting ducts,* which begin in the cortex and run downward into the medulla to become (5) the *medullary collecting ducts.* Urine passes from the renal pelvis to the *bladder,* where it is stored until it is eventually expelled from the body through the process of *micturition,* or urination.

Renal blood flow constitutes about 21 per cent of the cardiac output. Blood flows to each kidney through a renal artery, which branches progressively to form the *interlobar arteries, arcuate arteries, interlobular arteries,* and *afferent arterioles,* which lead to the glomerular capillaries, where filtration of fluid and solutes begins. The capillaries of each glomerulus coalesce to form an *efferent arteriole,* which leads to a second capillary network, the *peritubular capillaries,* which surround the tubules. The peritubular capillaries empty into the vessels of the venous system, which run parallel to the arteriolar vessels, and progressively form the *intralobular vein, arcuate vein, interlobar vein,* and *renal vein,* which leaves the kidney along the renal artery and ureter. The *vasa recta* are specialized peritubular capillaries that dip into the renal medulla and run parallel to the loops of Henle. The outer portion of the kidney, the renal cortex, receives most of the blood flow of the kidney;

only 1 to 2 per cent of the total renal blood flow passes through the vasa recta, which supply the renal medulla.

Two distinguishing features of the renal circulation are (1) the high rate of blood flow (about 1200 ml/min for a 70-kilogram man) relative to the tissue weight (about 300 grams for the two kidneys) and (2) the presence of two capillary beds, the glomerular and peritubular capillaries, which are arranged in series and separated by the efferent arterioles. The glomerular capillaries filter large amounts of fluid and solutes, most of which are reabsorbed from the renal tubules into the peritubular capillaries.

Renal blood flow is determined according to the pressure gradient across the renal vasculature and the total renal vascular resistance, as expressed by the following relationship:

$$\text{Renal blood flow} = \frac{(\text{Renal artery pressure} - \text{Renal vein pressure})}{\text{Total renal vascular resistance}}$$

The total renal vascular resistance is the sum of the resistances of the individual vascular segments, including the arteries, arterioles, capillaries, and veins. Most of the renal vascular resistance resides in three major segments: the interlobular arteries, afferent arterioles, and efferent arterioles.

GLOMERULAR FILTRATION IS THE FIRST STEP IN URINE FORMATION (p. 320)

The composition of glomerular filtrate is almost identical to that of plasma except that it has virtually no protein (only about 0.03 per cent). The glomerular filtration rate (GFR) is normally about 125 ml/min, or about 20 per cent of the renal plasma flow; thus, the fraction of renal plasma flow that is filtered (*filtration fraction*) averages about 0.2.

The GFR is determined according to the *net filtration pressure* across the glomerular capillaries and the *glomerular capillary filtration coefficient* (K_f), which is the product of the permeability and surface area of the capillaries:

$$\text{GFR} = K_f \times \text{Net filtration pressure}$$

The net filtration pressure is the sum of hydrostatic and colloid osmotic forces acting across the glomerular capillaries and includes (1) the hydrostatic pressure inside the capillaries, glomerular hydrostatic pressure (P_G), which is normally about 60 mm Hg and promotes filtration; (2) the hydrostatic pressure in Bowman's capsule outside the capillaries (P_B), which is normally 18 mm Hg and opposes filtration; (3) the colloid osmotic pressure of the glomerular capillary plasma proteins (π_G), which averages 33 mm Hg and opposes filtration; and (4) the colloid osmotic pressure of proteins in Bowman's capsule (π_B), which is near zero and therefore under normal conditions has little effect on filtration. Thus,

$$\text{Net filtration pressure} = P_G - P_B - \pi_G = 10 \text{ mm Hg}$$

$$\text{GFR} = K_f \times (P_G - P_B - \pi_B) = 125 \text{ ml/min}$$

Decreased glomerular capillary filtration coefficient (K_f) decreases GFR. Although changes in K_f have a proportional effect on GFR, this is not a primary mechanism for physiologic control of GFR. Nevertheless, in some diseases, such as uncontrolled hypertension and diabetes mellitus, GFR is reduced because of increased thickness of the glomerular capillary membrane, which reduces K_f, or because of severe damage to the capillaries and complete loss of capillary filtration surface area.

Increased Bowman's capsule pressure decreases GFR. Changes in Bowman's capsule pressure normally do not control GFR; however, in certain pathological states, such as urinary tract obstruction, Bowman's capsule pressure may increase to such a high level that GFR is reduced. For example, precipitation of calcium or uric acid may lead to "stones" that lodge in the urinary tract, often in the ureter, thereby obstructing urine flow and raising Bowman's capsule pressure.

Increased glomerular capillary colloid osmotic pressure decreases GFR. The two factors that influence glomerular capillary colloid osmotic pressure are: (1) the arterial colloid osmotic pressure and (2) the fraction of plasma filtered by the glomerular capillaries (filtration fraction). An increase in either the arterial colloid osmotic pressure or the filtration fraction will raise the glomerular capillary colloid osmotic pressure. Conversely, a decrease in the arterial plasma colloid osmotic pressure or the filtration fraction will reduce the glomerular colloid osmotic pressure. Because filtration fraction is the ratio of GFR to renal plasma flow, a decrease in renal plasma flow lowers the filtration fraction. Even with a constant glomerular hydrostatic pressure, decreased renal blood flow will tend to increase the glomerular colloid osmotic pressure and decrease the GFR.

Increased glomerular capillary hydrostatic pressure increases GFR. Glomerular hydrostatic pressure is determined by three variables, each of which is physiologically regulated:

- *Arterial pressure.* Increased arterial pressure tends to raise glomerular hydrostatic pressure and to increase GFR. However, this effect is normally buffered by *autoregulation,* which minimizes the effect of blood pressure on glomerular hydrostatic pressure.
- *Afferent arteriolar resistance.* Increased resistance of afferent arterioles decreases glomerular hydrostatic pressure and decreases GFR.
- *Efferent arteriolar resistance.* Increased efferent arteriolar resistance raises the resistance to outflow of the glomerular capillaries and raises the glomerular hydrostatic pressure, thereby tending to raise the GFR as long as the increased efferent resistance does not reduce renal blood flow to a great extent. With severe efferent constriction (e.g., more than a three-fold increase in resistance), the large decrease in renal blood flow more than offsets the increase in glomerular hydrostatic pressure and reduces GFR.

GLOMERULAR FILTRATION AND RENAL BLOOD FLOW ARE CONTROLLED BY NEUROHUMORAL SYSTEMS AND BY INTRARENAL MECHANISMS (p. 326)

The determinants of GFR that are most variable and subject to physiological control include glomerular hydrostatic pressure and glomerular capillary colloid osmotic pressure. These, in turn, are influenced by the sympathetic nervous system, hormones and autacoids (vasoactive substances released in the kidney), and other intrarenal feedback control mechanisms.

Sympathetic nervous system activation decreases GFR. Strong activation of the sympathetic nervous system constricts the renal arterioles and decreases renal blood flow and GFR. This effect seems to be most important in reducing GFR during severe, acute disturbances such as those elicited by the defense reaction, brain ischemia, or severe hemorrhage.

Hormones and autacoids control GFR and renal blood flow. Several hormones and autacoids can also influence GFR and renal blood flow, as follows:

- *Norepinephrine and epinephrine,* which are released from the adrenal medulla, constrict afferent and efferent arterioles and decrease GFR and renal blood flow.
- *Endothelin,* a peptide that is released from damaged vascular endothelial cells of the kidneys and other tissues, constricts renal arterioles and decreases GFR and renal blood flow.
- *Angiotensin II* constricts efferent arterioles to a greater extent than afferent arterioles and therefore tends to increase glomerular hydrostatic pressure and GFR while decreasing renal blood flow. Increased angiotensin II formation usually occurs with decreased arterial pressure or volume depletion, both of which tend to reduce GFR. In these instances, increased angiotensin II levels help to prevent decreases in GFR by constricting efferent arterioles.
- *Endothelium-derived nitric oxide (EDNO)* decreases renal vascular resistance and increases GFR and renal blood flow. EDNO is an autacoid that is released from the vascular endothelial cells throughout the body, and it appears to be important in the prevention of excessive vasoconstriction of the kidneys.
- *Prostaglandins* (especially PGE_2 and PGI_2) are not of major importance in the regulation of GFR and renal blood flow under normal conditions. Prostaglandins, however, may dampen the renal vasoconstrictor effects of sympathetic nerves or angiotensin II, especially the effects on the afferent arterioles. Blockade of prostaglandin synthesis (e.g., with aspirin and nonsteroidal anti-inflammatory drugs) may therefore cause significant decreases in GFR and renal blood flow, especially in patients whose extracellular fluid volume is reduced as a result of vomiting, diarrhea, dehydration, or diuretic therapy.

GFR and Renal Blood Flow Are Autoregulated During Changes in Arterial Pressure (p. 327)

In normal kidneys, a decrease in arterial pressure to as low as 75 mm Hg or an increase to as high as 160 mm Hg changes GFR by only a few percentage points; this relative constancy of GFR and renal blood flow is referred to as *autoregulation*. Autoregulation of GFR and renal blood flow is not perfect, but it prevents potentially great changes in GFR and, therefore, renal excretion of water and solutes that would otherwise occur with changes in blood pressure.

Tubuloglomerular feedback **is a key component of renal autoregulation.** This feedback has two parts— (1) an afferent arteriolar mechanism and (2) an efferent arteriolar mechanism—both of which depend on the special anatomic arrangement of the *juxtaglomerular complex*. The juxtaglomerular complex consists of *macula densa cells* in the initial portion of the distal tubule and *juxtaglomerular cells* in the walls of the afferent and efferent arterioles. When blood pressure is decreased, delivery of sodium chloride is decreased to the macula densa cells, which are capable of sensing this change. The decrease in sodium chloride concentration at the macula densa, in turn, causes two main effects: (1) a decrease in the resistance of the afferent arterioles, which raises glomerular hydrostatic pressure and GFR toward normal levels; and (2) an increase in renin release from the juxtaglomerular cells of the afferent and efferent arterioles, which causes an increase in angiotensin II formation. Angiotensin II then constricts efferent arterioles and increases glomerular hydrostatic pressure and GFR toward normal levels.

Myogenic mechanism **contributes to autoregulation of renal blood flow and GFR.** This mechanism refers to the intrinsic capability of blood vessels to constrict when blood pressure is increased. The constriction prevents the vessel from being overstretched and, by raising vascular resistance, helps to prevent excessive increases in renal blood flow and GFR when blood pressure rises. Conversely, with decreased blood pressure, the myogenic mechanism contributes to decreased vascular resistance.

Other Factors That Alter Renal Blood Flow and GFR

- *A high-protein diet* increases GFR and renal blood flow in part by stimulating growth of the kidneys but also by reducing renal vascular resistance. One mechanism that contributes to the effect of protein to raise GFR is tubuloglomerular feedback. A high-protein diet increases the release of amino acids into the blood, which are reabsorbed in the proximal tubule through co-transport with sodium. This in turn causes increased proximal tubule reabsorption of amino acids and sodium, decreased sodium chloride delivery to the macula densa, decreased afferent arteriolar resistance, and increased GFR.

- *Hyperglycemia,* as occurs in uncontrolled diabetes mellitus, may also increase renal blood flow and GFR through tubuloglomerular feedback because glucose, like amino acids, is co-transported with sodium in the proximal tubule.
- *Glucocorticoids* increase renal blood flow and GFR by reducing renal vascular resistance.
- *Fever* increases renal blood flow and GFR by reducing renal vascular resistance.
- *Aging* decreases renal blood flow and GFR mainly because of a reduction in the number of functional nephrons; renal blood flow and GFR decrease about 10 per cent during each decade of life after age 40.

27

Urine Formation by the Kidneys:
II. Tubular Processing of the Glomerular Filtrate

REABSORPTION AND SECRETION BY THE RENAL TUBULES (p. 331)

After the glomerular filtrate enters the renal tubules, it flows sequentially through the *proximal tubules, loops of Henle, distal tubules, collecting tubules,* and *collecting ducts* before it is excreted as urine. Along this course, some substances are reabsorbed from the tubules into the peritubular capillary blood, whereas others are secreted from the blood into the tubules. The urine that is formed and all of the substances in the urine represent the sum of three basic renal processes:

Urinary Excretion = Glomerular Filtration
　　　　　 − Tubular Reabsorption + Tubular Secretion

Tubular Secretion—The Net Movement of Solutes from Peritubular Capillaries into the Tubules

Some substances enter the tubules not only by glomerular filtration but also by secretion from the peritubular capillaries into the tubules via two steps: (1) simple diffusion of the substance from the peritubular capillaries into the renal interstitium and (2) movement of the substance across the tubular epithelium into the lumen through active or passive transport. Substances that are actively secreted into the tubules include *potassium* and *hydrogen ions* and certain *organic acids* and *organic bases.*

Reabsorption of Solutes in Water from the Tubules into the Peritubular Capillaries

For a substance to be reabsorbed, it must first be transported across the renal tubular epithelial membrane into the interstitial fluid and then through the peritubular capillary membrane back into the blood. Solutes can be transported either through the cell membranes *(transcellular route)* by active or passive transport or through the junctional spaces between the cells *(paracellular route)* by passive transport; water is transported through and between the epithelial cells by osmosis. After absorption into the interstitial fluids, water and solutes are transported through the peritubular capillary walls by *ultrafiltration (bulk flow)* that is mediated by hydrostatic and colloid osmotic forces. In contrast to the glomerular capillaries, which filter large amounts of fluid and solutes, the peritubular capillaries have a large reabsorptive force that rapidly moves fluid and solutes from the interstitium into the blood.

Reabsorption rates for different substances are highly variable. Some substances that are filtered, such as glucose and amino acids, are almost completely reabsorbed by the tubules, so that the urinary excretion rate is essentially zero (Table 27–1).

Most of the ions in the plasma, such as sodium, chloride, and bicarbonate, are also highly reabsorbed from the tubules, but their rates of reabsorption and urinary excretion vary depending on the needs of the body. The metabolic waste products, such as urea and creatinine, are poorly reabsorbed and excreted in relatively large amounts. Tubular reabsorption is highly selective, allowing the kidneys to regulate excretion of different substances independent of one another.

Active transport requires energy and can move solutes against an electrochemical gradient. Transport that is directly coupled to an energy source, such as hydrolysis of adenosine triphosphate (ATP), is termed *primary active transport.* A good example is the *sodium-potassium ATPase pump,* which plays a major role in reabsorption of sodium ions in many parts of the nephron. On the basal and lateral sides of the tubular epithelial cells, the *basolateral membrane* has an extensive sodium-potassium ATPase system that hydrolyzes ATP and uses the released energy to transport sodium ions out of the cell into the interstitium. At the same time, potassium is transported from the interstitium to the inside of the cell. This pumping of sodium out of the cell across the basolateral membrane favors passive diffusion of sodium into the cell across the *luminal membrane* (the side that faces the tubular lumen) and passive diffu-

TABLE 27-1 FILTRATION, REABSORPTION, AND EXCRETION RATES OF DIFFERENT SUBSTANCES BY THE KIDNEYS

	Amount Filtered	Amount Reabsorbed	Amount Excreted	% of Filtered Load Reabsorbed
Glucose (gm/day)	180	180	0	100
Bicarbonate (mEq/day)	4,320	4,318	2	>99.9
Sodium (mEq/day)	25,560	25,410	150	99.4
Chloride (mEq/day)	19,440	19,260	180	99.1
Urea (gm/day)	46.8	23.4	23.4	50
Creatinine (gm/day)	1.8	0	1.8	0

sion of potassium out of the cell into the tubular lumen.

In certain parts of the nephron, there are additional mechanisms for moving large amounts of sodium into the cell. In the proximal tubules, there is an extensive *brush border* on the luminal side of the membrane that multiplies the surface by 20-fold. There also are *sodium carrier proteins* that bind sodium ions on the luminal surface of the membrane and release them inside the cell, providing *facilitated diffusion* of sodium through the membrane into the cell. These sodium carrier proteins are also important for secondary active transport of other substances, such as glucose and amino acids.

Secondary active reabsorption of glucose and amino acids occurs through the renal tubular membrane. In secondary active transport, two or more substances interact with a specific membrane protein and are co-transported together across the membrane. As one of the substances (e.g., sodium) diffuses down its electrochemical gradient, the energy that is released is used to drive another substance (e.g., glucose) against its electrochemical gradient. Secondary active transport does not require energy directly from ATP or from other high-energy phosphate sources; rather, the source of the energy is that which is liberated by simultaneous facilitated diffusion of another transported substance down its own electrochemical gradient.

Transport maximum is displayed for actively transported substances. Many of the nutrients, such as glucose and amino acids, are reabsorbed through secondary active transport with sodium. In most instances, reabsorption of these substances displays a *transport maximum*, which refers to the maximum rate of reabsorption. When the filtered load of these substances exceeds the transport maximum, the excess amount is excreted. The *threshold* is the tubular load at which the transport maximum is exceeded in one or more nephrons, resulting in the appearance of that solute in the urine. The threshold usually occurs at a slightly lower tubular load than the transport maximum because not all nephrons have the same transport maximum and some nephrons excrete glucose before others have reached their transport maximum.

Passive water reabsorption by osmosis is coupled mainly to sodium reabsorption. When solutes are transported out of the tubule, through the primary or secondary active transport, their concentrations decrease in the tubule while they increase in the interstitium. This creates a concentration difference that causes osmosis of water in the same direction as that in which the solutes are transported, from the tubular lumen to the interstitium. Some parts of the renal tubule, especially the *proximal tubules, are highly permeable to water,* and reabsorption occurs so rapidly that there is only a small concentration gradient across the membrane. *In the ascending loops of Henle, however, water permeability is always low,* so that almost no water is reabsorbed despite a large osmotic gradient. *In the distal tubules, collecting tubules, and collecting ducts, water permeability depends on the presence or absence of antidiuretic hormone (ADH).* In the presence of ADH, these sections of the renal tubule are highly permeable to water.

Some solutes are reabsorbed by passive diffusion. When sodium, a positive ion, is reabsorbed through the tubular cell, negative ions such as *chloride* also tend to diffuse passively through the paracellular pathway (between the cells). Additional reabsorption of chloride also occurs because of a concentration gradient that develops when water is reabsorbed from the tubule by osmosis, thereby concentrating the chloride ions in the tubular lumen. Noncharged substances, such as *urea*, are also passively reabsorbed from the tubule because osmotic reabsorption of water tends to concentrate these solutes in the tubular lumen, favoring their diffusion into the renal interstitium. Urea and many other waste products do not permeate the tubule nearly as rapidly as water, allowing large amounts of these substances to be excreted in the urine.

REABSORPTION AND SECRETION ALONG DIFFERENT PARTS OF THE NEPHRON (p. 337)

The proximal tubules have a high capacity for reabsorption. Approximately 65 per cent of the filtered load of water, sodium, chloride, potassium, and several other electrolytes is reabsorbed in the proximal tubules. One important function of the proximal tubules, therefore, is to conserve substances that are needed by the body, such as glucose, amino acids, proteins, water, and electrolytes. In contrast, the proximal tubules are relatively impermeable to waste products of the body and reabsorb a much smaller percentage of the filtered load of the substances.

The loop of Henle has three functionally distinct segments—the *descending thin segment, ascending thin segment,* **and** *ascending thick segment.* The loop of Henle dips into the inner part of the kidney, the renal medulla, and plays an important role in allowing the kidney to form concentrated urine. *The descending thin loop of Henle is highly permeable to water,* which is rapidly reabsorbed from the tubular fluid into the hyperosmotic interstitium (osmolarity rises to 1200–1400 mOsm/liter in the inner renal medulla); approximately 20 per cent of the glomerular filtrate volume is reabsorbed in the thin descending loop of Henle, causing the tubular fluid to become hyperosmotic as it moves toward the inner renal medulla.

In the thin and thick segments of the ascending loop of Henle, water permeability is virtually zero, but large amounts of sodium, chloride, and potassium are reabsorbed, causing the tubular fluid to become dilute (hypotonic) as it moves back toward the cortex. At the same time, active transport of sodium chloride out of the thick ascending loop of Henle into the interstitium causes a very high concentration of these ions in the interstitial fluid of the renal medulla. As in the proximal tubule, reabsorption of sodium chloride in the loop of Henle is closely linked to activity of the sodium-potassium ATPase pump in the basolateral membrane. In addition, sodium chloride is rapidly transported across the luminal membrane by a *1-sodium, 2-chloride, 1-potassium co-transporter.* About 25 per cent of the filtered loads of sodium, chloride, and potassium are reabsorbed in the loop of Henle, mostly in the thick ascending limb. Considerable amounts of other ions, such as calcium, bicarbonate, and magnesium, are also reabsorbed in the thick ascending loop of Henle.

The early distal tubule dilutes the tubular fluid. The thick segment of the ascending limb empties into the distal tubule. The first portion of the distal tubule forms part of the *juxtaglomerular complex* that provides feedback control of the glomerular filtration rate (GFR) and blood flow in the same nephron. The next early portion of the distal tubule has many of the same characteristics as the ascending loop of Henle and avidly reabsorbs most of the ions but is virtually impermeable to water and urea. For this reason, it is referred to as the *diluting segment*; it also dilutes the tubular fluid. Fluid leaving this part of the nephron usually has an osmolarity of only about 100 mOsm/liter.

The late distal tubule and cortical collecting tubule are similar. The second half of the distal tubule and the cortical collecting tubule have similar functional characteristics. Anatomically, they are composed of two distinct cell types: (1) the *principal cells*, which absorb sodium and water from the lumen and secrete potassium into the lumen, and (2) the *intercalated cells*, which absorb potassium ions and secrete hydrogen ions into the tubular lumen.

The tubular membranes of both segments are almost completely impermeable to urea, and their permeability to water is controlled by the concentration of ADH. With high levels of ADH, these segments are highly permeable to water. The reabsorption of sodium and secretion of potassium by the principal cells are controlled by the hormone *aldosterone*. The secretion of hydrogen ions by the intercalated cells plays an important role in acid-base regulation of the body fluids (discussed later).

Medullary collecting ducts are final sites. Although the medullary collecting ducts reabsorb less than 10 per cent of the filtered water and sodium, they are the final site for processing of urine and therefore extremely important in determination of the final urine output of water and solutes. Some special characteristics of this tubular segment are as follows:

1. Its permeability to water is controlled by ADH; with high ADH levels, water is rapidly reabsorbed, thereby reducing urine volume and concentrating most solutes in the urine.
2. The medullary collecting duct is highly permeable to urea, allowing some of the urea in the tubule to be absorbed into the medullary interstitium and helping to raise the osmolality of the renal medulla, which contributes to the overall ability of the kidneys to form a concentrated urine.
3. It secretes hydrogen ions against a large concentration gradient, thereby playing a key role in acid-base regulation.

REGULATION OF TUBULAR REABSORPTION (p. 341)

Because it is essential to maintain a precise balance between tubular reabsorption and glomerular filtration, multiple nervous, hormonal, and local control mechanisms regulate the tubular reabsorption rate as well as the GFR. An important feature of tubular reabsorption is that excretion of water and solutes can be independently regulated, especially through hormonal controls.

Glomerulotubular balance represents the ability of the tubule to increase its reabsorption rate in response to a greater tubular load. If the GFR is increased, the absolute rate of tubular reabsorption is increased approximately in proportion to the rise in GFR. Glomerulotubular balance helps to prevent overloading of the more distal parts of the renal tubule when GFR increases; however, glomerulotubular balance does not completely prevent changes in GFR from altering urinary excretion.

Peritubular capillary and renal interstitial fluid physical forces alter tubular reabsorption. As the glomerular filtrate passes through the renal tubules, more than 99 per cent of the water and most of the solutes are reabsorbed—first into the renal interstitium and then into the peritubular capillaries. Of the 125 ml/min of fluid that is normally filtered by the glomerular capillaries, approximately 124 ml/min is reabsorbed into the peritubular capillaries.

Peritubular capillary reabsorption is regulated by hydrostatic and colloid osmotic pressures acting across the capillaries and by the capillary filtration coefficient (K_f), as shown in the following relation:

$$\text{Reabsorption} = K_f(P_c - P_{if} - \pi_c + \pi_{if})$$

where P_c is the peritubular capillary hydrostatic pressure, P_{if} is the interstitial fluid hydrostatic pressure, π_c is the colloid osmotic pressure of the peritubular capillary plasma proteins, and π_{if} is the colloid osmotic pressure of proteins in the renal interstitium. The two primary determinants of peritubular capillary reabsorption that are directly influenced by renal hemodynamic changes are the hydrostatic and colloid osmotic pressures of the peritubular capillaries. The peritubular capillary hydrostatic pressure is in turn influenced by (1) the arterial pressure and (2) the resistance of the afferent and efferent arterioles (Table 27–2).

The peritubular capillary colloid osmotic pressure is influenced by (1) the systemic plasma colloid osmotic pressure and (2) the *filtration fraction*, which is the GFR/renal plasma flow ratio; the higher the fil-

TABLE 27–2 FACTORS THAT CAN INFLUENCE PERITUBULAR CAPILLARY REABSORPTION

↑	$P_c \rightarrow$	↓ Reabsorption
		↓ $R_A \rightarrow$ ↑ P_c
		↓ $R_E \rightarrow$ ↑ P_c
		↑ Arterial Pressure $\rightarrow$ ↑ P_c
↑	$II_c \rightarrow$	↑ Reabsorption
		↑ $II_A \rightarrow$ ↑ II_c
		↑ FF $\rightarrow$ ↑ II_c
↑	$K_f \rightarrow$	↑ Reabsorption

P_c, peritubular capillary hydrostatic pressure; R_A and R_E, afferent and efferent arteriolar resistances, respectively; π_c, peritubular capillary colloid osmotic pressure; π_A, systemic plasma colloid osmotic pressure; FF, filtration fraction.

tration fraction, the greater the fraction of plasma that is filtered through the glomerular capillaries and, consequently, the more concentrated become the proteins in the plasma that remains behind. An increase in filtration fraction therefore tends to raise the peritubular capillary reabsorption rate.

Increased arterial pressure reduces tubular reabsorption. Even small increases in arterial pressure can raise the urinary excretion rates of sodium and water, phenomena that are referred to as *pressure natriuresis* and *pressure diuresis*, respectively. There are two primary mechanisms by which increased arterial pressure raises urinary excretion:

1. Increased arterial pressure causes slight elevations in renal blood flow and GFR; in normal kidneys, GFR and renal blood flow usually change less than 10 per cent between arterial pressures of 75 and 160 mm Hg because of the renal autoregulatory mechanisms discussed previously.
2. Increased arterial pressure raises peritubular capillary hydrostatic pressure, especially in the vasa recta of the renal medulla; this in turn decreases peritubular capillary reabsorption, which increases backleakage of sodium into the tubular lumen and thereby decreases the net sodium and water reabsorption and increases the urine output.
3. Increased arterial pressure also decreases angiotensin II formation, which greatly decreases sodium reabsorption by the renal tubules (discussed later).

Aldosterone increases sodium reabsorption and potassium secretion. Aldosterone, which is secreted by the adrenal cortex, acts mainly on the *principal cells* of the cortical collecting tubule to stimulate the sodium-potassium ATPase pump, which increases sodium reabsorption from the tubule and potassium secretion into the tubule. In the absence of aldosterone, as occurs with destruction or malfunction of the adrenals *(Addison's disease)*, there is marked loss of sodium from the body and accumulation of potassium. Conversely, excess aldosterone secretion, as occurs in patients with adrenal tumors *(Conn's syndrome)*, is associated with sodium retention and potassium depletion.

Angiotensin II increases sodium and water reabsorption. Angiotensin II, which may be the most powerful sodium-retaining hormone of the body, increases sodium and water reabsorption through three main effects:

1. Angiotensin II stimulates aldosterone secretion, which in turn increases sodium reabsorption.
2. Angiotensin II constricts the efferent arterioles which reduces peritubular capillary hydrostatic pressure and increases filtration fraction by reducing renal blood flow; both of these changes tend to increase the reabsorptive force at the peritubular capillaries and tubular reabsorption of sodium and water.
3. Angiotensin II directly stimulates sodium reabsorption, especially in the proximal tubules.

These multiple actions of angiotensin II cause marked sodium and water retention by the kidneys in circumstances associated with low blood pressure, low extracellular fluid volume, or both, such as during hemorrhage or loss of salt and water from the body fluids.

ADH increases water reabsorption. ADH, which is secreted by the posterior pituitary gland, increases water permeability of the distal tubules, collecting tubules, and collecting ducts, allowing these portions of the nephron to avidly reabsorb water and form a highly concentrated urine. These effects help the body to conserve water during circumstances such as dehydration, which greatly stimulates ADH secretion. In the absence of ADH, these portions of the nephrons are virtually impermeable to water, causing the kidneys to excrete large amounts of dilute urine.

Atrial natriuretic peptide decreases sodium and water reabsorption. Specific cells of the cardiac atria, when distended as a result of plasma volume expansion, secrete a peptide called *atrial natriuretic peptide*. Greater levels of this peptide inhibit reabsorption of sodium and water by the renal tubules, thereby increasing the excretion of sodium and water.

Parathyroid hormone increases calcium reabsorption and decreases phosphate reabsorption. Parathyroid hormone is one of the most important calcium- and phosphate-regulating hormones of the body. Its principal action in the kidneys is to increase reabsorption of calcium, especially in the thick ascending loop of Henle and distal tubule. Another action of parathyroid hormone is inhibition of phosphate reabsorption by the proximal tubule.

Sympathetic nervous system activation increases sodium reabsorption. Stimulation of the sympathetic nervous system constricts the afferent and efferent arterioles, thereby reducing GFR. At the same time, sympathetic activation directly increases sodium reabsorption in the proximal tubule and ascending loop of Henle and stimulates renin release and angiotensin II formation.

USE OF CLEARANCE METHODS TO QUANTIFY KIDNEY FUNCTION (p. 345)

Renal clearance is the volume of plasma that is completely cleared of a substance each minute. For given substance X, renal clearance is defined as the ratio of the excretion rate of substance X to its concentration in the plasma, as shown by the following relation:

$$C_X = (U_X \times V)/P_X$$

where C_X is renal clearance in ml/min, $U_X \times V$ is the excretion rate of substance X (U_X is the concentration of X in the urine, and V is urine flow rate in ml/min), and P_X is plasma concentration of X. Renal clearances can be used to quantify several aspects of kidney functions, including the rates of glomerular filtration, tubular reabsorption, and tubular secretion of different substances.

Renal clearance of creatinine or inulin can be used to estimate GFR. *Creatinine*, a by-product of skeletal muscle metabolism, is filtered at the glomerulus but is not reabsorbed or secreted appreciably by the tubules; therefore, the entire 125 milliliters of plasma that filters into the tubules each minute (GFR) is cleared of creatinine. This means that creatinine clearance is approximately equal to the GFR. For this reason, creatinine clearance is often used as an index of GFR. An even more accurate measure of GFR is the clearance of *inulin*, a polysaccharide that is not reabsorbed or secreted by the renal tubules.

Renal clearance of para-aminohippuric acid (PAH) can be used to estimate renal plasma flow. Some substances, such as PAH, are filtered and not reabsorbed by the tubules but are secreted into the tubules; therefore, the renal clearance of these substances is greater than GFR. In fact, about 90 per cent of the plasma flowing through the kidney is completely cleared of PAH, and renal clearance of PAH (C_{PAH}) can be used to estimate the renal plasma flow, as follows:

$$C_{PAH} = (U_{PAH} \times V)/P_{PAH} \cong \text{Renal Plasma Flow}$$

where U_{PAH} and P_{PAH} are urine and plasma concentrations of PAH, respectively, and V is the urine flow rate.

Filtration fraction is the ratio of GFR to renal plasma flow. If renal plasma flow is 650 ml/min and GFR is 125 ml/min, the filtration fraction is 125/650, or 0.19.

Tubular reabsorption or secretion can be calculated from renal clearances. For substances that are completely reabsorbed from the tubules (e.g., amino acids, glucose), the clearance rate is zero because the urinary secretion rate is zero. For substances that are highly reabsorbed (e.g., sodium), the clearance rate is usually less than 1 per cent of the GFR, or less than 1 ml/min. In general, waste products of metabolism, such as urea, are poorly reabsorbed and have relatively high clearance rates.

Tubular reabsorption rate is calculated as the difference between the rate of filtration of the substance (GFR × P_X) and the urinary excretion rate (U_X × V), as follows:

$$\text{Reabsorption}_X = (GFR \times P_X) - (U_X \times V)$$

If the excretion rate of a substance is greater than the filtered load, then the rate at which it appears in the urine represents the sum of the rate of glomerular filtration plus tubular secretion; secretion rate is therefore the difference between the rate of urinary excretion of a substance and the rate at which it is filtered, as follows:

$$\text{Secretion}_X = (U_X \times V) - (GFR \times P_X)$$

28

Regulation of Extracellular Fluid Osmolarity and Sodium Concentration

To function properly, cells must be bathed in extracellular fluid with a relatively constant concentration of electrolytes and other solutes. The total concentration of solutes in the extracellular fluid (the osmolarity) is determined by the amount of solute divided by the volume of the extracellular fluid. The most abundant solutes in the extracellular fluid are sodium and chloride; to a large extent, extracellular fluid osmolarity is determined by the amounts of extracellular sodium chloride and water, which are determined by the balances between intake and excretion of these substances.

In this chapter, we discuss the mechanisms that permit the kidney to excrete either dilute or concentrated urine, and therefore to regulate extracellular fluid sodium concentration and osmolarity, and the mechanisms that govern fluid intake.

THE KIDNEY EXCRETES EXCESS WATER BY FORMING A DILUTE URINE (p. 349)

When there is excess water in the body, the kidneys can excrete urine with an osmolarity as low as 50 mOsm/liter; conversely, when there is a deficit of water, the kidneys can excrete urine with a concentration as high as 1200 to 1400 mOsm/liter. Equally important, the kidneys can excrete a large volume of dilute urine or a small volume of concentrated urine without a major change in the rate of solute excretion.

Antidiuretic hormone controls urine concentration. When the osmolarity of the body fluids increases above normal, the posterior pituitary gland secretes more antidiuretic hormone (ADH), which increases the permeability of the distal tubules and collecting ducts to water, causing large amounts of water to be reabsorbed and decreasing urine volume without marked alteration of renal solute excretion.

When there is excess water in the body and extracellular fluid osmolarity is reduced, the secretion of ADH becomes decreased, thereby reducing the permeability of the distal tubules and collecting ducts to water and causing large amounts of dilute urine to be excreted.

Excretion of dilute urine is caused by decreased ADH and decreased water reabsorption. When the glomerular filtrate is formed, its osmolarity is about the same as that of plasma (300 mOsm/liter). As fluid flows through the proximal tubules, solutes and water are reabsorbed in equal proportions so that little change in osmolarity occurs. As fluid flows down the descending loop of Henle, water is reabsorbed and tubular fluid reaches equilibrium with the surrounding interstitial fluid, which is very hypertonic (osmolarity as high as 1200–1400 mOsm/liter). In the ascending limb of the loop of Henle, especially the thick segment, sodium, potassium, and chloride are avidly reabsorbed, but because this part of the tubule is impermeable to water, even in the presence of ADH, the tubular fluid becomes more dilute as it flows into the early distal tubule. *Regardless of whether ADH is present, fluid leaving the early distal tubule is hypo-osmotic, with an osmolarity of only about one third that of plasma.*

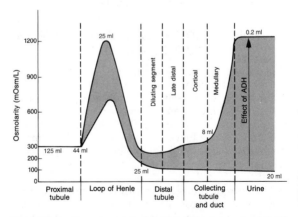

Figure 28–1 Changes in osmolarity of the tubular fluid as it passes through the different tubular segments in the presence of high levels of ADH and in the absence of ADH. (Numerical values indicate the approximate volumes in milliliters per minute or osmolarities in milliosmoles per liter of fluid flowing along the different tubular segments.)

As the dilute fluid of the early distal tubule passes into the late distal, convoluted tubule, cortical collecting ducts, and medullary collecting ducts, there is additional reabsorption of sodium chloride. In the absence of ADH, the tubule is relatively impermeable to water, and additional reabsorption of solutes causes the tubular fluid to become even more dilute, decreasing its osmolarity to as low as 50 mOsm/liter. This failure to reabsorb water and continued reabsorption of solutes lead to a large volume of dilute urine (Fig. 28–1).

THE KIDNEY CONSERVES WATER BY SECRETING A CONCENTRATED URINE (p. 350)

When there is a water deficit in the body and plasma osmolarity and ADH levels are elevated, the kidneys form a concentrated urine by continuing to excrete solutes while increasing water reabsorption and decreasing urine volume. *The two basic requirements for forming a concentrated urine are*

- *A high level of ADH*, which allows the distal tubules and collecting tubules to avidly reabsorb water
- *A high osmolarity of the renal medullary interstitial fluid.*

Tubular fluid flowing out of the loop of Henle is normally dilute, with an osmolarity of only about 100 mOsm/liter. The medullary interstitium outside the collecting tubules in the renal medulla is normally very concentrated with sodium and urea due to the operation of the *countercurrent multiplier*, which depends on the special permeability characteristics of the loop of Henle. As fluid flows into the distal tubules and finally into the collecting tubules and ducts, water is reabsorbed until tubular fluid osmolarity equilibrates with the surrounding medullary interstitial fluid osmolarity. This process leads to a highly concentrated urine with an osmolarity of 1200 to 1400 mOsm/liter when high ADH levels are present (see Fig. 28–1).

The countercurrent multiplier causes high osmolarity in the renal medulla. For the renal medulla to increase its osmolarity to a range of 1200 to 1400 mOsm/liter, the medullary interstitium must accumulate solutes in great excess of water. Once this has occurred, the high osmolarity is maintained by a balanced inflow and outflow of solutes and water in the medulla.

The major factors that contribute to the build-up of solute concentration in the renal medulla are

- Active transport of sodium ions and co-transport of potassium, chloride, and other ions out of the thick ascending limb of the loop of Henle into the medullary interstitium
- Active transport of ions from the collecting ducts into the medullary interstitium
- Diffusion of large amounts of urea from the inner medullary collecting ducts into the medullary interstitium
- Diffusion of only small amounts of water from the medullary collecting tubules into the interstitium, far less than the reabsorption of solutes into the medullary interstitium, and virtually no water diffusion into the medulla from the ascending loop of Henle.

Countercurrent exchange in the vasa recta preserves hyperosmolarity of the renal medulla. There are two special features of the *vasa recta*, which carry blood flow to the renal medulla, that help preserve the high solute concentrations:

1. Vasa recta blood flow is low, accounting for only 1 to 2 per cent of the total renal blood flow. This sluggish flow is sufficient to supply the metabolic needs of the tissues and helps to minimize solute loss from the medullary interstitium.
2. *The vasa recta serve as countercurrent exchangers*, minimizing washout of solutes from the medullary interstitium. This countercurrent exchange feature is due to the U shape of the vasa recta capillaries.

As blood descends into the medulla, it becomes progressively more concentrated because the vasa recta capillaries are highly permeable to water and solutes. However, as blood ascends back toward the cortex, it becomes progressively less concentrated as solutes diffuse back out into the medullary interstitium and water moves into the vasa recta. Although there is a large amount of fluid and solute exchange across the vasa recta, there is little net loss of solutes from the interstitial fluid.

QUANTIFYING RENAL URINE CONCENTRATION AND DILUTION WITH "FREE-WATER" AND OSMOLAR CLEARANCES (p. 358)

When the urine is dilute, water is excreted in excess of solutes. Conversely, when the urine is concentrated, solutes are excreted in excess of water. The rate at which solutes are cleared from the blood can be expressed as the *osmolar clearance* (C_{osm}); this is a measurement of the volume of plasma cleared of solutes each minute:

$$C_{osm} = (U_{osm} \times V)/P_{osm}$$

where U_{osm} is urine osmolarity, V is urine flow rate, and P_{osm} is plasma osmolarity.

The relative rates at which solutes in water are excreted can be assessed using the concept of *free-water clearance* (C_{H_2O}), which is defined as the difference between water excretion (urine flow rate) and osmolar clearance:

$$C_{H_2O} = V - C_{osm} = V - (U_{osm} \times V)/P_{osm}$$

The rate of free-water clearance is the rate at which solute-free water is excreted by the kidneys. *When free-water clearance is positive, excess water is being excreted by the kidneys; when free-water clearance is negative, excess solutes are being removed from the blood by the kidneys and water is being conserved.*

Disorders of Urinary Concentrating Ability

An impairment in the ability of the kidneys to concentrate the urine can occur with one or more of the following abnormalities:

- *Decreased secretion of the ADH,* which is referred to as *"central" diabetes insipidus.* This results in the inability to produce or release ADH from the posterior pituitary due to head injuries, infections, or congenital abnormalities.
- *Inability of the kidneys to respond to ADH,* a condition called *"nephrogenic" diabetes insipidus.* This abnormality can be caused by failure of the countercurrent mechanism to form a hyperosmotic renal medullary interstitium or failure of the distal and collecting tubules and collecting ducts to respond

to ADH. Many types of *renal diseases can impair the concentrating mechanism*, especially those that damage the renal medulla. In addition, *impairment of the function of the loop of Henle*, as occurs with diuretics that inhibit electrolyte reabsorption in that segment, can compromise urine concentrating ability. Marked *increases in renal medullary blood flow* can "wash out" some of the solutes in the renal medulla and reduce maximal concentrating ability. No matter how much ADH is present, maximal urine concentration is limited by the degree of hyperosmolarity of the medullary interstitium.

There also are certain drugs, such as lithium (to treat manic-depressive disorders) and tetracyclines (antibiotics to treat infections), that can impair the ability of the distal nephron segments to respond to ADH.

CONTROL OF EXTRACELLULAR FLUID OSMOLARITY AND SODIUM CONCENTRATION (p. 359)

The regulation of extracellular fluid osmolarity and sodium concentration are closely linked because sodium is the most abundant cation in the extracellular compartment. Plasma sodium concentration is normally regulated within close limits of 140 to 145 mEq/liter, with an average concentration of about 142 mEq/liter. Osmolarity averages about 300 mOsm/liter and seldom changes more than 2 to 3 per cent.

Although multiple mechanisms control the amount of sodium and water excretion by the kidneys, two primary systems are particularly involved in regulation of the concentration of sodium and osmolarity of extracellular fluid: (1) the osmoreceptor-ADH feedback system and (2) the thirst mechanism.

Osmoreceptor-ADH Feedback System

When osmolarity (plasma sodium concentration) increases above normal, the osmoreceptor-ADH feedback system operates as follows:

- Increased extracellular fluid osmolarity stimulates the osmoreceptor cells in the anterior hypothalamus, near the supraoptic nuclei, to send signals that are relayed to the posterior pituitary gland.

- Action potentials conducted to the posterior pituitary stimulate release of ADH, which is stored in secretory granules in the nerve endings.
- ADH, which is transported in the blood to the kidney, increases water permeability of the late distal tubules, cortical collecting tubules, and medullary collecting ducts.
- Increased water permeability in the distal nephron segments causes increased water reabsorption and excretion of a small volume of concentrated urine. This causes dilution of solutes in the extracellular fluid, thereby correcting the initial excessively concentrated extracellular fluid.

The opposite sequence of events occurs when the extracellular fluid becomes too dilute (hypo-osmotic).

ADH is synthesized in the supraoptic and paraventricular nuclei of the hypothalamus and released from the posterior pituitary. The hypothalamus contains two types of large neurons that synthesize ADH: (1) about five sixths of the ADH is synthesized in the *supraoptic nuclei*, and (2) about one sixth of the ADH is synthesized in the *paraventricular nuclei*. Both of these nuclei have axonal extensions to the posterior pituitary. Once ADH is synthesized, it is transported down the axons or the neurons that terminate in the posterior pituitary.

Secretion of ADH in response to an osmotic stimulus is rapid, so that plasma ADH levels can increase severalfold within minutes, thereby providing a rapid means of altering renal excretion of water.

Cardiovascular reflex stimulation of ADH release by decreased arterial pressure, decreased blood volume, or both. ADH release is also controlled by cardiovascular reflexes, including (1) the *arterial baroreceptor reflex* and (2) the *cardiopulmonary reflexes*, both of which were discussed in Chapter 18. Afferent stimuli carried by the *vagus* and *glossopharyngeal nerves* synapse with nuclei of the tractus solitarius, and projections from these nuclei relay signals to the hypothalamic nuclei that control ADH synthesis and secretion. Whenever blood pressure and blood volume are reduced, such as occurs during hemorrhage, increased ADH secretion through these reflex pathways causes increased fluid reabsorption by the

TABLE 28–1 REGULATION OF ADH SECRETION	
Increase ADH	**Decrease ADH**
↑ Plasma osmolarity	↓ Plasma osmolarity
↓ Blood volume	↑ Blood volume
↓ Blood pressure	↑ Blood pressure
Nausea	
Hypoxia	
Drugs:	Drugs:
Morphine	Alcohol
Nicotine	Clonidine (antihypertensive drug)
Cyclophosphamide	Haloperidol (dopamine blocker)

kidneys, helping to restore blood pressure and blood volume toward normal levels.

Although the usual day-to-day regulation of ADH secretion is effected mainly by changes in plasma osmolarity, large changes in blood volume, as occur during hemorrhage, also elicit marked increases in ADH levels.

Other stimuli cause ADH secretion. A summary of the various factors that can increase or decrease ADH secretion is shown in Table 28–1.

ROLE OF THIRST IN CONTROLLING EXTRACELLULAR FLUID OSMOLARITY AND SODIUM CONCENTRATION
(p. 361)

The kidneys minimize fluid loss through the osmoreceptor-ADH feedback system; however, adequate fluid intake is necessary to counterbalance fluid losses that normally occur through sweating and breathing and through the intestinal tract. Fluid intake is regulated by the thirst mechanism, which together with the osmoreceptor-ADH mechanism maintains precise control of extracellular fluid osmolarity and sodium concentration.

Many of the stimuli involved in controlling ADH secretion also increase thirst, which is defined as the conscious desire for water (see Table 28–2).

Two of the most important stimuli for thirst include increased extracellular fluid osmolarity and decreased extracellular fluid volume and arterial pressure. A third important stimulus for thirst is *angiotensin II*. Because angiotensin II is also stimu-

TABLE 28–2 CONTROL OF THIRST	
Increase Thirst	**Decrease Thirst**
↑ Osmolarity	↓ Osmolarity
↓ Blood volume	↑ Blood volume
↓ Blood pressure	↑ Blood pressure
↑ Angiotensin II	↓ Angiotensin II
Dryness of mouth	Gastric distention

lated by low blood volume and low blood pressure, its effect on thirst as well as its actions on the kidneys to decrease fluid excretion help to restore blood volume and blood pressure toward normal.

Other factors that influence water intake include dryness of the mouth and mucous membranes of the esophagus and the degree of gastric distention. These stimuli to the gastrointestinal tract are relatively short-lived, and the desire to drink is completely satisfied only when plasma osmolarity, blood volume, or both return to normal.

The ADH and thirst mechanisms operate together to control extracellular osmolarity. In the normal person, these two mechanisms work in parallel to precisely regulate extracellular fluid osmolarity and sodium concentration, despite the constant challenge of dehydration. Even with additional challenges, such as a high salt intake, these feedback mechanisms are able to keep plasma osmolarity reasonably constant. When either the ADH or thirst mechanism fails, the other ordinarily can still keep extracellular osmolarity and sodium concentration relatively constant, as long as there is sufficient fluid intake to balance the daily obligatory urine volume and water losses caused by respiration, sweating, and gastrointestinal losses. If both the ADH and thirst mechanisms fail simultaneously, however, either sodium concentration or osmolarity can be adequately controlled. In the absence of these mechanisms, there are no other feedback mechanisms in the body capable of precisely regulating plasma osmolarity.

Angiotensin II and aldosterone do not normally play a major role in controlling extracellular osmolarity and sodium concentration. As discussed in Chapter 27, angiotensin II and aldosterone are the two most important hormonal regulators of renal tubular sodium reabsorption. Despite the importance of these hormones in regulating sodium excretion, they do not have a major effect on plasma sodium concentration for two reasons:

1. Angiotensin II and aldosterone increase both sodium and water reabsorption by the renal tubules, leading to greater extracellular fluid volume and sodium *quantity* but little change in sodium *concentration*.
2. As long as the ADH and thirst mechanisms are functional, any tendency toward increased plasma sodium concentration is compensated for by increased water intake or increased ADH secretion, which tends to dilute the extracellular fluid back toward normal.

Under extreme conditions, associated with the complete loss of aldosterone secretion due to adrenalectomy or Addison's disease, there is a tremendous loss of sodium by the kidneys, which can lead decreased plasma sodium concentration; one of the reasons is that large losses of sodium are accompanied by severe volume depletion and decreased blood pressure, which can activate the thirst mechanism and lead to further dilution of plasma sodium concentration even though the increased water intake helps to minimize decreased body fluid volumes. There are extreme conditions during which plasma sodium concentration may change significantly, even with a functional ADH-thirst mechanism. Even so, the ADH-thirst mechanism is by far the most powerful feedback system in the body for controlling extracellular fluid osmolarity and sodium concentration.

29

Integration of Renal Mechanisms for Control of Blood Volume and Extracellular Fluid Volume; and Renal Regulation of Potassium, Calcium, Phosphate, and Magnesium

CONTROL OF EXTRACELLULAR FLUID VOLUME (p. 367)

In discussing control of extracellular fluid volume, we must consider factors that regulate the amount of sodium chloride in the extracellular fluid, because sodium chloride content of the extracellular fluid usually parallels extracellular fluid volume, provided the antidiuretic hormone (ADH)-thirst mechanisms are operative. In most cases, the burden of extracellular volume regulation is placed on the kidneys, which must adapt their excretion to match varying intakes of salt and water.

Sodium excretion is precisely matched with sodium intake under steady-state conditions. An important fact in considering overall control of sodium excretion—or excretion of any electrolyte—is that under steady-state conditions, a person must excrete almost precisely the amount of sodium ingested. Even with disturbances that cause major changes in renal excretion of sodium, the balance between intake and excretion is usually restored within a few days.

Sodium excretion is controlled by altering glomerular filtration or tubular reabsorption rates. The kidney alters sodium and water excretion by changing the rate of filtration, the rate of tubular reabsorption, or both, as follows:

Excretion
= Glomerular Filtration − Tubular Reabsorption

As discussed previously, glomerular filtration and tubular reabsorption are both regulated by multiple factors, including hormones, sympathetic activity, and arterial pressure. Normally, glomerular filtration rate (GFR) is about 180 liters/day, tubular reabsorption is 178.5 liters/day, and urine excretion is 1.5 liters/day. Small changes in either GFR or tubular reabsorption have the potential to cause large changes in renal excretion.

Tubular reabsorption and GFR are usually regulated precisely, so that excretion by the kidneys can be exactly matched to intake of water and electrolytes. Even with disturbances that alter GFR and tubular reabsorption, however, changes in urinary excretion are minimized by various buffering mechanisms. Two intrarenal buffering mechanisms are (1) *glomerulotubular balance*, which allows the renal tubules to raise their tubular reabsorption rates in response to increased GFR and filtered sodium load, and (2) *macula densa feedback*, in which increased sodium chloride delivery to the distal tubule, due to increased GFR or decreased proximal or loop of Henle sodium reabsorption, causes afferent arteriolar constriction and decreased GFR.

Because neither of these two intrarenal feedback mechanisms operates perfectly to restore urine output to normal, changes in GFR or tubular reabsorption can lead to significant changes in sodium and water excretion. When this happens, *systemic feedback mechanisms* come into play, such as changes in blood pressure and changes in various hormones, that eventually return sodium excretion to equal intake.

IMPORTANCE OF PRESSURE NATRIURESIS AND PRESSURE DIURESIS IN MAINTAINING BODY SODIUM AND FLUID BALANCE (p. 368)

One of the most powerful mechanisms for controlling blood volume and extracellular fluid volume and for maintaining sodium and fluid balance is the effect of blood pressure on sodium and water excretion (*pressure natriuresis* and *pressure diuresis*, respectively). As discussed in Chapter 19, this feedback between the kidneys and circulation also plays a dominant role in long-term blood pressure regulation.

Pressure diuresis refers to the effect of increased arterial pressure to raise urinary volume excretion, whereas pressure natriuresis refers to the increased sodium excretion that occurs with increased arterial pressure. Because pressure diuresis and natriuresis usually occur in parallel, we often refer to these mechanisms simply as *pressure natriuresis*.

Pressure natriuresis is a key component of the renal–body fluid feedback mechanism. During changes in sodium and fluid intake, this mechanism helps maintain fluid balance and minimizes changes in blood volume, extracellular fluid volume, and arterial pressure as follows:

1. An increase in fluid intake (assuming that sodium accompanies the fluid) above the level of urine output causes a temporary accumulation of fluid in the body and a small increase in blood volume and extracellular fluid volume.
2. An increase in blood volume raises mean circulatory filling pressure and cardiac output.
3. An increase in cardiac output raises arterial pressure, which raises urine output by way of pressure natriuresis. The steepness of the normal pressure natriuresis relation ensures that only a slight increase in blood pressure is required to raise urinary excretion severalfold.
4. An increase in fluid excretion balances the greater intake, and further accumulation of fluid is prevented.

The renal–body fluid feedback mechanism prevents continuous accumulation of salt and water in the body during increased salt and water intake. As long as kidney function is normal and pressure natriuresis is operating effectively, large increases in salt and water intake can be accommodated with only slight increases in blood volume, extracellular fluid volume, and arterial pressure. The opposite sequence of events occurs when fluid intake falls below normal.

DISTRIBUTION OF EXTRACELLULAR FLUID BETWEEN THE INTERSTITIAL SPACES AND VASCULAR SYSTEM (p. 370)

Ingested fluid and salt initially enter the blood but rapidly become distributed between the interstitial spaces and the plasma. Blood volume and extracellular fluid volume usually are controlled simultaneously and in parallel. There are conditions, however, that can markedly alter the distribution of extracellular fluid between the interstitial spaces and blood.

As discussed in Chapter 25, the principal factors that can cause loss of fluid from the plasma into the interstitial spaces (edema) include (1) increased capillary hydrostatic pressure, (2) decreased plasma colloid osmotic pressure, (3) increased permeability of the capillaries, and (4) obstruction of the lymphatic vessels.

NERVOUS AND HORMONAL FACTORS INCREASE THE EFFECTIVENESS OF RENAL–BODY FLUID FEEDBACK CONTROL (p. 370)

Nervous and hormonal mechanisms act in concert with pressure natriuresis to minimize changes in blood volume, extracellular fluid volume, and arterial pressure that occur in response to day-to-day challenges. Abnormalities of kidney function or of nervous and hormonal factors that influence the kidneys can, however, lead to serious changes in blood pressure and body fluid volumes (discussed later).

Sympathetic nervous system control of renal excretion involves the arterial baroreceptor and low-pressure stretch receptor reflexes. The kidneys receive extensive sympathetic innervation, and under some conditions, changes in sympathetic activity can alter renal sodium and water excretion and extracellular fluid volume. For example, when blood volume is reduced by hemorrhage, reflex activation of the sympathetic nervous system occurs because of decreased pressures in the pulmonary blood vessels and other low-pressure regions of the thorax and because of low arterial pressure. The increased sympathetic activity in turn has several effects to reduce sodium and water excretion: (1) renal vasoconstriction, which decreases GFR; (2) increased tubular reabsorption of salt and water; and (3) stimulation of renin release and increased formation of angiotensin II and aldosterone, both of which further elevate tubular reabsorption. All of these mechanisms together play an important role in the rapid restitution of the blood volume that occurs during acute conditions associated with reduced blood volume, low arterial pressure, or both.

Reflex decreases in renal sympathetic activity may contribute to rapid elimination of excess fluid in the circulation after ingestion of a meal that contained large amounts of salt and water.

Angiotensin II is a powerful controller of renal excretion. When sodium intake is increased above normal, renin secretion decreases and causes reduced angiotensin II formation, which has several effects on the kidney to raise tubular sodium reabsorption (see Chapter 27). Conversely, when sodium intake is reduced, increased levels of angiotensin cause sodium and water retention and oppose decreases in arterial pressure that would otherwise occur. Changes in activity of the renin-angiotensin system act as powerful amplifiers of the pressure natriuresis mechanism for maintaining stable blood pressures and body fluid volumes.

Although angiotensin II is one of the most powerful sodium- and water-retaining hormones in the body, *neither a decrease nor an increase in circulating angiotensin II has a large effect on extracellular fluid volume or blood volume in persons with an otherwise normal cardiovascular system.* The reason for this is that with large increases in angiotensin II levels, as occurs with a renin-secreting tumor in the kidney, there is only a transient sodium and water retention that raises arterial pressure; this quickly raises kidney output of sodium and water, thereby overcoming the sodium-retaining effects of angiotensin II and re-establishing balance between intake and output of sodium at a higher arterial pressure.

Conversely, *blockade of angiotensin II formation* with drugs, such as converting enzyme inhibitors, greatly increases the ability of the kidneys to excrete salt and water but does not cause a major change in extracellular fluid volume. After blockade of angiotensin II formation, there is a transient increase in sodium and water excretion, but this reduces arterial pressure, which helps to re-establish sodium balance. This effect of angiotensin II blockers has proved to be important in lowering blood pressure in hypertensive patients.

Aldosterone has a major role in controlling renal sodium excretion. The function of aldosterone in regulating sodium balance is closely related to that described for angiotensin II; with decreased sodium intake, the increased angiotensin II levels stimulate aldosterone secretion, which contributes to decreased urinary sodium excretion and the maintenance of sodium balance. Conversely, with high sodium intake, suppression of aldosterone formation decreases tubular sodium reabsorption, allowing the kidneys to secrete large amounts of sodium. Changes in aldosterone formation also aid the pressure natriuresis mechanism in maintaining sodium balance during variations in sodium intake.

However, *when there is excess aldosterone formation, as occurs in patients with tumors of the adrenal gland, the increased sodium reabsorption and decreased sodium excretion usually last only a few days* and extracellular fluid volume rises by only about 10 to 15 per cent, causing increased arterial pressure. When the arterial pressure rises sufficiently, the kidneys "escape" from the sodium and water retention (because of pressure natriuresis) and thereafter excrete amounts of sodium equal to the daily intake, despite continued high levels of aldosterone.

ADH controls renal water excretion. As explained previously, ADH plays an important role in allowing the kidneys to form a small volume of concentrated urine while excreting normal amounts of sodium. This effect is especially important during water deprivation. Conversely, when there is excess extracellular fluid volume, decreased ADH levels reduce reabsorption of water by the kidneys and help rid the body of excess volume.

Excessive levels of ADH, however, rarely cause large increases in arterial pressure or extracellular fluid volume. Infusion of large amounts of ADH into animals initially increases extracellular fluid volume by only 10 to 15 per cent. As the arterial pressure rises in response to this increased volume, much of the excess volume is excreted because of pressure diuresis; after several weeks, blood volume and extracellular fluid volume are elevated by no more than 5 to 10 per cent, and the arterial pressure is elevated by less than 10 mm Hg. High levels of ADH do not cause major increases in body fluid volume or arterial pressure, although *high ADH levels can cause severe reductions in extracellular sodium ion concentration*.

INTEGRATED RESPONSES TO CHANGES IN SODIUM INTAKE (p. 373)

The integration of different control systems that regulate sodium and fluid excretion can be summarized by examining the homeostatic responses to increases in dietary sodium intake. As sodium intake is increased, sodium output initially lags behind intake. This causes a slight increase in cumulative sodium balance and a slight increase in extracellular fluid volume. It is mainly the small increase in extracellular fluid volume that triggers various mechanisms in the body to raise the amount of sodium excretion. These mechanisms are as follows:

- *Activation of low-pressure receptor reflexes* that originate from the stretch receptors of the right atrium and pulmonary blood vessels; these reflexes inhibit sympathetic activity and angiotensin II formation, both of which tend to decrease tubular sodium reabsorption.
- *Increased secretion from the cardiac atria of atrial natriuretic peptide* (ANP), which reduces renal tubular sodium reabsorption
- *A small increase in arterial pressure,* which by itself promotes sodium excretion through *pressure natriuresis*
- *Suppression of angiotensin II formation,* caused by increased arterial pressure and extracellular volume expansion, decreases tubular sodium reabsorption by eliminating the normal effect of angiotensin II to raise sodium reabsorption. In addition, decreased angiotensin II reduces aldosterone secretion, further reducing sodium reabsorption.

The combined activation of natriuretic systems, such as ANP, and suppression of sodium- and water-retaining systems lead to increased excretion of sodium when sodium intake is raised. The opposite changes take place when sodium intake is reduced below normal levels.

CONDITIONS THAT CAUSE LARGE INCREASES IN BLOOD VOLUME AND EXTRACELLULAR FLUID VOLUME (p. 374)

Despite the powerful regulatory mechanisms that maintain blood volume and extracellular fluid volume at reasonably constant levels, there are abnormal conditions that can cause large increases in both of these variables. Almost all of these conditions result from circulatory abnormalities, including the following:

- *Heart diseases*—In congestive heart failure, blood volume may increase by 10 to 15 per cent and extracellular fluid volume sometimes increases by 200 per cent or more. The fluid retention by the kidneys helps to return the arterial pressure and cardiac output toward normal if the heart failure is not too severe. If the heart is greatly weakened, however, the arterial pressure will not be able to increase sufficiently to restore urine output to normal. When this occurs, the kidneys continue to retain volume until the person develops severe circulatory congestion and eventually dies of edema, especially pulmonary edema.

- *Increased capacity of the circulation*—Any condition that increases vascular capacity will also cause the blood volume to increase and fill this extra capacity. Examples of conditions associated with increased vascular capacity include *pregnancy* (due to increased vascular capacity of the uterus, placenta, and other enlarged organs) and *varicose veins*, which in severe cases may hold as much as an extra liter of blood.

CONDITIONS THAT CAUSE LARGE INCREASES IN EXTRACELLULAR FLUID VOLUME BUT WITH NORMAL BLOOD VOLUME (p. 374)

There are several pathophysiological conditions in which extracellular fluid volume becomes markedly increased but blood volume remains normal or even slightly decreased. These conditions are usually initiated by leakage of fluid and protein into the interstitium, which tends to decrease the blood volume. The kidneys' response to these conditions is similar to the response after hemorrhage—the kidneys retain salt and water in an attempt to restore blood volume toward normal. Two examples of this are as follows:

- *Nephrotic syndrome*, characterized by a loss of plasma proteins in the urine, reduces the plasma colloid osmotic pressure and causes the capillaries throughout the body to filter large amounts of fluid; this in turn causes edema and decreased plasma volume.
- *Liver cirrhosis*, characterized by decreased synthesis of plasma proteins by the liver; a similar sequence of events occurs in cirrhosis of the liver as in nephrotic syndrome, except that in liver cirrhosis the decreased plasma protein concentration results from destruction of the liver cells, rendering them unable to synthesize enough plasma proteins. Cirrhosis is also associated with fibrous tissue in the liver structures, which greatly impedes the flow of portal blood through the liver. This raises capillary pressure throughout the portal circulation and contributes to leakage of fluid and proteins into the peritoneal cavity, a condition called *ascites*.

REGULATION OF POTASSIUM EXCRETION AND POTASSIUM CONCENTRATION IN THE EXTRACELLULAR FLUID (p. 375)

Extracellular fluid potassium concentration normally is regulated precisely at about 4.2 mEq/liter, seldom rising or falling more than ±0.3 mEq/liter. A special difficulty in regulating potassium concentration is the fact that about 98 per cent of the total body potassium is contained in the cells and only 2 per cent is contained in the extracellular fluid. Failure to rapidly rid the extracellular fluid of potassium ingested each day could result in life-threatening *hyperkalemia* (increased plasma potassium concentration). A small loss of potassium from the extracellular fluid could cause severe *hypokalemia* in the absence of rapid compensatory responses.

Potassium excretion is regulated mainly by secretion in the distal and collecting tubules. Maintenance of potassium balance depends primarily on renal excretion because the amount in the feces is normally about 5 to 10 per cent of the potassium intake. Renal potassium excretion is determined by the sum of three processes: (1) the rate of potassium filtration (GFR multiplied by the plasma potassium concentration), (2) the rate of potassium reabsorption by the tubules, and (3) the rate of potassium secretion by the tubules. About 65 per cent of the filtered potassium is reabsorbed in the proximal tubule, and another 25 to 30 per cent is reabsorbed in the loop of Henle.

The normal day-to-day variation of potassium excretion, however, is regulated mainly by the secretion in the distal and collecting tubules rather than by changes in glomerular filtration or tubular reabsorption. In these tubular segments, potassium can at times be reabsorbed (e.g., during potassium depletion) or at other times secreted in large amounts depending on the needs of the body. With high potassium intakes, the required extra excretion of potassium is achieved almost entirely through increased secretion of potassium in the distal and collecting tubules.

Potassium secretion occurs in the *principal cells* of the late distal tubules and cortical collecting tubules. Secretion of potassium from the peritubular capillary blood into the tubular lumen is a three-step process, involving (1) passive diffusion of potassium from the blood to the interstitium, (2) active transport of potassium from the interstitium into the cell by the sodium-potassium ATPase pump at the basolateral membrane, and (3) passive diffusion of potassium from the cell interior to the tubular fluid. The primary factors that control potassium secretion by the principal cells include

- *Increased extracellular potassium concentration, which increases potassium secretion.* The mechanisms for this effect include stimulation of the sodium-potassium ATPase pump, an increase in the potassium gradient from the interstitial fluid to the tubular lumen, and the effect of greater potassium concentration to stimulate aldosterone secretion, which further stimulates potassium secretion.
- *Increased aldosterone concentration, which increases potassium secretion.* This effect is mediated through multiple mechanisms, including stimulation of the sodium-potassium ATPase pump and increase in the permeability of the luminal membrane for potassium.
- *Increased tubular flow rate, which increases potassium secretion.* The mechanism for the effect of high volume flow rate is as follows: When potassium is secreted into the tubular fluid, the luminal concentration of potassium increases, thereby reducing the driving force for potassium diffusion into the tubule. With an increased tubular flow rate, however, the secreted potassium is continuously flushed down the tubule, and the rise in tubular potassium concentration is minimized, thereby increasing net potassium secretion.
- *Acute increases in hydrogen ion concentration (acidosis), which decreases potassium secretion.* The mechanism for this effect is inhibition of the sodium-potassium ATPase pump by the elevated hydrogen ion concentration.

Aldosterone is the primary hormonal mechanism for regulating potassium ion concentration. There is a direct feedback by which aldosterone and potassium ion concentration are linked. This feedback mechanism operates as follows: Whenever extracellular fluid potassium concentration increases above normal, secretion of aldosterone is stimulated, which increases renal excretion of potassium, returning extracellular potassium concentration toward normal. The opposite changes take place when potassium concentration is too low.

CONTROL OF RENAL CALCIUM EXCRETION AND EXTRACELLULAR CALCIUM ION CONCENTRATION
(p. 380)

As with other substances, the intake of calcium must be balanced with the net loss of calcium over the long term. Unlike ions such as sodium and chloride, however, a large share of calcium excretion occurs in the feces. Only about 10 per cent of the ingested calcium normally is reabsorbed in the intestinal tract, and the remainder is excreted in the feces. Almost all of the calcium in the body (99 per cent) is stored in the bones, with only about 1 per cent in the intracellular fluid and 0.1 per cent in the extracellular fluid. Bones therefore act as large reservoirs for storing calcium and as sources of calcium when extracellular fluid calcium concentration tends to decrease *(hypocalcemia)*.

One of the most important regulators of bone uptake and release of calcium is *parathyroid hormone* (PTH). Decreased extracellular fluid calcium concentration promotes increased secretion of PTH, which acts directly on the bones to raise the reabsorption of bone salts (release of bone salts from the bones) and therefore release large amounts of calcium into the extracellular fluid. When calcium ion concentration is elevated *(hypercalcemia)*, PTH secretion decreases, and the excess calcium is deposited in the bones because of new bone formation.

The bones, however, do not have an inexhaustible supply of calcium. Over the long term, the intake of calcium must be balanced with calcium excretion by the gastrointestinal tract and kidneys. The most important regulator of calcium reabsorption at both of these sites is PTH; thus, PTH regulates plasma calcium concentration through three main effects: (1) by stimulating bone resorption; (2) by stimulating activation of vitamin D, which increases intestinal absorption of calcium; and (3) by directly increasing renal tubular calcium reabsorption. This is discussed in more detail in Chapter 79.

PTH controls renal calcium excretion. Calcium is not secreted by the renal tubules, and its excretion rate is therefore determined by the rate of calcium filtration and tubular reabsorption. One of the primary controllers of renal tubular calcium reabsorption is PTH. With increased levels of PTH, there is increased calcium reabsorption through the thick ascending loop of Henle and distal tubule, which reduces urinary excretion of calcium. Conversely, decreased PTH promotes calcium excretion by reducing reabsorption in the loop of Henle and distal tubules.

Greater plasma phosphate concentration stimulates PTH, which increases calcium reabsorption by the renal tubules and decreases calcium excretion.

Calcium reabsorption is also stimulated by *metabolic acidosis* and inhibited by *metabolic alkalosis*.

30

Regulation of Acid-Base Balance

Hydrogen ion concentration is precisely regulated.
Hydrogen ion concentration in the extracellular fluid is maintained at a very low level, averaging only .00000004 Eq/liter (40 nEq/liter). Normal variations are only about 3 to 5 nEq/liter. Because hydrogen ion concentration in the extracellular fluid is very low and because these small numbers are difficult to work with, hydrogen ion concentration is usually expressed in terms of pH units. The pH is the logarithm of the reciprocal of the hydrogen ion concentration $[H^+]$, expressed as equivalents per liter:

$$pH = \log \frac{1}{[H^+]} = -\log[H^+]$$

Arterial blood has a normal pH of 7.4, whereas the pH of venous blood and interstitial fluids is about 7.35. A person is considered to have *acidosis* when the arterial pH falls below 7.4 and to have *alkalosis* when the pH rises above 7.4. The lower limit of pH at which a person can live for more than a few hours is about 6.8, and the upper limit is about 8.0.

DEFENSES AGAINST CHANGES IN HYDROGEN ION CONCENTRATION: BUFFERS, LUNGS, AND KIDNEYS
(p. 386)

The body has three primary lines of defense against changes in hydrogen ion concentration in the body fluids:

- *The chemical acid-base buffer systems of the body fluids*, which immediately combine with acid or base to prevent excessive changes in hydrogen ion concentration
- *The respiratory system*, which regulates the removal of CO_2 and therefore carbonic acid (H_2CO_3) from the extracellular fluid; this mechanism operates

within seconds to minutes and acts as a second line of defense.

- *The kidneys*, which excrete either an alkaline or acidic urine, thereby adjusting the extracellular fluid hydrogen ion concentration toward normal during alkalosis or acidosis; this mechanism operates slowly but powerfully over a period of hours or several days to regulate acid-base balance.

BUFFERING OF HYDROGEN IONS IN THE BODY FLUIDS
(p. 387)

A buffer is any substance that can reversibly bind H^+. The general form of a buffering reaction is

$$\text{Buffer} + H^+ \leftrightarrows H \text{ Buffer}$$

In this example, free H^+ combines with the buffer to form a weak acid (H Buffer). When the H^+ concentration increases, the reaction is forced to the right and more H^+ binds to the buffer as long as available buffer is present. When the H^+ concentration decreases, the reaction shifts toward the left, and H^+ is released from the buffer.

Among the most important buffer systems in the body are the *proteins* of the cells and, to a lesser extent, the proteins of the plasma and interstitial fluids. The *phosphate buffer system* ($HPO_4^=/H_2PO_4^-$) is not a major buffer in the extracellular fluid but is important as an intracellular buffer and as a buffer in renal tubular fluid. The most important extracellular fluid buffer is the *bicarbonate buffer system* (HCO_3^-/PCO_2), primarily because the components of the system, CO_2 and HCO_3^-, are closely regulated by the lungs and kidneys, respectively.

The Bicarbonate Buffer System

The bicarbonate buffer system consists of a water solution that has two main ingredients: (1) a weak acid, H_2CO_3, and (2) a bicarbonate salt such as $NaHCO_3$. H_2CO_3 is formed in the body through the reaction of CO_2 with H_2O:

$$CO_2 + H_2O \underset{\text{Anhydrase}}{\overset{\text{Carbonic}}{\rightleftharpoons}} H_2CO_3$$

H_2CO_3 ionizes to form small amounts of H^+ and HCO_3^-:

$$H_2CO_3 \overset{\longrightarrow}{\longleftarrow} H^+ + HCO_3^-$$

The second component of the system, bicarbonate salt, occurs mainly as sodium bicarbonate ($NaHCO_3$) in the extracellular fluid. $NaHCO_3$ ionizes almost completely to form HCO_3^- and Na^+:

$$NaHCO_3 \overset{\longrightarrow}{\rightleftharpoons} Na^+ + HCO_3^-$$

Putting the entire system together, we have the following:

$$CO_2 + H_2O \leftrightharpoons H_2CO_3 \overset{\longrightarrow}{\longleftarrow} H^+ + \underbrace{HCO_3^-}_{\substack{+ \\ Na^+}}$$

When a strong acid is added to this buffer solution, the increased hydrogen ions are buffered by HCO_3^-:

$$\uparrow H^+ + HCO_3^- \rightarrow H_2CO_3 \rightarrow CO_2 + H_2O$$

The opposite reaction takes place when a strong base, such as sodium hydroxide (NaOH), is added to a bicarbonate buffer solution:

$$NaOH + H_2CO_3 \rightarrow NaHCO_3 + H_2O$$

In this case, the OH^- from the NaOH combines with H_2CO_3 to form additional HCO_3^-. The weak base $NaHCO_3$ replaces the strong base NaOH. At the same time, the concentration of H_2CO_3 decreases (because it reacts with NaOH), causing more CO_2 to combine with H_2O to replace the H_2CO_3:

$$\begin{array}{ccccc} CO_2 + H_2O & \rightarrow & H_2CO_3 & \rightarrow & \uparrow HCO_3^- + H^+ \\ + & & & & + \\ NaOH & & & & Na \end{array}$$

The net result is a tendency for the CO_2 levels to decrease, but the reduced CO_2 in the blood inhibits respiration and therefore decreases the rate of CO_2 expiration. The rise in blood HCO_3^- is compensated for by the rise in renal excretion of HCO_3^-.

The relationship of bicarbonate and carbon dioxide to pH is given by the Henderson-Hasselbalch equation. The following equation (the *Henderson-*

Hasselbalch equation) gives the relationship between the concentration of acid and base elements for the bicarbonate buffer system:

$$pH = 6.1 + \log \frac{HCO_3^-}{0.03 \times P_{CO_2}}$$

In this equation, CO_2 represents the acidic element, because it combines with water to form H_2CO_3, and HCO_3^- represents the basic element. HCO_3^- is expressed as mmol/liter, and P_{CO_2} is expressed as mm Hg. The greater the P_{CO_2}, the lower is the pH; the greater the HCO_3^-, the higher is the pH.

When disturbances of acid-base balance result from primary changes in extracellular HCO_3^-, these are referred to as *metabolic* acid-base disorders. Acidosis caused by a primary decrease in HCO_3^- concentration is termed *metabolic acidosis*, whereas alkalosis caused by a primary increase in HCO_3^- concentration is called *metabolic alkalosis*. Acidosis caused by an increase in P_{CO_2} is called *respiratory acidosis*, whereas alkalosis caused by a decrease in P_{CO_2} is called *respiratory alkalosis*.

RESPIRATORY REGULATION OF ACID-BASE BALANCE
(p. 390)

Because the lungs expel CO_2 from the body, rapid ventilation by the lungs decreases the concentration of CO_2 in the blood, which in turn decreases the carbonic acid (H_2CO_3) and H^+ concentration in the blood. Conversely, a decrease in pulmonary ventilation increases the concentration of CO_2 and H^+ in the blood.

Increased hydrogen ion concentration stimulates pulmonary ventilation. Not only does the pulmonary ventilation rate influence H^+ concentration by changing the P_{CO_2} of the body fluids, increases in H^+ concentration markedly stimulate pulmonary ventilation. As pH decreases from the normal value of 7.4 to the strong acidic value of 7.0, pulmonary ventilation increases to four to five times the normal rate. This in turn reduces the P_{CO_2} of the blood and returns H^+ concentration back toward normal. Conversely, if pH increases above normal, the respiration becomes depressed, and H^+ concentration increases toward normal. The respiratory system can return H^+ concentration and pH to about two thirds of normal within a few minutes after a sudden disturbance of acid-base balance.

Abnormalities of respiration can cause acid-base disturbances. Impairment of lung function, such as in severe *emphysema*, decreases the ability of the lungs to eliminate CO_2; this causes a build-up of CO_2 in the extracellular fluid and a tendency toward *respiratory acidosis*. The ability to respond to metabolic acidosis is impaired because the compensatory reductions in P_{CO_2} that would normally occur due to increased ventilation are blunted. Conversely, overventilation (rare) causes a reduction in P_{CO_2} and a tendency toward *respiratory alkalosis*.

RENAL CONTROL OF ACID-BASE BALANCE (p. 392)

The kidneys control acid-base balance by excreting either an acidic urine, which reduces the amount of acid in extracellular fluid, or a basic urine, which removes base from the extracellular fluid.

The overall mechanism by which the kidneys excrete acidic or basic urine is as follows: A large quantity of HCO_3^- is filtered continuously into the tubules; if HCO_3^- is excreted into the urine, base is removed from the blood. In contrast, a large quantity of H^+ is also secreted into the tubular lumen, thus removing acid from the blood. If more H^+ is secreted than HCO_3^- is filtered, there will be a net loss of acid from the extracellular fluid. Conversely, if more HCO_3^- is filtered than H^+ is secreted, there will be a net loss of base. In addition to secretion of H^+ and reabsorption of filtered HCO_3^-, the kidneys can generate new HCO_3^- from reactions that take place in the renal tubule. *The kidneys regulate extracellular fluid H^+ concentrations through three basic mechanisms: (1) secretion of H^+, (2) reabsorption of filtered HCO_3^-, and (3) production of new HCO_3^-.*

Secretion of Hydrogen Ions and Reabsorption of Bicarbonate Ions by the Renal Tubules

Hydrogen ion secretion and bicarbonate reabsorption occur in virtually all parts of the tubules except the descending and ascending thin limbs of the loop of Henle. Bicarbonate is not reabsorbed directly by the tubules; instead, bicarbonate is reabsorbed as a result of the reaction of secreted hydrogen ions with filtered bicarbonate ions in the tubular fluid under the influence of carbonic anhydrase in the tubular

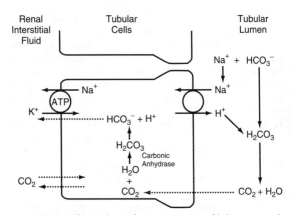

Figure 30–1 Cellular mechanisms for (1) active secretion of hydrogen ions into the renal tubule; (2) tubular reabsorption of bicarbonate by combination with hydrogen ions to form carbonic acid, which dissociates to form carbon dioxide and water; and (3) sodium ion reabsorption in exchange for the hydrogen ions secreted. This pattern of hydrogen ion secretion occurs in the proximal tubule.

epithelium. For each HCO_3^- reabsorbed, there must be an H^+ secreted.

In the proximal tubule, a thick ascending segment of the loop of Henle, and distal tubule, H^+ is secreted into the tubular fluid by sodium-hydrogen countertransport (Fig. 30–1). The secreted H^+ is consumed by reaction with HCO_3^-, forming H_2CO_3, which dissociates into CO_2 and H_2O. The CO_2 diffuses into the cell and is used to re-form H_2CO_3 and, eventually, HCO_3^-, which is then reabsorbed across the basolateral membranes of the tubules.

Normally, over 99 per cent of the filtered HCO_3^- is reabsorbed by the renal tubules, with about 95 per cent of this occurring in proximal tubules, loops of Henle, and early distal tubules.

In the distal and collecting tubules H^+ is secreted by primary active transport. The same basic mechanisms, however, are used for HCO_3^- reabsorption. Although the total amount of H^+ secreted in the late distal tubules and collecting ducts is not large, these segments are capable of increasing H^+ concentration to as much as 900-fold, which reduces the pH of the tubular fluid to about 4.5, the lower limit of pH that can be achieved in normal kidneys.

Bicarbonate ions are "titrated" against hydrogen ions in the tubules. Under normal conditions, the rate of tubular H^+ secretion is about 4400 mEq/day, and the rate of filtration of HCO_3^- is about 4320 mEq/day. The quantities of these two ions entering the tubules are almost equal, and they combine with each other to form CO_2 and H_2O; HCO_3^- and H^+ normally "titrate" each other in the tubules.

The titration process is not exact because there is usually a slight excess of H^+ in the tubules to be secreted into the urine. The excess H^+ (about 80 mEq/day) rids the body of nonvolatile acids produced by metabolism. Most of the H^+ is not excreted as free hydrogen ions but rather in combination with other urinary buffers, especially phosphate and ammonia.

In alkalosis, there is an excess of bicarbonate ions over hydrogen ions in the urine. Because the HCO_3^- cannot be reabsorbed unless it reacts with H^+, the excess HCO_3^- is left in the urine and eventually excreted, which helps to correct the alkalosis.

In acidosis, there are excess hydrogen ions, compared with bicarbonate ions, in the urine. This causes complete reabsorption of the filtered HCO_3^-, and the excess H^+ passes into the urine after combining with buffers in the tubules such as phosphate and ammonia. Thus, the basic mechanism by which the kidneys correct for acidosis or alkalosis is incomplete titration of H^+ against HCO_3^-, leaving one to pass into the urine and therefore be removed from the extracellular fluid.

COMBINATION OF EXCESS HYDROGEN IONS WITH PHOSPHATE AND AMMONIA BUFFERS IN THE TUBULES—A MECHANISM FOR GENERATING NEW BICARBONATE IONS (p. 395)

When H^+ is secreted in excess of HCO_3^- filtered into the tubular fluid, only a small part of the excess H^+ can be excreted in the urine in ionic form (H^+); the minimal urine pH is about 4.5, corresponding to an H^+ concentration of $10^{-4.5}$ mEq/liter, or 0.03 mEq/liter.

The excretion of large amounts of H^+ (greater than 500 mEq/day in severe acidosis) in the urine is accomplished primarily by combining the H^+ with buffers in the tubular fluid. The two most important buffers are *phosphate buffer* and *ammonia buffer*. For each H$^+$ secreted that combines with a nonbicarbonate buffer, a new HCO$_3$$^-$ is formed within the renal tubular cells and added to the body fluids.

Urinary phosphate buffer carries excess hydrogen ions into the urine and generates new bicarbonate. The phosphate buffer system is composed of HPO$_4$$^=$ and H$_2$PO$_4$$^-$. The H$^+$ remaining in the renal tubule in excess of that which reacts with HCO$_3$$^-$ can react with HPO$_4$$^=$ to form H$_2$PO$_4$$^-$, which can be excreted as a sodium salt (NaH$_2$PO$_4$). For each H$^+$ excreted with phosphate buffer, a new HCO$_3$$^-$ is generated in the renal tubule and reabsorbed. The HCO$_3$$^-$ that is generated in the tubular cell represents a net gain of HCO$_3$$^-$ by the blood rather than merely a replacement of filtered HCO$_3$$^-$.

In normal conditions, about 75 per cent of the filtered phosphate is reabsorbed, and only about 30 to 40 mEq/day is available for buffering H$^+$; therefore, much of the buffering of excess H$^+$ in the tubular fluid in severe acidosis occurs through the ammonia buffer system.

Ammonia is the most important urinary buffer in acidosis. The ammonia buffer system is composed of ammonia (NH$_3$) and ammonia ion (NH$_4$$^+$). Ammonia ion is synthesized from *glutamine*, which is actively transported into the cells of the proximal tubules, thick ascending limbs in the loop of Henle, and distal tubules. Once inside the cell, each molecule of glutamine is metabolized to form two NH$_4$$^+$ and two HCO$_3$$^-$. The NH$_4$$^+$ is secreted into the tubular lumen in exchange for sodium, and the HCO$_3$$^-$ moves across the basolateral membrane along with the reabsorbed sodium ion. For each molecule of glutamine metabolized, two NH$_4$$^+$ are secreted into the urine and two HCO$_3$$^-$ are reabsorbed into the blood. *The HCO$_3$$^-$ generated by this process constitutes new bicarbonate added to the blood.*

One of the most important features of the renal ammonia buffer system is that *renal glutamine metabolism is markedly stimulated by acidosis*, thereby increasing the formation of NH$_4$$^+$ and new HCO$_3$$^-$ to be used in hydrogen ion buffering.

QUANTIFICATION OF RENAL TUBULAR ACID SECRETION
(p. 397)

- *The total rate of hydrogen secretion* can be calculated as

$$H^+ \text{ Secretion Rate} = HCO_3^- \text{ Reabsorption Rate} \\ + \text{ Titratable Acid Excretion Rate} \\ + NH_4^+ \text{ Excretion Rate}$$

This assumes that almost all the H^+ secreted either combines with HCO_3^-, which is reabsorbed, or is excreted with phosphate (titratable acid) or ammonia buffer.

- The net acid excretion rate is calculated as

$$\text{Net Acid Excretion Rate} = \\ \text{Urinary Titratable Acid Excretion Rate} + \\ NH_4^+ \text{ Excretion Rate} - HCO_3^- \text{ Excretion Rate}$$

The reason we subtract HCO_3^- excretion is that loss of HCO_3^- is the same as the addition of H^+ to the blood. In acidosis, the net acid excretion rate increases markedly, thereby removing acid from the blood. The net acid excretion rate also equals the rate of new bicarbonate addition to the blood. *In acidosis, there is a net addition of bicarbonate back to the blood as more NH_4^+ and urinary titratable acid are excreted.* In alkalosis, titratable acid and NH_4^+ excretion drop to zero, whereas HCO_3^- excretion increases. *In alkalosis, there is a negative net acid secretion.*

Renal tubular hydrogen ion secretion is regulated by P_{CO_2} and extracellular [H^+]. In alkalosis, tubular secretion of H^+ must decrease to a level that is too low to achieve complete HCO_3^- reabsorption, enabling the kidneys to increase HCO_3^- excretion. In acidosis, tubular H^+ secretion must be sufficiently increased to reabsorb all the filtered HCO_3^- and still have enough H^+ to excrete large amounts of NH_4^+ and titratable acid, thereby contributing large amounts of new HCO_3^- to the blood.

The two most important stimuli for increasing H^+ secretion by the tubules in acidosis are (1) an increase in P_{CO_2} of the extracellular fluid and (2) an increase in hydrogen ion concentration of the extracellular fluid (decreased pH).

RENAL CORRECTION OF ACIDOSIS—INCREASED EXCRETION OF HYDROGEN IONS AND ADDITION OF NEW BICARBONATE IONS TO THE EXTRACELLULAR FLUID
(p. 398)

The condition of *acidosis* occurs when arterial pH falls below 7.4. If the decrease in pH is caused by a decrease in HCO_3^-, the condition is referred to as *metabolic acidosis*, whereas a decrease in pH caused by an increase in PCO_2 is referred to as *respiratory acidosis*.

Regardless of whether the acidosis is respiratory or metabolic, both conditions cause a decrease in the ratio of HCO_3^- to H^+ in the renal tubular fluid. This results in an excess of H^+ in the renal tubules, causing complete reabsorption of HCO_3^- and leaving still additional H^+ available to combine with the urinary buffers NH_4^+ and $HPO_4^=$. In acidosis, the kidneys reabsorb all of the filtered HCO_3^- and contribute new bicarbonate through the formation of NH_4^+ and titratable acid.

Metabolic acidosis results from a decrease in bicarbonate in the body fluids. The decreased extracellular fluid HCO_3^- concentration causes a decrease in glomerular filtration of HCO_3^-. The compensatory responses include stimulation of respiration, which eliminates CO_2 and returns pH toward normal. At the same time, renal compensation increases reabsorption of HCO_3^- and excretion of titratable acid and NH_4^+, which leads to the formation of new HCO_3^- and return of pH toward normal.

Some of the primary causes of metabolic acidosis are as follows:

- *Decreased renal tubular secretion of hydrogen ion or decreased reabsorption of bicarbonate*—This can occur as a result of a condition called *renal tubular acidosis* in which the kidneys are unable to secrete adequate amounts of H^+. As a result, large amounts of HCO_3^- are lost in the urine, causing a continued state of metabolic acidosis. *Chronic renal failure*, which occurs when kidney function declines markedly and H^+ is not adequately secreted by the tubules, also causes build-up of acids in the body fluids.

- *Formation of excess metabolic acids in the body*—An example of this is the metabolic acidosis that occurs in *diabetes mellitus* in which large amounts of acetoacetic acid are formed from metabolism of fats.

- *Ingestion of excess metabolic acids*—This can occur, for example, with ingestion of certain drugs such as *acetylsalicylics (aspirin)* and *methyl alcohol*, which are metabolized to form formic acid.

- *Excessive loss of base from the body fluids*—This most commonly occurs with severe *diarrhea* in which large amounts of gastrointestinal secretions, containing bicarbonate, are lost from the body.

Respiratory acidosis is caused by decreased ventilation, which increases Pco_2. A decrease in the pulmonary ventilation rate increases the Pco_2 of the extracellular fluid, causing a rise in H_2CO_3, H^+ concentration, and respiratory acidosis. As a compensation, increased Pco_2 stimulates H^+ secretion by the renal tubules, causing increased HCO_3^- reabsorption. The excess H^+ remaining in the tubular cells combines with buffers, especially ammonia, which leads to the generation of new HCO_3^- that is added back to the blood. These changes help return plasma pH toward normal.

Common causes of respiratory acidosis are pathological conditions that damage the respiratory centers or the ability of the lungs to effectively eliminate CO_2. For example, damage to the respiratory center in the medulla oblongata can cause respiratory acidosis. Obstruction of the passages of the respiratory tract, pneumonia, decreased pulmonary surface area, or any factor that interferes with the exchange of gases between the blood and alveolar membrane can cause respiratory acidosis.

RENAL CORRECTION OF ALKALOSIS—DECREASED TUBULAR SECRETION OF HYDROGEN IONS AND INCREASED EXCRETION OF BICARBONATE IONS (p. 399)

The condition of *alkalosis* occurs when arterial pH rises above 7.4. If the increase in pH results mainly from an increase in plasma HCO_3^-, it is called *metabolic alkalosis*, whereas alkalosis caused by a decrease in Pco_2 is called *respiratory alkalosis*.

The compensatory responses to alkalosis are basically opposite those of acidosis. In alkalosis, the ratio of HCO_3^- to CO_2 in the extracellular fluid increases, causing an increase in pH (a decrease in H^+ concentration). Regardless of whether the alkalosis is caused by metabolic or respiratory abnormalities, there still is an increase in the ratio of HCO_3^- to H^+ in the renal tubular fluid. The net effect is an excess of HCO_3^- that cannot be reabsorbed from the tubules and therefore is excreted in the urine. In alkalosis, HCO_3^- is removed from the extracellular fluid through renal excretion, which has the same effect as the addition of H^+ to the extracellular fluid.

Metabolic alkalosis results from increased HCO_3^- in the extracellular fluid. This causes an increase in the filtered load of HCO_3^-, which in turn results in an excess of HCO_3^- over H^+ in the renal tubular fluid. The excess HCO_3^- in the tubular fluid fails to be reabsorbed because it does not have sufficient H^+ with which to react and therefore is excreted in the urine. In metabolic acidosis, the primary compensations are increased renal excretion of HCO_3^- and decreased ventilation rate, which raises PCO_2.

Metabolic alkalosis is not nearly as common as metabolic acidosis, but some of the main causes are as follows:

- *Excess aldosterone secretion*—This promotes excessive reabsorption of sodium ions and at the same time stimulates the secretion of H^+ by the intercalated cells of the collecting tubules. This leads to increased secretion of H^+ by the kidneys, excessive production of HCO_3^- by the kidney, and therefore metabolic alkalosis.
- *Vomiting of gastric contents*—Vomiting of the gastric contents alone, without vomiting of the lower gastrointestinal contents, causes loss of HCl secreted by the stomach mucosa. The net result is a loss of acid from the extracellular fluid and the development of metabolic alkalosis.
- *Ingestion of alkaline drugs*—One of the most common causes of metabolic alkalosis is ingestion of drugs such as sodium bicarbonate for the treatment of gastritis or peptic ulcer.

Respiratory alkalosis is caused by increased ventilation, which decreases PCO_2. Respiratory alkalosis rarely occurs due to physical pathological conditions; however, a *psychoneurosis* can occasionally cause overbreathing to the extent that a person becomes alkalotic. A physiological respiratory alkalosis occurs when a person ascends to a *high altitude*. The low oxygen content of the air stimulates respiration, which causes an excessive loss of CO_2 and a development of mild respiratory alkalosis. The primary compensations are the chemical buffers of the body fluids and the ability of the kidneys to increase HCO_3^- excretion.

Table 30–1* shows the different types of acid-base disturbances and the characteristic changes in pH, hydrogen ion concentration, PCO_2, and bicarbonate ion concentration.

TABLE 30–1 CHARACTERISTICS OF PRIMARY ACID-BASE DISTURBANCES

	pH	H$^+$	Pco_2	HCO$_3^-$
Respiratory acidosis	↓	↑	⇑	↑
Respiratory alkalosis	↑	↓	⇓	↓
Metabolic acidosis	↓	↑	↓	⇓
Metabolic alkalosis	↑	↓	↑	⇑

* The primary event is indicated by the double arrow (⇑ or ⇓). Note that respiratory acid-base disorders are initiated by an increase or a decrease in Pco_2, whereas metabolic disorders are initiated by an increase or decrease in HCO$_3^-$.

31

Micturition, Diuretics, and Kidney Diseases

MICTURITION (p. 405)

Micturition is the process by which the urinary bladder empties when it becomes filled. This involves two main steps: (1) the bladder fills progressively until the tension in its walls rises above a threshold level, which elicits the second step; and (2) a nervous reflex, called the *micturition reflex*, occurs and empties the bladder or, if this fails, at least causes a conscious desire to urinate.

Physiologic Anatomy and Nervous Connections of the Bladder

The *ureters* carry the urine from the renal pelvis to the bladder, where they pass obliquely through the bladder wall before emptying into the bladder chamber. There are no major changes in the composition of the urine as it flows through the ureters to the bladder. Peristaltic contractions of the ureter, which are enhanced by parasympathetic stimulation, force the urine from the renal pelvis toward the bladder.

The urinary bladder is a smooth muscle chamber composed of two main parts: (1) *the body*, which is the major portion of the bladder in which urine collects, and (2) *the neck*, which is a funnel-shaped extension of the body that connects with the urethra.

The smooth muscle of the bladder is called the *detrusor muscle*. When the fibers contract, they can increase the pressure of the bladder to 40 to 60 mm Hg and therefore play a major role in emptying the bladder.

The bladder neck (posterior urethra) is composed of detrusor muscle interlaced with a large amount of elastic tissue. The muscle in this area is called the *internal sphincter*; its natural tone keeps the bladder from emptying until the pressure in the main part of the bladder rises above a critical threshold.

Beyond the posterior urethra, the urethra passes through the *urogenital diaphragm*, which contains a layer of muscle called the *external sphincter* of the bladder. This muscle is a voluntary skeletal muscle and can be used to consciously prevent urination even when involuntary controls are attempting to empty the bladder.

The pelvic nerves provide the principal nervous supply of the bladder. Coursing through the pelvic nerves, which connect with the spinal cord through the *sacral plexus*, are both *sensory nerve fibers* and *motor nerve fibers*. The sensory nerve fibers detect the stretch of the bladder wall and initiate reflexes that cause bladder emptying. The motor nerves transmitted to the pelvic nerves are *parasympathetic fibers*.

The Micturition Reflex Is a Spinal Cord Reflex

The micturition reflex is a single complete cycle of (1) a progressive and rapid increase in bladder pressure, (2) a period of sustained increase in bladder pressure, and (3) a return of the pressure to the basal tone of the bladder, as follows:

- Sensory signals from the bladder wall stretch receptors are conducted to sacral segments of the spinal cord through the pelvic nerves and then reflexively back to the bladder through the parasympathetic nerves by way of the pelvic nerves.
- Once the micturition reflex is sufficiently powerful, it causes another reflex that passes through the *pudendal nerves* to the external sphincter to inhibit this. If this inhibition is more potent than the voluntary constrictor signals to the external sphincter, urination will be controlled.
- The micturition reflex is an autonomic spinal cord reflex, but it can be inhibited or facilitated by centers in the brainstem, mainly the *pons* and several centers in the *cerebral cortex* that are mainly excitatory.

DIURETICS AND THEIR MECHANISMS OF ACTION
(p. 408)

A diuretic is a substance that increases the rate of urine volume output. Many diuretics also increase urinary excretion of solutes, especially sodium and chloride. Most diuretics that are used clinically act primarily by decreasing the rate of sodium chloride reabsorption in the renal tubules, which in turn causes *natriuresis* (increased sodium excretion) and *diuresis* (increased water output).

The most common clinical use of diuretics is to reduce extracellular fluid volume, especially in diseases associated with edema and hypertension. As discussed in Chapter 25, loss of sodium from the body mainly reduces extracellular fluid volume; therefore, diuretics are most often administered under clinical conditions in which extracellular fluid volume is expanded.

A balance between salt and water intake and renal output occurs during chronic diuretic therapy. Some diuretics can increase urine output by more than 20-fold within a few minutes after they are administered; however, the effect of most diuretics on renal output of salt and water subsides within a few days owing to activation of compensatory mechanisms initiated by decreased extracellular fluid volume. For example, reduced extracellular fluid volume decreases arterial pressure and glomerular filtration rate and increases renin secretion and angiotensin II formation. All these responses eventually override the effect of a diuretic on urine output so that in the steady state, urine output becomes equal to intake—but only after a reduction in extracellular fluid volume has occurred.

There are many diuretics available for clinical use, and they have different mechanisms of action and therefore inhibit tubular reabsorption at different sites along the renal nephron. The general classes of diuretics and their mechanisms of action are shown in Table 31–1.

KIDNEY DISEASES (p. 410)

Many kidney diseases can be divided into two main categories: (1) *acute renal failure*, in which the kidneys abruptly stop working entirely or almost entirely but may eventually recover nearly normal function, and (2) *chronic renal failure*, in which there is a progressive loss of function of nephrons that gradually decreases overall kidney function. Within these two general categories, there are many specific kidney diseases that can affect the blood vessels, glomeruli, tubules, renal interstitium, and parts of the urinary tract outside the kidney. In this chapter,

TABLE 31–1 CLASSES OF DIURETICS, MECHANISMS OF ACTION, AND TUBULAR SITES OF ACTION

Class of Diuretic	Example	Mechanisms of Action of Diuretic	Tubular Site of Action
Osmotic diuretics	Mannitol	Inhibits water and solute reabsorption by increasing osmolarity of tubular fluid	Mainly proximal tubule
Loop diuretics	Furosemide	Inhibits Na^+–K^+–Cl^- co-transport in luminal membrane	Thick ascending loop of Henle
Thiazide diuretics	Chlorothiazide	Inhibits Na^+–Cl^- co-transport in luminal membrane	Early distal tubules
Carbonic anhydrase inhibitors	Acetazolamide	Inhibits H^+ secretion and HCO_3^- reabsorption, which reduces Na^+ reabsorption	Proximal tubules
Competitive inhibitors of aldosterone	Spironolactone	Inhibits action of aldosterone on tubular receptor, decreases Na^+ reabsorption, and decreases K^+ secretion	Proximal tubules
Sodium channel blockers	Amiloride	Blocks entry of Na^+ into sodium channels of luminal membrane, decreases Na^+ reabsorption, and decreases K^+ secretion	Collecting tubules

we discuss physiological abnormalities that occur in a few of the most important types of kidney diseases.

Acute Renal Failure

The three main categories of acute renal failure include the following:

Prerenal acute failure is caused by decreased blood supply to the kidneys. This can be a consequence of heart failure, which reduces cardiac output and blood pressure, or conditions associated with diminished blood volume, such as severe hemorrhage. When blood flow to the kidney falls to less than 20 per cent of normal, the renal cells start to become hypoxic. Further decreases in flow, if prolonged, will cause damage or death to the renal cells. If the acute renal failure is not corrected, this type of failure can evolve into *intrarenal acute renal failure*.

Intrarenal acute renal failure results from abnormalities within the kidney itself, including those that affect the blood vessels, glomeruli, or tubules. *Acute glomerulonephritis* is a type of intrarenal acute renal failure caused by an abnormal immune reaction that causes an inflammation of the glomeruli. The acute inflammation usually subsides within about 2 weeks, but in some patients, many of the glomeruli are destroyed beyond repair. In a small percentage of patients, continued renal deterioration leads to progressive *chronic renal failure* (discussed next).

Other causes of intrarenal acute renal failure include acute *tubular necrosis*, which is caused by severe renal ischemia or toxins and medications that damage the tubular epithelial cells. If the damage is not too severe, some regeneration of the tubular epithelial cells can occur, and renal function can be restored.

Postrenal acute renal failure is caused by obstruction of the urinary collecting system anywhere from the calyces to the outflow from the bladder. The most important causes of obstruction of the urinary tract are *kidney stones*, which are the result of precipitation of calcium, urate, or cystine.

Chronic Renal Failure: Irreversible Decrease in the Number of Functional Nephrons (p. 412)

Serious clinical symptoms of chronic renal failure often do not occur until the number of functional nephrons falls to at least 70 per cent below normal. The maintenance of normal plasma concentrations of electrolytes and normal body fluid volumes occurs at the expense of systemic compensations, such as hypertension, which over the long term can lead to additional clinical problems.

In general, chronic renal failure, like acute renal failure, can occur because of disorders of the blood vessels, glomeruli, tubules, renal interstitium, and lower urinary tract. Despite the wide variety of diseases that can cause chronic renal failure, the end result is essentially the same—a decrease in the number of nephrons.

Chronic renal failure can initiate a vicious circle that leads to end-stage renal failure. In some cases, an initial insult to the kidney leads to progressive deterioration of renal function and further loss of nephrons to the point at which to survive, a person must be placed on dialysis treatment or must undergo transplantation with a functional kidney. This condition is referred to as *end-stage renal failure*.

The cause of this progressive injury is not known, but some investigators believe that it may be related in part to increased pressure or stretch in the remaining glomeruli that occurs as a result of adaptive vasodilation or increased blood pressure. The increased pressure and stretch of arterioles and glomeruli are believed to eventually cause *sclerosis* (replacement of normal tissue with connective tissue) of these vessels. These sclerotic lesions eventually obliterate the glomerulus, leading to a further reduction in kidney function and a slowly progressing vicious circle that terminates in end-stage renal failure. The most common causes of end-stage renal failure include *diabetes mellitus* and *hypertension*.

Some of the general causes of chronic renal failure are as follows:

- *Injury of the renal blood vessels*—Some of the most common causes of renal vascular injury include *atherosclerosis* of the larger renal arteries, *fibromuscular hyperplasia* of one or more of the large arteries, and *nephrosclerosis*, a condition caused by sclerotic lesions of the smaller vessels and glomeruli that is often a result of hypertension or diabetes mellitus.
- *Injury of the glomeruli*—One example of this is *chronic glomerulonephritis*, which can be the result of several diseases that cause inflammation and

damage to the glomeruli capillaries. In contrast to the acute form of this disease, chronic glomerulonephritis is a slowly progressive disease that may lead to irreversible kidney failure. It may be a primary kidney disease, occurring after acute glomerulonephritis, or it may be secondary to a systemic disease, such as *lupus erythematosus*.

- *Injury to renal interstitium*—Primary or secondary disease of the renal interstitium is referred to as *interstitial nephritis*. This can result from vascular, glomerular, or tubular damage that destroys individual nephrons, or it can involve primary damage to the renal interstitium caused by poisons, drugs, and bacterial infections. Renal interstitial injury caused by bacterial infection is called *pyelonephritis*. This infection can result from bacteria that reach the kidneys through the blood stream or, more commonly, ascension from the lower urinary tract through the ureters to the kidney. With long-standing pyelonephritis, invasion of the kidneys by bacteria not only causes damage to the renal interstitium but also results in progressive damage to the renal tubules, glomeruli, and other structures, eventually leading to the loss of functional nephrons.

Abnormal Nephron Function in Chronic Renal Failure

The loss of functional nephrons requires the surviving nephrons to excrete more water and solutes. The kidney normally filters about 180 liters of fluid each day at the glomerular capillaries and then transforms this filtrate to approximately 1.5 liters of urine as the fluid flows along successive nephron segments. Regardless of the number of functional nephrons, the kidney must excrete the same volume of urine (if intake is constant) to maintain fluid balance. The loss of functional nephrons therefore requires the surviving nephrons to excrete extra amounts of water and solutes to prevent serious accumulation of these substances in the body fluids. This is achieved by increasing the glomerular filtration rate or decreasing the tubular reabsorption rate in the surviving nephrons. These adaptations allow water and electrolyte balances to be maintained with very little change in extracellular volume or electrolyte composition, even in patients who have lost as much as 70 per cent of their nephrons.

In contrast to the electrolytes, many of the waste products in metabolism, such as urea and creatinine, accumulate almost in proportion to the number of nephrons that have been destroyed. These substances are not avidly reabsorbed by the renal tubules, and their excretion rate depends largely on the rate of glomerular filtration. If the glomerular filtration rate decreases, these substances accumulate in the body transiently, raising the plasma concentration until the filtered load (glomerular filtration rate × plasma concentration) and the excretion rate (urine concentration × urine volume) return to normal, which is the same rate at which the substance is either ingested or produced in the body.

Some substances, such as phosphate, urate, and hydrogen ions, are maintained near normal until the glomerular filtration rate falls below 20 to 30 per cent of normal. Plasma concentrations rise, thereafter, but not in proportion to the decline in the glomerular filtration rate.

Effects of Renal Failure on the Body Fluids—Uremia

The effect of renal failure on the body's fluids depends on the food and water intake and the degree of impairment of kidney function. With the assumption that intake remains relatively constant, important effects of renal failure include the following:

- *Water retention and development of edema*
- *An increase in extracellular fluid urea (uremia) and other nonprotein nitrogens (azotemia)*—The nonprotein nitrogens include urea, uric acid, and creatinine, and a few less important compounds. These, in general, are the end products of protein metabolism.
- *Acidosis*—This results from failure of the kidneys to rid the body of normal acidic products. The buffers of the body fluids can normally buffer 500 to 1000 millimoles of acid without lethal increases in the extracellular hydrogen ion concentration. Each day, however, the body normally produces about 50 to 80 millimoles more metabolic acid than metabolic alkali. Complete renal failure, therefore, leads to severe accumulation of acids in the blood within a few days.
- *Anemia*—If the kidneys are seriously damaged, they are unable to form adequate amounts of

erythropoietin, which stimulates the bone marrow to produce red blood cells.

- *Osteomalacia*—In prolonged kidney failure, adequate amounts of the active form of *vitamin D* are not produced, causing decreased intestinal absorption of calcium and decreased availability of calcium to the bones. These lead to the condition called osteomalacia, in which the bones are partially absorbed and become greatly weakened. Another important cause of demineralization of the bones in chronic renal failure is the *rise in serum phosphate concentration* that occurs because of the decreased glomerular filtration rate. The rise in serum phosphate level increases binding of phosphate with calcium in the plasma, decreasing the serum ionized calcium, which in turn stimulates *parathyroid hormone* secretion, increasing the release of calcium from bones and further demineralization.

UNIT
VI

Blood Cells, Immunity, and Blood Clotting

32

Red Blood Cells, Anemia, and Polycythemia

The major function of red blood cells is to transport hemoglobin, which in turn carries oxygen from the lungs to the tissues. Normal red blood cells are biconcave discs; the shapes can change remarkably as the cells pass through the capillaries. The normal red blood cell has a great excess of cell membrane in relation to the quantity of material it contains. Deformation of the cell does not stretch the membrane and consequently does not rupture the cell. The average number of red blood cells per cubic millimeter is $5,200,000 \pm 300,000$ in men and $4,700,000 \pm 300,000$ in women.

Red blood cells have the ability to concentrate hemoglobin. In normal individuals, the percentage of hemoglobin is almost always near the maximum level in each cell (about 34 gm/dL). The blood contains an average of 15 grams of hemoglobin per 100 milliliters (16 grams in men and 14 grams in women). Each gram of pure hemoglobin is capable of combining with approximately 1.39 milliliters of oxygen. In a normal person, more than 20 milliliters of oxygen can be carried in combination with the hemoglobin in each 100 milliliters of blood.

Genesis of blood cells. All circulating blood cells are derived from *pluripotential hemopoietic stem cells*. The pluripotential cells differentiate to form the peripheral blood cells. As these cells reproduce, a portion is exactly like the original pluripotential cells. These cells are retained in the bone marrow to maintain a constant supply. The early offspring of the stem cells cannot be recognized as different types of blood cells even though they already have been committed to a particular cell line; these cells are called *committed stem cells*. Different committed stem cells will produce different colonies of specific types of blood cells.

The growth and reproduction of the different stem cells are controlled by multiple proteins called *growth inducers*, which promote growth but not differentiation of the cells. This is the function of another set of proteins called *differentiation inducers*. Each of these inducers causes one type of stem cell to differentiate one or more steps toward the final type of adult blood cell. The formation of growth inducers and differentiation inducers is controlled by factors outside the bone marrow. In the case of red blood cells, exposure of the body to a low level of oxygen for a long period induces growth, differentiation, and production of greatly increased numbers of erythrocytes.

REGULATION OF RED BLOOD CELL PRODUCTION—THE ROLE OF ERYTHROPOIETIN (p. 426)

The total mass of red blood cells in the circulatory system is regulated within very narrow limits. An adequate number of red blood cells is always available to provide sufficient tissue oxygenation under normal conditions. Any condition that causes the quantity of oxygen that is transported in the tissues to decrease ordinarily increases the rate of red blood cell production. The principal factor that stimulates red blood cell production is the circulating hormone *erythropoietin*. Hypoxia is a potent stimulus for erythropoietin formation. In a normal individual, 80 to 90 per cent of the erythropoietin is formed in the kidneys; the remainder is formed mainly in the liver. The structure in the kidney in which the erythropoietin is formed is not known; one possibility is the renal tubular epithelial cell.

When both kidneys are surgically removed or destroyed by renal disease, the individual invariably becomes very anemic because the amount of erythropoietin formed in nonrenal tissues is sufficient to cause only one third to one half as many red blood cells to be formed as are needed by the body.

Vitamin B_{12} and folic acid are important for the final maturation of red blood cells. Both vitamin B_{12} and folic acid are essential to the synthesis of DNA. The lack of either of these vitamins results in a diminished quantity of DNA and, consequently, failure of nuclear maturation and division. In addition to failure to proliferate, the red blood cells become larger than normal, developing into *megaloblasts*. These cells have irregular shapes and flimsy cell membranes; they are capable of carrying oxygen normally, but their fragility causes them to have a short life span—one half to one third that of normal. Vitamin B_{12} or folic acid deficiency, therefore, causes *maturation failure* in the process of erythropoiesis.

A common cause of maturation failure is an inability to absorb vitamin B_{12} from the gastrointestinal tract. This often occurs in persons with *pernicious anemia*, a disease in which the basic abnormality is atrophic gastric mucosa. The parietal cells of the gastric gland secrete a glycoprotein called *intrinsic factor*, which combines with vitamin B_{12} to make it available for absorption by the gut. The intrinsic factor binds tightly with vitamin B_{12} and protects the vitamin from digestion by the gastrointestinal enzymes. The intrinsic factor–vitamin B_{12} complex binds to specific receptor sites on the brush border membranes of mucosal cells of the ileum. Vitamin B_{12} is then transported into the blood via the process of pinocytosis. Lack of intrinsic factor causes loss of much of the vitamin due to enzyme action in the gut and failure of absorption.

Formation of Hemoglobin

The synthesis of hemoglobin begins when the red blood cell is in the proerythroblast stage and continues into the reticulocyte stage, when the cell leaves the bone marrow and passes into the bloodstream. During the formation of hemoglobin, the *heme* molecule combines with a very long polypeptide chain called a *globin* to form a subunit of hemoglobin called a *hemoglobin chain*. Four hemoglobin chains bind together loosely to form the entire hemoglobin molecule.

The most important feature of the hemoglobin molecule is its ability to bind loosely and reversibly with oxygen. The oxygen atom binds loosely with one of the so-called coordination bonds of the iron atom in hemoglobin. When bound to the iron heme, oxygen is carried as molecular oxygen, composed of two oxygen atoms. Oxygen is released into the tissue fluids in the form of dissolved molecular oxygen rather than as ionic oxygen.

Iron Metabolism

Iron is important for the formation of hemoglobin, myoglobin, and other substances, such as the cytochromes, cytochrome oxidase, peroxidase, and catalase. The total average quantity of iron in the body is about 4 to 5 grams. About 65 per cent of this amount is in the form of hemoglobin. About 4 per cent is in the form of myoglobin, 1 per cent is in the form of the various heme compounds that promote intracellular oxidation, 0.1 per cent is combined with the protein *transferrin* in the blood plasma, and 15 to 30 per cent is stored mainly in the reticuloendothelial system and the liver parenchymal cells, principally in the form of *ferritin*.

Iron is transported and stored. When iron is absorbed from the small intestine it immediately combines with a beta globulin called *apotransferrin* to form *transferrin*, which is transported in the plasma. This iron is loosely bound. Excess iron in the blood is deposited in the liver hepatocytes and in the reticuloendothelial cells of the bone marrow. Once inside the cell cytoplasm, iron combines with the protein *apoferritin* to form *ferritin*. Varying quantities of iron can combine in clusters of iron radicals within ferritin.

When the quantity of iron in the plasma decreases to less than normal, iron is removed from ferritin quite easily and transported by transferrin in the plasma to the portions of the body where it is needed. A unique characteristic of the transferrin molecule is its ability to bind strongly with receptors in the cell membranes of the erythroblasts and bone marrow. Transferrin is ingested via endocytosis into the erythroblasts along with the bound iron. Transferrin delivers the iron directly to the mitochondria, where heme is synthesized.

When red blood cells have lived their life span and are destroyed, the hemoglobin released is ingested by cells of the monocyte-macrophage system. The free iron that is liberated can be stored in the ferritin pool or reused for formation of hemoglobin.

THE ANEMIAS (p. 431)

Anemia means a deficiency of red blood cells and can be caused by rapid loss of red blood cells or slow production of red blood cells.

- *Blood loss anemia* occurs after significant hemorrhage. The body is able to replace the plasma within 1 to 3 days; however, the concentration of red blood cells remains low. After a significant hemorrhage, a period of 3 to 4 weeks is required to return the number of red blood cells to normal levels.

- *Aplastic anemia* is the result of a nonfunctioning bone marrow. This can be due to exposure to gamma radiation or toxic industrial chemicals or an adverse reaction to drugs.
- *Megaloblastic anemia* is the result of a lack of vitamin B_{12}, folic acid, or intrinsic factor. Lack of these substances leads to very slow reproduction of the erythrocytes in the bone marrow. As a result, these erythrocytes grow into large, odd-shaped cells called *megaloblasts*.
- *Hemolytic anemia* is the result of fragile red blood cells that rupture as they pass through the capillaries. In hemolytic anemia, the number of red blood cells that form is normal or in excess of normal; however, because these cells are very fragile, their life span is very short. *Sickle cell anemia* is a type of hemolytic anemia that is caused by an abnormal composition of the globin chains of hemoglobin. When this abnormal hemoglobin is exposed to low concentrations of oxygen, it precipitates into long crystals inside the red blood cell. This causes the cell to have an abnormal sickle shape and to be very fragile.

POLYCYTHEMIA (p. 432)

Polycythemia is a condition in which the number of red blood cells in the circulation increases owing to hypoxia or genetic aberration. Individuals who live at high altitudes have a *physiologic polycythemia* as a result of the thin atmosphere. Polycythemia can also occur in individuals with cardiac failure because of the decreased delivery of oxygen to the tissues.

Polycythemia vera is a genetic aberration in the hemocytoblastic cell line. The blast cells continue to produce red blood cells even though too many blood cells are present in the circulation. The hematocrit can rise to 60 to 70 per cent.

Polycythemia greatly increases the viscosity of the blood; as a result, blood flow through the vessels is often sluggish.

33

Resistance of the Body to Infection

I. Leukocytes, Granulocytes, Monocyte-Macrophage System, and Inflammation

Our bodies have a special system for combating the different infectious and toxic agents to which we are continuously exposed. The leukocytes (white blood cells) are the mobile units of the protective system of the body. They are formed in the bone marrow and lymph tissue and transported in the blood to areas of inflammation to provide rapid and potent defense against any infectious agent that might be present. Six different types of leukocytes are normally found in the blood; the normal percentages are listed as follows:

- Polymorphonuclear neutrophils—62.0%
- Polymorphonuclear eosinophils—2.3%
- Polymorphonuclear basophils—0.4%
- Monocytes—5.3%
- Lymphocytes—30.0%.

The three types of polymorphonuclear cells have a granular appearance and are called *granulocytes,* or *"polys."*

The granulocytes and monocytes protect the body against invading organisms by ingesting them via the process of *phagocytosis.* The lymphocytes function mainly in connection with the immune system to attach to specific invading organisms and destroy them.

Genesis of leukocytes. Two lineages of white blood cells are formed from the *pluripotential hemopoietic stem cells*: the *myelocytic lineage* and the *lymphocytic lineage.* Granulocytes and monocytes are the products of the myelocytic lineage, whereas lymphocytes and plasma cells are the products of the lymphocytic lineage. The granulocytes and monocytes are formed only in the bone marrow. Lymphocytes and plasma cells are produced mainly in the various lymphoid organs, including the lymph glands, spleen, and thymus.

The life span of white blood cells varies. The main reason white blood cells are present in the blood is for transportation from the bone marrow or lymphoid tissue to areas of the body where they are needed. The life span of granulocytes released from the bone marrow is normally 4 to 5 hours in the circulating blood and an additional 4 to 5 days in the tissues. When there is serious tissue infection, the total life span is often shortened to only a few hours because the granulocytes proceed rapidly to the infected area, perform their function, and in the process are destroyed.

The monocytes also have a short transit time of 10 to 12 hours before they enter the tissues. Once in the tissues, they swell to a much larger size to become *tissue macrophages*, in which form they can live for months unless they are destroyed while performing phagocytic functions.

Lymphocytes enter the circulatory system continuously along with the drainage of lymph from the lymph nodes. After a few hours, they pass back into the tissue via diapedesis and re-enter the lymph to return to the blood again and again; thus, there is continuous circulation of the lymphocytes throughout the tissue. The lymphocytes have a life span of months or even years depending on the need of the body for these cells.

DEFENSIVE PROPERTIES OF NEUTROPHILS AND MACROPHAGES (p. 436)

It is mainly the neutrophils and monocytes that attack and destroy invading bacteria, viruses, and other injurious agents. The neutrophils are mature cells that can attack and destroy bacteria and viruses in the circulating blood. The blood monocytes are immature cells that have very little ability to fight infectious agents. Once they enter the tissue, they mature into tissue macrophages that are extremely capable of combating disease agents. Both the neutrophils and macrophages move through the tissues via ameboid motion when stimulated by products formed in inflamed areas. This attraction of the neutrophils and macrophages to the inflamed area is called *chemotaxis*.

One of the most important functions of the neutrophils and macrophages is *phagocytosis*. For obvious reasons, phagocytosis is very selective. Certain physical characteristics increase the chance for phagocytosis. Most natural structures in the tissue have smooth surfaces that resist phagocytosis; if the surface is rough, the likelihood of phagocytosis is increased. Most naturally occurring substances in the body have protective protein coats that repel phagocytosis. Dead tissues and most foreign particles frequently have no protective coat, which makes them subject to phagocytosis. The body also has specific means of recognizing certain foreign materials to which antibodies adhere; the binding of antibodies to foreign particles enhances phagocytosis.

Once a foreign particle has been phagocytized, lysosomes and other cytoplasmic granules immediately come in contact with the phagocytic vesicles and dump digestive enzymes and bactericidal agents into the vesicle. This digests the phagocytized particle.

INFLAMMATION AND FUNCTION OF NEUTROPHILS AND MACROPHAGES (p. 439)

When tissue injury occurs, multiple substances are released that cause secondary changes in the tissue. These substances increase local blood flow and permeability of the capillaries, which cause large quantities of fluid to leak into the interstitial spaces, migration of large numbers of granulocytes and monocytes into the tissues, and local swelling.

One of the first results of inflammation is to "wall off" the area of injury from the remaining tissues. The tissue spaces and lymphatics in the inflamed area are blocked by fibrinogen clots so that fluid barely flows through these spaces. This walling-off procedure delays the spread of bacteria or toxic products. The intensity of the inflammatory process is usually proportional to the degree of tissue injury. Staphylococci that invade the tissue liberate extremely lethal cellular toxins, which results in the rapid development of inflammation. Staphylococcal infections are characteristically walled off rapidly. By comparison, streptococci do not cause such intense local tissue destruction; therefore, the walling off develops slowly. As a result, streptococci have a far greater tendency to spread through the body and cause death than do staphylococci, even though staphylococci are far more destructive to the tissues.

Macrophage and Neutrophil Response to Inflammation
(p. 439)

The tissue macrophage is the first line of defense against invading organisms. Within minutes after inflammation begins, the macrophages present in the tissues immediately begin their phagocytic actions. Many sessile macrophages break loose from their attachments and become mobile in response to *chemotactic factors*. These macrophages migrate to the area of inflammation and contribute their activity.

Neutrophil invasion of the inflamed tissue is a second line of defense. Within the first hour or so after inflammation begins, large numbers of neutrophils invade the inflamed area as a result of products in the inflamed tissue that attract these cells and cause chemotaxis toward that area.

Within a few hours after the onset of severe acute inflammation, the number of neutrophils increases by as many as fourfold to fivefold. This *neutrophilia* is caused by inflammation products that are transported in the blood to the bone marrow, where they act to mobilize neutrophils from the marrow capillaries into the circulating blood. This process makes more neutrophils available to the inflamed tissue area.

A second macrophage invasion of the inflamed tissue is the third line of defense. Along with the invasion of neutrophils, monocytes from the blood enter the inflamed tissue and enlarge to become macrophages. The number of monocytes in the circulating blood is low, and the storage pool of monocytes in the bone marrow is much less than that of the neutrophils. The buildup of macrophages in inflamed tissue is much slower than that of neutrophils. After several days to several weeks, the macrophages become the dominant phagocytic cell in the inflamed area because of the increased bone marrow production of monocytes.

The fourth line of defense is greatly increased production of both granulocytes and monocytes by the bone marrow. This process results from stimulation of the granulocytic and monocytic progenitor cells of the marrow and takes 3 to 4 days for the newly formed granulocytes and monocytes to reach the stage of leaving the marrow area.

Many factors are involved in the feedback control of the macrophage and neutrophil response. More than two dozen factors have been implicated in the control of the macrophage-neutrophil response to inflammation. The five factors that are thought to play a dominant role are

1. *Tumor necrosis factor* (TNF)
2. *Interleukin-1* (IL-1)
3. *Granulocyte-monocyte colony stimulating factor* (GM-CSF)
4. *Granulocyte colony stimulating factor* (G-CSF)
5. *Monocyte colony stimulating factor* (M-CSF).

These five factors are formed by activated macrophages and T cells in the inflamed tissues. The main causes of the increased production of granulocytes and monocytes by the bone marrow are the three colony stimulating factors; this combination of TNF, IL-1, and colony stimulating factors provides a powerful feedback mechanism that begins with tissue inflammation and proceeds to the formation of defensive white blood cells and removal of the cause and inflammation.

Pus is formed. When the neutrophils and macrophages engulf large numbers of bacteria and necrotic tissue, essentially all the neutrophils and many of the macrophages eventually die. The combination of various portions of necrotic tissue, dead neutrophils, dead macrophages, and tissue fluid is commonly known as *pus*. When the infection has been suppressed, the dead cells and necrotic tissue in the pus gradually autolyze over a period of days and are absorbed into the surrounding tissues until most of the evidence of the tissue damage is gone.

Eosinophils are produced in large numbers in persons with parasitic infections. Most parasites are too large to be phagocytized. The eosinophils attach themselves to the surface of the parasites and release substances, such as hydrolytic enzymes, reactive forms of oxygen, and larvicidal polypeptides called *major basic proteins*; these substances kill many of the invading parasites.

The eosinophils normally constitute about 2 per cent of all the blood leukocytes. In addition to combating parasitic infections, eosinophils have a propensity to collect in tissues in which allergic reactions have occurred. The migration of the eosinophils to inflamed allergic tissue results from the release of eosinophil chemotactic factor from mast cells and basophils. The eosinophils are believed to detoxify some of the inflammation-inducing substances released by the mast cells and basophils and destroy allergen-antibody complexes, thus preventing the spread of the inflammatory process.

Basophils are circulating mast cells. Mast cells and basophils liberate heparin into the blood, which prevents blood coagulation. These cells release histamine as well as smaller quantities of bradykinin and serotonin, which contribute to the inflammation process. The mast cells and basophils play an important role in some types of allergic reactions. The IgE class of antibodies (those responsible for allergic reactions) has a propensity to become attached to mast cells and basophils. The resulting attachment of the allergic antigen to the IgE antibody causes the mast cells or basophils to rupture and release exceedingly large quantities of histamine, bradykinin, serotonin, heparin, slow-reacting substance of anaphylaxis, and lysosomal enzymes. These substances in turn cause local vascular and tissue reactions that are characteristic of allergic manifestation.

THE LEUKEMIAS (p. 442)

The leukemias are divided into two general types: *lymphogenous* and *myelogenous*. The lymphogenous leukemias are caused by uncontrolled cancerous production of lymphoid cells, which usually begins in a lymph node or other lymphogenous tissue and then spreads to other areas of the body. The myelogenous leukemias begin by the cancerous production of young myelogenous cells in the bone marrow and then spread throughout the body so that the white blood cells are produced by many extramedullary organs. Leukemic cells are usually nonfunctional, so they cannot provide the usual protection against infection associated with white blood cells.

Almost all leukemias spread to the spleen, lymph nodes, liver, and other regions that have a rich vascular supply regardless of whether the origin of the leukemia is in the bone marrow or lymph nodes. The rapidly growing cells invade the surrounding tissues, using the metabolic elements of these tissues, and subsequently cause tissue destruction via metabolic starvation.

34

Resistance of the Body to Infection
II. Immunity and Allergy

INNATE AND ACQUIRED IMMUNITY

Immunity is the ability to resist almost all types of organisms or toxins that damage the tissues of the body. Most organisms have *innate immunity*, which consists of general actions such as phagocytosis of bacteria, destruction of pathogens by acid secretions, digestive enzymes in the gastrointestinal tract, resistance of the skin to invasion, and certain chemicals in the blood that attach to foreign organisms or toxins and destroy them. *Acquired immunity* is the ability to develop extremely powerful protective mechanisms against specific invading agents such as lethal bacteria, viruses, toxins, and even foreign tissues from other organisms.

Acquired immunity is initiated by antigens. Two basic types of acquired immunity occur in the body. *Humoral immunity*, or *B cell immunity*, involves the development of circulating antibodies that are capable of attacking an invading agent. *Cell-mediated immunity*, or *T-cell immunity*, is achieved through the formation of large numbers of activated lymphocytes that are specifically designed to destroy the foreign agent.

Because acquired immunity does not occur until after the invasion by a foreign organism or toxin, the body must have some mechanism for recognizing the invasion. Each invading organism or toxin usually contains one or more specific chemical compounds that are different from all other compounds; these compounds are called *antigens*, and they initiate the development of acquired immunity.

For a substance to be antigenic, it usually must have a molecular weight of at least 8000. The process of antigenicity depends on the regular occurrence on the surface of the large molecules of molecular groups called *epitopes*; proteins and large

303

polysaccharides are almost always antigenic because they contain this type of stereochemical characteristic.

Lymphocytes are the basis of acquired immunity. Lymphocytes are found in the lymph nodes and in special lymphoid tissue such as the spleen, submucosal areas of the gastrointestinal tract, and bone marrow. Lymphoid tissue is distributed advantageously in the body to intercept invading organisms and toxins before the invaders can become too widespread.

There are two populations of lymphocytes; both are derived from *pluripotent hemopoietic stem cells* that differentiate to form lymphocytes. One population of lymphocytes is processed in the thymus gland; these are called *T lymphocytes*, and they are responsible for cell-mediated immunity. Another population of lymphocytes is processed in the liver during mid fetal life and in the bone marrow during late fetal life and after birth; these are called *B lymphocytes*, and they are responsible for humoral immunity.

The thymus gland preprocesses T lymphocytes. Lymphocytes divide rapidly and develop extreme diversity for reacting against different specific antigens in the thymus gland. The processed *T cells* leave the thymus and spread to lymphoid tissues throughout the body. Most of the preprocessing of the T lymphocytes in the thymus occurs shortly before and after birth. Removal of the thymus gland after this time diminishes but does not eliminate the T-lymphocyte system. Removal of the thymus several months before birth, however, prevents the development of all cell-mediated immunity.

The liver and bone marrow preprocess the B lymphocytes. Much less is known about the details or processing of B lymphocytes. In humans, B lymphocytes are known to be preprocessed in the liver during mid fetal life and in bone marrow during late fetal life and after birth. B lymphocytes differ from T lymphocytes; they actively secrete antibodies, which are large protein molecules capable of combining with and destroying substances. B lymphocytes have a greater diversity than T lymphocytes, forming millions, perhaps even billions, of antibodies with different specific reactivities. After processing, B lymphocytes migrate to lymphoid tissues throughout the body, where they lodge in locations near the T-lymphocyte areas.

When a specific antigen comes in contact with the T and B lymphocytes in the lymphoid tissue, a set of T and B lymphocytes becomes activated to form *activated T cells* and *activated B cells*, which subsequently form antibodies. The activated T cells and newly formed antibodies react specifically with the antigen that initiated their development and inactivate or destroy the antigen.

A preformed repertoire of lymphocytes awaits activation by an antigen. There are millions of different types of preformed T and B lymphocytes that are capable of responding to the appropriate antigen. Each of these preformed lymphocytes is capable of forming only one type of antibody or one type of T cell with a single type of specificity. Once the specific lymphocyte is activated by its antigen, it reproduces wildly, forming tremendous numbers of duplicate lymphocytes. If the lymphocyte is a B lymphocyte, the progeny will eventually secrete antibodies that circulate throughout the body. If the lymphocyte is a T lymphocyte, its progeny develop into *sensitized T cells* that are released into the blood, where they circulate through the tissue fluids throughout the body and back into the lymph. Each different set of lymphocytes that is capable of forming one specific antibody or activated T cell is called a *clone of lymphocytes*. The lymphocytes in each clone are identical, and all are derived from one progenitor lymphocyte of a specific type.

SPECIFIC ATTRIBUTES OF THE B-LYMPHOCYTE SYSTEM— HUMORAL IMMUNITY AND THE ANTIBODY (p. 448)

On entry of a foreign antigen, the macrophages in the lymphoid tissue phagocytize the antigen and present it to adjacent B lymphocytes. The previously dormant B lymphocytes specific for the antigen immediately enlarge and eventually become *plasma cells*. The plasma cells produce gamma globulin antibodies, which are secreted into the lymph and carried to the circulating blood.

The formation of memory cells enhances the immune response to subsequent antigen exposure. Some of the B lymphocytes formed during activation of the specific clone do not form plasma cells but instead form new B lymphocytes similar to those of the original clone. This causes the population of the specifically activated clone to become greatly enhanced. These B lymphocytes circulate throughout the body and inhabit all the lymphoid tissue but remain immunologically dormant until activated again by a new quantity of the same antigen. The cells of the expanded clone of lymphocytes are called *memory cells*. Subsequent exposure to the same antigen will cause a more rapid and potent antibody response because of the increased number of lymphocytes in the specific clone. The increased potency and duration of the secondary response are the reasons why *vaccination* is usually accomplished through the injection of antigen in multiple doses with periods of several weeks or months between injections.

Antibodies are gamma globulin proteins called *immunoglobulins*. All immunoglobulins are composed of combinations of light and heavy polypeptide chains. Each light and heavy chain has a *variable portion* and *constant portion*. The variable portion is different for each specific antibody; it is this portion that attaches to a particular type of antigen. The constant portion determines other properties of the antibody, such as diffusibility, adherence to structures within tissues, and attachment to the complement complex. There are five general classes of antibodies, each with a specific function: IgM, IgA, IgG, IgD, and IgE. The IgG class is the largest and constitutes about 75 per cent of the antibodies of a normal person.

Antibodies act by directly attacking the invader or activating the complement system, which subsequently destroys the invading organism. The antibodies can inactivate the invading agent directly in one of the following ways:

- *Agglutination*, in which multiple large particles with antigens on their surfaces, such as bacteria or red cells, are bound together in a clump.
- *Precipitation*, in which the molecular complex of soluble antigens and antibodies becomes so large that it is rendered insoluble.
- *Neutralization*, in which the antibodies cover the toxic sites of the antigenic agent.
- *Lysis*, in which antibodies are occasionally capable of causing rupture of an invading cell by directly attacking the cellular membranes.

Although antibodies have some direct effects in the destruction of the invaders, most of the protection afforded by antibodies comes from the amplifying effects of the complement system.

The complement system is activated by antigen-antibody reaction. *Complement* is a collective term that is used to describe a system of proteins normally present in the plasma that can be activated by the antigen–antibody reaction. When an antibody binds with an antigen, a specific reactive site on the *constant* portion of the antibody becomes uncovered, or activated. This activated antibody site binds directly with the C1 molecule of the complement system, setting into motion a *cascade* of sequential reactions. When complement is activated, multiple end products are formed. Several of these products aid in the destruction of the invading organism or neutralization of a toxin.

Complement can stimulate phagocytosis by both neutrophils and macrophages, cause rupture of the cell membranes of bacteria or other invading organisms, promote agglutination, attack the structure of viruses, promote chemotaxis of neutrophils and macrophages, and induce the release of histamine by mast cells and basophils, promoting vasodilatation and leakage of plasma, which in turn promote the inflammatory process. Activation of complement by an antigen–antibody reaction is called the *classical pathway*.

An *alternate pathway* for complement activation occurs without the mediation of an antigen–antibody reaction. The cell membranes of some invading micro-organisms contain large polysaccharide molecules that are capable of activating the complement cascade. The alternate pathway does not involve an antigen–antibody reaction and is capable of providing some protection against invasion before a person becomes immunized against such agents; it is one of the first lines of defense against invading micro-organisms.

SPECIAL ATTRIBUTES OF THE T-LYMPHOCYTE SYSTEM— ACTIVATED T CELLS AND CELL-MEDIATED IMMUNITY
(p. 451)

When macrophages present a specific antigen, T lymphocytes of the specific lymphoid clone proliferate, causing large numbers of activated T cells to be released in the same way antibodies are released by the activated B cells. These activated T cells pass into the circulation and are distributed throughout the body, where they circulate for months or even years. *T-lymphocyte memory cells* are formed in the same way that B memory cells are formed in the antibody system; on subsequent exposure to the same antigen, the release of activated T cells occurs far more rapidly and much more powerfully than in the first response.

Antigens bind with *receptor molecules* on the surface of the T cells in the same way that they bind with antibodies. These receptor molecules are composed of a variable unit similar to the variable portion of the humoral antibody, but the stem section of the receptor molecule is firmly bound to the cell membrane.

Several Types of T Cells with Different Functions

The three major groups of T cells are (1) *helper T cells,* (2) *cytotoxic T cells,* and (3) *suppresser T cells.* The function of each of these cell types is quite distinct.

Helper T cells are the most numerous type of T cell in the body. Helper T cells serve as regulators of virtually all immune functions. This task is accomplished through the formation of a series of protein mediators called *lymphokines* that act on other cells of the immune system and bone marrow. Helper T cells secrete the *interleukin-2* through *-6, granulocyte-monocyte colony stimulating factor,* and *interferon-γ.* In the absence of the lymphokines produced by the helper T cells, the remainder of the immune system is almost paralyzed. It is the helper T cells that are inactivated or destroyed by the *acquired immunodeficiency syndrome* virus, which leaves the body almost totally unprotected against infectious disease.

Helper T cells perform the following functions:

- *Stimulation of growth and proliferation of cytotoxic and suppresser T cells* through the actions of interleukin-2, -4, and -5
- *Stimulation of B cell growth and differentiation to form plasma cells and antibodies* mainly through the actions of interleukin-4, -5, and -6
- *Activation of the macrophage system*
- *Stimulation of helper T cells themselves*—Interleukin-2 has a direct positive feedback effect of stimulating activation of the helper T cells, which acts as an amplifier to further enhance the cellular immune response.

Cytotoxic T cells are capable of killing micro-organisms through a direct attack. For this reason, they are also called *killer cells*. Surface receptors on the cytotoxic T cells cause them to bind tightly to those organisms or cells that contain their binding-specific antigen. After binding, the cytotoxic T cells secrete *hole-forming proteins*, called *perforans*, that literally punch large holes in the membrane of the attacked cells. These holes disrupt the osmotic equilibrium of the cells, which leads to cell death. Cytotoxic T cells are especially important in destroying cells infected by viruses, cancer cells, or transplanted organ cells.

Suppresser T cells suppress the functions of both cytotoxic and helper T cells. It is believed that these suppresser functions serve the purpose of regulating the activities of the other cells so that excessive immune reactions that might severely damage the body do not occur.

IMMUNOLOGIC TOLERANCE (p. 452)

The immune system normally recognizes a person's own tissue as being completely distinct from that of invading organisms. It is believed that most of the phenomenon of *tolerance* develops during the processing of T lymphocytes in the thymus and B lymphocytes in the bone marrow. The mechanism of tolerance induction is not completely understood; however, it is thought that the continuous exposure to self-antigen in the fetus causes the self-reacting T and B lymphocytes to be destroyed.

Failure of the tolerance mechanism leads to auto-immune diseases in which the immune system attacks the tissues of the body, such as *rheumatic fever*, in which the body becomes immunized against the tissues of the joints and valves of the heart; *glomerulonephritis*, in which the body becomes immunized against the basement membrane of the glomeruli; *myasthenia gravis*, in which the body becomes immunized against the acetylcholine receptor proteins of the neuromuscular junction; and *lupus erythematosus*, in which the body becomes immunized against many different tissues.

ALLERGY AND HYPERSENSITIVITY (p. 453)

An important but undesirable side effect of immunity is the development of *allergy* or other types of *immune hypersensitivity*. Allergy can be caused by activated T cells and can cause skin eruptions, edema, or asthmatic attacks in response to certain chemicals or drugs. In some individuals, a resin in the poison ivy plant induces the formation of activated helper and cytotoxic T cells that diffuse into the skin and elicit a cell-mediated characteristic type of immune reaction to this plant.

Some types of allergies are caused by IgE antibodies; these antibodies are called *reagins*, or *sensitizing antibodies*, to distinguish them from the more common IgG antibodies. A special characteristic of IgE antibodies is their ability to bind strongly with mast cells and basophils, which causes the release of multiple substances that induce vasodilation, increased capillary permeability, and attraction of neutrophils and eosinophils. *Hives, hay fever*, and *asthma* can result from this mechanism.

35

Blood Groups; Transfusion; Tissue and Organ Transplantation

O-A-B BLOOD GROUPS (p. 457)

The antigens *type A* and *type B* occur on the surfaces of red blood cells in a large proportion of the population. These antigens, or *agglutinogens*, cause blood transfusion reactions. It is on the basis of the presence or absence of the agglutinogens on the red blood cells that blood is grouped for the purpose of transfusion. When neither A nor B agglutinogen is present, the blood group is *type O*. When only the type A agglutinogen is present, the blood group is *type A*. When only type B agglutinogen is present, the blood group is *type B*. When both type A and B agglutinogens are present, the blood group is *type AB*.

When type A agglutinogen is not present on a person's red blood cells, antibodies known as *anti-A agglutinins* develop in the plasma. When type B agglutinogen is not present on the red blood cells, antibodies known as *anti-B agglutinins* develop in the plasma. Type O blood contains both anti-A and anti-B agglutinins, and type A blood contains type A agglutinogens and anti-B agglutinins. Type B blood contains type B agglutinogens and anti-A agglutinins; type AB blood contains both type A and B agglutinogens but no agglutinins.

The agglutinins are gamma-globulins of the IgM and IgG immunoglobulin subclasses. The origin of the agglutinins in individuals who do not have the antigenic substance in their blood seems to result from the entry into the body of small numbers of group A and group B antigens in food and through contact with bacteria.

When bloods are mismatched so that anti-A or anti-B plasma agglutinins are mixed with red cells containing A or B agglutinogens, the red cells agglutinate into clumps. These clumps can plug small

311

blood vessels throughout the circulatory system. In some cases, the antibodies induce the lysis of red blood cells through activation of the complement system.

One of the most lethal effects of transfusion reactions is renal failure. The excess hemoglobin from the hemolyzed red cells leaks through the glomerular membranes into the renal tubules. Reabsorption of water from the tubules causes the concentration of hemoglobin to rise, resulting in hemoglobin precipitation and subsequent blockade of the tubules.

Rh BLOOD TYPES (p. 459)

The Rh system is another important factor in blood transfusion. In the Rh system, spontaneous occurrence of agglutinins almost never happens; instead, the individual must first be exposed to an Rh antigen, usually through transfusion of blood or pregnancy. When red blood cells containing Rh factor are injected into a person without the factor, anti-Rh agglutinins develop and reach a maximum concentration in about 2 to 4 months. On multiple exposures to the Rh factor, the Rh-negative person eventually becomes strongly sensitized to the Rh factor. The mismatch of Rh factor blood leads to agglutination and hemolysis.

Erythroblastosis fetalis is a disease of fetuses and newborn infants that is characterized by progressive agglutination and subsequent phagocytosis of red blood cells. In a typical case, the mother is Rh negative and the father is Rh positive. If the baby has inherited the Rh-positive antigen from the father and the mother has developed anti-Rh agglutinins in response to this antigen, these agglutinins can diffuse through the placenta into the fetal circulation and cause red blood cell agglutination.

TRANSPLANTATION OF TISSUES AND ORGANS (p. 460)

An *autograft* is the transplantation of tissues or whole organs from one part of the body to another. An *isograft* is the transplantation of an organ from one identical twin to another. An *allograft* is the transplantation of an organ from one human being to another. A *xenograft* is the transplantation of an organ from one species to another.

In the case of autografts and isografts, all cells in the transplanted organ contain virtually the same antigens and will survive indefinitely if provided with an adequate blood supply. In the case of allografts and xenografts, immune reactions almost always occur. These reactions cause the cells within the graft to die within 1 to 5 weeks after the transplantation unless specific therapy is given to prevent the immune reaction. When the tissues are properly "typed" and are similar between donor and recipient for their cellular antigens, successful long-term allograft survival can occur. Simultaneous drug therapy is needed to minimize the immune reactions.

Tissue typing is performed to identify the HLA complex of antigens. The most important antigens in graft rejection are a complex called the *HLA antigens*. Only six of these antigens are ever present on the cell surface of any one person, but there are more than 150 different types of HLA antigens; this number represents more than a trillion possible combinations. As a consequence, it is virtually impossible for two individuals, with the exception of identical twins, to have the same six HLA antigens.

The HLA antigens occur on white blood cells as well as on the cells of the tissues. Some of the HLA antigens are not severely antigenic; therefore, a precise match of antigens between donor and recipient is not essential to allograft survival, but the best results occur in the closest possible match between donor and recipient.

Prevention of graft rejection can be accomplished by suppressing the immune system with *(1) glucocorticoid hormones*; (2) various drugs that are toxic to the lymphoid system, such as *azathioprine* (Imuran); or *(3) cyclosporine*, which has a specific inhibitory effect on the formation of helper T cells. This drug is especially efficacious in blocking T-cell–mediated rejection reactions.

36

Hemostasis and Blood Coagulation

The term *hemostasis* means prevention of blood loss. Whenever a vessel is severed or ruptured, hemostasis is achieved through (1) vascular spasm, (2) formation of a platelet plug, (3) formation of a blood clot as a result of blood coagulation, and (4) eventual growth of fibrous tissue to close the rupture permanently.

- *Trauma to the blood vessel causes the wall of the blood vessel to constrict.* The constriction results from nervous reflexes, local myogenic spasms, and local humoral factors released from the traumatized tissue and blood platelets, such as the vasoconstrictor substance *thromboxane A_2.*
- *A platelet plug can fill a small hole in a blood vessel.* When platelets come in contact with a damaged vascular surface, they begin to swell and assume irregular forms, release granules containing multiple factors, which increase the adherence of the platelets (i.e., adenosine diphosphate), and form thromboxane A_2. The adenosine diphosphate and thromboxane act on nearby platelets to activate them, so that they adhere to the originally activated platelets, thus forming a platelet plug.
- *Formation of the blood clot is the third mechanism for hemostasis.* Clot formation begins to develop within 15 to 20 seconds if the trauma to the vascular wall has been severe and within 1 to 2 minutes if the trauma has been minor. Within 3 to 6 minutes after rupture of a vessel, the entire opening or the broken end of the vessel is filled with the clot (if the vessel opening was not too large). After 20 minutes to 1 hour, the clot retracts; this closes the vessel further. Once a blood clot has formed, it is invaded by fibroblasts, which subsequently form connective tissue throughout the clot.

MECHANISM OF BLOOD COAGULATION (p. 464)

Blood coagulation takes place in three essential steps:

- A complex of substances called *prothrombin activator* is formed in response to rupture or damage to the blood vessel.
- Prothrombin activator catalyzes the conversion of *prothrombin* into *thrombin*.
- The thrombin acts as an enzyme to convert *fibrinogen* into *fibrin threads* that enmesh platelets, blood cells, and plasma to form the clot.

Prothrombin is converted to thrombin. Prothrombin is an unstable plasma protein that can easily split into smaller compounds, one of which is thrombin. Prothrombin is formed continuously by the liver. If the liver fails to produce prothrombin, the concentration in the plasma falls too low to provide normal blood coagulation within 24 hours. *Vitamin K* is required by the liver for normal formation of prothrombin; therefore, either the lack of vitamin K or the presence of liver disease prevents normal prothrombin formation and results in bleeding tendencies.

Fibrinogen is converted to fibrin, and a clot forms. Fibrinogen is a high-molecular-weight protein formed in the liver. Because of its large molecular size, very little fibrinogen normally leaks through the capillary pores into the interstitial fluid. Thrombin is an enzyme that acts on the fibrinogen molecule to remove four low-molecular-weight peptides to form a molecule of *fibrin monomer*. The fibrin monomer molecule polymerizes with other fibrin monomer molecules to form the long fibrin threads that produce the reticulum of the clot. The newly formed fibrin reticulum is strengthened by a substance called *fibrin-stabilizing factor*, which is normally present in small amounts in the plasma. This substance is also released from platelets entrapped in the clot. Fibrin-stabilizing factor is an enzyme that causes covalent bonding between the fibrin monomer molecules and adjacent fibrin threads, thereby strengthening the fibrin meshwork.

In the initiation of coagulation, prothrombin activator is formed. Prothrombin activator can be formed in two basic ways: (1) by the *extrinsic pathway*, which begins with trauma to the vascular wall and surrounding tissue, and (2) by the *intrinsic path-*

way, which begins in the blood itself. Both pathways involve a series of beta-globulin plasma proteins. These blood clotting factors are proteolytic enzymes that induce the successive cascading reactions of the clotting process.

- *The extrinsic mechanism* for initiating the formation of prothrombin activator begins with trauma to the vascular wall or extravascular tissues and occurs according to the following three steps:
  1. *Release of tissue thromboplastin*—Traumatized tissue releases a complex of several factors called tissue thromboplastin; these factors include phospholipids from the membranes of the traumatized tissue and a lipoprotein complex that functions as a proteolytic enzyme.
  2. *Activation of Factor X to form activated Factor X*—The lipoprotein complex of tissue thromboplastin complexes with *blood coagulation Factor VII* and in the presence of tissue phospholipids and calcium ions acts enzymatically on Factor X to form activated Factor X.
  3. *Effect of activated Factor X to form prothrombin activator*—The activated Factor X complexes immediately with the tissue phospholipid released as part of the tissue thromboplastin and with *Factor V* to form a complex called prothrombin activator. Within a few seconds, this splits prothrombin to form thrombin, and the clotting process precedes as previously described. Activated Factor X is the actual protease that causes splitting of prothrombin to thrombin.
- *The intrinsic mechanism* for initiation of the formation of prothrombin activator begins with trauma to the blood or exposure of the blood to collagen in the traumatized vascular wall. This occurs via the following cascade of reactions:
  1. *Activation of Factor XII and release of platelet phospholipids*—Through trauma, Factor XII is activated to form a proteolytic enzyme called *activated Factor XII*. Simultaneously, the blood trauma damages the blood platelets, which causes the release of platelet phospholipids containing a lipoprotein called *platelet factor 3*, which plays a role in subsequent clotting reactions.
  2. *Activation of Factor XI*—The activated Factor XII acts enzymatically on Factor XI to activate Fac-

tor XI. This second step in the intrinsic pathway requires high-molecular-weight *kininogen*.

3. *Activation of Factor IX by activated Factor XI*— The activated Factor XI then acts enzymatically on Factor IX to activate this factor.

4. *Activation of Factor X*—The activated Factor IX, acting in concert with Factor VIII and with platelet phospholipids and Factor III from the traumatized platelets, activates Factor X. When either Factor VIII or platelets are in short supply, this step is deficient. Factor VIII is the factor that is missing in the person who has *classic hemophilia*. Platelets are the clotting factor lacking in the bleeding disease called *thrombocytopenia*.

5. *Activation of activated Factor X to form prothrombin activator*—This step in the intrinsic pathway is the same as the last step in the extrinsic pathway (i.e., activated Factor X combines with Factor V and platelets or tissue phospholipids to form the complex called prothrombin activator). The prothrombin activator in turn initiates cleavage of prothrombin to form thrombin, thereby setting into motion the final clotting process.

Calcium ions are required for blood clotting. Except for the first two steps in the intrinsic pathway, calcium ions are required for promotion of all the reactions; in the absence of calcium ions, blood clotting will not occur. Fortunately, the calcium ion concentration rarely falls sufficiently low to significantly affect the kinetics of blood clotting. When blood is removed, it can be prevented from clotting by reducing the calcium ion concentration below the threshold level for clotting. This can be accomplished through either deionization of the calcium via reaction with substances such as a *citrate ion* or precipitation of the calcium with substances such as an *oxalate ion*.

Prevention of Blood Clotting in the Normal Vascular System—The Intravascular Anticoagulants (p. 468)

The most important factors for the prevention of clotting in the normal vascular system are the (1) a *smoothness of the endothelium*, which prevents contact activation of the intrinsic clotting system; (2) a *layer of glycocalyx in the endothelium,* which repels the clotting factors and platelets; and (3) a *protein bound with the endothelial membrane* (called *thrombomodulin*), which binds thrombin. The thrombomodulin–thrombin complex also activates a plasma protein called *protein C*, which inactivates activated Factors V and VIII. When the endothelial wall is damaged, its smoothness and its glycocalyx–thrombomodulin layer are lost, which activates Factor XII and platelets and initiates the intrinsic pathway of clotting.

Agents that remove thrombin from blood, such as the fibrin threads that form during the process of clotting and an alpha-globulin called *antithrombin III*, are the most important anticoagulants in the blood. Thrombin becomes absorbed to the fibrin threads as they develop; this prevents the spread of thrombin into the remaining blood and prevents excessive spread of the clot. The thrombin that does not absorb to the fibrin threads combines with antithrombin III, which inactivates the thrombin.

Heparin increases the effectiveness of antithrombin III in removing thrombin. In the presence of excess heparin, the removal of thrombin from the circulation is almost instantaneous. *Mast cells* located in the pericapillary connective tissue throughout the body and the *basophils* of the blood produce heparin. These cells continually secrete small amounts of heparin that diffuse into the circulatory system.

Plasmin lyses blood clots. *Plasminogen* is a plasma protein that when activated becomes a substance called *plasmin*, a proteolytic enzyme that resembles trypsin. Plasmin digests the fibrin threads as well as other clotting factors. Plasminogen becomes trapped in the clot along with other plasma proteins.

The injured tissues and vascular endothelium slowly release a powerful activator called *tissue plasminogen activator* (t-PA) that converts plasminogen to plasmin and removes the clot. Plasmin not only destroys fibrin fibers but also functions as a proteolytic enzyme to digest fibrinogen and a number of other clotting factors. Small amounts of plasmin are continuously formed in the blood. The blood also contains another factor, *alpha$_2$-antiplasmin*, that binds with plasmin and causes inactivation; the rate of plasmin formation must rise above a certain critical level before it becomes effective.

CONDITIONS THAT CAUSE EXCESSIVE BLEEDING (p. 469)

Excessive bleeding can result from a deficiency of vitamin K, hemophilia, or thrombocytopenia (platelet deficiency). Vitamin K is necessary for the formation of five important clotting factors: *prothrombin, Factor VII, Factor IX, Factor X,* and *protein C.* In the absence of vitamin K, insufficiency of these coagulation factors can lead to a serious bleeding tendency.

Hemophilia is caused by a deficiency of Factor VIII or Factor IX and occurs almost exclusively in males. *Hemophilia A,* or *classic hemophilia,* is caused by a deficiency of Factor VIII and accounts for about 85 per cent of cases. The other 15 per cent of cases of hemophilia are the result of a deficiency of Factor IX. Both of these factors are transmitted genetically via the female chromosome as a recessive trait; a woman will almost never have hemophilia because at least one of her two X chromosomes will have the appropriate genes.

Thrombocytopenia is a deficiency of platelets in the circulatory system. People with thrombocytopenia have a tendency to bleed from small vessels or capillaries. As a result, small punctate hemorrhages occur throughout the body tissues. The skin of such a person displays many small, purplish blotches, giving the disease the name *thrombocytopenic purpura.*

THROMBOEMBOLIC CONDITIONS (p. 470)

An abnormal clot that develops in a blood vessel is called a *thrombus.* An *embolus* is a freely flowing thrombus. Emboli generally do not stop flowing until they come to a narrow point in the circulatory system. Thromboembolic conditions in human beings are usually the result of a roughened endothelial surface or a sluggish blood flow. The rough endothelium can initiate the clotting process. When blood flow is too slow, the concentration of procoagulant factors often rises high enough in a local area to initiate clotting.

ANTICOAGULANTS FOR CLINICAL USE (p. 471)

- *Heparin* is extracted from several animal tissues and prepared in almost pure form. Heparin increases the effectiveness of *antithrombin III.* The action of heparin in the body is almost instantaneous and at normal dosages (0.5 to 1 mg/kg) can increase clotting time from about 6 to 30 minutes or longer. If too much heparin is given, a substance called *protamine* can be administered, which combines electrostatically with heparin to cause its inactivation.

- *Coumarins* such as *warfarin* cause the plasma levels of prothrombin and Factors VIII, IX, and X to fall. Warfarin causes this effect by competing with vitamin K for reactive sites in the enzymatic processes for the formation of prothrombin and the other three clotting factors.

Respiration

Respiration

37

Pulmonary Ventilation

The respiratory system supplies oxygen to the tissues and removes carbon dioxide. The major functional events of respiration include (1) pulmonary ventilation, which is how air moves in and out of the alveoli; (2) diffusion of oxygen and carbon dioxide between the blood and alveoli; (3) transport of oxygen and carbon dioxide to and from the peripheral tissues; and (4) regulation of respiration. This chapter provides a discussion of pulmonary ventilation.

MECHANICS OF PULMONARY VENTILATION (p. 477)

Muscles That Cause Lung Expansion and Contraction

Lung volume increases and decreases as the thoracic cavity expands and contracts. The lungs are held to the thoracic wall as if glued, except that they can slide freely in the thoracic cavity. Any time the length or thickness of the thoracic cavity increases or decreases, simultaneous changes in lung volume occur. The lungs can be expanded and contracted in two ways: (1) by downward and upward movement of the diaphragm to lengthen or shorten the chest cavity and (2) by elevation and depression of the ribs to increase and decrease the anteroposterior diameter of the chest cavity.

- *Normal quiet breathing* is accomplished almost entirely with the diaphragm. During inspiration, contraction of the diaphragm pulls the lower surfaces of the lungs downward. During expiration, the diaphragm simply relaxes, and the elastic recoil of the lungs, chest wall, and abdominal structures compresses the lungs.
- During *heavy breathing*, the elastic forces are not sufficiently powerful to cause the rapid expiration required during heavy breathing. The extra force is achieved mainly through contraction of the ab-

dominal muscles, which pushes the abdominal contents upward against the bottom of the diaphragm.

Raising and lowering of the rib cage cause the lungs to expand and contract. In the natural resting position, the ribs slant downward, allowing the sternum to fall backward toward the vertebral column. When the rib cage is elevated, the ribs project almost directly forward so that the sternum also moves forward and away from the spine, increasing the anteroposterior thickness of the chest.

- *Muscles that raise the rib cage are muscles of inspiration*; these include the external intercostals, but others that help are (1) the sternocleidomastoid muscles, which lift upward on the sternum; (2) the anterior serrati, which lift many of the ribs; and (3) the scaleni, which lift the first two ribs.
- *Muscles that depress the rib cage are muscles of expiration*; these include (1) the abdominal recti, which pull downward on the lower ribs at the same time that they and the other abdominal muscles compress the abdominal contents upward toward the diaphragm; and (2) the internal intercostals.

Movement of Air In and Out of the Lungs—and the Pressures That Cause It

Pleural pressure is the pressure of the fluid in the narrow space between the lung pleura and chest wall pleura. The normal pleural pressure at the beginning of inspiration is about −5 centimeters of water, which is the amount of suction that is required to hold the lungs open to their resting level. During normal inspiration, the expansion of the chest cage pulls the surface of the lungs with still greater force and creates a still more negative pressure to an average of about −7.5 centimeters of water.

Alveolar pressure is the air pressure inside the lung alveoli. When the glottis is open and there is no movement of air, the pressures in all parts of the respiratory tree are equal to atmospheric pressure, which is considered to be 0 centimeter of water.

- *During inspiration*, the pressure in the alveoli decreases to about −1 centimeter of water. This slight negative pressure is sufficient to move about 0.5 liter of air into the lungs within the 2 seconds required for inspiration.

- *During expiration*, opposite changes occur: The alveolar pressure rises to about +1 centimeter of water, and this forces the 0.5 liter of inspired air out of the lungs during the 2 to 3 seconds of expiration.

Lung compliance is the change in lung volume for each unit change in transpulmonary pressure. Transpulmonary pressure is the difference in pressure between the alveolar pressure and pleural pressure. The normal total compliance of both lungs together in the average adult is about 200 milliliters per centimeter of water—every time the transpulmonary pressure increases by 1 centimeter of water, the lungs expand by 200 milliliters. Compliance is determined on the basis of the following elastic forces:

- *Elastic forces of the lung tissue* are determined mainly by the elastin and collagen fibers. In deflated lungs, the fibers are contracted; when the lungs are expanded, the fibers become stretched, thereby exerting elastic force to return to their natural state.
- *Elastic forces caused by surface tension* are much more complex. Surface tension within the alveoli, however, accounts for about two thirds of the total elastic forces in the normal lungs.

"Surfactant," Surface Tension, and Collapse of the Lungs
(p. 479)

Water molecules have a strong attraction for one another. This attraction is what holds raindrops together (i.e., there is a tight contractile membrane of water molecules around the entire surface of the raindrop). The water surface lining the alveoli is also attempting to contract. This attempts to force the air out of the alveoli through the bronchi, and in doing so, it causes the alveoli to attempt to collapse. The net effect is to cause an elastic contractile force of the entire lungs, which is called the *surface tension elastic force*.

Surfactant reduces the work of breathing (increases compliance) by decreasing alveolar surface tension. Surfactant is secreted by type II alveolar epithelial cells. Its most important component is the phospholipid dipalmitoylphosphatidylcholine, which is mainly responsible for reducing the surface tension. Surfactant is spread over the alveolar surface and reduces the surface tension to $\frac{1}{12}$ to one half of the surface tension of a pure water surface, depending on the concentration and orientation of the surfactant molecules on the surface.

Smaller alveoli have a greater tendency to collapse. Note from the following formula that the collapse pressure generated in the alveoli is inversely affected by the radius of the alveolus, which means that the smaller the alveolus, the greater becomes the collapse pressure:

$$\text{Pressure} = (2 \times \text{Surface Tension})/\text{Radius}$$

When the alveoli have one half of the normal radius, the collapse pressures are doubled.

Surfactant, "interdependence," and lung fibrous tissue are important for "stabilizing" the sizes of the alveoli. If some of the alveoli were small and others were large, theoretically the smaller alveoli would tend to collapse, decreasing their volume in the lungs. This loss of volume would cause expansion of the larger alveoli. There are several reasons why this instability of alveoli does not occur in the normal lung.

- *Interdependence*—The adjacent alveoli, alveolar ducts, and other air spaces tend to splint each other in such a way that a large alveolus usually cannot exist adjacent to a small alveolus because they share common septal walls; this is the *interdependence phenomenon*.
- *Fibrous tissue*—The lung is constructed of about 50,000 functional units, each of which contains one or a few alveolar ducts and their associated alveoli. All of these are surrounded by fibrous septa that act as additional splints.
- *Surfactant*—Surfactant reduces surface tension, allowing the interdependence phenomenon and fibrous tissue to overcome the surface tension effects. As an alveolus becomes smaller, the surfactant molecules on the alveolar surface are squeezed together, increasing their concentration; this reduces still further the surface tension.

PULMONARY VOLUMES AND CAPACITIES (p. 482)

Pulmonary volumes and capacities are measured with a spirometer. Figure 37–1 shows a recording

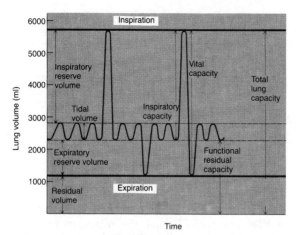

Figure 37–1 Diagram showing respiratory excursions during normal breathing and during maximal inspiration and maximum expiration.

for successive breath cycles at different depths of inspiration and expiration. The recording was made using an apparatus called a *spirometer*, which is a drum that is inverted in water with a tube extending from the air space in the drum to the mouth of the person being tested. As the person breathes in and out, the drum moves up and down, and a recording of the movement is made.

The pulmonary volumes added together equal the maximum volume to which the lungs can be expanded. The four pulmonary lung *volumes* are listed on the left in Figure 37–1:

- *Tidal volume* (VT) is the volume of air inspired or expired with each normal breath; it amounts to about 500 milliliters in the average young man.
- *Inspiratory reserve volume* (IRV) is the extra volume of air that can be inspired over and above the normal tidal volume; it is usually equal to about 3000 milliliters.
- *Expiratory reserve volume* (ERV) is the extra amount of air that can be expired by forceful expiration after the end of a normal tidal expiration; this normally amounts to about 1100 milliliters.
- *Residual volume* (RV) is the volume of air remaining in the lungs after the most forceful expiration; this is usually about 1200 milliliters.

Pulmonary capacities are combinations of two or more pulmonary volumes. In Figure 37–1 are listed the pulmonary capacities, which can be described as follows:

- *Inspiratory capacity* (IC) equals the tidal volume plus the inspiratory reserve volume. This is the amount of air (about 3500 milliliters) a person can breathe beginning at the normal expiratory level and distending the lungs to the maximum amount.
- *Functional residual capacity* (FRC) equals the expiratory reserve volume plus the residual volume. This is the amount of air that remains in the lungs at the end of normal expiration (about 2300 milliliters).
- *Vital capacity* (VC) equals the inspiratory reserve volume plus the tidal volume plus the expiratory reserve volume. This is the maximum amount of air a person can expel from the lungs after first filling the lungs to their maximum extent and then expiring to the maximum extent (about 4600 milliliters).
- *Total lung capacity* (TLC) is the maximum volume to which the lungs can be expanded with the greatest possible inspiratory effort (about 5800 milliliters); it is equal to the vital capacity plus the residual volume.

Interrelations among the pulmonary volumes and capacities are as follows:

$$VC = IRV + V_T + ERV$$
$$VC = IC + ERV$$
$$TLC = VC + RV$$
$$TLC = IC + FRC$$
$$FRC = ERV + RV$$

MINUTE RESPIRATORY VOLUME AND ALVEOLAR VENTILATION

The minute respiratory volume is the total amount of new air that is moved into the respiratory passages each minute. This is equal to the tidal volume multiplied by the respiratory rate. The normal tidal volume is about 500 milliliters, and the normal respiratory rate is about 12 breaths per minute; therefore, the minute respiratory volume normally averages about 6 liters/min.

Alveolar ventilation is the rate at which new air reaches the gas exchange areas of the lungs. During inspiration, some of the air never reaches the gas exchange areas but instead fills respiratory passages; this air is called *dead space air*. Because alveolar ventilation is the total volume of new air that enters the alveoli, it is equal to the respiratory rate multiplied by the amount of new air that enters the alveoli with each breath:

$$\dot{V}_A = Freq \cdot (V_T - V_D)$$

where $\dot{V}_A$ is the volume of alveolar ventilation per minute, Freq is the frequency of respiration per minute, V_T is the tidal volume, and V_D is the dead space volume. Thus, with a normal tidal volume of 500 milliliters, a normal dead space of 150 milliliters, and a respiratory rate of 12 breaths per minute, alveolar ventilation equals $12 \times (500 - 150)$, or 4200 ml/min.

There are three types of dead space air:

- *Anatomical dead space* is the air in the conducting airways that does not engage in gas exchange.
- *Alveolar dead space* is the air in the gas exchange portions of the lung that cannot engage in gas exchange; it is nearly zero in normal individuals.
- *Physiological dead space* is the sum of the anatomical dead space and the alveolar dead space (i.e., the total dead space air); it is equal to the anatomical dead space in normal individuals.

FUNCTIONS OF THE RESPIRATORY PASSAGEWAYS
(p. 486)

Trachea, Bronchi, and Bronchioles

Air is distributed to the lungs by way of the trachea, bronchi, and bronchioles. The trachea is called the first-generation respiratory passageway, and two main right and left bronchi are the second-generation respiratory passageways; each division thereafter is an additional generation. There are between 20 and 25 generations before the air finally reaches the alveoli.

The walls of the bronchi and bronchioles are muscular. In all areas of the trachea and bronchi not occupied by cartilage plates, the walls are composed mainly of smooth muscle. The walls of the bronchioles are almost entirely smooth muscle, except for the most terminal bronchiole (respiratory bronchiole), which has only a few smooth muscle fibers. Many obstructive diseases of the lung result from narrowing of the smaller bronchi and bronchioles, often because of excessive contraction of the smooth muscle itself.

The greatest resistance to air flow occurs in the larger bronchi, not in the small, terminal bronchioles. The reason for this high resistance is that there are relatively few bronchi in comparison with about 65,000 parallel terminal bronchioles, through each of which only a minute amount of air must pass. In disease conditions, the smaller bronchioles often do play a far greater role in determining air flow resistance for two reasons: (1) because of their small size, they are easily occluded; and (2) because they have a greater percentage of smooth muscle in the walls, they constrict easily.

Epinephrine and norepinephrine cause dilation of the bronchiole tree. Direct control of the bronchioles by sympathetic nerve fibers is relatively weak because few of these fibers penetrate to the central portions of the lung. The bronchial tree, however, is exposed to circulating norepinephrine and epinephrine released from the adrenal gland medullae. Both of these hormones, especially epinephrine because of its greater stimulation of *beta receptors*, cause dilatation of the bronchial tree.

The parasympathetic nervous system constricts the bronchioles. A few parasympathetic nerve fibers derived from the vagus nerves penetrate the lung parenchyma. These nerves secrete acetylcholine, which causes mild to moderate constriction of the bronchioles. When a disease process such as asthma has already caused some constriction, parasympathetic superimposed nervous stimulation often worsens the condition. When this occurs, administration of drugs that block the effects of acetylcholine, such as atropine, can sometimes relax the respiratory passages sufficiently to relieve the obstruction.

Mucous Coat of the Respiratory Passageways, and Action of Cilia to Clear the Passageways (p. 487)

All the respiratory passages are kept moist with a layer of mucus. The mucus is secreted in part by individual goblet cells in the epithelial lining of the passages and in part by small submucosal glands. In addition to keeping the surfaces moist, the mucus traps small particles out of the inspired air and keeps most of them from ever reaching the alveoli. The mucus itself is removed from the passages in the following manner.

The entire surface of the respiratory passages is lined with ciliated epithelium. Included in these passageways are the nose and lower passages down as far as the terminal bronchioles. About 200 cilia are on each epithelial cell. These cilia beat continually at a rate of 10 to 20 times per second, and the direction of their "power stroke" is always toward the pharynx (i.e., the cilia in the lungs beat upward, whereas those in the nose beat downward). This continual beating causes the coat of mucus to flow slowly, at a velocity of about 1 cm/min, toward the pharynx. The mucus and its entrapped particles are either swallowed or coughed to the exterior.

Respiratory Functions of the Nose (p. 487)

The nasal cavities warm, humidify, and filter the incoming air. These functions together are called the air conditioning function of the upper respiratory passageways.

- *Warming and humidifying the air*—Ordinarily, the temperature of the inspired air rises to within 1°F of body temperature and within 2 to 3 per cent of full saturation with water vapor before it reaches the trachea. When a person breathes air through a tube directly into the trachea (as through a tracheostomy), the cooling and especially the drying effect in the lower lung can lead to serious lung crusting and infection.
- *Filtering the air*—The hairs at the entrance to the nostrils are important for filtering out large particles. Much more important, however, is the removal of particles by turbulent precipitation. The air passing through the nasal passageways hits many obstructing vanes: the conchae, septum, and pharyngeal wall. When the particles strike the surfaces of the obstructions, they are entrapped in the mucous coating.

38

Pulmonary Circulation; Pulmonary Edema; Pleural Fluid

Special problems related to pulmonary hemodynamics have important implications for the gas exchange in the lungs. The present discussion is concerned specifically with these features of the pulmonary circulation.

PHYSIOLOGIC ANATOMY OF THE PULMONARY CIRCULATORY SYSTEM (p. 491)

The lung has three circulations—pulmonary, bronchial, and lymphatic.

- *Pulmonary circulation*—The pulmonary artery, which supplies blood to the lungs, is thin walled and distensible, giving the pulmonary arterial tree a large compliance, similar to that of the entire systemic arterial tree. This large compliance allows the pulmonary arteries to accommodate about two thirds of the stroke volume of the right ventricle. The pulmonary veins have distensibility characteristics similar to those of the veins in the systemic circulation.
- *Bronchial circulation*—Bronchial blood flow amounts to about 1 to 2 per cent of the total cardiac output. Oxygenated blood in the bronchial arteries supplies the connective tissue, septa, and large and small bronchi of the lungs. Because the bronchial blood empties into the pulmonary veins and bypasses the right heart, the right ventricular output is about 1 to 2 per cent less than the left ventricular output.
- *Lymphatic circulation*—Lymphatics are found in all the supportive tissues of the lungs. Particulate matter entering the alveoli is removed by way of lymphatic channels; plasma proteins leaking from the lung capillaries are also removed from the lung tissues, helping to prevent edema.

PRESSURES IN THE PULMONARY SYSTEM (p. 491)

Blood pressures in the pulmonary circulation are low compared with those in the systemic circulation.

- *Pulmonary artery pressure*—In the normal human being, the average systolic pulmonary arterial pressure is about 25 mm Hg, diastolic pulmonary arterial pressure is about 8 mm Hg, and mean pulmonary arterial pressure is about 15 mm Hg.
- *Pulmonary capillary pressure*—The mean pulmonary capillary pressure has been estimated through indirect means to be about 7 mm Hg. (The importance of this low capillary pressure in relation to fluid exchange is discussed subsequently.)
- *Left atrial and pulmonary venous pressures*—The mean pressure in the left atrium and the major pulmonary veins averages about 2 mm Hg in the recumbent human being.

The left atrial pressure can be estimated by measuring the pulmonary wedge pressure. Direct measurement of left atrial pressure is difficult because it requires passing a catheter through the left ventricle. The *pulmonary wedge pressure* can be measured by floating a balloon-tipped catheter through the right heart and pulmonary artery until the catheter wedges tightly in a smaller branch of the artery. Because all blood flow has been stopped in the blood vessels extending from the artery, an almost direct connection is made through the pulmonary capillaries with the blood in the pulmonary veins. The wedge pressure is usually only 2 to 3 mm Hg greater than the left atrial pressure. Wedge pressure measurements are used frequently for studying changes in left atrial pressure in persons with congestive heart failure.

BLOOD VOLUME OF THE LUNGS (p. 492)

The lungs provide an important blood reservoir. The pulmonary blood volume is about 450 milliliters, or about 9 per cent of the total blood volume. Under various physiological and pathological conditions, the quantity of blood in the lungs can vary from as little as one-half to two times normal. For instance, loss of blood from the systemic circulation by hemorrhage can be in part compensated for by an automatic shift of blood from the lungs into the systemic vessels.

Blood shifts between the pulmonary and systemic circulatory systems as a result of cardiac pathology. Left heart failure, mitral stenosis, or mitral regurgitation causes blood to dam up in the pulmonary circulation, increasing greatly the pulmonary vascular pressures. Because the volume of the systemic circulation is about nine times that of the pulmonary system, a shift of blood from one system to the other affects the pulmonary system greatly but usually has only mild effects elsewhere.

BLOOD FLOW THROUGH THE LUNGS AND ITS DISTRIBUTION (p. 493)

Pulmonary blood flow is nearly equal to cardiac output. Under most conditions, the pulmonary vessels act as passive, distensible tubes that enlarge with increasing pressure and narrow with decreasing pressure. For adequate aeration of the blood, it is important for the blood to be distributed to the segments of the lungs in which the alveoli are best oxygenated. This is achieved via the following mechanism.

Pulmonary blood flow distribution is controlled by alveolar oxygen. When the alveolar oxygen concentration decreases below normal, the adjacent blood vessels constrict, increasing the vascular resistance. This is opposite to the effect normally observed in systemic vessels, which dilate rather than constrict in response to low oxygen. This effect of a low oxygen level on pulmonary vascular resistance distributes blood flow where it is most effective (i.e., poorly ventilated alveoli receive relatively little blood flow).

The autonomic nervous system does not have a major function in normal control of pulmonary vascular resistance. Stimulation of the vagal fibers to the lungs causes a slight decrease in pulmonary vascular resistance, and stimulation of the sympathetics causes a slight increase in resistance; both effects are perhaps too slight to be more than marginally important. Sympathetic stimulation has a great effect in constricting the large pulmonary capacitative vessels, especially the veins. This large vessel constriction provides a means by which sympathetic stimulation can displace much of the extra blood in the lungs into other segments of the circulation when needed to combat low blood pressure.

EFFECT OF HYDROSTATIC PRESSURE GRADIENTS IN THE LUNGS ON REGIONAL PULMONARY BLOOD FLOW
(p. 493)

In the normal adult, the distance between the apex and base of the lungs is about 30 cm, which creates a 23 mm Hg difference in blood pressure. This pressure gradient has a large effect on blood flow in different regions of the lung.

Hydrostatic pressure gradients in the lung create three zones of pulmonary blood flow. The capillaries in the alveolar walls are distended by the blood pressure inside them, but simultaneously, they are compressed by the alveolar pressure outside. Any time the alveolar air pressure becomes greater than the capillary blood pressure, the capillaries close and blood flow stops. Under different normal and pathological lung conditions, any one of three possible zones of pulmonary blood flow can be found, as follows:

- *Zone 1 (top of lung)* has no blood flow during any part of the cardiac cycle because the local capillary pressure never rises higher than the alveolar pressure. In this zone, alveolar pressure > artery pressure > venous pressure; thus, the capillaries are pressed flat. Zone 1 does not occur during normal conditions; it can occur when pulmonary artery pressure is decreased (i.e., hemorrhage) and when alveolar pressure is increased (i.e., positive pressure ventilation).
- *Zone 2 (middle of lung)* has an intermittent blood flow that occurs during systole, when the artery pressure is greater than the alveolar pressure, but not during diastole, when the artery pressure is less than alveolar pressure. Zone 2 blood flow is thus determined by the difference between arterial and alveolar pressures.
- *Zone 3 (bottom of lung)* has a high, continuous blood flow because the capillary pressure remains greater than alveolar pressure during the entire cardiac cycle.

Pulmonary vascular resistance decreases during heavy exercise. During exercise, the blood flow through the lungs increases fourfold to sevenfold.

This extra flow is accommodated in the lungs in two ways: (1) by increasing the number of open capillaries, sometimes as much as threefold, and (2) by distending all the capillaries and increasing the rate of flow through each capillary by more than twofold. In the normal person, these two changes together decrease the pulmonary vascular resistance so much that the pulmonary arterial pressure rises very little, even during maximum exercise.

PULMONARY CAPILLARY DYNAMICS (p. 495)

The alveolar walls are lined with so many capillaries that in most places, the capillaries almost touch one another; therefore, the capillary blood flows in the alveolar walls as a "sheet" rather than through individual vessels.

Capillary Exchange of Fluid in the Lungs, and Pulmonary Interstitial Fluid Dynamics (p. 495)

The dynamics of fluid exchange through the lung capillaries are qualitatively the same as those for peripheral tissues. Quantitatively, however, there are important differences.

- *Pulmonary capillary pressure* is low (about 7 mm Hg) in comparison with a higher functional capillary pressure in the peripheral tissues (about 17 mm Hg).
- *Interstitial fluid pressure* is slightly more negative than in the peripheral subcutaneous tissue; values range from about -5 to -8 mm Hg.
- *Capillary permeability* is high, allowing extra amounts of protein to leak from the capillaries; therefore, the interstitial fluid colloid osmotic pressure is also high, averaging about 14 mm Hg, compared with an average of less than 7 mm Hg in many peripheral tissues.
- *The alveolar walls are thin.* The alveolar epithelium covering the alveolar surfaces is so weak that it ruptures when the interstitial pressure becomes greater than atmospheric pressure (i.e., more than 0 mm Hg), which allows dumping of fluid from the interstitial spaces into the alveoli.

The mean filtration pressure at the pulmonary capillaries is +1 mm Hg. This value is derived as follows:

- *Total outward force (29 mm Hg)*—Forces tending to cause movement of fluid out of the capillaries include capillary pressure (7 mm Hg), interstitial fluid colloid osmotic pressure (14 mm Hg), and interstitial fluid pressure (−8 mm Hg).
- *Total inward force (28 mm Hg)*—Only the plasma colloid pressure (28 mm Hg) tends to cause absorption of fluid into the capillaries.
- *Net mean filtration pressure (+1 mm Hg)*—Because the total outward force (29 mm Hg) is slightly greater than the total inward force (28 mm Hg), the net mean filtration pressure is slightly positive (29 − 28 = +1 mm Hg). This net filtration pressure causes a continual loss of fluid from the capillaries.

Negative pressure in the interstitial fluid helps to keep the alveoli "dry." Whenever extra fluid appears in the alveoli, it is simply sucked mechanically into the lung interstitium through small openings between the alveolar epithelial cells. The excess fluid is either carried away through the pulmonary lymphatics or absorbed into the pulmonary capillaries. Under normal conditions, the alveoli are kept in a "dry" state except for a small amount of fluid that seeps from the epithelium onto the lining surfaces of the alveoli to keep them moist.

Pulmonary Edema (p. 496)

Pulmonary edema occurs in the same way that peripheral edema occurs. The most common causes of pulmonary edema are as follows:

- *Left-sided heart failure* or mitral valvular disease causes a great increase in pulmonary capillary pressure and flooding of the interstitial spaces and alveoli.
- *Damage to the pulmonary capillary membrane* caused by infections or breathing of noxious substances produces rapid leakage of both plasma proteins and fluid out of the capillaries.

When the pulmonary interstitial fluid volume increases by more than 50 per cent, the alveolar epithelium ruptures, and fluid pours into the alveoli. The cause is simply that even the slightest positive pressure in the interstitial fluid spaces—probably even +1 mm Hg—seems to cause immediate rupture of the pulmonary alveolar epithelium. Except in the mildest cases of pulmonary edema, therefore, edema fluid always enters the alveoli; if this edema becomes sufficiently severe, it can cause death from suffocation.

Acute safety factors tend to prevent edema in the lungs. Before positive interstitial fluid pressure can occur and cause edema, all the following factors must be overcome: (1) normal negativity of the interstitial fluid pressure, (2) lymphatic pumping of fluid out of the interstitial spaces, and (3) increased osmosis of fluid into the pulmonary capillaries caused by decreased protein in the interstitial fluid when the lymph flow increases.

The pulmonary capillary pressure normally must rise to equal the plasma colloid osmotic pressure before significant pulmonary edema occurs. In the human being, who normally has a plasma colloid osmotic pressure of 28 mm Hg, the pulmonary capillary pressure must rise from the normal level of 7 mm Hg to more than 28 mm Hg to cause pulmonary edema, giving a safety factor against pulmonary edema of about 21 mm Hg.

The lymphatic system provides a chronic safety factor against pulmonary edema. When the pulmonary capillary pressure remains elevated chronically (for at least 2 weeks), the lungs become even more resistant to pulmonary edema. This is because the lymph vessels can expand greatly, increasing their capability for carrying fluid away from the interstitial spaces by perhaps as much as 10-fold. In a patient with chronic mitral stenosis, a pulmonary capillary pressure of 40 to 45 mm Hg has been measured without the development of significant pulmonary edema.

Lethal pulmonary edema can occur within hours. When the pulmonary capillary pressure does rise even slightly above the safety factor level, lethal pulmonary edema can occur within hours, and even within 20 to 30 minutes if the capillary pressure increases greatly. In acute left-sided heart failure, in which the pulmonary capillary pressure occasionally rises to 50 mm Hg, death frequently ensues within less than 30 minutes from acute pulmonary edema.

FLUIDS IN THE PLEURAL CAVITY (p. 497)

The lungs slide back and forth within the pleural cavity as they expand and contract during normal breathing. Small amounts of interstitial fluid transudate continually from the pleural membranes into the pleural space. These fluids contain proteins, which give the pleural fluid a mucoid characteristic allowing easy slippage of the moving lungs. The total amount of fluid in the pleural cavities is only a few milliliters. The pleural space—the space between the parietal and visceral pleurae—is called a potential space because it normally is so narrow that it is not obviously a physical space.

The lymphatic pump causes negative pressure in the pleural fluid. Because the recoil tendency of the lungs causes them to try to collapse, a negative force is always required on the outside of the lungs to keep the lungs expanded. This is provided by a negative pressure in the normal pleural space, which usually ranges from -4 to -8 mm Hg. The negativity of the pleural fluid keeps the normal lungs pulled tightly against the parietal pleura of the chest cavity, except for the extremely thin layer of mucoid fluid that acts as a lubricant.

Pleural effusion—the collection of large amounts of free fluid in the pleural space—is analogous to edema fluid in the tissues. The possible causes of effusion are the following:

- *Blockage of lymphatic drainage* from the pleural cavity allows excess fluid to accumulate.
- *Cardiac failure* causes excessively high peripheral and pulmonary capillary pressures, leading to excessive transudation of fluid into the pleural cavity.
- *Decreased plasma colloid osmotic pressure* allows excessive transudation of fluid from the capillaries.
- *Increased capillary permeability* caused by infection or any other source of inflammation of the pleural surfaces allows rapid dumping of both plasma proteins and fluid into the cavity.

39

Physical Principles of Gas Exchange; Diffusion of Oxygen and Carbon Dioxide Through the Respiratory Membrane

The diffusion of oxygen from the alveoli into the pulmonary blood and diffusion of carbon dioxide in the opposite direction occur through random molecular motion of molecules. The rate at which the respiratory gases diffuse is a much more complicated problem, requiring a deeper understanding of the physics of diffusion and gas exchange.

PHYSICS OF GAS DIFFUSION AND GAS PARTIAL PRESSURES (p. 501)

Respiratory gases diffuse from areas of high partial pressure to areas of low partial pressure. Respiratory physiology involves mixtures of gases, mainly of oxygen, nitrogen, and carbon dioxide. The rate of diffusion of each of these gases is directly proportional to the pressure caused by each gas alone, which is called the *partial pressure* of the gas. Partial pressures are used to express the concentrations of gases because it is the pressures that cause the gases to move via diffusion from one part of the body to another. The partial pressures of oxygen, carbon dioxide, and nitrogen are designated as PO_2, PCO_2, and PN_2, respectively.

The pressure of a gas in the air is calculated by multiplying its fractional concentration by the total pressure. Air has a composition of about 79 per cent nitrogen and about 21 per cent oxygen. The total pressure at sea level (atmospheric pressure) averages 760 mm Hg; therefore, 79 per cent of the 760 mm Hg is caused by nitrogen (about 600 mm Hg) and 21 per cent is caused by oxygen (about 160 mm Hg).

The partial pressure of nitrogen in the mixture is 600 mm Hg, and the partial pressure of oxygen is 160 mm Hg; the total pressure is 760 mm Hg, the sum of the individual partial pressures.

The pressure of a gas in a solution is determined not only by its concentration but also by its solubility coefficient. Some types of molecules, especially carbon dioxide, are physically or chemically attracted to water molecules, which allows far more of them to become dissolved without a build-up of excess pressure within the solution. The relationship between gas concentration and gas solubility is expressed by Henry's law:

$$\text{Pressure} = \frac{\text{Concentration of Dissolved Gas}}{\text{Solubility Coefficient}}$$

The vapor pressure of water at body temperature is 47 mm Hg. When air enters the respiratory passageways, water evaporates from the surfaces and humidifies the air. The pressure that the water molecules exert to escape from the surface is the vapor pressure of the water, which is 47 mm Hg at body temperature. Once the gas mixture has become fully humidified, the partial pressure of the water vapor in the gas mixture is also 47 mm Hg. This partial pressure is designated P_{H_2O}.

The rate of gas diffusion in a fluid (D) is affected by multiple factors. These factors are described as follows and expressed in the following single formula:

$$D \propto \frac{\Delta P \times A \times S}{d \times \sqrt{MW}}$$

- *Pressure difference* (ΔP)—The greater the difference in pressure between the two ends of a diffusion pathway, the greater is the rate of gas diffusion.
- *Cross-sectional area* (A)—The greater the cross-sectional area of the diffusion pathway, the greater is the total number of molecules to diffuse.
- *Gas solubility* (S)—The greater the solubility of the gas, the greater is the number of molecules available to diffuse for any given pressure difference.
- *Diffusion distance* (d)—The greater the distance that the molecules must diffuse, the longer it takes the molecules to diffuse the entire distance.
- *Molecular weight of gas* (MW)—The greater the velocity of kinetic movement of the molecules,

which is inversely proportional to the square root of the molecular weight, the greater is the rate of diffusion of the gas.

- *Temperature*—The temperature remains reasonably constant in the body and usually need not be considered.

The diffusion coefficient of a gas is directly proportional to its solubility and inversely proportional to its molecular weight. It is obvious from the above formula that the characteristics of the gas determine two factors of the formula—solubility and molecular weight—and together they determine the diffusion coefficient of the gas. That is, the diffusion coefficient is proportional to $S\sqrt{MW}$; also, the relative rates at which different gases at the same pressure levels diffuse are proportional to their diffusion coefficients. Considering the diffusion coefficient for oxygen to be 1, the relative diffusion coefficient for carbon dioxide is 20.3. The diffusion coefficients for nitrogen, carbon monoxide, and helium are less than that of oxygen.

COMPOSITION OF ALVEOLAR AIR—ITS RELATION TO ATMOSPHERIC AIR (p. 503)

The concentrations of gases in alveolar air are different than those in atmospheric air. These differences are shown in Table 39–1 **and explained as follows:**

1. Alveolar air is only partially replaced by atmospheric air with each breath.
2. Oxygen is constantly being absorbed from the alveolar air.
3. Carbon dioxide is constantly diffusing from the pulmonary blood into the alveoli.
4. Dry atmospheric air is humidified before it reaches the alveoli.

Water vapor dilutes the other gases in the inspired air. Table 39–1 (column 1) shows that atmospheric air is composed mostly of nitrogen and oxygen; it contains almost no carbon dioxide or water vapor. The atmospheric air becomes totally humidified as it passes through the respiratory passages. The water vapor at normal body temperature (i.e., 47 mm Hg) dilutes the other gases in the inspired air. The oxygen partial pressure decreases from 159 mm Hg in atmospheric air to 149 mm Hg in the

TABLE 39-1 PARTIAL PRESSURES OF RESPIRATORY GASES AS THEY ENTER AND LEAVE THE LUNGS (AT SEA LEVEL)

Partial Pressure (mm Hg)

	Atmospheric Air*		Humidified Air		Alveolar Air		Expired Air	
N_2	597.0	(78.62%)	563.4	(74.09%)	569.0	(74.9%)	566.0	(74.5%)
O_2	159.0	(20.84%)	149.3	(19.67%)	104.0	(13.6%)	120.0	(15.7%)
CO_2	0.3	(0.04%)	0.3	(0.04%)	40.0	(5.3%)	27.0	(3.6%)
H_2O	3.7	(0.50%)	47.0	(6.20%)	47.0	(6.2%)	47.0	(6.2%)
Total	760.0	(100.00%)	760.0	(100.00%)	760.0	(100.0%)	760.0	(100.0%)

* On an average cool, clear day.

humidified air, and the nitrogen partial pressure decreases from 597 to 563.4 mm Hg (see Table 39–1, column 4).

Alveolar air is renewed very slowly by atmospheric air. The amount of alveolar air replaced by new atmospheric air with each breath is only one seventh of the total, so that many breaths are required to completely exchange the alveolar air. This slow replacement of alveolar air prevents sudden changes in gas concentrations in the blood; makes the respiratory control mechanism much more stable than it would otherwise be; and helps prevent excessive increases and decreases in tissue oxygenation, tissue carbon dioxide concentration, and tissue pH when respiration is temporarily interrupted.

The alveolar oxygen concentration is controlled by the rate of oxygen absorption into the blood and the rate of entry of new oxygen into the lungs. Oxygen is continually being absorbed into the blood of the lungs, and new oxygen is continually being breathed into the alveoli from the atmosphere. The more rapidly oxygen is absorbed, the lower becomes its concentration in the alveoli. In comparison, the more rapidly new oxygen is breathed into the alveoli from the atmosphere, the higher its concentration.

Expired air is a combination of dead space air and alveolar air. The overall composition of expired air is determined by (1) the proportion of the expired air that is dead space air and (2) the proportion of the expired air that is alveolar air. When air is expired from the lungs, the first portion of this air (dead space air) is typical humidified air (see Table 39–1, column 4). Then, more and more alveolar air becomes mixed with the dead space air until all the dead space air has been washed out and only alveolar air is expired at the end of expiration. Normal expired air has approximate gas concentrations and partial pressures as shown (see Table 39–1, column 8).

DIFFUSION OF GASES THROUGH THE RESPIRATORY MEMBRANE (p. 505)

A respiratory unit is composed of a respiratory bronchiole, alveolar ducts, atria, and alveoli. There are about 300 million units in the two lungs. The alveolar walls are extremely thin, and within them is an almost solid network of interconnecting capillaries; the flow of blood in the alveolar wall has been described as a "sheet" of flowing blood. Gas exchange occurs through the membranes of all the terminal portions of the lungs, not merely in the alveoli themselves. These membranes are collectively known as the *respiratory membrane*, or the *pulmonary membrane*.

The respiratory membrane is composed of several different layers. The exchange of oxygen and carbon dioxide between the blood and alveolar air requires diffusion through the following different layers of the respiratory membrane:

- A layer of fluid lining the alveolus that contains surfactant
- The alveolar epithelium, which is composed of thin epithelial cells
- An epithelial basement membrane
- A thin interstitial space between the alveolar epithelium and the capillary membrane
- A capillary basement membrane that fuses in places with the epithelial basement membrane
- The capillary endothelial membrane.

The respiratory membrane is optimized for gas exchange:

- *Membrane thickness*—Despite the large number of layers, the overall thickness of the respiratory membrane averages about 0.6 micrometer except where there are cell nuclei.
- *Membrane surface area*—The total surface area of the respiratory membrane is about 70 square meters in the normal adult. This is equivalent to the floor area of a 25-by-30-foot room.
- *Capillary blood volume*—The capillary blood volume is 60 to 140 milliliters. By imagining that this small amount of blood is spread over the entire surface of a 25-by-30-foot floor, it is easy to understand the rapidity of respiratory exchange of gases.
- *Capillary diameter*—The average diameter of the pulmonary capillaries is about 5 micrometers; the red blood cell membrane usually touches the capillary wall so that oxygen and carbon dioxide need not pass through significant amounts of plasma as they diffuse between the alveolus and the red cell.

Multiple factors determine how rapidly a gas will pass through the respiratory membrane. These determining factors include the following:

- *Thickness of respiratory membrane*—The rate of diffusion through the membrane is inversely proportional to the membrane thickness. Edema fluid in the interstitial space and alveoli decreases diffusion because the respiratory gases must move not only through the membrane but also through this fluid. Fibrosis of the lungs can also increase the thickness of some portions of the respiratory membrane.
- *Surface area of respiratory membrane*—In emphysema, many of the alveoli coalesce, with dissolution of alveolar walls; this often causes the total surface area to decrease by as much as fivefold. During strenuous exercise, even the slightest decrease in surface area can be a serious detriment to respiratory exchange of gases.
- *Diffusion coefficient*—The diffusion coefficient for the transfer of each gas through the respiratory membrane depends on its solubility in the membrane and, inversely, on the square root of its molecular weight.
- *Pressure difference across respiratory membrane*—The difference between the partial pressure of gas in the alveoli and that of gas in the blood is directly proportional to the rate of gas transfer through the membrane.

Diffusing Capacity of the Respiratory Membrane (p. 508)

The diffusing capacity of the lungs for carbon dioxide is 20 times greater than that for oxygen. The ability of the respiratory membrane to exchange a gas between the alveoli and the pulmonary blood can be expressed in quantitative terms by its diffusing capacity, which is defined as the volume of a gas that diffuses through the membrane each minute for a pressure difference of 1 mm Hg. All the factors discussed that affect diffusion through the respiratory membrane can affect the diffusing capacity. The diffusing capacity of the lungs for oxygen when a person is at rest is about 21 ml/mm Hg/min. The diffusing capacity for carbon dioxide is about 20 times this value, or about 440 ml/mm Hg/min.

The diffusion capacity for oxygen increases during exercise. During exercise, the oxygenation of the blood is increased not only by greater alveolar ventilation but also by a greater capacity of the respiratory membrane for transmitting oxygen into the blood. During strenuous exercise, the diffusing capacity for oxygen can increase to about 65 ml/min/mm Hg, which is three times the diffusing capacity during resting conditions. This increase is caused by the following:

- *Increased surface area*—Opening up of closed pulmonary capillaries or extra dilatation of open capillaries increases the surface area for diffusion of oxygen.
- *Improved ventilation-perfusion ratio* ($\dot{V}A/\dot{Q}$)—Exercise improves the match between the ventilation of the alveoli and the perfusion of the alveolar capillaries with blood.

EFFECT OF THE VENTILATION-PERFUSION RATIO ON ALVEOLAR GAS CONCENTRATION (p. 509)

Even normally, and especially with many lung diseases, some areas of the lungs are well ventilated but have almost no blood flow, whereas other areas have excellent blood flow but little or no ventilation. In either of these conditions, gas exchange through the respiratory membrane is seriously impaired. A highly quantitative concept was developed to help understand respiratory exchange when there is imbalance between alveolar ventilation and alveolar blood flow; this concept is called the ventilation-perfusion ratio ($\dot{V}A/\dot{Q}$).

The $\dot{V}A/\dot{Q}$ **is the ratio of alveolar ventilation to pulmonary blood flow.** When $\dot{V}A$ (alveolar ventilation) is normal for a given alveolus and $\dot{Q}$ (blood flow) is normal for the same alveolus, then $\dot{V}A/\dot{Q}$ is also said to be normal. When the ventilation ($\dot{V}A$) is zero and there is still perfusion ($\dot{Q}$) of the alveolus, however, $\dot{V}A/\dot{Q}$ is zero. At the other extreme, when there is adequate ventilation ($\dot{V}A$) but zero perfusion ($\dot{Q}$), the ratio, $\dot{V}A/\dot{Q}$, is infinity.

The $\dot{V}A/\dot{Q}$ **can range from zero to infinity.**

- *When $\dot{V}A/\dot{Q}$ equals zero*, there is no alveolar ventilation, so that the air in the alveolus comes to equilibrium with the oxygen and carbon dioxide in the blood. Because the blood that perfuses the capillaries is venous blood, it is the gases in this blood that come to equilibrium with the alveolar gases; the alveolar P_{O_2} is 40 mm Hg, and the P_{CO_2} is 45 mm Hg.

- When $\dot{V}_A/\dot{Q}$ equals infinity, there is no capillary blood flow to carry oxygen away or to bring carbon dioxide to the alveoli. The alveolar air now becomes equal to the humidified inspired air (i.e., the air that is inspired loses no oxygen to the blood and gains no carbon dioxide). Because normal inspired and humidified air has a P_{O_2} of 149 mm Hg and a P_{CO_2} of 0 mm Hg, these are the partial pressures of these two gases in the alveoli.
- When $\dot{V}_A/\dot{Q}$ is normal, there is both normal alveolar ventilation and normal alveolar capillary blood flow; exchange of oxygen and carbon dioxide is nearly optimal; and alveolar P_{O_2} is normally at a level of 104 mm Hg, which lies between that of the inspired air (149 mm Hg) and that of venous blood (40 mm Hg). Likewise, alveolar P_{CO_2} lies between the two extremes; it is normally 40 mm Hg, in contrast to 45 mm Hg in venous blood and 0 mm Hg in inspired air.

Concept of "Physiologic Shunt" (When $\dot{V}_A/\dot{Q}$ Is Less Than Normal) (p. 510)

The greater the physiologic shunt, the greater is the amount of blood that fails to be oxygenated as it passes through the lungs. Whenever $\dot{V}_A/\dot{Q}$ is below normal, a fraction of the venous blood passes through the pulmonary capillaries without becoming oxygenated. This fraction is called *shunted blood.* Some additional blood flows through the bronchial vessels rather than through the alveolar capillaries (normally, about 2 per cent of the cardiac output); this, too, is nonoxygenated, shunted blood. The total amount of shunted blood flow per minute is called the *physiologic shunt,* which can be calculated with the following equation:

$$\frac{\dot{Q}_{PS}}{\dot{Q}_T} = \frac{C_{i_{O_2}} - C_{a_{O_2}}}{C_{i_{O_2}} - C\bar{v}_{O_2}}$$

where $\dot{Q}_{PS}$ is the physiologic shunted blood flow per minute, $\dot{Q}_T$ is cardiac output per minute, $C_{i_{O_2}}$ is the arterial oxygen concentration if $\dot{V}_A/\dot{Q}$ is "ideal," $C_{a_{O_2}}$ is the measured arterial oxygen concentration, and $C\bar{v}_{O_2}$ is the measured mixed venous oxygen concentration.

Concept of "Physiologic Dead Space" (When $\dot{V}A/\dot{Q}$ Is More Than Normal) (p. 511)

When the physiologic dead space is great, much of the work of ventilation is wasted because some of the ventilated air never reaches the blood. When ventilation of some of the alveoli is great but the alveolar blood flow is low, there is far more available oxygen in the alveoli than can be transported away by the flowing blood; the ventilation of these alveoli is said to be wasted. The ventilation of the anatomical dead space areas of the respiratory passageways is also wasted. The sum of these two types of wasted ventilation is called the physiologic dead space and is calculated with the Bohr equation:

$$\frac{VD_{phys}}{VT} = \frac{PaCO_2 - P\bar{E}CO_2}{PaCO_2}$$

where VD_{phys} is the physiologic dead space, VT is the tidal volume, $PaCO_2$ is the partial pressure of carbon dioxide in arterial blood, and $P\bar{E}CO_2$ is the average partial pressure of carbon dioxide in all expired air.

Abnormalities of $\dot{V}A/\dot{Q}$ (p. 511)

The $\dot{V}A/\dot{Q}$ is high at the top and low at the bottom of the lung. Blood flow and ventilation both increase from the top to the bottom of the lung, but blood flow increases more progressively. $\dot{V}A/\dot{Q}$ is, therefore, higher at the top of the lung than at the bottom. In both extremes, inequalities of ventilation and perfusion decrease the effectiveness of the lung for exchange of oxygen and carbon dioxide. During exercise, however, the blood flow to the upper part

TABLE 39–2 $\dot{V}A/\dot{Q}$ CHARACTERISTICS AT THE TOP AND BOTTOM OF THE LUNG

Area of Lung	Ventilation	Perfusion (Blood Flow)	$\dot{V}A/\dot{Q}$	Local Alveolar PO_2	Local Alveolar PCO_2
Top	Low	Lower	Highest	Highest	Lowest
Bottom	High	Higher	Lowest	Lowest	Highest

of the lung increases markedly, so that far less physiologic dead space occurs, and the effectiveness of gas exchange approaches optimum. The differences in ventilation and perfusion at the top and bottom of the upright lung and their effect on the regional PO_2 and PCO_2 are summarized in Table 39–2.

The $\dot{V}A/\dot{Q}$ may be increased or decreased in chronic obstructive lung disease. Most chronic smokers develop bronchial obstruction, which can cause alveolar air to become trapped with resultant emphysema. The emphysema in turn causes many of the alveolar walls to be destroyed; thus, two abnormalities occur in smokers to cause abnormal $\dot{V}A/\dot{Q}$.

- *Anatomical dead space (low $\dot{V}A/\dot{Q}$)*—Because many of the small bronchioles are obstructed, the alveoli beyond the obstructions are unventilated.
- *Physiologic dead space (high $\dot{V}A/\dot{Q}$)*—In areas where the alveolar walls have been destroyed but there is still alveolar ventilation, the ventilation is wasted because of inadequate blood flow.

40

Transport of Oxygen and Carbon Dioxide in the Blood and Body Fluids

Oxygen is transported principally in combination with hemoglobin to the tissue capillaries, where it is released for use by the cells. In the tissue cells, oxygen reacts with various foodstuffs to form large quantities of carbon dioxide. This in turn enters the tissue capillaries and is transported back to the lungs. This chapter presents the physical and chemical principles of oxygen and carbon dioxide transport in the blood and body fluids.

PRESSURES OF OXYGEN AND CARBON DIOXIDE IN THE LUNGS, BLOOD, AND TISSUES (p. 513)

The PO_2 of pulmonary blood rises to equal that of alveolar air within the first third of the capillary. The PO_2 averages 104 mm Hg in the alveolus, whereas the PO_2 of venous blood entering the capillary averages only 40 mm Hg. The initial pressure difference that causes oxygen to diffuse into the pulmonary capillary is $104 - 40$ mm Hg, or 64 mm Hg. The PO_2 rises to equal that of alveolar air by the time the blood has moved a third of the distance through the capillary, becoming almost 104 mm Hg.

The pulmonary capillary blood becomes almost fully saturated with oxygen, even during strenuous exercise. During strenuous exercise, oxygen utilization may increase by 20-fold, and the higher cardiac output reduces the time that the blood stays in the pulmonary capillaries to less than one-half normal. The blood, however, is still almost fully saturated with oxygen when it leaves the pulmonary capillaries for the following reasons:

- *Increased diffusing capacity*—As discussed in Chapter 39, the diffusing capacity for oxygen increases almost threefold during exercise; this results mainly from greater capillary surface area and more nearly ideal ventilation-perfusion ratio in the upper part of the lungs.

- *Transit time safety factor*—Normally, the blood becomes almost fully saturated with oxygen by the time it has passed through the first one third of the pulmonary capillary. In exercise, even with the shortened time of exposure in the capillaries, the blood still can become fully oxygenated.

Bronchial venous "shunt" flow decreases the arterial PO_2 from a capillary value of 104 mm Hg to an arterial value of 95 mm Hg. About 2 per cent of the blood that enters the left atrium has passed directly from the aorta through the bronchial circulation. This blood flow represents "shunt" flow because it is shunted past the gas exchange areas and its PO_2 value is typical of venous blood, about 40 mm Hg. This blood mixes with oxygenated blood from the lungs; this mixing of the bloods is called *venous admixture of blood.*

Tissue PO_2 is determined by the rate of oxygen transport to the tissues and the rate of oxygen utilization by the tissues. The PO_2 in the initial portions of the capillaries is 95 mm Hg, and the PO_2 in the interstitial fluid surrounding the tissue cells averages 40 mm Hg. This pressure difference causes oxygen to diffuse rapidly from the blood into the tissues, and the PO_2 of the blood leaving the tissue capillaries is also about 40 mm Hg. Two main factors can affect the tissue PO_2:

- *Rate of blood flow*—If the blood flow through a particular tissue becomes increased, greater quantities of oxygen are transported into the tissue in a given period, and the tissue PO_2 becomes correspondingly increased.
- *Rate of tissue metabolism*—If the cells use more oxygen for metabolism than normal, the interstitial fluid PO_2 tends to be reduced.

Carbon dioxide diffuses in a direction exactly opposite that of oxygen. There is, however, one major difference between the diffusion of carbon dioxide and that of oxygen: carbon dioxide can diffuse about 20 times as rapidly as oxygen. The pressure differences that cause carbon dioxide diffusion are far less than the pressure differences that cause oxygen diffusion.

TRANSPORT OF OXYGEN IN THE BLOOD (p. 516)

About 97 per cent of the oxygen is carried to the tissues in chemical combination with hemoglobin. The remaining 3 per cent is carried to the tissues in the dissolved state in the water of the plasma and cells. Hemoglobin combines with large quantities of oxygen when the PO_2 is high and then releases the oxygen when the PO_2 level is low. When blood passes through the lungs, where the blood PO_2 rises to 95 mm Hg, hemoglobin picks up large quantities of oxygen. As it passes through the tissue capillaries, where the PO_2 falls to about 40 mm Hg, large quantities of oxygen are released from the hemoglobin. The free oxygen then diffuses to the tissue cells.

The oxygen-hemoglobin dissociation curve shows the percent saturation of hemoglobin plotted as a function of PO_2. The oxygen-hemoglobin dissociation curve shown in Figure 40–1 demonstrates a progressive rise in the percentage of the hemoglobin that is bound with oxygen as the blood PO_2 increases, which is called the *per cent saturation of the hemoglobin*. Note the following features in the curve:

- *When the PO_2 is 95 mm Hg (arterial blood)*, the hemoglobin is about 97 per cent saturated with oxygen and the oxygen content is about 19.4 milliliters per 100 milliliters of blood; an average of nearly four molecules of oxygen are bound to each molecule of hemoglobin.
- *When the PO_2 is 40 mm Hg (mixed venous blood)*, the hemoglobin is 75 per cent saturated with oxygen

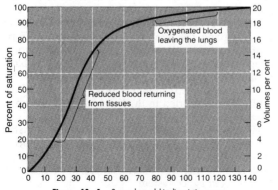

Figure 40–1   Oxygen-hemoglobin dissociation curve.

and the oxygen content is about 14.4 milliliters per 100 milliliters of blood; an average of three molecules of oxygen are bound to each molecule of hemoglobin.

- When the P_{O_2} is 25 mm Hg (mixed venous blood during moderate exercise), the hemoglobin is 50 per cent saturated with oxygen and the oxygen content is about 10 milliliters per 100 milliliters of blood; an average of two molecules of oxygen are bound to each molecule of hemoglobin.

The sigmoid shape of the oxygen-hemoglobin dissociation curve results from stronger binding of oxygen to hemoglobin as more molecules of oxygen become bound. Each molecule of hemoglobin can bind four molecules of oxygen. After one molecule of oxygen has bound, the affinity of hemoglobin for the second molecule is increased, and so forth. The affinity for the fourth molecule of oxygen is the greatest. Note that the affinity for oxygen is high in the lungs where the P_{O_2} value is about 95 mm Hg (flat portion of the curve) and low in the peripheral tissues where the P_{O_2} value is about 40 mm Hg (steep portion of the curve) (see Fig. 40–1).

The maximum amount of oxygen transported by hemoglobin is about 20 milliliters of oxygen per 100 milliliters of blood. In a normal person, each 100 milliliters of blood contains about 15 grams of hemoglobin, and each gram of hemoglobin can bind with about 1.34 milliliters of oxygen when it is 100 per cent saturated ($15 \times 1.34 = 20$ milliliters of oxygen per 100 milliliters of blood). The total quantity of oxygen bound with hemoglobin in normal arterial blood is about 97 per cent, however, so that about 19.4 milliliters of oxygen are carried in each 100 milliliters of blood. The hemoglobin in venous blood leaving the peripheral tissues is about 75 per cent saturated with oxygen, so that the amount of oxygen transported by hemoglobin in venous blood is about 14.4 milliliters of oxygen per 100 milliliters of blood. About 5 milliliters of oxygen is, thereby, normally transported to the tissues in each 100 milliliters of blood.

Hemoglobin functions to maintain a constant P_{O_2} in the tissues. Although hemoglobin is necessary for the transport of oxygen to the tissues, it performs another major function essential to life as a *tissue oxygen buffer* system.

- *Under basal conditions*, the tissues require about 5 milliliters of oxygen from each 100 milliliters of blood. For the 5 milliliters of oxygen to be released, the PO_2 must fall to about 40 mm Hg. The tissue PO_2 level normally does not rise to 40 mm Hg because the oxygen needed by the tissues is at that level not released from the hemoglobin; therefore, the hemoglobin sets the tissue PO_2 level at an upper limit of about 40 mm Hg.
- *During heavy exercise*, oxygen utilization increases to as much as 20 times normal. This can be achieved with little further decrease in tissue PO_2—down to a level of 15 to 25 mm Hg—because of the steep slope of the dissociation curve and the increase in tissue blood flow caused by the decreased PO_2 (i.e., a small fall in PO_2 causes large amounts of oxygen to be released).

When atmospheric oxygen concentration changes markedly, the buffer effect of hemoglobin still maintains almost constant tissue PO_2. The PO_2 of alveolar air may vary greatly—from 60 to more than 500 mm Hg—and still the PO_2 in the tissue does not vary more than a few millimeters from normal, demonstrating the tissue oxygen buffer function of the blood hemoglobin.

- *When the alveolar PO_2 falls to 60 mm Hg*, the arterial hemoglobin is still 89 per cent saturated with oxygen (see Fig. 40–1), and the tissues still remove about 5 milliliters of oxygen from each 100 milliliters of blood. To remove oxygen, the PO_2 of the venous blood falls to 35 mm Hg, only 5 mm Hg below the normal value; thus, the tissue PO_2 level hardly changes, despite the fall in alveolar PO_2 from 104 to 60 mm Hg.
- *When the alveolar PO_2 rises to 500 mm Hg*, the maximum oxygen saturation of hemoglobin can never rise above 100 per cent, which is only 3 per cent above the normal level of 97 per cent. The blood still loses several millimeters of oxygen to the tissues, which automatically reduces the PO_2 of the capillary blood to a value only a few millimeters greater than the normal level of 40 mm Hg.

The oxygen-hemoglobin dissociation curve is shifted to the right in metabolically active tissues in which temperature, PCO_2, and hydrogen ion concentration are increased. The oxygen-hemoglobin dissociation curve shown (see Fig. 40–1) is for normal, average blood. A shift in the curve to the right occurs when the affinity for oxygen is low, facilitating the unloading of oxygen. Note that for any given value of PO_2, the percent saturation with oxygen is low when the curve is shifted to the right; the shift to the right is beneficial in exercising muscles and other instances in which oxygen delivery is poor or metabolic activity is increased. The oxygen-hemoglobin dissociation curve is also shifted to the right as an adaptation to chronic hypoxemia associated with life at high altitude. Chronic hypoxemia increases the synthesis of 2,3-diphosphoglycerate, a factor that binds to hemoglobin and decreases the affinity for oxygen.

Carbon monoxide interferes with oxygen transport because it has about 250 times the affinity of oxygen for hemoglobin. Carbon monoxide combines with hemoglobin at the same point on the hemoglobin molecule as does oxygen and can therefore displace oxygen from the hemoglobin. Because it binds with about 250 times as much tenacity as oxygen, relatively small amounts of carbon monoxide can tie up a large portion of the hemoglobin, making it unavailable for oxygen transport. A patient with severe carbon monoxide poisoning can be helped by the administration of pure oxygen because oxygen at high alveolar pressures displaces carbon monoxide from its combination with hemoglobin more effectively than does oxygen at low atmospheric pressures.

TRANSPORT OF CARBON DIOXIDE IN THE BLOOD
(p. 520)

Under resting conditions, about 4 milliliters of carbon dioxide are transported from the tissues to the lungs in each 100 milliliters of blood. Approximately 70 per cent of the carbon dioxide is transported in the form of bicarbonate ions, 23 per cent is transported in combination with hemoglobin and plasma proteins, and 7 per cent is transported in the dissolved state in the fluid of the blood.

- *Transport in the form of bicarbonate ions (70 per cent)*— Dissolved carbon dioxide reacts with water inside red blood cells to form carbonic acid. This reaction is catalyzed by an enzyme in the red blood cells called *carbonic anhydrase*. Most of the carbonic acid immediately dissociates into bicarbonate ions and hydrogen ions; the hydrogen ions in turn combine with hemoglobin. Many of the bicarbonate ions diffuse from the red blood cells into the plasma while chloride ions diffuse into the red blood cells to take their place, a phenomenon called the *chloride shift*.

- *Transport in combination with hemoglobin and plasma proteins (23 per cent)*—Carbon dioxide reacts directly with amine radicals of the hemoglobin molecules and plasma proteins to form the compound carbaminohemoglobin (CO_2Hgb). This combination of carbon dioxide with the hemoglobin is a reversible reaction that occurs with a loose bond, so that the carbon dioxide is easily released into the alveoli where the P_{CO_2} is lower than that in the tissue capillaries.

- *Transport in the dissolved state (7 per cent)*—The amount of carbon dioxide dissolved in the fluid of the blood at 45 mm Hg is about 2.7 milliliters per 100 milliliters of blood (2.7 volumes per cent). The amount dissolved at 40 mm Hg is about 2.4 milliliters, or a difference of 0.3 milliliter. Therefore, only about 0.3 milliliter of carbon dioxide is transported in the form of dissolved carbon dioxide by each 100 milliliters of blood; this represents about 7 per cent of all carbon dioxide that is transported.

41

Regulation of Respiration

The nervous system adjusts the rate of alveolar ventilation to maintain the arterial blood oxygen pressure (PO_2) and carbon dioxide pressure (PCO_2) at relatively constant levels under a variety of conditions. This chapter describes the operation of this regulatory system.

RESPIRATORY CENTER (p. 525)

The respiratory centers are composed of three main groups of neurons.

- *The dorsal respiratory group* generates inspiratory action potentials in a steadily increasing ramp-like fashion and thus is responsible for the basic rhythm of respiration. It is located in the distal portion of the medulla and receives input from peripheral chemoreceptors and other types of receptors via the vagus and glossopharyngeal nerves.
- *The pneumotaxic center*, which is located dorsally in the superior portion of the pons, helps to control the rate and pattern of breathing. It transmits inhibitory signals to the dorsal respiratory group and thus controls the filling phase of the respiratory cycle. Because it limits inspiration, it has a secondary effect to increase respiratory rate.
- *The ventral respiratory group*, which is located in the ventrolateral part of the medulla, can cause either expiration or inspiration, depending on which neurons in the group are stimulated. It is inactive during normal quiet breathing but is important for stimulating the abdominal expiratory muscles when high levels of respiration are required.

The Hering-Breuer reflex prevents overinflation of the lungs. This reflex is initiated by nerve receptors located in the walls of the bronchi and bronchioles that detect the degree of stretch of the lungs. When the lungs become overly inflated, the receptors send signals through the vagi into the dorsal respiratory group that affects inspiration much the same as inhibitory signals from the pneumotaxic center. That is, when the lungs become overly inflated, the stretch receptors activate an appropriate feedback response that "switches off" the inspiratory ramp and thus stops further inspiration; this is called the Hering-Breuer inflation reflex.

CHEMICAL CONTROL OF RESPIRATION (p. 527)

The ultimate goal of respiration is to maintain proper concentrations of oxygen, carbon dioxide, and hydrogen ions in the tissues. Excess carbon dioxide or hydrogen ions mainly stimulate the respiratory center itself, causing increased strength of inspiratory and expiratory signals to the respiratory muscles. Oxygen, in contrast, acts on peripheral chemoreceptors located in the carotid and aortic bodies, and these in turn transmit appropriate nervous signals to the respiratory center for control of respiration.

Increased P_{CO_2} or hydrogen ion concentration stimulates a chemosensitive area of the respiratory center. The sensor neurons in the chemosensitive area are especially excited by hydrogen ions; however, hydrogen ions do not easily cross the blood-brain or blood–cerebrospinal fluid barrier. For this reason, changes in blood hydrogen ion concentration have little acute effect in stimulating the chemosensitive neurons compared with carbon dioxide, even though carbon dioxide is believed to stimulate these neurons secondarily by changing the hydrogen ion concentration. Carbon dioxide diffuses into the brain and reacts with the water of the tissues to form carbonic acid. This in turn dissociates into hydrogen ions and bicarbonate ions; the hydrogen ions then have a potent direct stimulatory effect.

Increased blood carbon dioxide concentration has a potent acute effect to stimulate respiratory drive but only a weak chronic effect. The excitation of the respiratory center by carbon dioxide is greatest during the first few hours of increased carbon dioxide concentration in the blood but then gradually declines over the next 1 to 2 days. This decline is caused by the following:

- The kidneys return the hydrogen ion concentration back toward normal after the carbon dioxide first increases the hydrogen ion concentration. The kidneys increase the blood bicarbonate, which binds with hydrogen ions in the blood and cerebrospinal fluid and reduces their concentration.
- More importantly, the bicarbonate ions diffuse through the blood-brain and blood–cerebrospinal fluid barriers and combine directly with the hydrogen ions around the respiratory neurons.

PERIPHERAL CHEMORECEPTOR SYSTEM FOR CONTROL OF RESPIRATORY ACTIVITY—ROLE OF OXYGEN IN RESPIRATORY CONTROL (p. 529)

Oxygen is unimportant for direct control of the respiratory center. Changes in oxygen concentration have virtually no direct effect on the respiratory center to alter respiratory drive, but when the tissues do lack oxygen, the body has a special mechanism for respiratory control that is located in the peripheral chemoreceptors, outside the brain respiratory center. This mechanism responds when the blood oxygen falls too low, mainly below a PO_2 value of 60 to 70 mm Hg.

Special nervous chemical receptors, called chemoreceptors, are important for detecting changes in arterial PO_2. Chemoreceptors also respond to changes in PCO_2 and hydrogen ion concentrations. The following two types of chemoreceptors transmit nervous signals to the respiratory center to help regulate respiratory activity:

- The *carotid bodies* are located in the bifurcations of the common carotid arteries; their afferent nerve fibers pass through Hering's nerves to reach the glossopharyngeal nerves and then the dorsal respiratory area of the medulla.
- The *aortic bodies* are located along the arch of the aorta; their afferent nerve fibers pass through the vagi and reach the dorsal respiratory area.

The oxygen lack stimulus is often counteracted by decreases in blood PCO_2 and hydrogen ion concentrations. When a person breathes air that has too little oxygen, it decreases the blood PO_2 and excites the carotid and aortic chemoreceptors, thereby increasing respiration. The increased respiration removes carbon dioxide from the lungs and decreases both blood P_{CO_2} and hydrogen ion concentrations. These two changes severely depress the respiratory center so that the final effect of increased respiration in response to low P_{O_2} is mostly counteracted. The effect of low arterial P_{O_2} on alveolar ventilation is far greater under some other conditions, including the following:

- *Pulmonary disease*—In pneumonia, emphysema, or other conditions that prevent adequate gas exchange through the pulmonary membrane, too little oxygen is absorbed into the arterial blood, and at the same time the arterial P_{CO_2} and hydrogen ion concentrations remain near normal or are increased because of poor transport of carbon dioxide through the membrane. The oxygen lack stimulus continues to increase ventilation.
- *Acclimatization to low oxygen*—When climbers ascend a mountain over a period of days rather than a period of hours, they can withstand far lower atmospheric oxygen concentrations. The reason is that the respiratory center loses about four fifths of its sensitivity to changes in arterial P_{CO_2} and hydrogen ions, and the low oxygen can drive the respiratory system to a much higher level of alveolar ventilation.

REGULATION OF RESPIRATION DURING EXERCISE
(p. 532)

In strenuous exercise, the arterial P_{O_2}, P_{CO_2}, and pH values remain almost exactly normal. Strenuous exercise can increase oxygen consumption and carbon dioxide formation by as much as 20-fold, but alveolar ventilation ordinarily increases almost exactly in step with the higher level of metabolism through two mechanisms:

- *Collateral impulses*—The brain, on transmitting impulses to the contracting muscles, is believed to transmit collateral impulses into the brain stem to excite the respiratory center.
- *Body movements*—During exercise, movements of the arms and legs are believed to increase pulmonary ventilation by exciting joint and muscle proprioceptors that in turn transmit excitatory impulses to the respiratory center.

Chemical factors can also play a role in the control of respiration during exercise. When a person exercises, the nervous factors usually stimulate the respiratory center by the proper amount to supply the extra oxygen requirements for the exercise and to blow off the extra carbon dioxide. Occasionally, however, the nervous signals are either too strong or too weak in their stimulation of the respiratory center; then, the chemical factors play a significant role in bringing about the final adjustment in respiration required to keep the carbon dioxide and hydrogen ion concentrations of the body fluids as nearly normal as possible.

OTHER FACTORS THAT AFFECT RESPIRATION (p. 533)

- *Voluntary control*—One can hyperventilate or hypoventilate to such an extent that serious derangements in PCO_2, pH, and PO_2 can occur in the blood. The nervous pathway for voluntary control passes directly from the cortex and other higher centers to the spinal neurons that drive the respiratory muscles.
- *Irritant receptors in the airways*—The epithelium of the trachea, bronchi, and bronchioles is supplied with sensory nerve endings called pulmonary irritant receptors that are stimulated by some irritants that enter the respiratory airways. They cause coughing, sneezing, and possibly also bronchial constriction in diseases such as asthma and emphysema.
- *Lung "J receptors"*—A few sensory nerve endings occur in the alveolar walls in juxtaposition to the pulmonary capillaries, and they are called J receptors. They are especially stimulated when the pulmonary capillaries become engorged with blood or when pulmonary edema occurs in conditions such as congestive heart failure.
- *Brain edema*—The activity of the respiratory center may be depressed or even inactivated by acute brain edema resulting from concussion.
- *Anesthesia*—Perhaps the most prevalent cause of respiratory depression and respiratory arrest is overdose with anesthetics or narcotics.

Cheyne-Stokes breathing is a respiratory disorder most commonly characterized by periodic breathing. With Cheyne-Stokes breathing, the patient breathes deeply for a short time and then breathes slightly or not at all for a short time, with the cycle repeating itself again and again. The period of overbreathing decreases the carbon dioxide level in the pulmonary blood (and increases the oxygen level), but it takes several seconds for the pulmonary blood to reach the respiratory center so that the person has continued to overventilate for a few extra seconds. When the changed pulmonary blood does reach the respiratory center, it becomes depressed, and the cycle repeats itself. Two causes of Cheyne-Stokes breathing are the following:

- *Slow blood flow*—Delayed transport of blood from the lungs to the brain allows changes in blood gases to continue for many more seconds than usual. This type of Cheyne-Stokes breathing often occurs in patients with severe cardiac failure because the left heart is large and blood flow is slow.
- *Increased negative feedback*—A change in arterial P_{O_2} or P_{CO_2} causes far greater change in ventilation than normally occurs. This type of Cheyne-Stokes breathing is found mainly in patients with brain damage; the respiratory drive is turned off for a few seconds, and an increase in blood P_{CO_2} turns it back on with great force.

42

Respiratory Insufficiency — Pathophysiology, Diagnosis, Oxygen Therapy

The diagnosis and treatment of respiratory disorders require an understanding of the basic physiological principles of respiration and gas exchange. Pulmonary disease can result from inadequate ventilation and from abnormalities of gas exchange in the lungs or transport from the lungs to the tissues.

METHODS FOR STUDYING RESPIRATORY ABNORMALITIES (p. 537)

The most fundamental tests of pulmonary performance are determinations of the blood PO_2, PCO_2, and pH. It is often important to make these measurements rapidly as an aid in determining the appropriate therapy for acute respiratory distress or acute abnormalities of acid-base balance. Measuring devices for pH, PCO_2, and PO_2 are built into the same apparatus, and all these measurements can be made within a minute or so with one small sample of blood; thus, changes in the blood gases and pH can be followed almost moment by moment.

MEASUREMENT OF MAXIMUM EXPIRATORY FLOW (p. 538)

A forced expiration is the simplest test of lung function. Figure 42–1B shows the instantaneous relationship between pressure and flow when the patient expires with as much force as he or she can after having inspired as much air as possible. Thus, expiration begins at total lung capacity (TLC) and ends at residual volume (RV) (see Fig. 42–1B). The middle curve shows the maximum expiratory flow at all lung volumes in a normal person. Note that

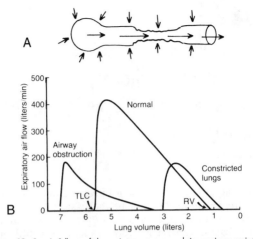

Figure 42–1 *A*, Collapse of the respiratory passageway during maximum expiratory effort, an effect that limits the expiratory flow rate. *B*, Effect of two respiratory abnormalities—constructed lungs and airway obstruction—on the maximum expiratory flow-volume curve.

the expiratory flow reaches a maximum value of over 400 liters/min at a lung volume of 5 liters and then decreases progressively as lung volume decreases. An important aspect of the curve is that the expiratory flow reaches a maximum value beyond which the flow cannot be increased further with additional effort (i.e., the descending portion of the curve representing the maximum expiratory flow is effort independent).

The maximum expiratory flow is limited by dynamic compression of the airways. Figure 42–1*A* shows the effect of pressure applied to the outsides of the alveoli and passageways caused by compression of the chest cage. The arrows indicate that the same amount of pressure is applied to the outsides of both the alveoli and bronchioles. Not only does this pressure force air from the alveoli into the bronchioles, but it also tends to collapse the bronchioles at the same time, which will oppose the movement of air to the exterior. Once the bronchioles have become almost completely collapsed, further expiratory force can still greatly increase the alveolar pressure, but it can also increase the degree of bronchiolar collapse and airway resistance by an equal amount, thus preventing a further rise in flow. Beyond a critical degree of expiratory force, a maximum expiratory flow has been reached.

The maximum expiratory flow-volume curve is useful in determining whether an obstructive or a restrictive lung disease is present. Figure 42–1*B* shows a normal maximum flow-volume curve along with curves generated from patients with obstructive lung disease or restrictive lung disease.

- *Restrictive lung disease*—The flow-volume curve in a restrictive lung disease (e.g., interstitial fibrosis) is characterized by low lung volumes and slightly higher-than-normal expiratory flow rates at each lung volume.
- *Obstructive lung disease*—The flow-volume curve in obstructive lung diseases (e.g., chronic bronchitis, emphysema, asthma) is characterized by high lung volumes and lower-than-normal expiratory flow rates. The curve may also have a "scooped-out" appearance.

PHYSIOLOGICAL PECULIARITIES OF SPECIFIC PULMONARY ABNORMALITIES (p. 539)

Obstructive lung disease is characterized by increased resistance to air flow and high lung volumes. Patients with obstructive lung disease find it easier to breathe at high lung volumes because this tends to increase the caliber of the airways (greater radial traction) and thus decrease the resistance to airflow. Mechanisms of airway obstruction include the following:

- *The airway lumen* may be partially obstructed by excessive secretions (chronic bronchitis), edema fluid, or aspiration of food or fluids.
- *The airway wall smooth muscle* may be contracted (asthma) or thickened because of inflammation and edema (asthma, bronchitis), or the mucus glands may be hypertrophied (chronic bronchitis).
- *Outside the airway*, the destruction of lung parenchyma may decrease radial traction, causing the airways to be narrowed (emphysema). In addition, swollen lymph nodes or neoplasm may compress the airways.

Restrictive lung disease is characterized by low lung volumes. Patients with restrictive lung disease find it easier to breathe at low lung volumes because it is difficult to expand the lungs. Expansion of the lung may be restricted for the following reasons:

- Abnormal lung parenchyma in which excessive fibrosis increases lung elasticity (e.g., pulmonary fibrosis, silicosis, asbestosis, tuberculosis)
- Problems with the pleura (e.g., pneumothorax, pleural effusion)
- Neuromuscular problems (e.g., polio, myasthenia gravis).

Chronic Pulmonary Emphysema

The term *pulmonary emphysema* **literally means excess air in the lungs.** Chronic pulmonary emphysema, however, signifies a complex obstructive and destructive process of the lungs and is usually a consequence of long-term smoking. The following pathophysiological events contribute to its development:

- *Chronic infection*—The bronchi and bronchioles are irritated. The irritant (usually smoke) deranges the normal protective mechanisms of the airways: (1) cilia may be partially paralyzed so that mucus cannot be moved easily out of the passageways, (2) secretion of excess mucus exacerbates the condition, and (3) alveolar macrophages can be inhibited.
- *Airway obstruction*—The infection, excess mucus, and inflammatory edema of the bronchiolar epithelium combine to cause chronic obstruction of many of the smaller airways.
- *Destruction of alveolar walls*—The obstruction of the airways makes it especially difficult to expire, causing entrapment of air in the alveoli and overstretching of the alveoli. This, combined with lung infection, causes marked destruction of the alveolar cells.

The physiological effects of chronic emphysema are extremely varied. They depend on the severity of the disease and the relative degree of bronchiolar obstruction versus lung parenchymal destruction. Chronic emphysema usually progresses slowly over many years. Among the different abnormalities are the following:

- *Increased airway resistance*—This is caused by bronchiolar obstruction and greatly increases the work of breathing. Expiration is especially difficult because the force on the outside of the lung compresses the bronchioles, which further increases their resistance.
- *Decreased diffusing capacity*—This is caused by the marked loss of alveolar walls and reduces the ability of the lungs to oxygenate the blood and remove carbon dioxide.
- *Abnormal ventilation-perfusion ratio* ($\dot{V}_A/\dot{Q}$)—In the same lungs, areas of the lung with bronchiolar obstruction have a very low $\dot{V}_A/\dot{Q}$ (physiological shunt), resulting in poor aeration of blood, and other areas with loss of alveolar walls have a very high $\dot{V}_A/\dot{Q}$ (physiological dead space), resulting in wasted ventilation.
- *Increased pulmonary vascular resistance*—Loss of alveolar walls decreases the number of pulmonary capillaries. The loss of capillaries causes the pulmonary vascular resistance to increase, in turn causing pulmonary hypertension.

Pneumonia (p. 540)

The term *pneumonia* includes any inflammatory condition of the lung in which alveoli are filled with fluid and blood cells. A common type of pneumonia is bacterial pneumonia, caused most frequently by pneumococci. This disease begins with infection in the alveoli; the pulmonary membrane becomes inflamed and highly porous, so that fluid and even red and white blood cells pass out of the blood into the alveoli. The infected alveoli become progressively filled with fluid and cells, and the infection spreads by extension of bacteria from alveolus to alveolus. Eventually, large areas of the lungs, sometimes whole lobes or even a whole lung, become "consolidated," which means that they are filled with fluid and cellular debris.

Atelectasis (p. 541)

Atelectasis **is the collapse of the alveoli.** The following are two common causes:

- *Airway obstruction*—This type of atelectasis usually results from (1) blockage of small bronchi with mucus and (2) bronchial obstruction caused by a mucous plug or solid object. The air trapped beyond the block is absorbed, causing alveolar collapse. If the lung cannot collapse, negative pressure develops in the alveoli, causing edema fluid to collect.

- *Lack of surfactant*—Surfactant decreases the alveolar surface tension, which helps to prevent alveolar collapse. In hyaline membrane disease (also called respiratory distress syndrome), however, the quantity of surfactant secreted by the alveoli is greatly depressed. As a result, the surface tension of the alveolar fluid is increased, causing the lungs to collapse or become filled with fluid.

Asthma

Asthma is an obstructive lung disease. The usual cause is hypersensitivity of the bronchioles to foreign substances in the air. The allergic reaction produces (1) localized edema in the walls of the small bronchioles as well as secretion of thick mucus into the bronchiolar lumens and (2) spasm of the bronchiolar smooth muscle. The airway resistance increases greatly.

The asthmatic person can usually inspire adequately but has great difficulty expiring. Clinical measurements show greatly reduced maximum expiratory rate and timed expiratory volume; this results in dyspnea, or "air hunger." The functional residual capacity and residual volume of the lung are increased during the asthmatic attack because of the difficulty in expiring air. Over a period of years, the chest cage becomes permanently enlarged, causing a "barrel chest," and the functional residual capacity and residual volume become permanently increased.

Tuberculosis

In tuberculosis, the tubercle bacilli cause a peculiar tissue reaction in the lungs, including (1) invasion of the infected region by macrophages and (2) walling off of the lesion by fibrous tissue to form the so-called tubercle. Tuberculosis in its late stages causes many areas of fibrosis and reduces the total amount of functional lung tissue. These effects cause the following:

- *Increased "work" by respiratory muscles* to cause pulmonary ventilation because of reduced vital capacity and breathing capacity.
- *Reduced total respiratory membrane surface area and increased thickness of the respiratory membrane* causes decreased pulmonary diffusing capacity.
- *Abnormal $\dot{V}_A/\dot{Q}$* further reduces the pulmonary diffusing capacity.

HYPOXIA AND OXYGEN THERAPY (p. 542)

Hypoxia can result from multiple causes. The following is a descriptive classification of the causes of hypoxia:

- Inadequate oxygenation of the lungs because of extrinsic reasons
  1. Deficiency of oxygen in atmosphere
  2. Hypoventilation (neuromuscular disorders)
- Pulmonary disease
  1. Hypoventilation due to increased airway resistance or decreased pulmonary compliance
  2. Uneven alveolar ventilation-perfusion ratio
  3. Diminished respiratory membrane diffusion
- Venous-to-arterial shunts ("right-to-left" cardiac shunts)
- Inadequate oxygen transport by the blood to the tissues
  1. Anemia or abnormal hemoglobin
  2. General circulatory deficiency
  3. Localized circulatory deficiency (peripheral, cerebral coronary vessels)
  4. Tissue edema
- Inadequate tissue capability of using oxygen
  1. Poisoning of cellular enzymes
  2. Diminished cellular metabolic capacity because of toxicity, vitamin deficiency, or other factors

The classic example of an inability of tissues to use oxygen is cyanide poisoning. Cyanide blocks the action of the enzyme cytochrome oxidase to such an extent that the tissues simply cannot utilize oxygen even though plenty is available. Deficiencies of some oxidative enzymes or other elements in the tissue oxidative system can lead to this type of hypoxia. A special example occurs in the disease beriberi, in which several important steps in the tissue utilization of oxygen and formation of carbon dioxide are compromised because of vitamin B deficiency.

Oxygen Therapy in the Different Types of Hypoxia
(p. 542)

Oxygen therapy is of great value in certain types of hypoxia but of almost no value in other types. Recalling the basic physiological principles of the different types of hypoxia, one can readily decide when oxygen therapy will be of value and, if so, how valuable.

- *Atmospheric hypoxia*—Oxygen therapy can correct the depressed oxygen level in the inspired gases and therefore provide 100 per cent effective therapy.
- *Hypoventilation hypoxia*—A person breathing 100 per cent oxygen can move five times as much oxygen into the alveoli with each breath as when breathing normal air. Again, here oxygen therapy can be extremely beneficial.
- *Hypoxia caused by impaired alveolar membrane diffusion*—Essentially the same result occurs as in hypoventilation hypoxia because oxygen therapy can increase the PO_2 in the lungs from a normal value of about 100 mm Hg to as high as 600 mm Hg. This raises the oxygen diffusion gradient between the alveoli and the blood by more than 800 per cent.
- *Hypoxia caused by oxygen transport deficiencies*—In hypoxia caused by anemia, abnormal hemoglobin transport of oxygen, circulatory deficiency, or physiological shunt, oxygen therapy is of less value because oxygen is already available in the alveoli. Instead, the problem is deficient transport of oxygen to the tissues. Extra oxygen can be transported in the dissolved state in the blood when alveolar oxygen is increased to the maximum level; this extra oxygen may be the difference between life and death.
- *Hypoxia caused by inadequate tissue use of oxygen.* In this type of hypoxia, there is abnormality neither of oxygen pickup by the lungs nor of transport to the tissues. Instead, the tissue metabolic enzyme system is simply incapable of utilizing the oxygen that is delivered. It is therefore doubtful that oxygen therapy is of any measurable benefit.

HYPERCAPNIA (p. 543)

Hypercapnia **means excess carbon dioxide in the body fluids.** When the alveolar P_{CO_2} rises above about 60 to 75 mm Hg, the person is breathing about as rapidly and deeply as he or she can, and air hunger, or *dyspnea*, becomes severe. As the P_{CO_2} rises to 80 to 100 mm Hg, the person becomes lethargic and sometimes even semicomatose. Anesthesia and death can result when the P_{CO_2} rises to 120 to 150 mm Hg. At the higher levels of P_{CO_2}, the excess carbon dioxide now begins to depress respiration rather than stimulate it, thus causing a vicious circle.

Cyanosis **means bluish skin.** It is caused by deoxygenated hemoglobin in the skin blood vessels, especially in the capillaries. This deoxygenated hemoglobin has a dark blue-purple color. In general, definite cyanosis appears whenever the arterial blood contains more than 5 grams of deoxygenated hemoglobin in each 100 milliliters of blood. A person with anemia almost never becomes cyanotic because there is not enough hemoglobin for 5 grams of it to be deoxygenated in the arterial blood. In comparison, in a person with excess red blood cells, the great excess of available hemoglobin leads frequently to cyanosis, even under otherwise normal conditions.

Dyspnea **means mental anguish associated with an inability to ventilate sufficiently to satisfy the demand for air.** A common synonym is *air hunger*. At least three factors often enter into the development of the sensation of dyspnea, as follows:

- *Hypercapnia*—A person becomes dyspneic from a buildup of carbon dioxide in body fluids.
- *Work of breathing*—The work of breathing associated with the attainment of normal blood gases after hypercapnia or hypoxic conditions frequently gives the person a sensation of dyspnea.
- *State of mind*—Dyspnea may be experienced even with normal respiratory function because of an abnormal state of mind; this is called *neurogenic* or *emotional dyspnea*.

UNIT

VIII

Aviation, Space, and Deep Sea Diving Physiology

VII

43

Aviation, High Altitude, and Space Physiology

Technological advancements have made it increasingly more important to understand the effects of altitude and low gas pressures as well as several other factors—acceleratory forces, weightlessness, and so forth—on the human body. This chapter discusses each of these problems.

EFFECTS OF LOW OXYGEN PRESSURE ON THE BODY
(p. 549)

A decrease in barometric pressure is the basic cause of high-altitude hypoxia. Note on Table 43–1 that as altitude increases, barometric pressure decreases, and PO_2 decreases proportionately. The alveolar PO_2 is also reduced by carbon dioxide and water vapor.

- *Carbon dioxide*—The alveolar PCO_2 falls from a sea level value of 40 mm Hg to lower values as altitude increases. In the acclimatized person with a fivefold increase in ventilation, the PCO_2 can be as low as 7 mm Hg because of higher respiration.
- *Water vapor pressure*—In the alveoli, it remains at 47 mm Hg as long as the body temperature is normal, regardless of altitude.

Carbon dioxide and water vapor pressure reduce the alveolar oxygen. The barometric pressure is 253 mm Hg at the top of 29,028-foot Mount Everest; 47 mm Hg of this must be water vapor, leaving 206 mm Hg for other gases. In the acclimatized person, 7 mm of the 206 mm Hg must be carbon dioxide, leaving 199 mm Hg. If there were no use of oxygen by the body, one fifth of this 199 mm Hg would be oxygen and four fifths would be nitrogen, or the PO_2 in the alveoli would be 40 mm Hg. Some of this

TABLE 43–1 EFFECTS OF ACUTE EXPOSURE TO LOW ATMOSPHERIC PRESSURES ON ALVEOLAR GAS CONCENTRATIONS AND ARTERIAL OXYGEN SATURATION*

Altitude (ft)	Barometric Pressure (mm Hg)	P_{O_2} in Air (mm Hg)	Breathing Air			Breathing Pure Oxygen		
			P_{CO_2} in Alveoli (mm Hg)	P_{O_2} in Alveoli (mm Hg)	Arterial Oxygen Saturation (%)	P_{CO_2} in Alveoli (mm Hg)	P_{O_2} in Alveoli (mm Hg)	Arterial Oxygen Saturation (%)
0	760	159	40 (40)	104 (104)	97 (97)	40	673	100
10,000	523	110	36 (23)	67 (77)	90 (92)	40	436	100
20,000	349	73	24 (10)	40 (53)	73 (85)	40	262	100
30,000	226	47	24 (7)	18 (30)	24 (38)	40	139	99
40,000	141	29				36	58	84
50,000	87	18				24	16	15

* Numbers in parentheses are acclimatized values.

alveolar oxygen would be absorbed into the blood, leaving an alveolar PO_2 of about 35 mm Hg.

Breathing pure oxygen increases arterial oxygen saturation at high altitudes. Table 43–1 shows arterial oxygen saturation while breathing air (column 6) and while breathing pure oxygen (column 9).

- *Breathing air*—Up to an altitude of about 10,000 feet, the arterial oxygen saturation remains at least as high as 90 per cent; it falls progressively until it is only about 70 per cent at 20,000 feet and much less at still higher altitudes.
- *Breathing pure oxygen*—When pure oxygen is breathed, the space in the alveoli formerly occupied by nitrogen becomes occupied by oxygen. At 30,000 feet, an aviator could have an alveolar PO_2 as high as 139 mm Hg instead of the 18 mm Hg he or she would have when breathing air.

One of the most important effects of hypoxia is decreased mental proficiency. Associated deficiencies include decreases in judgment, memory, and performance of discrete motor movements. Other acute effects of hypoxia are drowsiness, lassitude, muscle fatigue, sometimes headache, occasionally nausea, and sometimes euphoria. All of these progress to a stage of twitchings or seizures above 18,000 feet and, above 23,000 feet, coma occurs in the unacclimatized person.

A person remaining at high altitudes for days, weeks, or years becomes more and more acclimatized to the low PO_2. Acclimatization makes it possible for the person to work harder without hypoxic effects or to ascend to still higher altitudes. The principal means by which acclimatization comes about are the following:

- Increased pulmonary ventilation
- Increased red blood cells
- Increased diffusing capacity of the lungs
- Increased vascularity of the tissues
- Increased ability of the cells to use oxygen despite the low PO_2.

Pulmonary ventilation can increase fivefold in the acclimatized person but only 65 per cent in the unacclimatized person. Acute exposure to a hypoxic environment increases alveolar ventilation to a maximum of about 65 per cent above normal. If the person remains at a very high altitude for several days, the ventilation gradually increases to an average of about five times normal (400 per cent above normal).

- *Acute increase in pulmonary ventilation*—The immediate 65 per cent increase in pulmonary ventilation on rising to a high altitude blows off large quantities of carbon dioxide, reducing the P_{CO_2} and increasing the pH of the body fluids. Both these changes inhibit the respiratory center and thereby oppose the effect of low P_{O_2} to stimulate the peripheral respiratory chemoreceptors in the carotid and aortic bodies.
- *Chronic increase in pulmonary ventilation*—The acute inhibition fades away within 2 to 5 days, allowing the respiratory center to respond with full force, and the ventilation increases by about fivefold. The decreased inhibition results mainly from a reduction in bicarbonate ion concentration in the cerebrospinal fluid as well as the brain tissues. This in turn decreases the pH in the fluids surrounding the chemosensitive neurons of the respiratory center, thus increasing the activity of the center.

Hematocrit and blood volume increase during acclimatization. Hypoxia is the principal stimulus for an increase in red blood cell production. In full acclimatization to low oxygen, the hematocrit rises from a normal value of 40 to 45 to an average of about 60, with a proportionate increase in hemoglobin concentration. In addition, the blood volume increases, often by 20 to 30 per cent, resulting in a total rise in circulating hemoglobin of 50 or more per cent. This increase in hemoglobin and blood volume begins after 2 weeks, reaching half development in a month or so, and full development only after many months.

The pulmonary diffusing capacity can increase as much as threefold after acclimatization. The normal diffusing capacity for oxygen through the pulmonary membrane is about 21 ml/mm Hg/min. The following factors contribute to the threefold increase after acclimatization:

- *Increased pulmonary capillary blood volume* expands the capillaries and increases the surface area through which oxygen can diffuse into the blood.
- *Increased lung volume* expands the surface area of the alveolar membrane.

- *Increased pulmonary arterial pressure* forces blood into greater numbers of alveolar capillaries—especially in the upper parts of the lungs, which are poorly perfused under usual conditions.

Chronic hypoxia increases the number of capillaries in the tissues. The cardiac output often increases as much as 30 per cent immediately after a person ascends to high altitude but then decreases back toward normal as the blood hematocrit increases, so that the amount of oxygen transported to the tissues remains about normal. The number of capillaries in the tissues increases, especially in animals born and bred at high altitudes. The greater capillarity is especially marked in active tissues. For instance, the capillary density in the right ventricular muscle increases greatly because of the combined effects of hypoxia and excess work load on the right ventricle caused by pulmonary hypertension at high altitude.

The work capacity of all the muscles is greatly decreased in hypoxia. This includes not only the skeletal muscles but also the cardiac muscle, so that even the maximum level of cardiac output is reduced. In general, the work capacity is reduced in direct proportion to the decrease in maximum rate of oxygen uptake that the body can achieve. Even at a very high altitude, naturally acclimatized natives can achieve a daily work output that is almost equal to that of normal persons at sea level. Well-acclimatized lowlanders can almost never achieve this result.

A person who remains at a high altitude too long can develop *chronic mountain sickness.* The following effects contribute to the development of mountain sickness: (1) red blood cell mass and hematocrit become exceptionally high, (2) pulmonary arterial pressure increases even more than normal, (3) right side of the heart becomes greatly enlarged, (4) peripheral arterial pressure begins to fall, (5) congestive heart failure ensues, and (6) death often follows unless the person is removed to a lower altitude. The causes of this sequence of events are probably threefold:

- The red blood cell mass becomes so great that the blood viscosity increases several-fold; this now actually decreases tissue blood flow, so that oxygen delivery also begins to decrease.

- The pulmonary arterioles become exceptionally vasospastic because of lung hypoxia. Because all the alveoli are hypoxic, all the arterioles become constricted, and the pulmonary arterial pressure rises excessively, causing the right side of the heart to fail.
- The pulmonary arteriolar spasm diverts blood flow through nonalveolar pulmonary vessels, causing an excess of pulmonary shunt blood flow where the blood is not oxygenated.

A small percentage of people who ascend rapidly to high altitudes develop *acute mountain sickness*. The sickness begins from within a few hours to about 2 days after the ascent and may cause death if the individual is not administered oxygen or moved to a low altitude. The following two events frequently occur:

- *Acute cerebral edema*—Hypoxia vasodilates the cerebral blood vessels, increasing capillary pressure and in turn causing fluid to leak into the cerebral tissues. The cerebral edema can lead to severe disorientation and other effects that are related to cerebral dysfunction.
- *Acute pulmonary edema*—The cause is still unknown, but the following sequence has been suggested: The hypoxia causes pulmonary arterioles to constrict, but this is greater in some parts of the lungs than in others, causing the blood to flow through still unconstricted pulmonary vessels. The capillary pressure in these areas of the lungs becomes especially high, and local edema occurs.

EFFECTS OF ACCELERATORY FORCES ON THE BODY IN AVIATION AND SPACE PHYSIOLOGY (p. 552)

Pilots are exposed to positive and negative acceleratory forces—*G*. When a pilot is sitting in his or her seat, the force pressing the body against the seat results from the pull of gravity and is equal to his or her weight. The intensity of this force is said to be +1 G because it is equal to the pull of gravity.

- *Positive G force*—If the force with which a pilot presses against the seat becomes five times the normal weight during "pull-out" from a dive, the force acting on the seat is +5 G.

- *Negative G force*—If the airplane goes through an outside loop so that the pilot is held down by the seat belt, negative G is applied to the body; if the force with which he or she is thrown against the belt is equal to the weight of the body, the negative force is −1 G.

The most important effect of centrifugal acceleration is on the circulatory system. The reason that the circulatory system is so affected is that blood is mobile and can be translocated by centrifugal forces. Acceleration greater than 4 to 6 G causes "blackout" of vision within a few seconds and unconsciousness soon afterward. The following major cardiovascular effects often occur in response to positive G:

- *Cardiac output decreases*—The blood is centrifuged toward the lower part of the body. The more the blood pressure increases in the lower body, the more the blood "pools" there, and the less the cardiac output becomes.
- *Arterial blood pressure decreases*—The systolic and diastolic arterial pressures in the upper body fall greatly for the first few seconds after acceleration begins. They then recover somewhat within another 10 to 15 seconds. This secondary recovery is caused mainly by activation of baroreceptor reflexes.

The effects of negative G on the body are less dramatic acutely but possibly more damaging permanently than the effects of positive G. Negative acceleratory forces of −4 to −5 G cause intense hyperemia of the head with transient psychotic disturbances resulting from brain edema. Greater negative G forces can cause cerebral blood pressure to reach 300 to 400 mm Hg. The vessels inside the cranium, however, are somewhat protected by the cerebrospinal fluid because it is centrifuged toward the head at the same time that blood is centrifuged toward the cranial vessels. The greatly increased pressure of this fluid acts as a cushioning buffer on the outside of the brain to prevent vascular rupture. Because the eyes are not protected by the cranium, intense hyperemia can occur, causing them to become temporarily blinded with "redout."

Specific procedures and apparatus protect aviators against centrifugal acceleratory forces. Tightening the abdominal muscles and leaning forward to compress the abdomen can prevent some of the pooling of blood in the large vessels of the abdomen. Special "anti-G" suits prevent pooling of blood in the lower abdomen and legs. The simplest of these applies positive pressure to the legs and abdomen by inflating compression bags as the G increases.

WEIGHTLESSNESS IN SPACE (p. 554)

Physiological problems exist with weightlessness. Most of the problems that occur appear to be related to three effects of weightlessness: (1) motion sickness during the first few days of travel, (2) translocation of fluids within the body because of the failure of gravity to cause hydrostatic pressures, and (3) diminished physical activity because no strength of muscle contraction is required to oppose the force of gravity. The physiological consequences of prolonged periods in space are the following:

- Decreased blood volume
- Decreased red blood cell mass
- Decreased muscle strength and work capacity
- Decreased maximum cardiac output
- Loss of calcium and phosphate from the bones as well as loss of bone mass.

The physiological consequences of prolonged weightlessness are similar to those experienced by people who lie in bed for an extended period of time. For this reason, extensive exercise programs are carried out during prolonged space laboratory missions, and most of the effects mentioned are greatly reduced, except for some bone loss. In previous space laboratory expeditions in which the exercise program had been less vigorous, astronauts had severely decreased work capacities for the first few days after returning to earth. They also had a tendency to faint when they stood up during the first day or so after return to gravity because of their diminished blood volume and perhaps diminished responses of the arterial pressure control mechanisms. Even with the exercise program, fainting continues to be a problem after a prolonged expedition.

44

Physiology of Deep Sea Diving and Other Hyperbaric Conditions

The pressure around a diver increases greatly as he or she descends beneath the sea. Air must be supplied also under high pressure exposing the blood in the lungs to extremely high alveolar gas pressure, a condition called hyperbarism. These high pressures can cause tremendous alterations in the body physiology.

As one descends beneath the sea, the pressure increases and the gases are compressed to smaller volumes.

- *Increase in pressure.* A column of sea water 33 feet deep exerts the same pressure at its bottom as all the atmosphere above the earth. A person 33 feet beneath the ocean surface is, therefore, exposed to a pressure of 2 atmospheres: the first atmosphere of pressure caused by the air above the water and the second atmosphere by the weight of the water itself. At 66 feet the pressure is 3 atmospheres (and so forth, in accord with Table 44–1).
- *Decrease in volume.* If a bell jar at sea level contains 1 liter of air, the volume will have been compressed to one-half liter at 33 feet beneath the sea where the pressure is 2 atmospheres; at 8 atmospheres (233 feet) the volume is one-eighth liter. The volume to which a given quantity of gas is compressed is inversely proportional to the pressure, as shown in Table 44–1. This is the physical principle called *Boyle's law.*

EFFECT OF HIGH PARTIAL PRESSURES OF GASES ON THE BODY (p. 557)

Nitrogen narcosis can occur when nitrogen pressure is high. When a diver remains beneath the sea for an hour or more and is breathing compressed air, the depth at which the first symptoms of mild

TABLE 44–1 EFFECT OF SEA DEPTH ON PRESSURE AND ON GAS VOLUMES		
Depth (feet)	**Atmospheres (s)**	**Volume (liters)**
Sea level	1	1.0000
33	2	0.5000
66	3	0.3333
100	4	0.2500
133	5	0.2000
166	6	0.1667
200	7	0.1429
300	10	0.1000
400	13	0.0769
500	16	0.0625

narcosis appear is about 120 feet. At this level, the diver begins to exhibit joviality and seems to lose many of his or her cares. At 150 to 200 feet, he or she becomes drowsy. At 200 to 250 feet, strength wanes considerably. Beyond 250 feet, the diver usually becomes almost useless as a result of nitrogen narcosis if he or she remains at these depths too long. Nitrogen narcosis has characteristics similar to those of alcohol intoxication, and for this reason it has frequently been called "raptures of the depths."

The amount of oxygen transported in the blood increases greatly at extremely high PO_2. As the pressure rises progressively into the thousands of mm Hg, a large portion of the total oxygen is then dissolved, rather than bound with hemoglobin. If the PO_2 in the lungs is about 3000 mm Hg (4 atmospheres pressure), the total amount of oxygen dissolved in the blood's water is 9 ml/100 ml of blood. As this blood passes through the tissue capillaries and as the tissues use their normal amount of oxygen, the PO_2 of the blood is about 1200 mm Hg, instead of the normal 40 mm Hg. Thus, once the alveolar PO_2 rises above a critical level, the hemoglobin-oxygen buffer mechanism (discussed in Chapter 40) is no longer capable of keeping the tissue PO_2 in the normal safe range between 20 and 60 mm Hg.

The brain is especially susceptible to acute oxygen poisoning. Exposure to 4 atmospheres pressure of oxygen ($PO_2 = 3040$ mm Hg) causes seizures followed by coma in most people after 30 minutes. The seizures often occur without warning and, for obvious reasons, are likely to be lethal to divers submerged beneath the sea. Other symptoms encountered in acute oxygen poisoning include nausea, muscle twitching, dizziness, vision disturbance, irritability, and disorientation. Exercise greatly increases a diver's susceptibility to oxygen toxicity, causing symptoms to appear much earlier and with far greater severity than at rest.

Nervous system oxygen toxicity is caused by "oxidizing free radicals." Molecular oxygen (O_2) must first be converted into an "active" form before it can oxidize other chemical compounds. There are several forms of active oxygen; they are called oxygen free radicals. One of the most important of these is the superoxide free radical O_2^-, and another is the peroxide radical in the form of hydrogen peroxide.

- *Normal tissue PO_2.* Even when the tissue PO_2 is normal (40 mm Hg), small amounts of free radicals are continually being formed from the dissolved molecular oxygen. The tissues also contain enzymes that remove these free radicals, especially peroxidases, catalases, and superoxide dismutases. At a normal tissue PO_2, the oxidizing free radicals are removed so rapidly that they have little or no effect in the tissues.
- *High tissue PO_2.* Above about 2 atmospheres, the tissue PO_2 increases greatly and large amounts of oxidizing free radicals literally swamp the enzyme systems for removing them. One of the principal effects of the oxidizing free radicals is to oxidize the polyunsaturated fatty acids of the membranous structures of the cells, and another effect is to oxidize some of the cellular enzymes, thus damaging severely the cellular metabolic systems. The nervous tissues are especially susceptible because of the high lipid content.

Chronic oxygen poisoning causes pulmonary disability. A person can be exposed to 1 atmosphere pressure of oxygen almost indefinitely without developing the acute oxygen toxicity of the nervous system. After only 12 or so hours of 1 atmosphere oxygen exposure, lung passageway congestion, pulmonary edema, and atelectasis begin to develop. The reason for this effect in the lungs and not in other tissues is that the air spaces of the lungs are directly exposed to the high oxygen pressure.

Depth alone does not increase the carbon dioxide partial pressure in the alveoli. Depth does not increase the rate of carbon dioxide production in the body. As long as the diver continues to breathe a normal tidal volume, he or she continues to expire the carbon dioxide as it is formed, maintaining the alveolar carbon dioxide partial pressure at a normal value.

When a person breathes air under high pressure for a long time, the amount of nitrogen dissolved in the body fluids becomes great. The blood flowing through the pulmonary capillaries becomes saturated with nitrogen to the same high pressure as that in the breathing mixture. Over several hours, enough nitrogen is carried to all the tissues of the body to saturate the tissues also with dissolved nitrogen. Because nitrogen is not metabolized by the body, it remains dissolved until the nitrogen pressure in the lungs decreases, at which time the nitrogen is then removed by the reverse respiratory process. This removal takes hours to occur, which is the source of multiple problems collectively called "decompression sickness."

Decompression sickness results from formation of nitrogen bubbles in the tissues. If large amounts of nitrogen have dissolved in a diver's body and then he or she suddenly comes back to the surface of the sea, significant quantities of nitrogen bubbles can develop in the body fluids either intracellularly or extracellularly and cause minor or serious damage, depending on the number and size of bubbles formed; this is "decompression sickness." Exercise hastens the formation of bubbles during decompression because of increased agitation of the tissues and fluids. Synonyms for decompression sickness include bends, compressed air sickness, caisson disease, diver's paralysis, and dysbarism.

Most of the symptoms of decompression sickness are caused by gas bubbles blocking blood vessels in the different tissues. At first, only the smallest vessels are blocked by minute bubbles, but as the bubbles coalesce, progressively larger vessels are affected. Tissue ischemia and sometimes tissue death are the result.

- *Joint pain.* In about 89 per cent of people with decompression sickness, the symptoms are pain in the joints and muscles of the legs or arms. The joint pain accounts for the term "bends" that is often applied to this condition.

- *Nervous system symptoms.* In 5 to 10 percent of those with decompression sickness, nervous system symptoms occur, ranging from dizziness in about 5 percent to paralysis or collapse and unconsciousness in 3 per cent. The paralysis may be temporary, but in some instances, the damage is permanent.
- *The "chokes."* About 2 percent of those with decompression sickness develop "the chokes," caused by massive numbers of microbubbles plugging the capillaries of the lungs; this is characterized by serious shortness of breath, often followed by severe pulmonary edema and, occasionally, death.

Tank decompression is used for treatment of decompression sickness. The diver is put into a pressurized tank, and the pressure is then lowered gradually back to normal atmospheric pressure. Tank decompression is even more important for treating people in whom symptoms of decompression sickness develop minutes or even hours after they have returned to the surface. In this case, the diver is recompressed immediately to a deep level. Then decompression is carried out over a time period several times as long as the usual period of decompression.

Helium is often used to replace nitrogen in the gas mixture for very deep dives. Helium is used for the following three reasons:

- It has only about ⅕ the narcotic effect of nitrogen.
- About half as much volume of helium dissolves in the body tissues as nitrogen, and the volume that does dissolve diffuses out of the tissues several times as rapidly as does nitrogen, thus reducing the problem of decompression sickness.
- The low density of helium keeps the airway resistance for breathing at a minimum, which is extremely important because highly compressed nitrogen is so dense that airway resistance can become extreme, sometimes making the work of breathing beyond endurance.

HYPERBARIC OXYGEN THERAPY (p. 561)

Hyperbaric oxygen can have valuable therapeutic effects in several important clinical conditions. The oxygen is usually administered at P_{O_2} of 2 to 3 atmospheres of pressure. It is believed that the same oxidizing free radicals responsible for oxygen toxicity are also responsible for the therapeutic benefits. Some of the conditions in which hyperbaric oxygen therapy has been especially beneficial are the following:

- *Gas gangrene.* The bacteria that cause this condition, clostridial organisms, grow best under anaerobic conditions and stop growing at oxygen pressures greater than about 70 mm Hg. Hyperbaric oxygenation of the tissues can frequently stop the infectious process entirely and thus convert a condition that formerly was almost 100 per cent fatal into one that is cured in most instances of early treatment.
- *Leprosy.* Hyperbaric oxygenation might have almost as dramatic an effect in curing leprosy as in curing gas gangrene—also because of the susceptibility of the leprosy bacillus to destruction by high oxygen pressures.
- *Other conditions.* Hyperbaric oxygen therapy has also been valuable in the treatment of decompression sickness, arterial gas embolism, carbon monoxide poisoning, osteomyelitis, and myocardial infarction.

UNIT

IX

The Nervous System

A. General Principles and Sensory Physiology

45

Organization of the Nervous System; Basic Functions of Synapses and Transmitter Substances

GENERAL DESIGN OF THE NERVOUS SYSTEM (p. 565)

This involves both sensory (input) and motor (output) systems and is rather simple, although the actual substrate for many of its activities is extremely complex. The fundamental unit of operation in the nervous system is the *neuron*, which typically consists of a cell body (*soma*), several *dendrites*, and a single *axon*. Although most neurons exhibit these same three components, there is enormous variability in the specific morphology of individual neurons throughout the brain. It is estimated that the nervous system is composed of more than 100 billion neurons.

Much of the activity in the nervous system arises from mechanisms that stimulate *sensory receptors* located at the distal termination of a *sensory neuron*. Signals then travel over peripheral nerves to reach the spinal cord, and from there information is transmitted throughout the brain. Incoming sensory messages are processed and integrated with information stored in various pools of neurons in the brain, so that the resulting signals can be used to generate an appropriate *motor response*.

The motor division of the nervous system is responsible for controlling a variety of bodily activities such as contraction of skeletal and smooth muscles and secretion by exocrine and endocrine glands. Actually, only a relatively small proportion of the sensory input received by the brain is employed to generate an immediate motor response; much of it is discarded as irrelevant to the function at that moment. A large share of sensory input is stored in the form of *memory*.

397

Information stored as memory can become part of the processing mechanism used to manage subsequent sensory input. The brain is able to compare new sensory experiences with those stored in memory and in this way develop successful strategies to form a motor output.

FUNCTION OF THE CENTRAL NERVOUS SYSTEM (p. 569)

This is based on interactions that occur between neurons at specialized junctions called synapses. At a site of termination, an axon typically forms a number of branches that exhibit small dilated segments called *synaptic terminals* or *synaptic boutons*. The synaptic bouton is apposed to, but separated from, an adjacent postsynaptic structure (dendrite or soma) by a narrow space (10 to 30 nanometers) called the *synaptic cleft*. The synaptic boutons contain a variety of organelles including numerous mitochondria and exhibit an aggregation of relatively small spheroidal *synaptic vesicles* (40-nanometer diameter) that contain molecules of a chemical *neurotransmitter* agent. When released from the axon terminal, this transmitter agent binds to receptors on the postsynaptic neuron and alters its permeability to certain ions.

The two types of synapses that are most prevalent in the brain are chemical and electrical synapses. The overwhelming majority are *chemical synapses*. One neuron, the *presynaptic element*, releases a transmitter agent that binds to the *postsynaptic neuron*, which is then excited or inhibited. The transmission of signals at chemical synapses is typically "one way," from presynaptic axon terminal to the postsynaptic dendrite or soma.

The least common type of synapse (in mammals) is the *electrical synapse*; these consist of gap junctions that form low-resistance channels between the presynaptic and postsynaptic elements. At these synapses, various ions can freely move between the two related neurons, thereby mediating rapid transfer of signals that can spread through large pools of neurons.

When a synaptic bouton is invaded by an action potential, the transmitter agent is released into the synaptic cleft, where it can diffuse to and bind with specific receptors located in the membrane of the adjacent postsynaptic dendrite or soma. The excitatory or inhibitory action of the transmitter agent is actually determined by the response of the postsynaptic receptors.

Neurotransmitter Release Is Calcium Dependent (p. 570)

- When invaded by an action potential, the large number of voltage-gated *calcium channels* in the surface membrane of the synaptic bouton are opened, and calcium moves *into* the terminal.
- The inward calcium flux is coupled to the movement of synaptic vesicles to release sites at the membrane (presynaptic) nearest the postsynaptic structure. The vesicles fuse with the presynaptic membrane and release their transmitter agent into the synaptic cleft by a process called *exocytosis*. The quantity of transmitter released is directly related to the quantity of calcium entering the terminal.

Action of a Neurotransmitter: Determined by Its Postsynaptic Receptor (p. 570)

Receptors are complex proteins with (1) a *binding domain* that protrudes into the synaptic cleft and (2) an *ionophore* that extends through the membrane to protrude into the interior of the postsynaptic structure. The ionophore can be an ion channel that is specific for the passage of a certain ion, or it can form a *second messenger activator*. In both cases, the receptors are linked to chemically gated or *ligand-gated ion channels*.

- *Ligand-gated ion channels* can be *cationic* (passing sodium, potassium, or calcium ions) or *anionic* (passing mainly chloride ions).
- In general, ligand-gated channels that allow sodium to enter the postsynaptic neuron are *excitatory*, whereas channels that allow chloride to enter (or potassium to exit) the postsynaptic neuron are *inhibitory*. When activated, these channels typically open and close within fractions of a millisecond; therefore, these mechanisms provide for very rapid interaction between neurons.
- *Second messenger activators* are most commonly *G-proteins* that are attached to a portion of the receptor that protrudes into the postsynaptic element. When the receptor is activated, a portion of the G-protein is released and able to move within the cytoplasm of the postsynaptic neuron (as a

"second messenger"), where it performs one of four possible activities: (1) opening a specific membrane channel for an ion species such as sodium or potassium and allowing it to remain open for a longer period of time than is generally noted with a typical ligand-gated channel; (2) activating cyclic AMP or cyclic GMP, which in turn can stimulate specific metabolic machinery in the neuron; (3) activating certain enzymes, which then initiate a specific biochemical reaction in the postsynaptic neuron; and (4) activating gene transcription resulting in protein synthesis that may alter the metabolism or morphology of the cell. All of these activities are especially well suited to induction of relatively long-term changes in the excitability, biochemistry, or functional activity of the postsynaptic neuron.

Chemical Substances Function as Neurotransmitters
(p. 572)

At present, more than 50 substances have been reported to fulfill the criteria as neurotransmitters. Generally, these substances can be divided into two groups: *small-molecule transmitters* and *neuroactive peptides* as listed in Tables 45–1 and 45–2.

Small-molecule, rapidly acting transmitters are synthesized and packaged into synaptic vesicles in the axon terminal. The effect of these agents on the postsynaptic membrane is brief in duration (1 millisecond or less) and is typically involved with ion channel opening or closure. In some instances, the small-molecule agents can stimulate receptor-activated enzymes and thereby alter the metabolic oper-

---

TABLE 45–1 SMALL-MOLECULE, RAPIDLY ACTING NEUROTRANSMITTERS

Class I	*Class III: Amino Acids*
Acetylcholine	γ-Aminobutyric acid
Class II: The Amines	Glycine
Norepinephrine	Glutamate
Epinephrine	Aspartate
Dopamine	*Class IV*
Serotonin	Nitric oxide
Histamine	

TABLE 45–2 NEUROPEPTIDE, SLOWLY ACTING NEUROTRANSMITTERS

A. Hypothalamic-Releasing Hormones	C. Peptides That Act on Gut and Brain
Thyrotropin-releasing hormone	Leucine enkephalin
Leutinizing hormone–	Methionine enkephalin
	Substance P
releasing hormone	Gastrin
Somatostatin (growth hormone inhibitory factor)	Cholecystokinin
	Vasoactive intestinal polypeptide (VIP)
B. Pituitary Peptides	Neurotensin
β-Endorphin	Insulin
α-Melanocyte-stimulating hormone	Glucagon
Prolactin	D. From Other Tissues
Leutinizing hormone	Angiotensin II
Thyrotropin	Bradykinin
Growth hormone	Carnosine
Vasopressin	Sleep peptides
Oxytocin	Calcitonin
ACTH	

ation of the postsynaptic neuron. The synaptic vesicles used by these neurotransmitters are *recycled* at the axon terminal (i.e., after fusion with the presynaptic membrane at the synaptic active site, newly formed vesicles are released from the axon terminal membrane more peripherally and subsequently replenished with the transmitter agent). *Acetylcholine* is one of the typical small-molecule transmitters; it is synthesized from acetyl coenzyme A and choline in the presence of the enzyme *choline acetyltransferase*. The latter substance is actually synthesized in the soma and delivered to synaptic boutons via axonal transport mechanisms. When acetylcholine is released from vesicles into the synaptic cleft, it binds to receptors on the postsynaptic membrane. Within milliseconds, it is broken down into acetate and choline by the enzyme *acetylcholinesterase*, which is also present in the synaptic cleft. As a general rule, the small-molecule transmitters are rapidly inactivated shortly after they bind to their receptors. In this example, choline is actively transported back into the synaptic bouton for subsequent re-use in the synthesis of acetylcholine.

Neuropeptides form the second group of transmitter agents, and these substances are typically synthesized in the soma as integral components of large proteins. These large molecules are cleaved in the cell body and packaged into vesicles in the Golgi apparatus either as the active peptidergic agent or as a precursor of the neuroactive substance. The vesicles are delivered to the axon terminals, and the transmitter is released into the synaptic cleft as described. Commonly, however, smaller amounts of the neuroactive peptide are released compared with the small-molecule transmitters, and their vesicles do not appear to be recycled. An important feature of the neuropeptides is that the duration of their action is more prolonged than that of the small-molecule agents. The peptides can alter ion channel function, modify the metabolism of the cell, or modify gene expression, and these actions can be sustained for minutes, hours, days, or presumably even longer.

In most instances, neurons use only one neurotransmitter agent; however, more examples are being reported in which a small-molecule substance and neuropeptide are *co-localized* in a single synaptic bouton. How the neuron might coordinate the use of the two substances remains to be established.

Certain Electrical Events Are Characteristic of Excitatory Synaptic Interactions (p. 574)

- The neuronal membrane exhibits a characteristic *resting membrane potential* of about -65 millivolts. Moving this potential to a more positive value (depolarization) makes the cell more excitable, whereas changing the potential to a more negative value (hyperpolarization) makes the cell less excitable.
- At rest, the concentrations of various ions measured external or internal to the cell membrane are different. Extracellular sodium concentration is much higher than its intracellular concentration, whereas just the opposite is true for potassium. The distribution of chloride ions is similar to that of sodium ions, although the concentration gradient is less.
- Recall that the Nernst potential for an ion is that electrical potential that opposes movement of that ion down its concentration gradient.

Nernst Potential (in millivolts)

$$= \pm 61 \times \log \frac{\text{(Ion Concentration } Inside\text{)}}{\text{(Ion Concentration } Outside\text{)}}$$

For sodium, the Nernst potential is +61 millivolts. Because the resting membrane potential in neurons is approximately −65 millivolts, one might expect sodium to move into the cell at rest. It does not, however, because the voltage-gated sodium channels are closed. A small amount does "leak" in, while potassium "leaks" out, but a sodium/potassium pump exchanges sodium ions for potassium ions and moves sodium back out and potassium back into the cell, thereby maintaining the resting potential.

- The neuronal membrane at rest is maintained at about −65 millivolts, mainly due to the fact that it is much more permeable to *potassium* ions than to sodium ions. As a result, large numbers of positively charged potassium ions move out of the cell, leaving behind an abundance of negatively charged ion species, and the interior becomes negatively charged with respect to the extracellular environment. The interior of the soma (and dendrites) consists of a highly conductive fluid environment with essentially no electrical resistance. Changes in electrical potential that occur in one part of the cell can easily spread throughout the neuron.

- When a transmitter-receptor interaction results in the opening of *ligand-gated sodium channels* in the postsynaptic membrane, sodium enters the postsynaptic neuron, and the membrane potential moves in the positive direction (i.e., toward the Nernst potential for sodium [+61 millivolts]). This new, more positive local potential is called an *excitatory postsynaptic potential (EPSP)*. If the membrane potential of the neuron is moved above *threshold* at the axon initial segment, an action potential will be generated by the postsynaptic neuron. The action potential is said to be initiated at the axon initial segment because this region has been shown to contain approximately seven times the number of voltage-gated membrane channels than are found in other parts of the neuron. In most instances, the simultaneous discharge of *many* axon terminals is required to bring the postsynaptic neuron to threshold; this is called *summation,* and this concept is discussed further.

Electrical Events Are Characteristic of Inhibitory Synaptic Interactions (p. 577)

- Neurotransmitters that selectively open *ligand-gated chloride channels* are the basis for the production of an *inhibitory postsynaptic potential (IPSP)*.
- The Nernst potential for chloride is −70 millivolts. In general, this is more negative than the postsynaptic neuron resting membrane potential. As a result, chloride ions move into the cell, the membrane potential becomes more negative (hyperpolarized), and the cell is less excitable (inhibited). Similarly, if a transmitter selectively opens potassium channels, positively charged potassium ions exit the cell, also making the interior more negative.
- Another inhibitory mechanism involves "short-circuiting," or shunting, the membrane potential. In some neurons, the resting membrane potential closely approximates the chloride Nernst potential. In this situation, when chloride channels open, there is no *net* inward or outward movement of chloride ions; instead, chloride ions move bidirectionally in a rapid manner. Under these conditions, if a sodium-induced EPSP occurs, the extensive ongoing chloride flux through the open chloride channels means that a higher-than-normal sodium influx (5 to 20 times greater) must occur to move the membrane away from the chloride Nernst potential. This makes the cell less excitable during the period of chloride ion flux.

EPSPs and IPSPs Are Summated Over Time and Space (p. 578)

- *Temporal summation* occurs when a second postsynaptic potential (EPSP or IPSP) arrives before the membrane has returned to its resting level. Because a typical postsynaptic potential may last for about 15 milliseconds and ion channels are open for about 1 millisecond (or less), there usually is more than sufficient time for several channel openings to occur over the course of a single postsynaptic potential. The effects of these two potentials are additive (summed over time).
- *Spatial summation* occurs when a number of axon terminals over the surface of a neuron are simultaneously active. Their aggregate effects are summed, and the combined postsynaptic potential

will be greater than any one individual potential. Commonly, the magnitude of a single EPSP might be only 0.5 to 1 millivolt, which is far less than the 10 to 20 millivolts that are frequently required to reach threshold. Spatial summation enables the combined EPSP to be of sufficient magnitude to exceed threshold.

- At any given point in time, a neuron is combining the affects of *all* the EPSPs and IPSPs that are occurring over its surface. As a consequence, the postsynaptic neuron might (1) become more excitable and possibly generate an action potential or increase its firing rate and (2) become less excitable and not initiate any action potentials or decrease its level of firing.

Dendrites Perform Special Functions in the Excitation and Inhibition of Neurons (p. 579)

Since the dendritic surface forms such a large proportion of the total surface of the neuron, it is estimated that 80 to 95 per cent of all synaptic boutons terminate on dendritic elements. Dendrites contain a relatively small number of voltage-gated ion channels in their surface membranes and therefore are not able to *propagate* action potentials. They can support the spread of electrical current by *electrotonic conduction*, but this mode of transmission is subject to decay (decrement) over time and space. The EPSPs (or IPSPs) that arise at distal points on the dendritic tree may diminish to such a low level by the time they reach the soma and axon initial segment that there is insufficient current to bring the neuron to threshold. Conversely, synapses on proximal dendrites or soma have more influence over the initiation of action potentials because they are simply closer to the axon initial segment. The synaptic potentials will not decrement to a level below that necessary to reach threshold.

Firing Rate of a Neuron: Related to Its State of Excitation (p. 580)

Many factors contribute to the determination of the threshold of a neuron, so that this functional characteristic varies among neurons. Some neurons are inherently more excitable than others (i.e., it takes less current to reach threshold), whereas others fire at a more rapid rate once threshold is exceeded. The firing rate of a neuron is directly related to the degree to which threshold is exceeded; the farther above threshold, the greater the firing rate, although there is an upper limit.

SYNAPTIC TRANSMISSION EXHIBITS SPECIAL CHARACTERISTICS (p. 581)

- When synapses are repetitively stimulated at a rapid rate, the response of the postsynaptic neuron can diminish over time, and the synapse is said to be *fatigued*. This decreased responsiveness is mainly thought to be the result of increased buildup of calcium within the synaptic bouton and inability to rapidly replenish the supply of neurotransmitter agent.
- When repetitive (tetanic) stimulation is applied to an excitatory synapse and is followed by a brief rest period, subsequent activation of that synapse may require less current and produce an enhanced response; this is called *post-tetanic facilitation*.
- The pH of the extracellular synaptic environment influences the excitability of the synaptic function. More acidic values *increase* excitability, whereas more alkaline levels *decrease* synaptic activity.
- A *decrease* in the supply of oxygen diminishes synaptic activity.
- The effects of drugs and chemical agents on neuronal excitability are diverse, complex, and variable. For example, caffeine clearly increases the excitability of many neurons, whereas strychnine can *indirectly* increase the activity of neurons by inhibiting certain populations of inhibitory interneurons
- The transmission of current across a synapse requires a certain amount of time that varies from one neuronal pool to another. This is called *synaptic delay*, and it is influenced by (1) the time required to release the transmitter, (2) the time required to diffuse across the synaptic cleft, (3) the time required by the transmitter to bind to its receptor, (4) the time required by the receptor to carry out its action, and (5) the time required for ions to diffuse into the postsynaptic cell and alter the membrane potential.

46

Sensory Receptors; Neuronal Circuits for Processing Information

SENSORY RECEPTORS (p. 583)

Five Basic Types of Nervous System Receptors

- *Mechanoreceptors* detect physical deformation of the receptor membrane or the tissue immediately surrounding the receptor.
- *Thermoreceptors* detect changes in the temperature (warm or cold) of the receptor.
- *Nociceptors* detect the presence of physical or chemical damage to the receptor or the tissue immediately surrounding it.
- *Photoreceptors* (electromagnetic receptors) detect light (photons) striking the retina.
- *Chemoreceptors* are responsible for taste and smell, O_2 and CO_2 levels in the blood, and osmolality of tissue fluids.

Sensory receptors are highly sensitive to one particular type of stimulus (modality)—the "labeled line" principle. Once activated, a receptor initiates action potentials in its associated sensory fiber, which then conveys these impulses into the spinal cord in the form of a labeled line in a peripheral nerve. These impulses or action potentials are similar in all sensory fibers. They may exhibit qualitative differences in amplitude or frequency, but an action potential initiated by a painful stimulus is not perceived as uniquely distinguishable from an action potential initiated by any other receptor or *sensory modality*.

What does allow us to differentiate one type of sensation from another is the location in the nervous system where the fiber leads or terminates. Each fiber or collection of neurons linked by related sensory fibers is referred to as a "labeled line." Action potentials, for example, traveling along the fibers and neurons that compose the *anterolateral system* (spinothalamic tract) are perceived as pain, whereas action potentials carried over the dorsal column–medial lemniscal system are perceived as touch or pressure.

SENSORY RECEPTORS (p. 584)

These transduce a physico-chemical stimulus into a nerve impulse. When activated by the appropriate stimulus, a local current is generated at the receptor; this is referred to as a receptor potential. Regardless of whether the stimulus is mechanical, chemical, or physical (e.g., heat, cold, and light), the transduction process results in a change in the ionic permeability of the receptor membrane and consequently a change in the potential difference across this membrane. The maximum receptor potential amplitude of about 100 millivolts is achieved when the membrane sodium permeability is at its maximum level.

The sensory fiber linked to each receptor exhibits "threshold phenomena." Only when the receptor potential exceeds a set value (threshold) is a self-propagating action potential initiated in the fiber. The receptor potential is a graded potential, meaning that it will diminish over time and space.

The receptor potential is proportional to stimulus intensity (p. 585). As stimulus intensity increases, subsequent action potentials usually increase in *frequency*. The receptor potential amplitude may change substantially with a relatively small stimulus change but then will increase only minimally with greater stimulus intensity.

Sensory receptors adapt to their stimuli either partially or completely over time (p. 586). This adaptation occurs by one of two different mechanisms. First, the physico-chemical properties of the receptor may be altered by the stimulus; for example, when a pacinian corpuscle is initially deformed (and its membrane permeability increased), the fluid within its concentric lamellae redistributes the applied pressure. This redistribution is reflected as a decrease in membrane permeability, and the receptor potential diminishes or adapts. Second, a process of *accommodation* may occur in the sensory fiber itself. Although poorly understood, this may involve a gradual "inactivation" of sodium channels over time.

Receptors are classified as slowly adapting or rapidly adapting (p. 587). *Slowly adapting receptors* continue to transmit signals with little change in frequency as long as the stimulus is present. For this reason, they are called "tonic receptors" and are able to signal stimulus strength for extended periods of time. Some examples are muscle spindles, Golgi tendon organs, pain receptors, baroreceptors, and chemoreceptors. *Rapidly adapting receptors* are activated only when stimulus intensity changes. These receptors are referred to as "rate receptors" or "movement detectors." The pacinian corpuscle is the best example of the latter receptor category, along with the receptor of the semicircular canal and joint receptor (proprioceptor).

PHYSIOLOGICAL CLASSIFICATION OF RECEPTORS
(p. 587)

Two different schemes have been devised to classify the entire range of peripheral nerve fibers:

- In the more general scheme of classification, all peripheral fibers are divided into types A and C, with type A fibers subdivided into four categories (Fig. 46–1). This scheme is based on the diameter and conduction velocity of each fiber, with Type Aα being the largest and most rapidly conducting variety.
- A second scheme, which was devised mainly by sensory physiologists, distinguishes five main categories that are also based on fiber diameter and conduction velocity.

Intensity of a Stimulus (p. 588)

This is represented in sensory fibers using the features of spatial and temporal summation. Commonly, a single sensory nerve trunk in a peripheral nerve contains several fibers, each of which is related to a variable number of receptors (more than 100 in the case of free nerve endings in the skin) at its distal termination. The aggregate of all the receptors and fibers related to a single nerve defines the *receptive field* of that nerve. An intense stimulus that extends to the entire receptive field will activate all the fibers in the sensory nerve trunk, whereas a less intense stimulus will activate proportionally fewer fibers.

Different gradations of stimulus intensity are signaled by involving a variable number of "parallel" fibers in the same nerve (spatial summation) or by

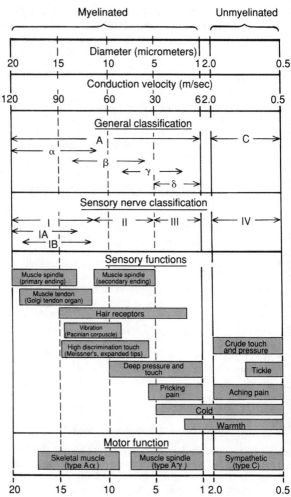

Figure 46–1 Physiological classifications and functions of nerve fibers.

changing the frequency of impulses traveling in a single fiber (temporal summation).

TRANSMISSION AND PROCESSING OF SIGNALS IN NEURONAL POOLS (p. 589)

Any aggregate of neurons, such as the cerebral cortex or thalamus, or an individual nucleus in the thalamus can be referred to as a *neuronal pool*. Typically, each neuronal pool has a set of several inputs (*afferents*), its *receptive field*, and one or several "targets" to which it projects via a set of organized efferent axons.

Afferent systems can provide either *threshold or subthreshold stimulation* to the neuronal pool. Threshold stimulation will obviously raise the membrane potential above firing levels in several cells and generate action potentials. In other cells, the membrane potential may be slightly depolarized but not quite sufficient to reach threshold (subthreshold). These cells are said to be *facilitated*, or they are more excitable because the postsynaptic potentials that are smaller than normal will bring the cell to threshold and fire action potentials.

In some neuronal pools, *divergence* of incoming signals is a common feature. This divergence may take one of two different forms. In an *amplification* mechanism, an input fiber may branch to contact many neurons in the pool, and these postsynaptic cells in turn project in a unified manner to one or a restricted number of targets. In the other form of divergence, the activated neurons in the pool project to multiple, unrelated targets.

The processing in neuronal pools might use the mechanism of *convergence*. Multiple inputs from a single afferent system may terminate on a single neuron in the pool. Alternatively, convergence can result when input signals from multiple sources reach a single neuron in the pool.

On the afferent side, a single neuron or pool of neurons can give rise to both excitatory and inhibitory output signals. A single efferent axon may provide excitatory output to one neuron in the next (postsynaptic) pool that is itself an excitatory (relay) neuron, or it may synapse with an inhibitory interneuron in the next pool, which might then inhibit relay neurons in the postsynaptic pool. This is called *feedforward inhibition*.

Signal processing in neuronal pools can involve a *reverberating circuit* or *oscillating circuits*. In these circuits, the output axons of the pool give rise to collateral branches that synapse with *excitatory* interneurons located *within* the pool. These excitatory interneurons then provide feedback to the same output neurons of the pool, thus leading to a self-propagating sequence of signals. The excitatory postsynaptic potentials produced by the excitatory interneurons can be facilitatory or actually stimulate firing by the pool output neurons. The latter situation is the substrate for a neuronal cell group that emits a *continuous train* of efferent signals (p. 592). Some neuronal pools generate a *rhythmical output signal* (e.g., the respiratory centers in the medullary reticular formation) (p. 593); this function uses a reverberating circuit.

Extensive and Diverse Connectivity in the Nervous System
(p. 593)

These can produce functional instability in the brain when operations go awry. One of the most obvious examples of this instability is an *epileptic seizure*. Two mechanisms are used by the nervous system to combat functional instability.

- The most prominent of these mechanisms is *feedback inhibition*. In this circuit, the output of a neuronal pool activates *inhibitory interneurons* located within the pool, and these cells then provide inhibitory feedback to the main output neurons of the pool. Such a circuit forms an internally regulated "brake" on the output of the pool. When the brake fails, as occurs in a seizure, the pool output fires in an uncontrolled manner.
- The second method to limit instability is called *synaptic fatigue*. The actual substrate for this feature is not well understood; it may have a molecular basis, such as a decrease in the uptake or utilization of calcium. Alternatively, it may be related to a more long-term change in receptor sensitivity involving the process of receptor number (sensitivity) up-regulation or down-regulation, which is known to occur in the brain.

47

Somatic Sensations

I. General Organization; The Tactile and Position Senses

The somatic senses can be divided into three main components:

1. *Mechanoreception* includes both tactile and position (proprioceptive) sensations.
2. *Thermoreception* detects increases or decreases in temperature.
3. *Nociception* detects tissue damage or the release of specific pain-mediating molecules.

The sensory modalities conveyed over the somatic sensory systems include discriminative (precisely localized) touch, crude (poorly localized) touch, pressure, and vibration, and the senses of static body position and body movement that are collectively referred to as *proprioception*. *Exteroceptive* sensations are those that originate from stimulation of body surface structures, such as the skin and subcutaneous tissues, as well as from deeper structures, such as the muscle, fascia, and tendons. In contrast, sensory signals that arise from internal organs (endodermally derived structures) are called *visceral* sensations.

DETECTION AND TRANSMISSION OF TACTILE SENSATIONS (p. 595)

Even though touch, pressure, and vibration are often classified as separate and distinct sensations, they are detected by the same general class of tactile receptors: the *mechanoreceptors*. At least six different types of mechanoreceptors are classified as *tactile* receptors:

- *Free nerve endings* are found in varying densities in all areas of the skin as well as the cornea of the eye.
- *Meissner's corpuscle* is an encapsulated, rapidly adapting receptor found in the nonhairy (glabrous) areas of skin such as the fingertips and lips, areas that are particularly sensitive to even the lightest touch stimulation.
- *Merkel's discs* (known as expanded tip receptors) are found in glabrous skin but are also found in moderate numbers in hairy skin surfaces. These receptors are relatively slowly adapting and are thought to signal the continuous touch of objects against the skin.
- *Hair end-organs* (peritrochal endings) are entwined about the base of each hair on the body surface. They are rapidly adapting and detect the movement of objects over the skin surface, which displaces the hairs.
- *Ruffini's end-organs* are encapsulated endings that are located in the skin and deeper tissues as well as joint capsules. They exhibit very little adaptation and thus signal continuous touch and pressure applied to the skin or movement around the joint where they are located.
- *Pacinian corpuscles* are present in the skin and deeper tissues such as fascia. They adapt very rapidly and are thought to be especially important for detecting vibration or other rapid changes in the mechanical state of the tissues.

Most of these categories of tactile receptors transmit signals over relatively large myelinated fibers that exhibit rapid conduction velocities. In contrast, the free nerve endings are linked to small myelinated fibers and unmyelinated type C fibers that conduct at relatively slow velocities.

Each of the tactile receptors is also involved in the detection of vibration. Pacinian corpuscles detect the most rapid vibratory stimuli (30 to 800 cycles per second) and are linked to the large, rapidly conducting myelinated fibers. Low-frequency vibration (up to about 80 cycles per second) stimulate Meissner's corpuscles and the other tactile receptors, which generally transmit at relatively slow conduction velocities and are less rapidly adapting than the pacinian corpuscles.

The sense of a tickle or an itch is related to very sensitive, rapidly adapting free nerve endings in the superficial layers of the skin that mainly transmit over type C fibers. The function of this sensory modality is presumably to call attention to light skin irritations that can be relieved by movement or scratching, a stimulus that appears to override the itch signals.

SOMATOSENSORY PATHWAYS IN THE CENTRAL NERVOUS SYSTEM (p. 597)

The main pathways for the transmission of somatosensory signals are the dorsal column–medial lemniscal and anterolateral systems. With a few exceptions, sensory information carried by nerve fibers from the body surface (exclusive of the face) enter the spinal cord through dorsal roots. Once in the central nervous system, the signals are segregated into one of two pathways. Signals that originated at thermoreceptors and nociceptors are processed along the anterolateral system and are described in Chapter 48. Signals that arise from mechanoreceptors travel in the dorsal column–medial lemniscal (DC-ML) system. These modalities include discriminative touch, vibration, and proprioception. In a similar manner, somatosensory information from the face is carried mainly in branches of the trigeminal nerve; when such fibers enter the brain stem, they also segregate into two pathways: one is specialized for processing pain, temperature, and crude touch, and the other carries discriminative touch, vibration, and proprioception.

TRANSMISSION IN THE DORSAL COLUMN–MEDIAL LEMNISCAL SYSTEM (p. 598)

The anatomy of the dorsal column–medial lemniscal system is characterized by a high degree of somatotopic (spatial) organization as follows:

Primary Sensory Neurons. The central processes of dorsal root ganglion primary sensory neurons that enter the spinal cord through the medial division of the dorsal root entry zone are the larger, myelinated fibers that carry signals related to discriminative touch, vibration, and proprioception. On entering the cord, some of these fibers form local synapses in the gray matter, whereas many simply pass into the dorsal column area and ascend without synapsing until they reach the *dorsal column nuclei* in the caudal medulla. Here, fibers carrying information from the lower extremities synapse in the nucleus gracilis, whereas those from the upper extremities terminate in the nucleus cuneatus.

Dorsal Column Nuclei. Axons of cells in the cuneate and gracile nuclei form the *medial lemniscus*, which crosses the midline in the caudal medulla as the sensory decussation. This fiber bundle continues rostrally to the thalamus, where the axons terminate in the ventrobasal complex, mainly the ventrolateral posterior nucleus (VPL). Axons of VPL neurons then enter the posterior limb of the internal capsule and project to the *primary somatosensory cortex* (SI) in the postcentral gyrus.

Medial Lemniscal Pathway. The fibers of the DC-ML system exhibit a high degree of somatotopic organization *(spatial orientation)*. Fibers carrying signals from the lower extremity pass upward through the medial portion of the dorsal column, terminate in the gracile nucleus, and form the ventral and lateral portion of the medial lemniscus. They eventually terminate laterally in VPL; neurons here project to the most medial part of SI, on the medial wall of the hemisphere. Information from the upper extremity travels in the lateral part of the dorsal column, terminates in the cuneate nucleus, and enters the dorsal and medial portion of the medial lemniscus. These fibers synapse in the medial part of VPL and finally reach the arm territory of SI in the hemisphere contralateral to the body surface, where the signals actually originated. Throughout the system, there is a point-to-point relationship between the origin in the periphery and the termination in SI.

Somatosensory Signals from the Face. Tactile somatosensory signals from the face travel in the trigeminal nerve and enter the brain stem at midpontine levels, where the primary sensory fibers terminate in the principal trigeminal sensory nucleus. From here, axons cross the midline and course rostrally, adjacent to the medial lemniscus, and eventually terminate medially in a portion of the ventrobasal complex, the ventral posteromedial nucleus (VPM). This system of fibers is comparable to the DC-ML system and conveys similar types of somatosensory information from the face.

Somatosensory Areas of the Cerebral Cortex. The postcentral gyrus comprises the primary somatosensory cortex, which corresponds to Brodmann's areas 3, 1, and 2. A second somatosensory area (SII) is much smaller than SI and is located just posterior to the face region of SI bordering on the lateral fissure. Within SI, the segregation of body parts is maintained so that the face region is ventrally located nearest the lateral fissure, the upper extremity continues medially and dorsally from the face region and extends toward the convexity of the hemisphere, and the lower extremity projects onto the medial surface of the hemisphere. In fact, there is a complete but separate body representation in areas 3, 1, and 2. Within each of these body representations, there is an *unequal volume* of cortex devoted to each body part. Those body surfaces with a high density of sensory receptors are represented by larger areas in the cortex than those with a relatively low density of receptors.

Functional Anatomy of the Primary Somatosensory Cortex

- Contains six different horizontally arranged cellular layers numbered from I to VI beginning with layer I at the cortical surface (p. 600). The most characteristic is layer IV because it receives the important projections from VPL and VPM of the ventrobasal thalamus. From here, information is spread dorsally into layers I to III and ventrally to layers V and VI.
- Contains an array of vertically organized columns of neurons that extend through all six layers (p. 601). These are *functionally determined columns* that vary in width from 0.3 to 0.5 millimeter and are estimated to contain about 10,000 neurons each. In the most anterior part of area 3 in SI, the vertical columnar arrays are concerned with muscle afferents, whereas in the posterior part of area 3, they process cutaneous input. In area 1, the vertical columns process additional cutaneous input, whereas in area 2, they are concerned with pressure and proprioception.

The functions of the primary and association somatosensory areas can be inferred from studies of patients with lesions in these areas as follows.

- Lesions that involve *primary somatosensory cortex* result in the (1) inability to precisely localize cutaneous stimuli on the body surface; some crude localizing ability may be retained; (2) inability to judge degrees of pressure, or the weight of objects touching the skin; and (3) inability to identify objects by touch or texture *(astereognosis)*.

- Lesions that involve Brodmann's areas 5 and 7 will damage the *association cortex for somatic sensation*. Common signs and symptoms include the (1) inability to recognize objects that have a relatively complex shape or texture when palpated with the contralateral hand and (2) loss of the awareness of the contralateral side of the body *(hemineglect)* (this symptom is most acute with lesions in the nondominant parietal lobe); and (3) when feeling an object, patients will explore only the side that is ipsilateral to their lesion and ignore the contralateral side *(amorphosynthesis)*.

Characteristics of Signal Transmission and Processing in the Dorsal Column–Medial Lemniscal System (p. 602)

The *receptive field* of an SI cortical neuron is determined by the combination of primary sensory neurons, dorsal column nuclear neurons, and thalamic neurons that provide afferent projections to that SI neuron.

Two-point discrimination is used to test the functional status of the DC-ML system. This method is frequently used to determine the individual's ability to distinguish two simultaneously applied cutaneous stimuli as two separate "points" *(two-point discrimination)*. This capability varies substantially over the body surface. On the fingertips and lips, two points of stimulation as close together as 1 to 2 millimeters can be distinguished as separate points, whereas on the back, the two points must be separated by at least 30 to 70 millimeters. This function depends on the central processing elements in the DC-ML pathway to recognize that the two excitatory signals generated peripherally are separate and nonoverlapping.

Lateral inhibition is a mechanism used throughout the nervous system to "sharpen" signal transmission. This process uses inhibition of the input from the peripheral portion of a receptive field to better define the boundaries of the excited zone. In the DC-ML system, lateral inhibition occurs at the level of the dorsal column nuclei and in thalamic nuclei.

The **DC-ML system is particularly effective in sensing** *rapidly changing* **and** *repetitive stimuli,* **and this is the basis for** *vibratory sensation.* This capability resides in the rapidly adapting pacinian corpuscles, which are able to detect vibrations up to 700 cycles per second, and in Meissner's corpuscles, which detect somewhat lower frequencies such as 200 cycles per second and below.

The awareness of body position or body movement is called *proprioceptive sensation.* The sense of body movement is also called the *kinesthetic sense* or *dynamic proprioception.* A combination of tactile, muscle, and joint capsule receptors is used by the nervous system to produce the sense of proprioception. For the movement of small body parts such as the fingers, tactile receptors in the skin and joint capsules are thought to be most important in determining the proprioceptive signal. For complex movements of the upper or lower limbs in which some joint angles are increasing and others are decreasing, muscle spindles are an important determinant of proprioceptive sensation. At the extremes of joint angulation, the stretch imposed on ligaments and deep tissues around the joint can activate pacinian corpuscles and Ruffini endings. The latter, being rapidly adapting receptors, are probably responsible for detection of the rate of change in movement.

Transmission of Poorly Localized Tactile Input: Anterolateral System (p. 605)

Signals traveling on small myelinated fibers and unmyelinated C fibers can arise from tactile receptors (typically free-nerve endings) in the skin. This information is transmitted along with pain and temperature signals in the anterolateral portion of the spinal cord white matter. As discussed in Chapter 48, the anterolateral system extends to the ventrobasal thalamus as well as to the intralaminar and posterior thalamic nuclei. Although some painful stimuli are fairly well localized, the precise point-to-point organization in the DC-ML system and relative diffusiveness of the anterolateral system probably account for the less effective localizing ability of the latter system.

48

Somatic Sensations

II. Pain, Headache, and Thermal Sensations

Pain is mainly a protective mechanism for the body because it is not a pure sensation but rather a response to tissue injury that is created within the nervous system.

PAIN SENSATION: FAST AND SLOW PAIN CLASSIFICATION (p. 609)

Fast pain is felt within about 0.1 second after the stimulation, whereas *slow pain* is felt 1 second or more after the painful stimulation. Slow pain is usually associated with tissue damage and can be referred to as burning pain, aching pain, or chronic pain.

All pain receptors are free nerve endings. They are found in largest number and density in the skin, periosteum, arterial walls, joint surfaces, and dura and its reflections inside the cranial vault.

THREE TYPES OF STIMULI (p. 609)

Pain receptors are activated by

- *Mechanical* and *thermal* stimuli, tend to elicit *fast pain*.
- *Chemical* stimuli tend to produce *slow pain*, although this is not always the case. Some of the more common chemical agents that elicit pain sensations include bradykinin, serotonin, histamine, potassium ion, acid, acetylcholine, and proteolytic enzyme. The tissue concentration of these substances appears to be directly related to the degree of tissue damage and in turn the perceived degree of painful sensation. In addition, prosta-

421

glandins and substance P enhance the sensitivity of pain receptors but do not directly excite them.

● *Pain receptors adapt very slowly or essentially not at all.* In some instances, the activation of these receptors becomes progressively *greater* as the pain stimulus continues; this is called *hyperalgesia.*

TWO SEPARATE PATHWAYS OF PAIN TRANSMISSION
(p. 610)

Fast pain signals elicited by mechanical or thermal stimuli are transmitted over Aδ fibers in peripheral nerves at velocities between 6 and 30 m/sec. In contrast, the slow, chronic type of pain signals are transmitted over type C fibers at velocities ranging from 0.5 to 2 m/sec. As these two types of fibers enter the spinal cord through dorsal roots, they are segregated such that Aδ fibers primarily excite neurons in lamina I of the dorsal horn, whereas C fibers synapse with neurons in the substantia gelatinosa. The latter cells then project deeper into the gray matter and activate neurons mainly in lamina V but also in lamina VI and VII. The neurons that receive Aδ fiber input (fast pain) form the *neospinothalamic tract*, whereas those that receive C fiber input form the *paleospinothalamic tract.*

The neospinothalamic tract is used in pain localization. Axons from neurons in lamina I that form the neospinothalamic tract cross the midline close to their origin and ascend the white matter of the spinal cord as part of the anterolateral system. Some of these fibers terminate in the brain stem reticular formation, but most project all the way to the ventral posterolateral nucleus (VPL) of the thalamus (ventrobasal thalamus). From here, thalamic neurons project to the primary somatosensory (SI) cortex. This system is primarily used in the localization of pain stimuli.

Activity in the paleospinothalamic system may impart the unpleasant perception of pain. The *paleospinothalamic pathway* is the phylogenetically older of the two pain pathways. The axons of cells in lamina V, like those from lamina I, cross the midline near their level of origin and ascend in the anterolateral system. The axons of lamina V cells terminate almost exclusively in the brain stem rather than in the thalamus. In the brain stem, these fibers reach the reticular formation, superior colliculus, and periaqueductal gray. A system of ascending fibers, mainly from the reticular formation, proceed rostrally to the intralaminar nuclei and posterior nuclei of the thalamus, as well as to portions of the hypothalamus. Pain signals transmitted over this pathway are typically localized only to a major part of the body. For example, if the stimulus originates in the hand, it may be localized to the upper extremity.

- The role of SI cortex in pain perception is not entirely clear. Complete removal of SI cortex does not eliminate the perception of pain. Such lesions do, however, interfere with the ability to interpret the quality of pain and determine its precise location.
- The fact that the brain stem reticular areas and the intralaminar thalamic nuclei that receive input from the paleospinothalamic pathway are part of the brain stem activating or alerting system may explain why individuals with chronic pain syndromes have difficulty sleeping.

BRAIN AND SPINAL CORD: INTERNAL PAIN SUPPRESSION SYSTEM (p. 613)

There is tremendous variability in the degree to which individuals react to painful stimuli; this is in large part due to the existence of a mechanism for pain suppression (analgesia) that resides within the central nervous system. This pain suppression system consists of three major components:

- The *periaqueductal gray* of the mesencephalon and rostral pons receives input from the ascending pain pathways in addition to descending projections from the hypothalamus and other forebrain regions.
- The *nucleus raphe magnus* (serotonin) and *nucleus paragigantocellularis* (noradrenaline) in the medulla receive input from the periaqueductal gray and project to neurons in the spinal cord dorsal horn.
- In the dorsal horn, *enkephalin interneurons* receive input from descending serotonergic raphe magnus axons, and the latter form direct synaptic contact with incoming pain fibers. This is called *primary afferent depolarization* (or presynaptic inhibition), and the effect is to interrupt the transmission of the pain signals over these incoming primary sensory fibers. This effect is thought to be mediated by calcium channel blockade in the membrane of the sensory fiber terminal. Other serotonergic fi-

bers exert more conventional postsynaptic inhibitory effects at their synapses with the dorsal horn neurons that process pain signals. The mechanism used by descending noradrenergic axons from the medullary reticular formation is not as well understood, but activity in this system also leads to the interruption of pain transmission at the level of the dorsal horn.

The Role of the Periaqueductal Gray in Pain Perception

Neurons in the periaqueductal gray and nucleus raphe magnus (but not the noradrenergic medullary reticular neurons) have *opiate receptors* on their surface membranes. When stimulated by exogenously administered opioid compounds (analgesics) or by endogenous opioid neurotransmitter agents (endorphins and enkephalins) found in the brain, the pain suppression circuitry is activated; this leads to reduced pain perception.

Pain Sensation: Inhibited by Certain Types of Tactile Stimulation

Activation of the large, rapidly conducting tactile sensory fibers of the dorsal roots appears to suppress the transmission of pain signals in the dorsal horn, probably through lateral inhibitory circuits. Although poorly understood, such circuitry probably explains the relief of pain achieved through the simple maneuver of rubbing the skin in the vicinity of a painful stimulus.

Electrical Stimulation: Relief of Pain

Stimulation of electrodes implanted over the spinal cord dorsal columns or stereotaxically positioned in the thalamus or periaqueductal gray has been utilized to reduce chronic pain. The level of stimulation can be regulated upward or downward by the patient to more effectively manage the pain suppression.

Referred Pain: Distant from the Stimulus

Most frequently, referred pain involves pain signals originating in an internal (visceral) organ or tissue. The mechanism is not well understood but is thought to be caused by visceral pain fibers that may synapse with neurons in the spinal cord, which also receive pain input from cutaneous areas seemingly unrelated to the visceral stimulation site. A common example is pain from the heart wall being *referred* to the surface of the left side of the jaw and neck or the left arm. Rather than associating the pain with the heart, the patient perceives the pain sensation as coming from the face or arm. This implies that visceral afferent signals from the heart converge on the same spinal cord neurons that receive cutaneous input from the periphery (or the convergence may occur in the thalamus).

In other instances, leakage of gastric secretions from a perforated or an ulcerated gastrointestinal tract may directly stimulate pain endings in the peritoneum and lead to severe painful sensations in the body wall. The pain may localize to the dermatomal surface related to the embryonic location of the visceral structure. In addition, spasms in the muscular wall of the gut or distention of a muscular wall of an organ such as the urinary bladder may lead to painful sensations.

Pain from an internal organ such as an inflamed appendix may be experienced in two locations. If the involved appendix actually touches the parietal peritoneum, pain may be felt in the wall of the right lower abdominal quadrant. In addition or alternatively, pain can be referred to the region around the umbilicus owing to the termination of visceral pain fibers in the T-10 or T-11 segment of the spinal cord, which receives cutaneous input from those dermatomes.

Clinical Abnormalities of Pain and Other Sensations
(p. 616)

- *Hyperalgesia* involves a heightened sensitivity to painful stimuli. Local tissue damage or local release of certain chemicals can lower the threshold for activation of pain receptors and the subsequent generation of pain signals.
- Interruption of the blood supply or damage to the ventrobasal thalamus (somatosensory region) may cause the *thalamic pain syndrome*. This is initially characterized by a loss of all sensation over the contralateral body surface. After a few weeks to months, sensations may return, but they are

poorly localized and nearly always painful. Eventually, a state is reached in which even minor skin stimulation can lead to excruciatingly painful sensations; this is known as *hyperpathia*.

- Viral infection of a dorsal root ganglion or cranial nerve sensory ganglion may lead to segmental pain and severe skin rash in the area subserved by the affected ganglion; this is known as *herpes zoster (shingles)*.

- Severe lancinating pain may occur in the cutaneous distribution of one of the three main branches of the trigeminal nerve (or glossopharyngeal nerve); this is called *tic douloureux or trigeminal neuralgia* (or glossopharyngeal neuralgia). In some instances, this is caused by the pressure of a blood vessel compressing the surface of the trigeminal nerve within the cranial cavity; often, it can be surgically corrected.

- The *Brown-Séquard syndrome* is caused by extensive damage to either the right or left half of the spinal cord, as with hemisection. A characteristic set of somatosensory deficits ensues. Transection of the anterolateral system results in a loss of pain and temperature sensation *contralaterally* that usually begins just a few segments caudal to the level of the lesion. On the side *ipsilateral* to the lesion, there is a loss of dorsal column sensations beginning at about the level of the lesion and extending through all levels caudal to the lesion. If the lesion involves several segments of the cord, there may be a loss of *all* sensation ipsilaterally in those dermatomes that correspond to the location of the cord lesion. The patients will, of course, also exhibit motor deficits.

Headache results when pain from deeper structures is referred to the surface of the head. The source of the pain stimuli may be intracranial or extracranial; in this chapter, we focus on intracranial sources. The brain itself is insensitive to pain, and when somatosensory structures are damaged, the patient experiences the sensation of tingling or *pins and needles*. The exceptions, as described previously, are tic douloureux and thalamic pain syndrome.

Pressure on the venous sinuses and stretching of the dura or blood vessels and cranial nerves passing through the dura lead to the sensation of headache. When structures above the tentorium cerebelli are affected, pain is referred to the frontal portion of the head. Involvement of structures below the tentorium results in occipital headaches.

Meningeal inflammation typically produces pain involving the entire head. When a small volume of cerebrospinal fluid is removed (as little as 20 milliliters) and the patient is not recumbent, gravity will cause the brain to "sink"; this leads to stretching of meninges, vessels, and cranial nerves and results in a diffuse headache. The headache that follows excessive drinking is thought to be due to the direct toxic irritation of alcohol on the meninges. Constipation may also cause headache as a result of the direct toxic effects of circulating metabolic substances or circulatory changes related to the loss of fluid into the gut.

Although the mechanism is still not completely understood, *migraine headaches* are thought to be the result of vascular phenomena. Prolonged unpleasant emotions or anxiety produces spasm in brain arteries and leads to local ischemia in the brain. This may result in prodromal visual or olfactory symptoms. As a result of the prolonged spasm and ischemia, the muscular wall of the vessel loses its capability to maintain normal tone. The pulsation of circulating blood alternately stretches (dilates) and relaxes the vessel wall, and this stimulates the pain receptors in the vascular wall or in the meninges surrounding the entry points of vessels into the brain or cranium. The result is an intense headache. Other theories are being investigated, and a number of new and effective treatments should be available.

THERMAL SENSATIONS

Changes in Temperature: Three Types of Receptors
(p. 619)

- *Pain receptors* are stimulated only by extreme degrees of cold or warmth. In this case, the perceived sensation is one of pain, not temperature.
- Specific *warmth receptors* have not yet been identified, although their existence is suggested by psychophysical experiments; at present, they are simply regarded as free nerve endings. Warmth signals are transmitted over type C sensory fibers.
- The *cold receptor* has been identified as a small nerve ending, with tips that protrude into the basal aspect of basal epidermal cells. Signals from these receptors are transmitted over Aδ type sen-

sory fibers. There are 3 to 10 times as many cold receptors as warmth receptors, and their density varies from 15 to 25 per square centimeter on the lips to 3 to 5 receptors per square centimeters on the fingers.

Cold and Warmth Receptors: 7°C to 50°C (p. 619)

Temperatures below 7°C and above 50°C activate pain receptors; both of these extremes are perceived similarly as very painful and not cold or warm. The peak temperature for activation of cold receptors is about 24°C, whereas the warmth receptors are maximally active at about 45°C. Both cold and warm receptors can be stimulated with temperatures in the range of 31°C to 43°C.

When the cold receptor is subjected to an abrupt temperature decrease, it is strongly stimulated initially, but then the generation of action potentials falls off dramatically after the first few seconds. The decrease in firing, however, progresses more slowly after the next 30 minutes or so. The cold and warm receptors respond to *steady state temperature* as well as to *changes in temperature*. This explains why a cold outdoor temperature "feels" so much colder at first as one emerges from a warm environment.

The stimulatory mechanism in thermal receptors is believed to be related to the change in metabolic rate in the nerve fiber induced by the temperature change. For every 10°C temperature change, there is a twofold change in the rate of intracellular chemical reactions.

The density of thermal receptors on the skin surface is relatively small; therefore, temperature changes that affect only a small surface area are not as effectively detected as are temperature changes that affect a large skin surface area. If the entire body is stimulated, a temperature change as small as 0.01°C can be detected. Thermal signals are transmitted through the central nervous system in parallel with pain signals.

UNIT

X

The Nervous System

B. The Special Senses

49

The Eye

I. The Optics of Vision

PHYSICAL PRINCIPLES OF OPTICS (p. 623)

- Light travels through transparent objects at a *slower velocity* than it does through air. The *refractive index* of a transparent substance is the ratio of its velocity in air to its velocity in the transparent object.
- The direction that light travels is always *perpendicular* to the plane of the wave front. When a light wave passes through an angulated surface, it is bent (refracted) at some angle if the refractive indices of the two media are different. The angle depends on the refractive index of the barrier material and the angle between the two surfaces.

Refractive Principles and Their Application to Lenses

- A *convex* lens focuses light rays. Light rays that pass through the lens perimeter are bent (refracted) toward those (to make themselves perpendicular to the wave front) that pass through the central region. The light rays are said to *converge*.
- A *concave* lens diverges light rays. At the lens perimeter, light waves are refracted, so that they will travel perpendicular to the wave front or interface, and are bent away from those passing through the central region. This is called *divergence*.
- The *focal length* of a lens is the distance beyond a convex lens at which parallel light rays converge to a single point.
- Each point source of light in front of a convex lens is focused on the opposite side of the lens in line with the lens center; that is, the object appears to be upside down and reversed from left to right.

- The more a lens bends light rays, the greater is its refractive power. The unit of measure for refractive power is the *diopter*. A spherical (or convex) lens that converges parallel light rays to a point 1 meter beyond the lens has a refractive power of +1 diopter. If the light rays are bent twice as much, then the diopter is +2.

THE OPTICS OF THE EYE (p. 626)

The eye is optically equivalent to a photographic camera. It has a lens and variable aperture (pupil), and the retina corresponds to the film. If all the refractive surfaces of the eye are algebraically added, the eye may be represented as a "reduced eye." In this case, a single refractive surface exists with its central point 17 mm in front of the retina, and the eye has a total refractive power of 59 diopters when totally accommodated for distant vision. Most of the refractive power of the eye is actually provided by the cornea. The total refractive power of the lens alone is only about 20 diopters. The lens system of the eye focuses an inverted, upside-down image on the retina. We perceive the image as right-side up because the brain has "learned" that this is the correct, or normal, orientation.

Accommodation **depends on a change in the shape of the lens and allows the eye to focus on a near object.** When shifting the gaze from a far to a near object, the process of *accommodation* involves (1) making the lens more convex, (2) narrowing the pupillary diameter, and (3) producing adduction (vergence) of both eyes. When the lens is in a "relaxed" state with no tension exerted on the edges of its capsule, it assumes a nearly spherical shape because of its own intrinsic elastic properties. When the inelastic zonule fibers attached to the lens perimeter become taut and are pulled radially by their attachment to the *inactive ciliary muscle (and ciliary body)*, the lens is relatively flat or less convex. When the ciliary muscle is *activated* by postganglionic parasympathetic fibers in the oculomotor nerve, the circular fibers of the ciliary muscle contract, producing a sphincter-like action that relaxes the tension on the zonule fibers and allows the lens to become more convex because of its own inherent elasticity. This increases its refractive capability and allows the eye to focus on near objects. At the same time, the sphincter pupillae muscle is activated, the pupil constricts, and the two eyes are medially deviated.

Presbyopia is the loss of accommodation by the lens. As an individual ages, the lens begins to lose its intrinsic elastic properties and becomes less responsive and unable to focus on near objects. This condition is known as *presbyopia*, and it is corrected with reading glasses designed to magnify near objects or with bifocals in which one lens (the upper portion) is designed to enhance distance vision and the second lens (the lower portion) has greater refractive capability to improve near vision.

The diameter of the pupil (iris) is also a factor in accommodation. The greater the diameter, the more light enters the eye. The amount of light that enters is proportional to the area of the pupil opening or to the square of the pupillary diameter. In addition, greater depth of focus is achieved by decreasing the pupillary opening. Squinting (narrowing the pupil opening) improves the sharpness of the image by increasing the focal plane.

Refractive Errors (p. 628)

Correction with Various Types Of Eye Glasses

- *Emmetropia* refers to the normal eye. When the ciliary muscle is completely relaxed, all distant objects are in sharp focus on the retina.
- *Hyperopia*, also known as *farsightedness*, is due to an eyeball that is too short from top to bottom. In this situation, light rays are not sufficiently bent, and they come to focus behind the retina; this condition is corrected with a convex lens.
- *Myopia*, also known as *nearsightedness*, is due to an eye that is elongated from front to back. In this situation, there is too much refraction by the lens when the ciliary muscle is relaxed, and the distant object is focused in front of the retina; this condition is corrected with a concave lens, which decreases refraction by producing divergence of the entering light rays.
- *Astigmatism* is caused by substantial differences in curvature over different planes through the eye. For example, the curvature in a vertical plane through the eye may be much less than the curvature through a horizontal plane. As a result, light rays entering from the eye in different directions

are focused at different points. This condition requires a cylindrical lens for correction.

- *Cataracts* are caused by an opacity that forms in a portion of the lens. The affected portion of the lens no longer allows the passage of light rays, and vision is diminished. The treatment of choice at present is to remove the lens and substitute an artificial lens implant.

- *Keratoconus* is a condition that results from the formation of an oddly shaped cornea with a prominent bulge on one side. If the bulge is extreme, no single glass lens can correct for the severity of the refractive problem. The best solution is a contact lens, which adheres to the surface of the cornea and is held in place by a film of tear fluid. This lens is ground to compensate for the bulge in the cornea so that the anterior surface of the contact lens becomes a far more uniform and effective refractive surface.

Visual Acuity (p. 631)

Sharpest within the Foveal Region of the Retina

The *fovea* is made up entirely of cone photoreceptors, each of which has a diameter of about 1.5 micrometers. Normal visual acuity in humans allows the discrimination of two points of light as being distinct when they are separated by at least 25 seconds of arc on the retina.

The fovea is normally about 0.5 millimeters in diameter. Maximal acuity occurs in less than 2 degrees of the visual field. The reduction in acuity outside the foveal region is due in part to the presence of rod photoreceptors intermixed with cones and to the linkage of some rod and cone receptors to the same ganglion cells.

The test chart for visual acuity is usually placed 20 feet from the individual being tested. If the letters of a particular size can be recognized at a distance of 20 feet, the individual is said to have 20/20 vision. If the individual can see only at 20 feet letters that should be visible at 200 feet, that individual has 20/200 vision.

Depth Perception (p. 631)

Distinguishing the Distance of an Object from the Eye

Knowing the size of an object allows the brain to calculate its distance from the eye. If an individual looks at a distant object without moving the eyes, no *moving parallax* is apparent. However, if the head is moved from side to side, close objects move rapidly across the retina, whereas distant objects move very little or not at all.

Binocular vision also aids in determining the distance of an object. Because the eyes are typically about 2 inches apart, an object placed 1 inch in front of the bridge of the nose is seen by a small part of the peripheral retina in the left and right eyes. In contrast, the image of an object at 20 feet falls on closely corresponding points in the middle of each retina. This type of binocular parallax (stereopsis) provides the ability to accurately judge distances from the eyes.

Ophthalmoscope: To View the Retina (p. 632)

This instrument allows the retina of the observed eye to be illuminated by means of an angled mirror or prism and a small bulb. The observer positions the instrument to view the subject's retina through the subject's pupil. If either the subject's eyes or the examiner's eyes are not emmetropic, refraction can be adjusted using a series of movable lenses within the instrument.

Two Types of Intraocular Fluid of the Eye (p. 632)

- *Vitreous humor* lies between the lens and retina and is actually more of a gelatinous body than a liquid. Substances can diffuse through the vitreous, but there is little movement or flow in this liquid.
- *Aqueous humor* is a watery fluid secreted by the epithelial lining of ciliary processes on the ciliary body at a rate of 2 to 3 microliters per minute. This fluid migrates between the ligaments supporting the lens and through the pupil into the anterior chamber of the eye (between the lens and cornea). From here, the fluid flows into the angle between the cornea and iris and then through a trabecular meshwork to enter the canal of Schlemm, which empties directly into extraocular veins.

Intraocular pressure is normally about 15 mm Hg, with a range of 12 to 20 mm Hg. A tonometer is usually used to measure intraocular pressure. This device consists of a small footplate that is placed on the anesthetized cornea. A small force is applied to the footplate, which displaces the cornea inward, and the distance of inward displacement is calibrated in terms of intraocular pressure. This pressure is determined by the resistance to movement of fluid through the trabecular meshwork leading into the canal of Schlemm. The rate of fluid entering the canal normally is equal to the rate of fluid formation by the ciliary body epithelium. When debris from intraocular hemorrhage or infection accumulates in the trabecular meshwork, reticuloendothelial cells in the vicinity can phagocytize this material and maintain the flow of aqueous humor.

Glaucoma is a condition in which intraocular pressure can reach dangerously high levels in the range of 60 to 70 mm of Hg. As the pressure rises above 20 to 30 mm of Hg, axons of retinal ganglion cells that form the optic nerve are compressed to the extent that axonal flow is interrupted. This may cause permanent injury to the parent neuron. It is also possible that compression of the central retinal artery may lead to neuronal death in the retina. Glaucoma can be treated with eyedrops that reduce the secretion of aqueous humor or increase its absorption. If drug therapy fails, surgical procedures are implemented to open the trabecular spaces or to drain the trabecular meshwork directly into subconjunctival spaces outside the eyeball.

50

The Eye

II. Receptor and Neural Function of the Retina

ANATOMY AND FUNCTION OF THE STRUCTURAL ELEMENTS OF THE RETINA (p. 637)

The retina is composed of 10 cellular layers or boundaries. These are listed as follows in sequence, beginning with the most external layer (most distant from the center of the eyeball):

1. Pigment layer
2. Layer of rods and cones
3. Outer limiting membrane
4. Outer nuclear layer
5. Outer plexiform layer
6. Inner nuclear layer
7. Inner plexiform layer
8. Ganglionic layer
9. Layer of optic nerve fibers
10. Inner limiting membrane.

When light passes through the lens system of the eye, it first encounters the inner limiting membrane, optic nerve fibers, and ganglion cell layer and then continues through the remaining layers to finally reach the receptors, rods, and cones. The *fovea* is a specialized region of approximately 1 square millimeter in the central region of the retina. In the very center of the fovea is an area 0.3 millimeter in diameter called the *central fovea*. This is the region of maximal visual acuity; it is here that the photoreceptor layer contains only cones. In addition, the subjacent retinal layers—all the way to the optic nerve fibers and blood vessels—are displaced laterally to enable more direct access to the receptor elements.

Each photoreceptor consists of (1) an outer segment, (2) an inner segment, (3) a nuclear region, and (4) a synaptic body or terminal. The receptors are

437

referred to as *rods* or *cones,* depending primarily on the shape of the outer segment. The cones are generally wider than rods, although in the fovea, both types are relatively slender and not easily distinguished on the basis of their morphology.

The light-sensitive photopigment *rhodopsin* is found in the rod outer segment, whereas a similar material called color-sensitive pigment *photopsin,* or cone pigment, is found in the cone outer segment. These photopigments are proteins that are incorporated into a stacked array of membranous discs in the receptor outer segment, which actually represent infolding of the photoreceptor outer cell membrane. This is not readily apparent in the distal portion of the rod outer segment, however, in which the membranous discs become secondarily detached from and entirely contained within the limiting membrane of the outer segment.

The inner segments of the rods and cones are essentially indistinguishable and contain the cytoplasmic components and organelles common to other neuronal cell bodies. Individual photoreceptor nuclei are continuous with their own inner segment, but the outer limiting membrane of the retina forms an incomplete separation or boundary between the layer of inner segments and the layer of photoreceptor nuclei (outer nuclear layer).

The *synaptic body* contains elements such as mitochondria and synaptic vesicles that are typically found in axon terminals in the brain. The black pigment *melanin* in the pigment layer reduces light reflection throughout the globe of the eye and thus performs a function similar to the black color inside the bellows of a camera. The importance of this pigment is best illustrated by its absence in albino individuals. Because of the large amount of reflection inside the globe of the eye, albinos rarely exhibit better than 20/100 visual acuity. The pigment layer also stores large quantities of vitamin A, which is used in the synthesis of visual pigments.

The *central retinal artery* provides the blood supply only to the innermost layers of the retina (ganglion cell axons to inner nuclear layer). The outermost layers of the retina receive their blood supply by diffusion from the highly vascularized choroid, which is situated between the sclera and retina.

When an individual has a traumatic *retinal detachment,* the line of separation occurs between the neural retina and pigment epithelium. Because of the independent blood supply to the inner layers of the retina via the central retinal artery, the retina can survive for several days and may resist functional degeneration if surgically returned to its normal apposition to the pigment epithelium.

PHOTOCHEMISTRY OF VISION (p. 640)

Rhodopsin-Retinal Cycle and Excitation of Rod Photoreceptors

Rhodopsin is decomposed by light energy. The rod photopigment rhodopsin is concentrated in the portion of the outer segment that protrudes into the pigment layer. This substance is a combination of the protein *scotopsin* and the carotenoid pigment *retinal* or, more specifically, 11-*cis* retinal. When light energy is absorbed by rhodopsin, the retinal portion is transformed into the all-*trans* configuration, and the retinal and scotopsin components begin to separate. In a series of reactions that occur extremely rapidly, the retinal component is converted to lumirhodopsin, metarhodopsin I, metarhodopsin II, and, finally, scotopsin, and all-*trans* retinal is cleaved. During this process, metarhodopsin II is believed to elicit the electrical changes in the rod membrane that lead to subsequent impulse transmission through the retina.

Rhodopsin reformation occurs. In the first stage of the reformation of rhodopsin, all-*trans* retinal is converted to the 11-*cis* configuration, which then immediately combines with scotopsin to form rhodopsin. There also is a second pathway leading to the formation of rhodopsin that involves the conversion of *all-trans* retinal into all-*trans* retinol, which is a form of vitamin A. The retinol is enzymatically converted to 11-*cis* retinol and then to 11-*cis* retinal, which is able to combine with scotopsin to form rhodopsin. If there happens to be excess retinal in the retina, it is converted to vitamin A, thus reducing the total amount of rhodopsin in the retina. Night blindness occurs in vitamin A–deficient individuals because rods are the photoreceptors maximally used under relatively dim lighting conditions, and the formation of rhodopsin is dramatically decreased because of the absence of vitamin A. This condition can actually be reversed in about 1 hour or less with an intravenous injection of vitamin A.

Rhodopsin Activation: Hyperpolarization and Rod Membrane Potential (p. 641)

Rod photoreceptors behave quite differently from other neural receptor elements. In the dark (in the absence of photic stimulation), rod outer segment membranes are "leaky" to sodium; that is, sodium ions *enter* the outer segment and alter its membrane potential from the typical level of -70 to -80 millivolts observed in sensory receptors to a more positive value of -40 millivolts. This is known as a sodium current or the "dark current," and it actually causes a small amount of transmitter release in the dark. When light strikes the rod outer segment, rhodopsin molecules undergo the series of reactions as described; this *decreases* the conductance of sodium into the outer segment and diminishes the dark current. Some sodium ions continue to be pumped out through the cell membrane, and this loss of positive ions causes the interior to become more negative; the membrane potential becomes more negative and is said to *hyperpolarize*. The flow of transmitter is halted.

When light strikes a photoreceptor, the transient hyperpolarization in rods reaches a peak in about 0.3 second and lasts for more than 1 second. In addition, the magnitude of the receptor potential is proportional to the *logarithm* of the light intensity. This has great functional significance because it allows the eye to discriminate light intensity through a range many thousand times as great as would otherwise be possible. This function is the result of an extremely sensitive chemical cascade that amplifies the stimulatory effects by about 1 million-fold, as follows. Activated rhodopsin (metarhodopsin II) acts as an enzyme to activate many molecules of *transducin*, a protein also found in the outer segment disc membrane. The activated transducin in turn activates *phosphodiesterase*, an enzyme that immediately hydrolyzes many molecules of *cyclic GMP*. The loss of cyclic GMP results in the closure of many sodium channels, which is then accompanied by an increasingly more negative (hyperpolarized) membrane potential. Within about 1 second, metarhodopsin II is inactivated, and the entire cascade reverses: the membrane potential becomes more depolarized as sodium channels are reopened, and sodium again enters the outer segment as the dark current is reestablished. Cone photoreceptors behave similarly, but the amplification factor is 30 to 300 times less than that in the rods.

Photochemistry, Color Vision, Cone Photoreceptors
(p. 642)

As in the rods described, the photochemical transduction process in cones involves an opsin and a retinal. In cones, the opsin is called *photopsin*, which has a different chemical composition than rhodopsin, whereas the retinal component is exactly the same as that in rods. There are three types of cones, each of which is characterized by a different photopsin that is maximally sensitive to a particular wavelength of light in the blue, green, or red portion of the light spectrum.

Light or Dark Adaptation: Retinal Sensitivity (p. 642)

If exposed to bright light for long periods of time, large proportions of the photochemicals in both the rods and cones are depleted, and much of the retinal is converted to vitamin A. As a result, the overall sensitivity to light is reduced; this effect is called *light adaptation*. Conversely, when an individual remains in the dark for a long period of time, the opsins and retinal are converted back into light-sensitive pigments. In addition, vitamin A is converted into retinal providing even more photosensitive pigment; this is the process of *dark adaptation*. The latter process occurs about four times as rapidly in cones as in rods, but cones exhibit less sensitivity change in the dark. Cones cease adapting after only a few minutes. The more slowly adapting rods continue the process for minutes to hours, and their sensitivity increases over a very broad range.

Adaptation can also occur through changes in pupillary size. This change can be on the order of 30-fold within a fraction of a second. Neural adaptation can also take place in the circuits that exist within the retina and brain. If light intensity increases, transmission from bipolar cell to horizontal cell to amacrine and ganglion cell may also increase. Although the latter form of adaptation is less substantial than pupillary changes, neural adaptation, like pupillary adaptation, occurs very rapidly.

The value of light and dark adaptive processes is that they provide the eye with the ability to change its sensitivity by as much as 500,000- to 1 million-fold. This can be appreciated when entering a dark room from a brightly lit environment. The sensitivity of the retina is low because it is light adapted and little can be seen in the dark room. As dark adaptation occurs, vision in the dark improves. The intensity of sunlight is estimated to be 10 billion times greater than light intensity on a starlit night, yet the eye can function to some degree under both conditions because of its enormous adaptive range.

COLOR VISION (p. 643)

Color Detecton: A Three-Color Mechanism

Spectral sensitivity of the three types of cones is based on the light absorption curves for the three different cone pigments. All visible color (other than blue, green, or red) is the result of the combined stimulation of two or more types of cones. The nervous system then interprets the ratio of activity of the three types of cones as a color. Approximately equal stimulation of blue, green, and red cones is interpreted as white light.

Changing the color of the light that illuminates a scene does not substantially alter the hues of color in the scene; this is called *color constancy*, and it is believed that the mechanism for this phenomenon resides in the primary visual cortex.

When a particular type of cone is missing from the retina, some colors cannot be distinguished from others. An individual who lacks red cones is called a *protanope*. The overall spectrum is shortened at the long wavelength end by the absence of red cones. Red-green color blindness is a genetic defect in males but is transmitted by the female. Genes on the female X chromosome code for the respective cones. This defect rarely occurs in females because they have two X chromosomes, so they almost always have one normal copy of the gene.

NEURAL FUNCTIONS OF THE RETINA (p. 645)

Neural Circuits: Six Types of Cells

- *Photoreceptors* consist of rod and cone outer segments and inner segments in the photoreceptor layer, a cell body in the outer nuclear layer, and a synaptic terminal in the outer plexiform layer.

- *Horizontal cells*, *bipolar cells*, and *amacrine cells* receive synaptic input in the outer plexiform layer, have the somata in the inner nuclear layer, and form presynaptic contacts in the inner plexiform layer.
- *Ganglion cells* receive synaptic input in the inner plexiform layer, have a soma in the ganglion cell layer, and give rise to axons that course within the optic nerve fiber layer.
- *Interplexiform cells* transmit signals in the opposite direction—from the inner plexiform to the outer plexiform layer.

In the fovea, the pathway from a cone to a ganglion cell is relatively direct and can involve a receptor, bipolar cell, and ganglion cell. Horizontal cells can be involved in the outer plexiform layer, whereas amacrine cells are active in the inner plexiform layer. More peripherally in the retina, in which rod photoreceptors are most abundant, the input from several photoreceptors can converge on a single bipolar neuron, which may have an output only to an amacrine cell that then projects to a ganglion cell. This represents the pure rod vision pathway. Horizontal and amacrine cells can provide the lateral connectivity.

Neurotransmitters present in the retina include *glutamate* used by rods and cones and *GABA*, *glycine*, *dopamine*, *acetylcholine*, and *indolamines* utilized by amacrine cells. The transmitter that is used by horizontal, bipolar, or interplexiform cells is unclear.

Beginning with the photoreceptors, transmission of signals up to the ganglion cell layer occurs only by *electrotonic conduction* (graded potentials) and not by action potentials. The ganglion cell is the only retinal neuron that is capable of generating an action potential; this ensures that signals in the retina accurately reflect illumination intensity, and it gives retinal neurons more flexibility in their response characteristics.

Lateral Inhibition, Contrast, and Horizontal Cells (p. 647)

Horizontal cell processes connect laterally with photoreceptor synaptic terminals and bipolar cell dendrites. The photoreceptors that lie within the center of a light beam are maximally stimulated, whereas those at the periphery are inactivated by horizontal cells that are themselves activated by the light beam. It is said that the *surround* is inhibited, whereas the central region is *excited*, although these terms may not be precisely correct. This is the basis for the enhancement of visual contrast. Amacrine cells may also contribute to contrast enhancement through their lateral projections in the inner plexiform layer. Interestingly, although horizontal cells may have axons, amacrine cells do not, and therefore their physiological properties are very complex.

Some Bipolar Cells and Light (p. 647)

Some bipolar cells depolarize when their related photoreceptor or photoreceptors are stimulated by light, whereas others hyperpolarize. There are two possible explanations for this observation; one is that the two bipolar cells simply respond differently to glutamate release by the photoreceptor. One bipolar cell is excited by glutamate, and the other type of bipolar cell is inhibited. Another explanation is that one type of bipolar may receive direct (excitatory) input from the photoreceptor, and the other type may receive indirect inhibitory input from a horizontal cell. Having some bipolar cells excited and others inhibited may also contribute to the lateral inhibition scheme.

Amacrine Cells (p. 647)

Nearly 30 different types of amacrine cells have been identified by morphological or histochemical means. Some amacrine cells respond vigorously at the onset of a visual stimulus, others respond at the offset, and still others respond at both the onset and offset. Another type responds only to a moving stimulus. Because of the variety of neurotransmitters used by this class of cell, there can be no generalization regarding its effect on its target neuron.

Three General Classes of Ganglion Cells (p. 648)

There are about 1.6 million ganglion cells in the retina, yet it is estimated that 100 million rods are present with 3 million cones. This means that an average of 60 rods and 2 cones converge on each retinal ganglion cell. The population of ganglion cells is divided into W-, X-, and Y-type cells.

- *W-type ganglion cells* constitute nearly 40 per cent of the entire pool, are small, and have a somal diameter of 10 micrometers; they transmit action potentials at the relatively slow velocity of 8 m/sec. They receive most of their input from rod photoreceptors (via bipolar and amacrine cells) and exhibit a relatively broad dendritic field. These cells appear to be especially sensitive to movement in the visual field, and because of their dominant input from rods, they are probably responsible for dark-adapted vision.
- *X-type ganglion cells* are somewhat more numerous than W cells and represent about 55 per cent of all ganglion cells. They have a somal diameter of 10 to 15 micrometers and conduct at velocities of about 14 m/sec. These cells exhibit relatively small dendritic fields and therefore represent discrete locations in the visual field. Each X cell receives input from at least one cone photoreceptor, so this class of cell is probably responsible for color vision.
- *Y-type ganglion cells* are the largest; they exhibit somal diameters of up to 35 micrometers and conduct at velocities of about 50 m/sec. As might be predicted, their dendritic spread is broad. They are the fewest in number, however, and constitute only about 5 per cent of the total pool. These cells respond rapidly to changes anywhere in the visual field (either intensity or movement) but are not capable of specifying with accuracy where the change occurred.

Ganglion Cells and Continuous Activity (p. 648)

It is the ganglion cell axons that form the optic nerve fibers. Even when unstimulated, they transmit action potentials at rates varying between 5 and 40 per second. Visual signals are thus superimposed on this background or spontaneous level of firing.

Many ganglion cells are particularly sensitive to changes in light intensity. Some cells will respond with more firing when light intensity increases. Others may increase their firing when light intensity decreases. These effects are due to the presence of depolarizing and hyperpolarizing bipolar cells. The responsiveness to light intensity fluctuation is equally well developed in the peripheral and foveal regions of the retina.

Contrast Levels and Lateral Inhibition (p. 649)

Ganglion cells are said to be responsive to contrast borders rather than to absolute levels of illumination. When photoreceptors are activated by flat, diffuse light, depolarizing bipolar cells provide excitatory output, but at the same time, hyperpolarizing bipolar and horizontal cells can produce inhibitory output. When a light stimulus has contrast (sharp borders or edges between light and dark), at the light-dark border, one photoreceptor in the light is activated and a signal is transmitted through its bipolar cell to a ganglion cell, which then increases its firing. A neighboring photoreceptor in the dark is inactivated, which means that no signal is transmitted through its bipolar-ganglion cell line. At the same time, a horizontal cell turned on by the activated photoreceptor will inhibit its laterally adjacent bipolar cells, while horizontal cells that are not activated by the lateral photoreceptor in the relatively dark zone will not inhibit the photoreceptor in the light region, thereby further contributing to excitation through the active photoreceptor-bipolar-ganglion cell line. The dark is made "darker," and the light is made "lighter" (i.e., contrast is enhanced).

Ganglion Cell Activity: Color Signals (p. 649)

Some ganglion cells are stimulated by all three types of cone photoreceptors, and such a ganglion cell is thought to signal "white" light. Most ganglion cells, however, are stimulated by light of one wavelength and inhibited by another. For example, red light may excite and green may inhibit a particular ganglion cell; this is called a *color-opponent mechanism* and is thought to be the process employed to differentiate color. Because the substrate for such a process is present in the retina, recognition and perception of color may actually begin in the retina at the level of the primary sensory receptive element.

51

The Eye

III. Central Neurophysiology of Vision

THE MAIN VISUAL PATHWAY (FROM THE RETINA TO THE DORSAL LATERAL GENICULATE NUCLEUS; TO THE PRIMARY VISUAL CORTEX) (p. 651)

Axons of retinal ganglion cells form the *optic nerve*. Axons that originate from the nasal retina cross at the *optic chiasm*, whereas those from the temporal retina pass through the lateral aspect of the chiasm without crossing. Retinal axons continue posterior to the chiasm as the *optic tract*, and the majority terminate in the *dorsal lateral geniculate nucleus*. From here, the axons of geniculate neurons proceed further posteriorly as the *geniculocalcarine (optic) radiations* and terminate in the *primary visual (striate) cortex*. In addition, retinal axons extend to other regions of the brain including the (1) suprachiasmatic nucleus (control of circadian rhythms), (2) pretectal nuclei (for pupillary light reflexes), (3) superior colliculus (control of rapid eye movements), and (4) ventral lateral geniculate nucleus.

Functions of the Dorsal Lateral Geniculate Nucleus (p. 651)

The dorsal lateral geniculate nucleus (DLGN) is a laminated structure that consists of six concentrically arranged layers. The most internal layer is layer 1, whereas the most superficial is layer 6. Retinal axons that terminate in the DLGN arise from the contralateral nasal retina and the ipsilateral temporal retina and thus carry point-to-point information from the contralateral visual field. The contralateral nasal fibers terminate in layers 1, 4, and 6, whereas the ipsilateral temporal fibers terminate in layers 2, 3, and 5. Information from the two eyes remains segregated in the DLGN, as does input from X and Y retinal ganglion cells. The Y cell input terminates in layers 1 and 2, which are termed the *magnocellular layers* because they contain relatively large neurons. This is a rapidly conducting pathway that is color blind and does not carry effective localizing information because there are so few Y cells in the retina. Layers 3 to 6 are termed the *parvicellular layers* because they contain relatively small neurons that receive input from X cells that transmit color information; because of the large population of X retinal ganglion cells, they convey accurate point-to-point localizing signals. Information from the retina is processed along at least two parallel pathways.

The processing of information in the DLGN is also influenced by input from the cerebral cortex and the mesencephalon, namely, the locus ceruleus. Both of these pathways mediate either direct or indirect inhibitory influence on DLGN neurons.

ORGANIZATION AND FUNCTIONS OF THE VISUAL CORTEX (p. 652)

The *primary visual cortex*, or area 17 of Brodmann, is also referred to as V-1. It is located on the medial surface of the hemisphere lining both walls of the calcarine sulcus near the occipital pole. It receives visual input from each eye and contains the representation of the entire contralateral visual field, with the lower visual field contained in the upper bank of the calcarine sulcus and the upper visual field located in the lower bank. The macular portion of the retina is represented posteriorly near the occipital pole, whereas more peripheral retinal input reaches more anterior, concentrically arranged territories.

Secondary visual cortex (also called *visual association cortex*) is located on the lateral surface of the hemisphere in rostral portions of the occipital lobe and caudal parts of parietal and temporal cortices. This territory corresponds to area 18 of Brodmann and is known as V-2. As many as 12 visual areas have been described, but little is known about their contribution to visual function.

The primary visual cortex has a layered structure (p. 653). Like all other areas of the neocortex, the primary visual cortex is organized into six horizontally arranged layers. The Y-type incoming geniculate fibers terminate most heavily in a subdivision of layer IV called IVca, whereas the X-type fibers terminate primarily in layers IVa and IVcb.

There is also a vertical, columnar organization within V-1. A vertical array of neurons is approximately 50 micrometers in width and extends through the entire thickness of the cortex from pial surface to the underlying subcortical white matter. As thalamic input terminates in layer IV, signals are spread by local circuits upward and downward in the column.

Interspersed among the vertical columns are *color blobs*. These aggregates of neurons respond specifically to color signals mediated by surrounding cortical columns.

Visual signals from the two eyes remain segregated through projections from the DLGN to V-1. The cells in one vertical column in layer IV are primarily responsive to input from one eye, whereas neurons in the next adjacent column are preferentially responsive to the other eye; these are called *ocular dominance columns*.

Processing in the Primary Visual Cortex: Signals and Two Pathways in Visual Association Cortex (p. 654)

Neuronal linkages that follow a more *dorsal* stream from V-1 into the rostrally adjacent area 18 (V-2) and then on to the parietal cortex carry information concerning precise localization of the visual image in space and the gross form of the image, and whether or not it is moving. This information seems to be carried in the Y cell pathway and permits a determination of the location ("where") of the object at any moment—and whether it is moving or not.

Conversely, a more *ventral pathway* from V-1 into the adjacent V-2 and temporal association cortex seems to carry information that is necessary for the analysis of visual details that are conveyed along the X cell pathway: information that might be used to recognize textures and letters, along with the color of objects. In general, this information might be used to determine "what" the object is and its meaning.

NEURONAL ACTIVITY IN OCCIPITAL CORTEX DURING ANALYSIS OF THE VISUAL IMAGE (p. 654)

The visual cortex detects the orientation of lines and borders. We discuss in Chapter 50 that a major function of the visual system involves the detection of contrast—particularly the edges formed by lines and borders. Neurons in layer IV of V-1 called *simple cells* are maximally responsive to lines or edges that are aligned in a preferred orientation. For a line of any orientation, there are certain neurons that are most active and others that are some degree less than maximally active.

Other cells in V-1, called *complex cells*, are responsive to lines or edges with a preferred orientation, but the line can be displaced laterally or vertically for some defined distance.

A third class of cell called the *hypercomplex cell* is primarily located in visual association areas. These cells detect lines or edges that have a specific length, specific angulated shape, or other relatively complex feature.

Neurons of various types in the visual cortex participate in some circuits that are *serially* organized as well as pathways in which information is transmitted in a *parallel* manner. Both of these categories of functional organization are important to normal vision.

Neuronal Detection of Color (p. 655)

Color is detected by means of color contrast. Often, color is contrasted with a white portion of the scene; this is the basis for the *color constancy* concept discussed in Chapter 50. Color contrast is detected by an opponent process in which some colors excite certain neurons and inhibit others. The simple cells in V-1 contribute to this process, whereas higher-order visual cortical neurons are involved in more complex color contrast discrimination.

Removal of the primary visual cortex causes loss of conscious vision. Individuals may still be able to react "reflexively" to changes in light intensity, movement in the visual scene, and gross patterns of light stimuli. This activity is mainly due to activity in subcortical visual centers such as the superior colliculus.

Testing the Visual Fields: Perimetry (p. 655)

The visual field is the area seen by an eye; it is divided into a nasal (medial) portion and a temporal (lateral) portion. The process of testing the visual field of each eye independently is called *perimetry*. The subject fixates a single point in the center of the visual field while a second small spot is moved in and out of the visual field, and the subject acknowledges its location.

A *blind spot* exists in that portion of the visual field occupied by the *optic disc*. A blind spot in any other portion of the visual field is called a *scotoma*. In *retinitis pigmentosa*, portions of the retina degenerate, and excessive melanin pigment is deposited in these areas. This process usually begins in the peripheral retina and then spreads centrally.

Effects of lesions in the optic pathways on fields of vision. Interruption of the crossing fibers in the optic chiasm causes a visual field loss in the temporal portion of the visual field of each eye; this is called *bitemporal heteronymous hemianopsia*. Section of one optic tract leads to a loss of the nasal visual field in the ipsilateral eye and the temporal field contralaterally; this condition is called *contralateral homonymous hemianopsia*. A lesion involving the optic radiations in one hemisphere produces a similar defect. These two lesions can be differentiated by the presence or absence of the pupillary light reflexes. If the reflexes are preserved, the lesion is in the optic radiations; if they are lost, the lesion must involve the optic tracts that carry retinal signals to the pretectal region.

Eye Movements and Their Control (p. 656)

For a visual scene to be interpreted correctly, the brain must be able to move the eyes into position to properly view it. Eye movement is provided by three pairs of muscles: the medial and lateral recti, superior and inferior recti, and superior and inferior oblique muscles. These muscles are innervated by motoneurons in the nuclei of the third, fourth, and sixth cranial nerves. The activity of these motoneurons is influenced by a variety of areas in the brain, including cells in the frontal, parietal, and occipital lobes; brain stem reticular formation; superior colliculus; cerebellum; and vestibular nuclei. Three general categories of eye movements are considered: fixation, saccadic, and pursuit.

Fixation **involves moving the eyes to bring a discrete portion of the visual field into focus on the fovea**. Voluntary fixation is controlled by the frontal eye fields, Brodmann's area 8, and an area in the occipital lobe that represents a portion of secondary visual cortex (area 19). It is thought that the occipital eye field "locks" the eyes on a fixation point, whereas the frontal eye field "unlocks" the eyes from the fixation point and allows them to be directed to a different target. Small flicking and drifting movements of the eyes may be necessary to keep an object on the region of the central fovea. These movements are probably directed by the superior colliculus.

Saccadic movement of the eyes is a mechanism of successive fixation points. When the eyes rapidly jump from one object to another, each jump is called a *saccade*. These movements are very rapid, and the brain suppresses the visual image during the movement so that one typically is not conscious of the point-to-point movement. As one reads, the eyes make several saccadic movements for each line.

Pursuit movements occur when the eyes fixate a moving target. The control system for such movements involves the transmission of visual information to the cerebellum by various routes. The brain then computes the trajectory of the target and activates the appropriate motoneurons to move the eyes to keep the target in focus on the fovea.

The superior colliculi are mainly responsible for turning the eyes and head toward a visual stimulus. If the visual cortex is damaged, the head can still be turned toward a visual stimulus in the periphery of the visual field. This ability is lost, however, when the superior colliculus is destroyed. The visual field is mapped in the superior colliculus independent of a similar map in the visual cortex. This activity is thought to be mediated by input through the Y-type retinal ganglion cells (and perhaps also W-type cells). The superior colliculus also directs turning of the head and body toward a visual stimulus through its descending projections in the medial longitudinal fasciculus. Interestingly, other sensory inputs, such as audition and somatosensation, are funneled through the superior colliculus and its descending connections such that the superior colliculus performs a global integrating function with respect to orientation of the eyes and body toward various stimulus points.

AUTONOMIC CONTROL OF ACCOMMODATION AND PUPILLARY APERTURE (p. 659)

Parasympathetic fibers to the eye originate in the Edinger-Westphal nucleus and course via the oculomotor nerve to the ciliary ganglion, where postganglionic fibers originate and extend to the eye with ciliary nerves. *Sympathetic fibers* originate in the intermediolateral cell column of the spinal cord and pass to the superior cervical ganglion. Postganglionic sympathetic fibers course on the carotid and ophthalmic arteries to eventually reach the eye.

When the fixation point of the eyes changes, the focusing power of the lens is adjusted in the proper direction by the appropriate activation of the autonomic innervation of the ciliary and sphincter pupillae muscles in each eye. Chromatic aberration also appears to be used to adjust the lens; for example, red light focuses slightly posterior to blue light because the lens bends blue light rays more than red light rays.

When the eyes focus from far to near (or vice versa), they must also converge. This involves bilateral activation of the medial rectus muscles in each eye. The areas of the brain that control pupillary changes and convergence are sufficiently separated because lesions may disrupt one function but not the other. For example, an Argyll-Robertson pupil is one that does not exhibit normal light reflexes but will accommodate. Such a pupil is common in individuals afflicted with syphilis.

52

The Sense of Hearing

THE TYMPANIC MEMBRANE AND THE OSSICULAR SYSTEM (p. 663)

Sound from the Tympanic Membrane to the Cochlea

The tympanic membrane is cone shaped; attached to its center is the handle of the *malleus*, which is the first in the series of bony elements that make up the ossicular chain. The *incus* is attached to the malleus by ligaments so that the two bones move together when the tympanic membrane moves the malleus. At its other end, the incus articulates with the *stapes*, which in turn is attached to the oval window of the membranous labyrinth. The malleus is also attached to the *tensor tympani* muscle; this keeps the tympanic membrane taut.

Impedance matching between sound waves in air and sound waves in the cochlear fluid is mediated by the ossicular chain. The amplitude of stapes movement at the oval window is only three fourths as large as the movement of the handle of the malleus. The ossicular chain does not amplify sound waves by increasing the *movement* of the stapes as is commonly believed; instead, the system actually increases the *force* of the movement by about 1.3-fold. Because the area of the tympanic membrane is so great relative to the surface area of the oval window (55 versus 3.2 square millimeters), the lever system multiplies the pressure of the sound wave exerted against the tympanic membrane by a factor of 22. The fluid within the membranous labyrinth has far greater inertia than air, so that the pressure amplification added by the ossicular chain is necessary to cause vibration in the fluid. The tympanic membrane and ossicles together provide *impedance matching* between sound waves in air and sound vibrations in the fluid of the membranous labyrinth. In the absence of a functioning ossicular chain, normal sounds are barely perceptible.

Contraction of the stapedius and tensor tympani muscles attenuates sound conduction. When extremely loud sounds are transmitted through the ossicular chain, there is a reflex damping of the malleus by the *stapedius* muscle, which acts as an antagonist to the tensor tympani. In this way, the rigidity of the ossicular chain is increased, and the conduction of sound, particularly at lower frequencies, is greatly reduced. Interestingly, this same mechanism is used to diminish the sensitivity to one's own speech.

Sound Through Bone Conduction (p. 664)

Because the cochlea is entirely embedded in bone, vibration of the skull can stimulate the cochlea. When a tuning fork is vibrated and applied to the skull on the forehead or mastoid region, a humming sound can be heard. In general, however, even relatively loud sounds in the air do not have sufficient energy to enable effective hearing through bone conduction.

THE COCHLEA (p. 664)
Functional Anatomy

The cochlea consists of three tubes coiled side by side. The *scala vestibuli* and *scala media* are separated by the vestibular membrane (Reissner's membrane); the scala media and the scala tympani are separated by the basilar membrane. The *organ of Corti* lies on the surface of the basilar membrane and within the scala media. The roof of the organ of Corti is formed by the tectorial membrane. At the end of the cochlea opposite the round and oval windows, the scala vestibuli is continuous with the scala tympani at the helicotrema. The overall stiffness of the basilar membrane is 100 times less at the helicotrema than it is near the oval window. The stiffest portion near the oval window is most sensitive to high-frequency vibrations, whereas the more compliant end near the helicotrema is sensitive to low-frequency vibrations.

Transmission of Sound Waves (p. 666)

When a sound wave strikes the tympanic membrane, the ossicles are set into motion and the footplate of the stapes is pushed into the membranous labyrinth at the oval window. This initiates a wave that travels along the basilar membrane toward the helicotrema.

Vibration patterns are induced by different sound frequencies. The pattern of vibration initiated in the basilar membrane is different for the various sound frequencies. Each wave is relatively weak at its outset but becomes strongest at the portion of the basilar membrane that has a resonant frequency equal to that of the sound wave. The wave essentially dies out at this point and does not affect the remainder of the basilar membrane. In addition, the velocity of the traveling wave is greatest near the oval window and then gradually decreases as it proceeds toward the helicotrema.

Vibration patterns are induced by different sound amplitudes. The maximal amplitude of vibration for sound frequency is spread in an organized way over the surface of the basilar membrane. For example, maximal vibration for an 8000 cycle-per-second (Hertz or Hz) sound occurs near the oval window, whereas for a 200-Hz sound, it occurs near the helicotrema. The principal method for sound discrimination is the "place" of maximal vibration on the basilar membrane for that sound.

Organ of Corti

Nerve Impulses in Response to Vibration of the Basilar Membrane

The receptor cells of the organ of Corti consist of two types: the *inner* and the *outer hair cells*. There is a single row of about 3500 inner hair cells and three or four rows of about 12,000 outer hair cells. Nearly 95 per cent of the eighth cranial nerve sensory fibers that innervate the cochlea form synaptic contact with inner hair cells. The cell bodies of the sensory fibers are found in the spiral ganglion, which is located within the bony modiolus (the center) that serves as support for the basilar membrane at one end. The central processes of these ganglion cells enter the brain stem in the rostral medulla to synapse in the cochlear nuclei.

Vibration of the basilar membrane excites the hair cells. The apical surface of the hair cells gives rise to many stereocilia and a single kinocilium, which project upward into the overlying tectorial membrane. When the basilar membrane vibrates, the hair cell cilia, which are embedded in the tectorial membrane, are bent in one direction and then in the other. It is this movement that mechanically opens ion channels and leads to depolarization of the hair cell.

Even though there are far more outer than inner cells, the majority of sensory fibers receive synapses from the inner hair cells. If outer hair cells are damaged, however, a significant amount of hearing loss ensues. It has thus been proposed that the outer hair cells in some way regulate the sensitivity of the inner hair cells.

Hair cell receptor potentials activate auditory nerve fibers. The approximately 100 cilia protruding from the apical surface of the hair cells progressively increase in length from the region of the attachment of the basilar membrane to the modiolus. The longest of these cilia is referred to as a kinocilium because it does not appear to move. When the stereocilia are bent toward the kinocilium, potassium channels in the cilial membrane are opened, potassium enters, and the hair cell is depolarized. Just the reverse occurs when the cilia move away from the kinocilium: the hair cell is hyperpolarized. The fluid bathing the cilia and apical surface of the hair cells is *endolymph*. This watery fluid is different from the *perilymph* in the scala vestibuli and scala tympani, which, like extracellular fluid, is *low* in potassium and *high* in sodium. The endolymph is secreted by the stria vascularis, a specialized epithelium in the wall of the scala media, and it is *high* in potassium and *low* in sodium. The electrical potential across the endolymph is about +80 millivolts; this is called the *endocochlear potential*. The interior of the hair cell, however, is about −70 millivolts. The potential difference across the membrane of the cilia and apical surface of the hair cells is about 150 millivolts; this greatly increases their sensitivity.

Sound Frequency and the "Place Principle" (p. 668)

The nervous system determines sound frequency by the point of maximal stimulation along the basilar membrane. Sounds at the high-frequency end of the spectrum will maximally stimulate the basal end near the oval window. Low-frequency stimulation maximally stimulates the apical end near the helicotrema. Sound frequencies below 200 Hz, however, are discriminated differently. These frequencies cause synchronized *volleys* of impulses at the same frequency in the eighth cranial nerve, and cells in the cochlear nuclei that receive input from these fibers can distinguish the different frequencies.

Loudness of Sound

1. As the sound becomes louder, the amplitude of vibration in the basilar membrane increases, and hair cells are activated more rapidly.
2. With increased amplitude of vibration, more hair cells are activated, and spatial summation enhances the signal.
3. Outer hair cells are activated by large-amplitude vibrations; somehow, this signals the nervous system that the sound has surpassed a certain level that delimits high intensity.

The auditory system can discriminate between a soft whisper and a loud noise that might represent as much as a 1 trillion-fold increase in sound energy. The intensity scale is compressed by the brain to provide the wide range of sound discrimination.

Because of the wide range in sound sensitivity, intensity is expressed as the logarithm of the actual intensity. The unit of sound intensity is the *bel*, and sound levels are most often expressed in 0.1-bel units or as 1 *decibel*.

The threshold for hearing in the human is different at varying intensities. For example, a 3000-Hz tone can be heard at an intensity level of 70 decibels, whereas a 100-Hz tone can be heard only if the intensity is increased to a level 10,000 times as great.

The range of hearing is typically listed as 20 to 20,000 Hz, but, again, the intensity level is significant because at a level of 60 decibels, the frequency range is only 500 to 5000 Hz. To hear the full range of sound, the intensity level must be very high.

CENTRAL AUDITORY MECHANISMS
Anatomy of Central Auditory Pathways (p. 669)

Primary sensory fibers from the spiral ganglion enter the brain stem and terminate in the *dorsal* and *ventral cochlear nuclei*. From there, signals are sent to the contralateral (and ipsilateral) *superior olivary nucleus*, where cells give rise to fibers that enter the lateral lemniscus, which then terminates in the *inferior colliculus*. Cells in the inferior colliculus project to the *medial geniculate nucleus* of the thalamus; from there, signals are transmitted to the primary auditory cortex, the *transverse temporal gyrus of Heschel*. It is important to understand that (1) beginning with the output from the cochlear nuclei, signals are transmitted bilaterally through central pathways with a contralateral predominance; (2) collaterals from central pathways synapse in the brain stem reticular formation; and (3) spatial representations of sound frequency (tonotopic organization) are found at many levels within the various cell groups of the central auditory pathways.

Role of the Primary Auditory Cortex in Hearing (p. 670)

The primary auditory cortex is a relatively small region of the temporal lobe that corresponds to Brodmann areas 41 and 42. Surrounding these areas is area 22, a portion of which is considered secondary auditory cortex.

At least six separate *tonotopic representations (maps)* of sound frequency have been described in primary auditory cortex. The question of why these separate maps exist is unanswered at the moment, but it is presumed that each region selects a particular feature of sound or sound perception and performs an analysis of that feature.

Bilateral destruction of the primary auditory cortex does not eliminate the ability to detect sound; it does, however, cause two specific deficits. There is an inability to recognize certain *sound patterns*, and there is a distinct problem in *localizing sounds* in the environment. Lesions in secondary auditory cortex interfere with the ability to interpret the *meaning* of particular sounds. This is particularly true for spoken words and is referred to as a *receptive aphasia*.

Mechanism for Sound Localization (p. 671)

The superior olivary nucleus is divided into *medial* and *lateral* subdivisions. The lateral subnucleus determines sound direction by detecting the difference in sound intensity transmitted from the two ears. The medial subnucleus localizes sound by detecting the difference in the time of arrival of sound to the two ears. The input to individual cells in the latter nucleus is segregated so that signals from the right ear reach one dendritic system of a given cell, while input from the left ear synapses with a separate dendritic system on the same neuron.

Centrifugal Projections in the Auditory System

Each processing level in the central auditory pathway gives rise to descending or retrograde fibers that project back toward the cochlear nuclei as well as the cochlea itself. These centrifugal connections are more pronounced in the auditory system than in any other sensory pathway. It is speculated that these connections allow one to selectively attend to particular sound features.

Common Hearing Abnormalities

Hearing difficulties can be assessed with an *audiometer*, which allows specific sound frequencies to be individually delivered to each ear. When a patient has *nerve deafness*, both air and bone sound conductions are affected, and damage usually involves one or more of the neural components of the auditory system. When only air conduction is affected, damage to the ossicular chain is usually the cause. This is often due to repeated middle ear infections.

53

The Chemical Senses—Taste and Smell

The sense of taste is mainly a function of the *taste buds*, but the sense of smell contributes substantially to taste perception. The texture of food as sensed by tactile receptors in the mouth also contributes to the taste experience.

THE PRIMARY SENSATIONS OF TASTE (p. 675)

At present, receptors for 13 different chemical substances are identified; these include the following:

2 sodium receptors
2 potassium receptors
1 chloride receptor
1 adenosine receptor
1 hydrogen ion receptor

1 inosine receptor
2 sweet receptors
2 bitter receptors
1 glutamate receptor.

For practical purposes, the activities of these receptors have been grouped into four categories called the *primary sensations of taste*; they include *sour, salty, sweet,* and *bitter*.

- *Sour taste* is caused by acidic substances, and the taste intensity is proportional to the logarithm of the hydrogen ion concentration.
- *Salty taste* is mainly contributed to by the cations of ionized salts, but some salts activate additional receptors; this explains the slight difference among salty tasting items.
- *Sweet taste* is the result of activation of several receptor types, including sugars, glycols, alcohols, aldehydes, and many others, most of which are organic chemicals.
- *Bitter taste* also is caused by the activation of several different receptors associated with organic chemicals. Two of the more common substances in this category are long-chain, nitrogen-containing items and alkaloids. The latter category in-

cludes medicinal compounds such as quinine, caffeine, strychnine, and nicotine. A strong bitter taste often causes a substance to be rejected; this is related to the fact that dangerous toxins found in some plants are alkaloids.

Taste Threshold (p. 676)

To be recognized as salty, a substance need only be 0.01 M, whereas for quinine to be perceived as bitter, its concentration need only be 0.000008 M. This correlates with the notion that bitter serves a protective function against dangerous alkaloids; thus, its sensitivity is high. Some individuals are "taste blind" for certain substances. This is probably due to the normal variation one sees in the presence of or number of certain classes of receptors.

Taste Buds and Their Function

A taste bud is composed of about 50 modified epithelial cells, some of which serve a supporting function, the *sustentacular cells*, and others that are the actual *receptor cells*. The latter are continuously replaced by mitotic division from the surrounding epithelial cells. The life span of a taste cell in lower mammals is about 10 days, but it is unknown for humans. The apical surfaces of the taste cells are arranged around a *taste pore*. Microvilli or taste hairs protrude from the pore into the mouth cavity and provide the receptor surface for taste molecules. Intertwined among the cell bodies are the sensory nerve fibers that form postsynaptic elements and respond to activity in taste cells.

The 3000 to 10,000 taste buds in the adult are found on three types of *papillae* of the tongue. *Fungiform* papillae are found on the anterior two thirds of the tongue; *circumvallate* papillae form a V-shaped configuration on the posterior one third of the tongue; and *foliate* papillae are found along the lateral margins of the tongue. A small number of taste buds are also found on the palate, tonsils, and epiglottis and in the proximal esophagus. Each taste bud typically responds to only one of the four primary taste substances, except when an item is present in a very high concentration; then, it may stimulate more than one receptor type.

Like other receptors, taste cells produce a *receptor potential*. The application of the substance to which it is sensitive causes the taste cell to be depolarized. The degree of depolarization correlates with the concentration of the taste substance. The binding of the taste substance to the receptor opens ion-specific channels that allow sodium to enter the cell. At the first stimulation, the taste substance elicits a rapid response in the associated sensory fibers that then adapts to a lower level within a few seconds. The taste substance is washed away from the receptor by saliva.

Transmission of Taste Signals in the Central Nervous System (p. 677)

Taste fibers from the anterior two thirds of the tongue first travel in branches of the trigeminal nerve and then join the chorda tympani, a branch of the facial nerve. Taste sensation from the posterior third of the tongue is carried by fibers in the glossopharyngeal nerve, whereas any taste fibers from the epiglottis or other areas course within branches of the vagus nerve. From their entry into the brain stem, all taste fibers are funneled into the *solitary tract* and eventually synapse in the rostral portion of the *nucleus of the solitary tract*. From here, axons pass rostrally in rather ill-defined pathways to the *ventromedial nucleus* of the thalamus and then onto the cerebral cortex in the ventral region of the postcentral gyrus, which curls into the lateral fissure. This is the primary pathway for taste recognition.

In addition to the cortical pathway for taste perception, *taste reflexes* involve fibers that course from the solitary tract directly to the superior and inferior salivatory nuclei, which contain preganglionic parasympathetic neurons for the eventual activation of saliva secretion by the submandibular, sublingual, and parotid glands. Although some of the adaptive qualities of taste are the result of activity at the receptor level, most taste adaptation apparently occurs through central mechanisms that are not well defined at present.

Taste Preferences and Control of Diet

In animals, taste preferences change with the needs of the body for certain substances. These mechanisms are most likely present in modified forms in humans as well, and these functions are probably the result of activity in modifiable circuits in the central nervous system. For example, if a person becomes ill after eating a particular substance, an aversion may develop to that substance; conversely, we associate certain pleasant responses to other substances that we learn to enjoy.

THE SENSE OF SMELL (p. 678)

In humans, the sense of smell is probably the least understood sense—perhaps because it is largely a subjective phenomenon. Compared with some animals, in humans, it is poorly developed.

Olfactory Epithelium

The receptor surface for smell is located in the upper part of the nasal cavity and typically exhibits a surface area of only about 2.4 square centimeters. Olfactory receptor cells are bipolar neurons derived from the central nervous system. There are usually about 100 million of these cells in each individual interspersed with a much smaller number of sustentacular cells. The apical surface of the receptor cell exhibits a knob that emits 6 to 12 olfactory hairs or cilia, which contain the actual receptors and project into the mucus present on the epithelial surface. Spaced among the receptor cells are the glands of Bowman, which secrete mucus onto the epithelial surface.

Stimulation of the Olfactory Receptor Cells

Odorant molecules diffuse into the mucus and bind to receptor proteins that protrude from the surface of the olfactory cilia. The receptor protein is linked to a cytoplasmic G-protein. On activation, the *alpha subunit* of the G-protein separates and activates adenyl cyclase, which in turn leads to the formation of cyclic AMP. Sodium channels are then activated by cyclic AMP, and sodium ions enter the cell and depolarize it, leading to the production of action potentials in the olfactory sensory fibers. The utility of this depolarization method is that it can multiply the excitatory effect of a weak odorant molecule concentration and greatly enhance the sensitivity of the system.

Like the taste system, the intensity of olfactory stimulation is proportional to the logarithm of stimulus strength. The receptors adapt about 50 per cent in the first second and thereafter adapt very little and very slowly. Although most odors appear to adapt to extinction within 1 or 2 minutes, this is not a physiological process at the level of the receptor but rather a function of central mechanisms that alter perception. This may be one of the functions of the large number of centrifugal fibers that course from olfactory regions of the brain back into the olfactory bulb.

Search for the Primary Sensations of Smell (p. 680)

As many as 100 different smell sensations have been reported, and they have been narrowed down to seven primary odor sensations: *camphoraceous*, *musky*, *floral*, *peppermint*, *ethereal*, *pungent*, and *putrid*.

Smell, even more than taste, is associated with pleasant or unpleasant affective qualities. The threshold for some odorant molecules is extremely low, on the order of one 25 billionth of a milligram. The range of sensitivity, however, is only 10 to 50 times that of the threshold level, which is relatively low compared with other sensory systems.

Transmission of Smell Signals into the Central Nervous System

The olfactory bulb lies over the cribriform plate of the ethmoid bone, which separates the cranial and nasal cavities. The olfactory nerves pass through perforations in the cribriform plate and enter the olfactory bulb, where they terminate in relation to glomeruli. This is a tangled knot of mitral and tufted cell dendrites and olfactory nerve fibers. Mitral and tufted cell axons leave the olfactory bulb via the olfactory tract and enter specialized regions of the cortex without first passing through the thalamus.

The *medial olfactory area* is best represented by the septal nuclei, which project to the hypothalamus and other regions that control behavior. This system is thought to be involved in primitive functions such as licking, salivating, and other feeding-related behavior.

The *lateral olfactory area* is composed of the prepiriform, pyriform, and cortical amygdaloid regions. From here, signals are directed to less-primitive limbic structures, such as the hippocampus. This apparently is the system that associates certain odors with specific behavioral responses.

Another, phylogenetically newer, pathway actually projects to the dorsomedial thalamic nucleus and then onto the orbitofrontal cortex.

Fibers that originate in the brain course centrifugally to reach granule cells in the olfactory bulb. The latter cells inhibit mitral and tufted neurons of the bulb; in this way, the ability to distinguish different odors is sharpened.

UNIT
XI

The Nervous System

C. Motor and Integrative Neurophysiology

54

Motor Functions of the Spinal Cord; The Cord Reflexes

The spinal cord is often relegated to a role secondary to that of the brain when nervous system functions are analyzed. Circuits exist within the spinal cord, however, that process sensory information and are capable of generating complex motor activity. In addition, it is clear that even the most advanced and complicated functions involving the control of movement performed by the brain cannot be implemented if the spinal cord and its direct connections with skeletal muscles are not intact.

COMPONENTS OF THE SPINAL CORD: MOTOR FUNCTION
(p. 686)

Anterior horn motor neurons are present at all levels of the cord and give rise to axons that exit the cord via its ventral roots and then pass distally in peripheral nerves to innervate skeletal striated muscles. A motor neuron and all the muscle fibers it innervates are referred to collectively as a *motor unit*.

Spinal cord ventral horn motor neurons are of two varieties: alpha and gamma motor neurons. The largest are the *alpha motor neurons*, which give rise to myelinated axons that average about 14 micrometers in diameter and conduct action potentials very rapidly. The *gamma motor neurons* are much smaller and give rise to smaller axons that average about 5 micrometers in diameter and conduct action potentials at a slower velocity than the alpha motor neurons.

A third cell type that contributes to motor and sensory functions in the spinal cord is the *interneuron*. There are several different varieties of these cells; they are about 30 times more numerous than motor neurons and highly excitable, and they may have spontaneous firing rates as high as 1500 per second. The interneurons actually receive the bulk of synaptic input that reaches the spinal cord as either incoming sensory information or signals descending from higher centers in the brain.

The *Renshaw cell* is a particular variety of interneuron that receives input from collateral branches of motor neuron axons and then, via its own axonal system, provides inhibitory connections with the same or neighboring motor neurons. This suggests that the motor system, like the sensory systems, uses the mechanism of lateral inhibition to focus or sharpen its signals. Other interneurons are responsible for interconnection with one or several adjacent segments of the cord in an ascending or a descending direction; the latter cells are called *propriospinal neurons*.

MUSCLE SENSORY RECEPTORS: MUSCLE SPINDLES, GOLGI TENDON ORGANS

Receptor Function of the Muscle Spindle (p. 687)

The sensory feedback from skeletal muscles includes (1) the current length of the muscle and (2) the current tension in the muscle. The length value is derived from a *muscle spindle*, whereas tension is signaled by a *Golgi tendon organ*.

A *muscle spindle* is 3 to 10 millimeters in length and consists of 3 to 12 thin intrafusal muscle fibers that are actually striated muscle fibers. Each is attached at its distal ends to the associated extrafusal skeletal muscle. The central region of each intrafusal fiber is devoid of actin-myosin contractile elements and instead forms a capsule containing several nuclei. When the nuclei are arranged more or less linearly, the fiber is called a *nuclear chain fiber*; when nuclei are simply aggregated or clumped in the central region, the fiber is called a *nuclear bag fiber*. Typically, a muscle spindle will contain one to three nuclear bag fibers and three to nine nuclear chain fibers. The distally located contractile elements of each intrafusal fiber are innervated by relatively small, *gamma motor neuron axons*.

Two types of sensory fibers are associated with muscle spindle intrafusal fibers. One is called the *primary ending*, or the annulospiral ending. The primary ending is a type Ia myelinated primary sensory fiber with an average diameter of 17 micrometers and a rapid conduction velocity of 70 to 120 m/ sec. Typically, a spindle will also have at least one type II, secondary or flower spray–ending that exhibits an average diameter of 8 micrometers, is lightly myelinated, and conducts at a slower velocity than type Ia fibers. The primary ending wraps itself around the central (nuclear) region of both a nuclear bag and a nuclear chain intrafusal fiber, whereas the secondary ending forms numerous small terminal branches that cluster around the nuclear region of only the nuclear chain intrafusal fibers.

Dynamic and Static Responses of the Muscle Spindle

When the central region of a spindle is *slowly* stretched, the number of impulses in both the primary and secondary endings increases in proportion to the degree of stretch; this is called the *static response*. Because the nuclear chain fibers are innervated by both the primary and secondary sensory fibers, the static response is thought to be mediated by these intrafusal fibers.

When the length of a spindle is *suddenly* increased, the primary sensory fiber exhibits a vigorous response. This is called the *dynamic response*, and it appears to signal the rate of change in length. Because most nuclear bag fibers are mainly associated with primary endings, it is assumed that they are responsible for the dynamic response.

Control of the Static and Dynamic Responses by Gamma Motor Neurons

Gamma motor neurons are divided into two categories on the basis of the type of intrafusal fiber they innervate. Gamma motor neurons distributing to nuclear bag fibers are called *dynamic*, whereas those distributing to nuclear chain fibers are called *static*. Stimulation of a dynamic gamma motor neuron enhances only the dynamic response, whereas stimulation of static gamma motor neurons enhances the static response.

Muscle spindles exhibit a continuous or background level of activity that can be modulated upward (increased firing) or downward (decreased firing) as necessary for the ongoing muscle activity.

Stretch Reflex: Muscle Spindle Activation (p. 689)

Type Ia sensory fibers enter the spinal cord through the dorsal roots and give rise to branches that either terminate in the cord near their level of entry or ascend to the brain. Those that terminate in the cord synapse directly (monosynaptic) with alpha motor neurons in the ventral horn that innervate extrafusal fibers in the *same* muscle in which the primary sensory fibers originated. This circuitry is the substrate for the *stretch reflex*. This reflex actually has two components: a dynamic phase that occurs while the spindle is stretched and a static phase that occurs when the muscle has stopped increasing in length and reaches a new static length. An important function of the stretch reflex is its *damping effect* on oscillatory or jerky movements. In the absence of normally functioning spindle sensory mechanisms, an unusual repetitive contraction of muscles appears *(clonus)*.

Muscle Spindles Function During Voluntary Movement
(p. 690)

Approximately 31 per cent of the axons distributing to any given muscle are from gamma motor neurons. When signals are transmitted from the motor cortex or other control centers, both alpha and gamma motor neurons are *co-activated*. The stimulation of gamma motor neurons during contraction of a muscle maintains the sensitivity of the spindle and prevents it from going "slack" and stopping its output. The gamma motor neuron system is most strongly influenced by descending projections from the facilitatory regions of the brain stem reticular formation, which in turn are influenced by output from the cerebellum, basal ganglia, and cerebral cortex.

Muscle Spindle System and Body Position During Tense Action

During movement that requires substantial force generation, the gamma motor neurons act on muscle spindles located in muscles on both sides of a joint and are activated in tandem by the brain stem reticular formation. The result is enhancement of muscle tone on both sides of the joint; this provides a stabilizing effect on movement at that joint.

Clinical Applications for the Stretch Reflex

The physician can assess the general state of reflex activity by testing the stretch reflex at a number of key joint locations. For example, tapping on the patellar tendon at the knee stretches spindles in the quadriceps and normally elicits a reflex contraction of that muscle group (stretch reflex), which produces a knee jerk. The physician notes the characteristics of the reflex response and can make diagnostic inferences concerning possible abnormalities in the muscles or nervous system. A reflex that is too strong or too brisk can indicate one type of problem, whereas a reflex that is weak or absent may suggest other types of problems.

Clonus, an alternate contraction of the agonist and antagonistic muscles crossing a joint, is a sign of abnormal stretch reflex function. This sign is often prominent at the ankle, where a rapid and maintained dorsiflexion induced by the examiner might elicit sustained jerking movements (alternate flexion and extension) of the foot at the ankle joint. It is a sign that the spinal cord circuits that mediate the stretch reflex are not being properly influenced by the descending projections from the brain.

Golgi Tendon Reflex: Control of Muscle Tension

The Golgi tendon organ is an encapsulated receptor through which a small bundle of muscle tendon fibers pass just before their bony insertion. Sensory fibers intermingle with and entwine the tendon fibers and are stimulated when the *tension* imposed by muscle contraction is increased. Like the muscle spindle, the tendon organ responds vigorously when the tendon is undergoing stretch (dynamic response) and then settles down to a steady state level that is proportional to the degree of tension (static response).

Signals from the tendon organ are conducted through large myelinated type Ib fibers that conduct nearly as rapidly as the type Ia fibers from the muscle spindles. On entering the cord, these fibers form branches with some fibers terminating locally on the pool of interneurons and others entering a long ascending pathway. Local inhibitory interneurons link the tendon organ input to alpha motor neurons that innervate those muscles with which the tendon organ is associated. In contrast to muscle spindle input, which excites its related motor neurons, the tendon organ produces *inhibition* of the motor neurons to which it is connected. This negative feedback prevents injury to the muscle when it exceeds its upper limit of tension. In addition to this protective function, the tendon organ reflex serves to redistribute the muscle load more evenly over a larger number of muscle fibers. In addition, via their ascending projections, the tendon organs provide input to the cerebellum and motor areas of the cerebral cortex that will be used by these centers in the control of movement.

WITHDRAWAL OR FLEXOR REFLEX (p. 692)

The *withdrawal* (*flexor*) *reflex* is elicited by pain receptors, usually those located in the skin. The muscles that are activated are those required to remove the body part from the painful stimulus. Typically, these are flexor muscles in the limbs, but the reflex is not limited to these muscles. The sensory fibers that carry these signals terminate on the pool of spinal cord interneurons, most of which provide excitatory input to the appropriate ventral horn motor neurons, whereas others inhibit motor neurons that innervate antagonistic muscles. The latter mechanism is called *reciprocal inhibition*.

CROSSED EXTENSOR REFLEX (p. 693)

The *crossed extensor reflex* often occurs in conjunction with the flexor reflex. Removal of a limb from a painful stimulus may require support from one or more body parts. For example, withdrawal of the foot might necessitate that the other foot support the entire body. In this situation, interneurons that receive the incoming pain signal from one foot can project across the midline to excite the appropriate contralateral motor neurons to support the body; often, these are extensor motor neurons. It also is possible, if the lower extremity is initially affected by the pain stimulus, for impulses to spread to more rostral cord levels by propriospinal neurons that synapse with motor neurons, innervating upper extremity musculature that might be needed to stabilize the body.

REFLEXES OF POSTURE AND LOCOMOTION (p. 694)

Postural and Locomotor Reflexes of the Spinal Cord

In experimental animals in which the spinal cord has been isolated from the remainder of the brain via a cervical level transection, certain reflex motor patterns are released from the normal descending control mechanisms from the brain.

- Pressure on a foot pad causes the limb to be extended against the applied pressure. In some animals, when held in place on all four limbs, this reflex can generate sufficient muscle force to support the entire body. This reflex is called the *positive supportive reaction*.
- Similarly, when an animal with a cervical cord transection is placed on its side, it tries to raise itself to a standing position, although this maneuver is rarely successful. This is called the *cord righting reflex*.
- If a cord-transected animal is suspended on a treadmill so that each of the limbs can touch the surface of the treadmill, all four limbs will move in a synchronous and coordinated manner as if the animal were trying to walk on the treadmill.
- These observations indicate that circuits intrinsic to the spinal cord are capable of generating movements in a single extremity, a pair of extremities, or all four extremities. This circuitry involves connections between flexor and extensor motor neurons in a single cord segment, across the midline, and rostrally and caudally through the propriospinal system.

SPINAL SHOCK SYMPTOMS

When the spinal cord is transected, all cord functions below the lesion become substantially depressed; this is referred to as *spinal shock*. This condition may persist for a few hours, days, or weeks. It is thought to represent a period during which the excitability of spinal neurons is dramatically reduced due to the loss of all descending projections. As is the case in other areas of the nervous system, the affected neurons gradually regain their excitability as they reorganize and adapt to the new levels of reduced synaptic input. In some instances, the affected neurons actually become hyperexcitable; this produces motor abnormalities such as spasm, tremor, or cramping that may be difficult to overcome.

Some of the more common symptoms that appear during spinal shock include the following:

- *Arterial blood pressure may fall significantly*, indicating that the output of the *sympathetic* nervous system is completely interrupted. This response usually recovers rapidly if there is no structural damage to the sympathetic nervous system as a result of the process that produced the spinal cord transection.

- *All skeletal muscle reflexes are nonfunctional.* In humans, 2 weeks to several months may be required for reflex activity to return to normal. If the transection is incomplete and some descending pathways remain intact, some reflexes may actually become hyperactive.
- *Sacral autonomic reflexes that regulate bladder and bowel function may be suppressed for several weeks.*

55

Cortical and Brain Stem Control of Motor Function

Essentially, each voluntary movement that an individual consciously decides to make has at least some component controlled by the cerebral cortex. However, not all movement is "voluntary," and much of the control over muscles and their coordinated activity involves a variety of brain centers, including the basal ganglia, cerebellum, brain stem, and spinal cord, that work in concert with areas of the cerebral cortex.

MOTOR CORTEX AND CORTICOSPINAL SYSTEM (p. 699)

Primary Motor Cortex

The primary motor cortex is located in the frontal lobe within the gyrus immediately anterior to the central sulcus called the *precentral gyrus* or *Brodmann's area 4*. Many years ago, during neurosurgical procedures in humans, Penfield and Rasmussen discovered that stimulation of points in the precentral gyrus led to movement or activation of muscles in various parts of the body. They observed that muscle activation was *somatotopically* organized in this gyrus so that stimulation of the most lateral portion caused activation of head and neck muscles; activation of the middle portion led to movement in the hand, arm, or shoulder; and stimulation in the medial portion of the gyrus caused activation of trunk and lower extremity muscles. At some stimulation points, individual muscles were activated, whereas at others, a group of muscles were activated.

Premotor Area (p. 700)

Immediately anterior to the lateral portion of the primary motor cortex is the *premotor cortex*. This cortex forms a portion of *Brodmann's area 6* and contains a somatotopically organized map of the body musculature. Stimulation in this cortex, however, typically produces movements that involve groups of muscles. For example, the arm and shoulder may be activated to place the hand in position to perform a certain task.

Supplementary Motor Area (p. 700)

The *supplementary motor area* is located within the medial portion of area 6, on the dorsal convexity and medial wall of the hemisphere just anterior to the lower extremity portion of the precentral gyrus. Stimulation in this area requires greater intensity and typically causes bilateral muscle activation, usually involving the upper extremities.

Other Cortical Areas Specialized for Motor Control

- *Broca's area (motor speech area)* lies just anterior to the face portion of the primary motor cortex near the sylvian fissure. Activity in this area engages the musculature needed to convert simple vocal utterances into whole words and complete sentences.
- The *frontal eye field (Brodmann's area 8)* also lies just anterior to the precentral gyrus but somewhat more dorsal than Broca's area. This cortical region controls the conjugate eye movements required to shift gaze from one object to another.
- A *head rotation area* associated with the frontal eye field is functionally linked to area 8 and serves to enable movements of the head correlated with eye movement.
- An *area related to the control of fine movements of the hand* is located within the premotor cortex just anterior to the hand region of area 4. When this area is damaged, the muscles of the hand are not paralyzed, but certain hand movements are lost; this is called *motor apraxia.*

Corticospinal Tract (Pyramidal Tract)

Primary Output Pathway from the Motor Cortex (p. 701)

The corticospinal tract mainly originates from the primary motor cortex (30 per cent) and premotor cortex (30 per cent), whereas the remainder is divided among several other areas, including the primary somatosensory cortex (postcentral gyrus), supplementary cortex, parietal lobe areas, and portions of the cingulate gyrus. After leaving the cortex, axons of this tract enter the posterior limb of the internal capsule and pass caudally through the brain stem to the ventral surface of the medulla, where they are contained in the medullary pyramids. At the junction of medulla and spinal cord, a majority of the fibers cross the midline to enter the lateral funiculus of the spinal cord and form the *lateral corticospinal tract*, which extends throughout the length of the cord. The fibers that do not cross continue as far as the thoracic spinal cord in the *ventral corticospinal tract*.

The largest fibers in the pyramidal tract are about 16 micrometers in diameter and are believed to originate from the giant cells of Betz found in the precentral gyrus. There are approximately 34,000 Betz cells, and the total number of fibers in the corticospinal tract is about 1 million, so that the large fibers represent only about 3 per cent of the entire tract.

Other Fiber Pathways from the Motor Cortex

In addition to projections to the spinal cord, branches of pyramidal tract fibers reach many other areas, including the caudate and putamen, red nucleus, reticular formation, basilar pontine nuclei, and inferior olive. The projections to the red nucleus may provide an alternate pathway for the motor cortex to influence the spinal cord via the rubrospinal tract, if corticospinal axons are damaged at a level caudal to the red nucleus.

Incoming Pathways to the Motor Cortex

It is important to consider the areas of the brain that provide *input* to the motor areas that give rise to the corticospinal system; these are surrounding areas of cortex in the same and contralateral hemisphere, including the somatosensory cortex as well as fibers from a variety of thalamic nuclei that carry information from the ascending somatosensory pathways, cerebellum, basal ganglia, and reticular activating system.

Activity in the Corticospinal and Rubrospinal Tracts and Spinal Cord Neurons (p. 703)

Like neurons in the visual cortex, those in the motor cortex are organized into vertical modules. Each vertical unit may control the activity of a synergistic group of muscles or an individual muscle. Approximately 50 to 100 pyramidal neurons must be activated simultaneously or in rapid succession to cause muscle contraction. Often, if a strong signal is needed to cause initial muscle activation, a weaker signal is then able to maintain the contraction for longer periods. The substrate for this function may involve two populations of corticospinal neurons: *dynamic neurons* produce high output for short time periods and may specify the development of the proper force needed to initiate the movement, whereas *static neurons* fire a less intense signal at a slower rate to maintain the force of contraction. Interestingly, the red nucleus also exhibits neurons with dynamic and static properties, with the dynamic variety outnumbering their counterpart in the cortex, and the static variety being proportionally less than those in the cortex.

Somatosensation and Precision of Muscle Contraction
(p. 704)

The signals that arise in muscle spindles, Golgi tendon organs, and skin near joints when movement occurs are relayed to the motor cortex and influence the output of that motor cortex. In general, the somatosensory input tends to enhance the activity of motor cortex. For example, as an object is grasped by the fingers, compression of the skin by the object tends to cause further excitement of the muscles and tightening of the fingers around the object.

Corticospinal Fibers: Spinal Motor Neurons

Large numbers of corticospinal fibers terminate in the cervical and lumbosacral enlargements of the spinal cord; this probably reflects the control over the muscles of the upper and lower extremities that is exerted by this system. Most of the cortical input is focused on the pool of spinal interneurons, but apparently some corticospinal axons synapse directly with ventral horn motor neurons. The corticospinal system may carry *command signals* that activate patterns of movement with a composition determined by aggregates of spinal interneurons. Similarly, it is not necessary for corticospinal signals to directly inhibit the action of antagonist muscles.

This can be accomplished by activating the intrinsic cord circuits that produce reciprocal inhibition.

Lesions of the Corticospinal Tract

"Stroke" Syndrome (p. 705)

A *stroke* is caused by a ruptured blood vessel that bleeds into the brain or by thrombosis of a vessel that produces local ischemia in neighboring brain tissue. When either event involves the primary motor cortex (origin of the corticospinal tract), the resulting motor deficits are characterized by the loss of voluntary control of discrete movements involving the distal portions of the extremities, particularly the fingers and hands. This does not necessarily mean that the muscles are completely paralyzed but rather that the control of fine movements is lost. Postural movements or gross positioning of the limbs may not be affected. Hemorrhagic or ischemic cortical strokes usually involve more territory than just the primary motor cortex. When the tissue damage extends beyond the primary cortex and involves neurons that project to the caudate, putamen, or reticular formation, characteristic symptoms occur such as hyper-reflexia, hypertonia, and spasticity.

ROLE OF THE BRAIN STEM IN CONTROLLING MOTOR FUNCTION

Support of the Body Against Gravity—Roles of the Reticular and Vestibular Nuclei (p. 706)

The pontine and medullary areas of the reticular formation function in opposition to one another through their contributions to the reticulospinal system. The pontine levels tend to excite antigravity muscles, whereas medullary levels inhibit them. Pontine levels are strongly activated by ascending somatosensory fibers, vestibular nuclei, and cerebellar nuclei. When unopposed by medullary levels, the excitation of antigravity muscles is sufficiently strong to support the body. In comparison, the inhibitory influence derived from the medullary reticulospinal fibers is strongly influenced by input from the cerebral cortex and red nucleus. The pontine and medullary systems can be selectively activated or inactivated to produce the desired excitation of or inhibition of antigravity muscles.

Role of the Vestibular Nuclei in Controlling Antigravity Muscles (p. 706)

The lateral vestibular nucleus transmits excitatory signals (mainly by way of the lateral vestibulospinal tract) that strongly excite antigravity muscles. This system is influenced most strongly by the vestibular sensory apparatus and uses the antigravity muscles to maintain balance.

Decerebrate Rigidity After Brain Stem Lesions (p. 707)

When the brain stem is sectioned at about midcollicular levels, leaving the reticulospinal and vestibulospinal tracts intact, a condition develops known as *decerebrate rigidity*. This is characterized by hyperactivity in the antigravity muscles, primarily in the neck, trunk, and extremities. The activation of the antigravity muscles is unopposed because the corticospinal and rubrospinal tracts have been sectioned along with the cortical activation of the medullary reticulospinal fibers. Although the cortical drive on the pontine reticulospinal system has also been interrupted, there is sufficient activation remaining from other excitatory inputs, such as the ascending somatosensory pathways and cerebellar nuclei. Examination of the antigravity muscles reveals that their stretch reflexes are greatly enhanced, and they are said to exhibit *spasticity*. It is believed that the descending influence from the pontine reticulospinal fibers affects primarily the gamma motor neurons. This is substantiated in animal experiments in which section of the dorsal roots in such a situation eliminates the hyperactivity in the antigravity muscles. The enhanced activation in these muscles is dependent on the action of gamma motor neuron input to muscle spindles and the resultant increased activity of Ia primary afferent fibers.

VESTIBULAR SENSATIONS AND EQUILIBRIUM (p. 707)

Vestibular Sensory Apparatus

The sensory organs for the vestibular sense are located in a system of bony chambers within the petrous portion of the temporal bone. Each bony enclosure houses a membranous chamber or tubular structure that contains the sensory hair cells and the terminal ends of primary sensory fibers of the eighth cranial nerve that lead into the brain. The membranous structures include the three semicircular canals or ducts and two larger chambers: the utricle and saccule.

Utricle and Saccule and Orientation of the Head Relative to Gravity

Within each utricle and saccule is a small specialized structure called the *macula*. It is a flattened area approximately 2 millimeters in diameter that lies in the horizontal plane on the inferior surface of the utricle and in the vertical plane within the saccule. The surface of each macula is covered by a gelatinous layer in which are imbedded calcium carbonate crystals called *statoconia*.

The macula contains supporting cells and sensory hair cells that exhibit cilia that protrude upward into the gelatinous layer. Each cell has 50 to 70 stereocilia and one large kinocilium. The latter is always the tallest cilium and is positioned off to one side of the apical surface of the hair cell. The stereocilia become progressively shorter toward the side opposite the kinocilium. Minute filaments connect the tip of each cilium to the next adjacent one and serve to open ion channels in the cilial membrane, which is bathed in endolymphatic fluid. When the stereocilia are bent toward the kinocilium, ion channels are opened, ions enter the cell from the endolymph, and the cell is depolarized. Conversely, movement of the stereocilia away from the kinocilium results in closure of membrane channels and hyperpolarization of the cell. In each macula, groups of hair cell cilia are oriented in specific directions so that some are stimulated and others are inhibited with head movement in any direction. The brain recognizes patterns of excitation and inhibition in the sensory fibers and translates that pattern into head orientation.

Semicircular Canals (p. 708)

The three membranous semicircular canals are named the *anterior*, *posterior*, and *lateral* canals; each is oriented at right angles to the others so that they represent the three planes in space. The lateral canal is in the true horizontal plane when the head is tilted forward 30 degrees, whereas the anterior and posterior canals are both in the vertical plane, with the anterior canal angled forward at 45 degrees and the posterior canal angled 45 degrees posteriorly. The sensory epithelium within each canal is formed by an *ampulla*, which is composed of ciliated sensory hair cells capped by a small crest called the *crista ampullaris* that in turn protrudes into an overlying gelatinous mass, the *cupula*. Each canal contains *endolymph*, which is free to move with rotation of the head; as it does, the cupula is deflected along with the cilia, which protrude into it from the hair cells. Movement in one direction is depolarizing; movement in the opposite direction is hyperpolarizing.

Utricle and Saccule: Static Equilibrium (p. 709)

The utricle and saccule are sensitive to *linear acceleration* (but not linear velocity). When the head accelerates in any plane relative to gravity, the statoconia shift and displace hair cell cilia in a specific direction, which depolarizes some cells and hyperpolarizes others.

Semicircular Canals: Head Rotation

When the head begins to rotate (angular acceleration), the endolymph in the canals, because of its inertia, tends to remain stationary and produces relative endolymph flow opposite that of head rotation. The cupula is deflected; the cilia are displaced; and the hair cells are depolarized or hyperpolarized, depending on the direction of cupula deflection. If the head rotation persists in the same direction, the endolymph attains the same direction and velocity as head rotation; the cupula is no longer deflected; and the hair cells are not stimulated. When the rotation stops, there again is flow of endolymph relative to the cupula (in the direction of rotation); some hair cells depolarize; and others hyperpolarize. The semicircular canals do not serve to maintain equilibrium but rather signal the beginning (or the end) of head rotation and thus are "predictive" in function.

Vestibular Reflex Actions (p. 710)

- Sudden changes in head orientation result in postural adjustments caused by activation of receptors in the utricle, saccule, or semicircular canals. The activation of motor responses is achieved via projections from the vestibular nuclei to the lateral vestibulospinal tract.

- When head orientation changes, the eyes must be moved to maintain a stable image on the retina. This correction is accomplished through connections from the semicircular canals to the vestibular nuclei, which then control motor neurons of the third, fourth, and sixth cranial nerves via projections through the medial longitudinal fasciculus.
- Proprioceptors in muscles and joints of the neck provide input to the vestibular nuclei, which counteracts the sensation of malequilibrium when the neck is bent.
- Input from the visual system, which signals a slight shift in the position of an image on the retina, is effective in maintaining equilibrium when the vestibular system is damaged.

Central Connections of the Vestibular System

The vestibular nuclei are richly interconnected with components of the brain stem reticular formation. These pathways are used to regulate eye movements via the medial longitudinal fasciculus and to control posture in the trunk and limbs in conjunction with the vestibulospinal tracts. The former connections function to maintain the eyes on a target when head orientation changes. The perception of head and body movement is achieved through vestibular input to the thalamus, which then projects to the cerebral cortex. Relatively little is known about the anatomy and function of this pathway.

The vestibular system also maintains extensive projections to and receives projections from the cerebellum. The cerebellar flocculonodular lobe is related to semicircular canal function and, when affected by lesions, causes a loss of equilibrium during rapid changes in the direction of head motion. The uvula of the cerebellum plays a similar role in regard to static equilibrium.

56

The Cerebellum, Basal Ganglia, and Overall Motor Control

THE CEREBELLUM AND ITS MOTOR FUNCTIONS (p. 715)

The cerebellum is especially vital to the control of rapid movements. Damage to the cerebellum does not usually produce muscle paralysis but instead produces an inability to use the affected muscles in a rapid, smooth, and coordinated manner.

Anatomical Features of the Cerebellum

The cerebellum consists of a three-layer cortex that surrounds four pairs of centrally located nuclei. The surface cortex exhibits numerous folds called *folia*, which are similar to the gyri of the cerebral cortex. The cerebellar cortex is divided into three major subdivisions: *anterior*, *posterior*, and *flocculonodular lobes*. The anterior and posterior lobes are further divided in the sagittal plane into a midline portion, the *vermis*; a slightly more lateral portion with ill-defined borders, the *intermediate zone*; and most lateral, the large *lateral hemispheres*.

The vermis and intermediate zones contain a somatotopic map of the body surface that reflects peripheral sensory input from muscle, tendon, joint capsule, and some cutaneous receptors.

The lateral hemispheres receive input primarily from the cerebral cortex via the basilar pontine nuclei, and only portions of each hemisphere exhibit a *fractured somatotopic organization*. Some body regions are thereby spatially segregated from their adjoining parts. For example, a lower limb territory might be located adjacent to a portion of the face. Some body regions are represented in more than one location.

The nuclei of the cerebellum include the medial or *fastigial nucleus*; *globose* and *emboliform* nuclei, which are collectively referred to as *interposed nuclei*; and lateral, or *dentate*, nucleus. The output of these nuclei is directed to the cerebral cortex via the thalamus and to the brain stem.

Input (Afferent) Pathways to the Cerebellum (p. 717)

- The largest afferent projection, the *pontocerebellar system*, originates from cells of the basilar pontine nuclei. It is this pathway that transmits signals that originate in nearly all regions of the cerebral cortex.
- The *olivocerebellar projection* originates from cells in the inferior olivary nuclei.
- *Spinocerebellar fibers* originate in the spinal cord or medulla.
- *Reticulocerebellar fibers* originate from a variety of cell groups in the brain stem.
- *Vestibular fibers* originate from the vestibular nuclei and vestibular sensory apparatus.

Output (Efferent) Signals from the Cerebellum

- The midline portions (vermis) of the cerebellar cortex project to the fastigial (medial) nucleus and then to the vestibular nuclei and reticular formation.
- The cortex of the intermediate zone projects to the globose and emboliform nuclei (interposed nuclei) and then to the ventrolateral and ventral anterior thalamic nuclei. From the thalamus, signals are transmitted to the cerebral cortex and basal ganglia.
- The lateral hemispheres project to the dentate (lateral) cerebellar nucleus and then to the ventrolateral and ventral anterior thalamic nuclei, which project to the cerebral cortex.

Functional Neuronal Circuits of the Cerebellum

The three layers of the cerebellar cortex (beginning nearest the pial surface) are the *molecular, Purkinje cell,* and *granular* layers. The fundamental circuit through the cerebellar cortex that is repeated approximately 30 million times is shown in Figure 56–1. The principal cell type is the Purkinje cell; it receives input to its fan-shaped dendritic tree located in the molecular layer. This input comes from two main sources: (1) *climbing fibers* that originate from cells of the inferior olivary complex and (2)

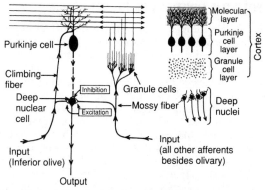

Figure 56–1   The left side of this figure shows the basic neuronal circuit of the cerebellum, with excitatory neurons shown in red and the Purkinje cell (an inhibitory neuron) shown in black. To the right is shown the physical relation of the deep cerebellar nuclei to the cerebellar cortex with its three layers.

parallel fibers that represent the axons of granule cells. The granule cells receive synaptic input from *mossy fibers*, which are formed by all the other cerebellar afferent systems. Another class of afferent fibers that apparently form synaptic contact with Purkinje cells—the *multilayered fibers*—have been shown to originate from biogenic amine cell groups, such as the locus ceruleus and other brain stem nuclei as well as portions of the hypothalamus.

The fundamental cerebellar circuit is completed by the axon of the Purkinje cell that forms synaptic contact in one of the cerebellar nuclei, although a few Purkinje axons extend into the vestibular nuclei. The transmission of signals through the fundamental circuit is influenced by three additional considerations.

1. Purkinje cells and cerebellar nuclear cells exhibit a high level of background activity that can be modulated upward or downward.
2. The cells of the central nuclei receive direct excitatory input from climbing fibers and most mossy fiber systems, whereas the input from Purkinje cells is inhibitory.
3. Within the cerebellar cortex, three other types of inhibitory interneurons (basket, stellate, and Golgi cells) also influence the transmission of signals through the fundamental circuit.

POTENTIAL FUNCTIONS OF THE CEREBELLUM (p. 720)

The cerebellum has a turn-on/turn-off function. During nearly every movement, certain muscles must be rapidly turned on and then quickly turned off. Because mossy and climbing fiber afferents can form direct excitatory contact with cerebellar nuclear cells (the cerebellar output neurons), it is possible that such connections establish the *turn-on* signal. In contrast, mossy and climbing fiber afferents also pass through the cerebellar cortex, where they can activate Purkinje cells that will inhibit cerebellar nuclear neurons and in this way specify the *turn-off* signal. Such a theory has some merit because cerebellar lesions are known to produce an inability to perform rapid alternating movements (e.g., pronation-supination of the wrist), which are known as *dysdiadochokinesia*.

Purkinje cells may learn to correct motor errors. It has been proposed that the role of the climbing fiber input to a Purkinje cell is to modify the sensitivity of that cell to parallel fiber input. The climbing fiber activity is more vigorous when a mismatch occurs between the anticipated result of a movement and its actual result. Gradually, as the movement is practiced, the mismatch declines, and the climbing fiber activity begins to return to its previous level. During the time of increased climbing fiber activity, the Purkinje cell can become more or less responsive to parallel fiber input.

The vestibulocerebellum joins with the brain stem and spinal cord to regulate equilibrium and posture (p. 721). The vestibulocerebellum is a combination of the flocculus and nodulus of the cerebellum and certain vestibular nuclei of the brain stem. It is believed that the role of these brain components is to calculate from the rate and direction of movement—that is, where the body will be in the next few milliseconds. This computation is the key to obtainment of the next sequential movement or maintenance of equilibrium. Because the vestibulocerebellar circuitry is associated mainly with axial and girdle muscles, this system seems to be primarily involved in setting and maintaining the posture appropriate for a movement.

The spinocerebellum is involved in the control of distal limb movements (p. 722). The spinocerebellum consists of the intermediate zone of the anterior and posterior lobes plus most of the vermis of the anterior and posterior lobes; it is the portion of the cerebellar cortex that receives the bulk of the ascending spinal cord projections (spinocerebellar and cuneocerebellar tracts), particularly the input from muscle spindles, Golgi tendon organs, and joint capsules. It also receives input from the cerebral cortex via the pontine nuclei, so it has information concerning intended movements as well as ongoing movements.

This part of the cerebellum may be involved in damping movements. For example, when an arm is moved, momentum develops and must be overcome for the movement to stop. When lesions affect the spinocerebellum, overshoot develops; that is, the arm might extend past the target in one direction, and then as a correction is made, the arm may overshoot in the opposite direction. This is sometimes interpreted as an *intention* or *action tremor*.

Very rapid movements, such as the finger movements of a touch-typist, are called *ballistic* movements. This implies that the entire movement is preplanned to go into motion, travel a specific distance, and then come to a stop. Saccadic eye movements are ballistic movements. These types of movements are disrupted when the spinocerebellum is damaged. The movement is slow to be initiated, its force development is weak, and it is slow to be terminated; this results in overshoot or *past-pointing*.

The cerebrocerebellum is involved with the planning, sequencing, and timing of movement (p. 723). The lateral cerebellar hemispheres receive the bulk of their input from the cerebral cortex via the pontine nuclei and essentially do not receive any projections directly from the spinal cord. The "plan" of an intended, sequential movement is thought to be transmitted from the premotor and sensory cortex to the basilar pons and then to the cerebellar nuclei and cortex of the lateral hemisphere. Interestingly, it has been reported that activity in the dentate nucleus reflects the movement that *will be* performed and not the ongoing movement.

When the lateral hemisphere is damaged, the timing of sequential movements is lost; that is, a succeeding movement may begin too early or too late, and complex movements such as writing or running are uncoordinated and do not progress in an orderly sequence from one movement to the next. The timing function involved in estimating the progression of auditory and visual phenomena may also be disrupted. For example, an individual can lose the ability to predict on the basis of sound or sight how rapidly an object is approaching.

Clinical Abnormalities of the Cerebellum (p. 724)

Dysmetria and ataxia—Movements that overshoot or undershoot the intended target; the effect is called *dysmetria*, and the abnormal movements are described as ataxic.

Past pointing—Failure of movement signal to be terminated at the proper time; the limb continues past or beyond its intended target.

Dysdiadochokinesia—Inability to perform rapid alternating movements; the switch that shifts from flexion to extension (or vice versa) is not timed properly.

Dysarthria—Speech defect that involves inappropriate progression from one syllable to the next; this is slurred speech in which some syllables are held and others are dropped too quickly.

Intention tremor—A type of tremor that is present only when a voluntary movement is attempted and that intensifies as the limb nears its target.

Cerebellar nystagmus—In effect, a tremor of the eyes when attempting to fixate on a point in the periphery of the visual field.

Hypotonia—Decreased muscle tone in the affected muscles, accompanied by diminished reflexes.

BASAL GANGLIA AND THEIR MOTOR FUNCTIONS
(p. 725)

The term *basal ganglia* refers to the combination of brain region that includes the *caudate nucleus, putamen, globus pallidus, substantia nigra*, and *subthalamic nucleus*. These structures are located deep within the core of each cerebral hemisphere.

The Role of the Basal Ganglia in Executing Patterns of Motor Activity

The circuits that interconnect the structures composing the basal ganglia are intricate and extremely complex. These connections are schematically represented in Figure 56–2.

In general, functions that involve movement are primarily linked to the putamen rather than to the

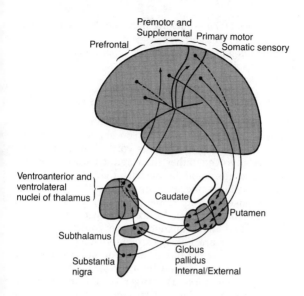

The Putamen Circuit

Figure 56-2 Putamen circuit through the basal ganglia for subconscious execution of learned patterns of movement.

caudate nucleus. Signals initiated in the premotor and supplementary cortex are transmitted to the putamen and then onto the globus pallidus. The latter structure has internal and external subdivisions that project to different locations. The external segment is reciprocally linked with the subthalamic nucleus, whereas the internal segment projects to the thalamus and substantia nigra. Motor nuclei in the thalamus that receive pallidal input project back to premotor and primary motor regions of the cortex. In addition to complex connectivity, the basal ganglia synaptic milieu contain an unusually diverse variety of neurotransmitter agents, and individual neurons of the putamen and caudate may actually express more than one neurotransmitter agent. Consequently, lesions of the basal ganglia give rise to a wide variety of clinical signs and symptoms, as follows:

Globus pallidus lesion—Writhing movements of the hand and arm or face, called *athetosis*
Subthalamic lesion—Flailing movements of an extremity, called *hemiballismus*

Putamen lesion—Flicking movements of the hands or face, called *chorea*

Substantia nigra dopamine cell degeneration—Parkinson's disease.

Cognitive Control of Motor Patterns (p. 726)

Like the putamen, the caudate nucleus receives dense projections from the cerebral cortex; in this case, however, primarily the cortical association areas are involved rather than motor cortex. The output from the caudate nucleus that is sent to the globus pallidus internal segment and thalamus eventually makes its way to the prefrontal, premotor, and supplementary motor cortices; thus, it appears that the caudate may function in the control of motor patterns that are linked to memory of previous experience. An example is a situation in which an individual is confronted by a threat. First, he recognizes the situation as dangerous on the basis of prior experience. Then, a judgment is made to take some course of action in response to the circumstances. When judgment or memory of past experience is associated with movement, it is likely that circuits through the caudate nucleus are involved in controlling the actions.

Basal Ganglia and the Timing and Scale of the Intensity of Movement (p. 727)

Two important parameters of any movement are the *speed* and *size* of the movement; these are called *timing* and *scaling* functions. Both of these features are disrupted in patients who have basal ganglia lesions, particularly those with lesions that involve the caudate nucleus. This correlates well with the fact that the posterior parietal cortex (especially in the nondominant hemisphere) is the locus for the spatial coordinates of the body and its relationship with the external environment. This part of the cortex projects heavily to the caudate nucleus.

Parkinson's Disease and Huntington's Disease and Basal Ganglia Lesions (p. 728)

Parkinson's disease may be caused by the loss of dopamine-secreting nerve fibers. Parkinson's disease is characterized by (1) the presence of *rigidity* in many muscle groups, (2) a *tremor present at rest* when no voluntary movement is under way, and (3) difficulty in initiating movement (referred to as *akinesia*). Much of this symptomatology is thought to be linked to a progressive loss of the dopamine-producing cells in the substantia nigra. These neurons are known to project diffusely throughout the caudate and putamen, and the severity of symptoms seems to be proportional to the degree of cell loss in the substantia nigra. The question of why these neurons degenerate remains unanswered at present.

There are several methods for treatment of Parkinson's disease. Because the cell loss results in diminished levels of dopamine, *a dopamine precursor, L-DOPA*, can be administered to increase dopamine availability. This substance will cross the blood-brain barrier, but dopamine itself will not. There are two major problems with this treatment: (1) not all L-DOPA consistently reaches the brain because tissues outside the central nervous system are capable of producing dopamine, and (2) as more and more neurons degenerate in the substantia nigra, the required dosage of L-DOPA changes.

- *L-Deprenyl* is an inhibitor of monoamine oxidase, the substance that breaks down dopamine after its release in the brain. It also appears to slow the degeneration of substantia nigra cells; it can be combined with L-DOPA to increase the availability of dopamine.
- *Transplantation of fetal substantia nigra neurons into the caudate and putamen* has been tried in an attempt to increase dopamine levels, but it has shown only limited success. The transplanted cells remain viable for only a few months, and the use of aborted fetal tissue creates a potential ethical dilemma. Cultured cell lines (e.g., fibroblasts) that have been genetically altered to produce dopamine are beginning to show promise as a fetal transplantation alternative.
- A procedure called *pallidotomy* is also beginning to show positive results. It has been reasoned that the motor deficits seen in patients with Parkinson's disease are the result of abnormal signals transmitted from the globus pallidus to the thalamus. Although the direct effects of dopamine loss appear to be restricted to the caudate and putamen, the output of the latter cell groups in the form of axons projecting to the globus pallidus is

still functional but presumably altered in a major way. One approach has been to position an electrode in the globus pallidus near its output pathways and make a destructive lesion that interrupts the projection to the thalamus. Surprisingly, this is not a technically difficult surgical procedure, and the results thus far look very good. One slight modification has been tried; this is to implant a stimulating electrode in the globus pallidus rather than making a destructive lesion. When activated by the patient, signals are generated by the electrode that disrupt the flow of impulses from the pallidum to the thalamus; the effect is much the same as a lesion.

Huntington's disease is a genetically transmitted disorder (autosomal dominant). Typically Huntington's disease does not appear until the fourth or fifth decade of life. It is characterized by choreiform (flicking) movements at certain joints that gradually progress to the point of involving much of the body. Severe dementia also gradually appears in tandem with the motor deficits. The neural substrate for this disorder is less well understood than that of Parkinson's disease. It is thought to involve a loss of GABA neurons in the caudate and putamen and perhaps a loss of acetylcholine neurons in several parts of the brain, including the cerebral cortex. The gene responsible for this defect has been isolated and traced to the short arm of chromosome 4. This determination should eventually facilitate the development of gene therapy for this disorder.

SUMMARY OF THE ENTIRE MOTOR CONTROL SYSTEM
(p. 729)

- *Spinal cord level*—Organized in the spinal cord are patterns of movement that involve nearly all muscles in the body. These range from the relatively simple withdrawal reflex to coordinated movement of all four extremities.
- *Brain stem (hindbrain) level*—With regard to somatomotor function, neurons in the brain stem play a major role in the control of reflexive eye movements that involve the vestibular sensory apparatus. In addition, the brain stem mediates control over posture and balance as influenced by the vestibular system and plays a major role in regulating muscle tone via gamma motor neurons.

- *Corticospinal system*—The output of the motor cortex is delivered to the spinal cord over this vast network of fibers. In general, the motor areas of the cortex can devise a unique and specific motor program that is then sent to the spinal cord, and various muscle groups are activated; alternatively, the cortex may select from among the set of motor patterns defined by intrinsic spinal cord circuitry.
- *Cerebellum*—The cerebellum functions at several levels within the motor control hierarchy. At the spinal cord level, it can facilitate stretch reflexes, so that the ability to manage an unexpected load change or perturbation is enhanced. In the brain stem, the cerebellum is interconnected with the vestibular system to aid in the regulation of posture, equilibrium, and eye movements. The output of the cerebellum is directed primarily to the thalamus and cerebral cortex to provide accessory motor commands or to program in advance the progression from a rapid movement in one direction to a rapid movement in the opposite direction.
- *Basal ganglia*—These neurons and associated cell groups function with motor areas of the cortex to control learned patterns of movement and multiple sequential movements designed to accomplish self-generated or internally guided tasks. Included in this function are the modifications to the motor program needed to regulate the speed and size of the movement—the timing and scaling functions.

57

The Cerebral Cortex; Intellectual Functions of the Brain; and Learning and Memory

PHYSIOLOGIC ANATOMY OF THE CEREBRAL CORTEX
(p. 733)

The cerebral cortex consists of a relatively thin layer of neurons ranging from 2 to 5 millimeters in thickness with a total surface area of approximately one fourth of a square meter and containing about 100 billion neurons.

Most cortical neurons fall into one of three categories: (1) *granular* (or *stellate*), (2) *fusiform*, and (3) *pyramidal*. The granule cells are short-axon, local circuit neurons that use *glutamate (excitatory)* or *GABA (inhibitory)* as neurotransmitters. In contrast, fusiform and pyramidal neurons have long axons that project at some distance from the cortex. Fusiform cells project to the thalamus, whereas pyramidal neurons project to other locations in the same or opposite hemisphere and to a variety of subcortical locations, such as the red nucleus, basilar pons, and spinal cord.

The neurons of the cerebral cortex are organized into six horizontal layers. Layer IV receives incoming sensory signals from the thalamus, whereas neurons in layer V give rise to long subcortical projections to the brain stem and spinal cord. Corticothalamic fibers originate from cells in layer VI. The corticothalamic interconnections are most significant because damage to the cortex alone seems to result in less dysfunction than occurs when both cortex and thalamus are damaged. Layers I, II, and III are specialized to receive input from and project to other parts of the cortex in the same or opposite hemispheres.

FUNCTIONS OF SPECIFIC CORTICAL AREAS (p. 734)

Studies have clearly shown that many areas of the cerebral cortex are specialized for specific functions. Some areas, called *primary cortex*, have direct connections with the spinal cord for the control of movement, whereas other primary regions receive sensory input from various thalamic nuclei that represent each of the special senses (except olfaction) and somatosensation. Secondary cortical areas are called *association cortex*, and they serve to interconnect various portions of the cortex in the same or opposite hemisphere.

Association Areas of Cortex (p. 735)

- *Parieto-occipito-temporal area* includes (1) the postero-parietal area that contains the spatial coordinates for all parts of the contralateral side of the body as well as all contralateral extrapersonal space; (2) the area for language comprehension called *Wernicke's area*, which lies in the superior temporal gyrus; (3) the area for the initial processing of visual language (reading) in the angular gyrus; and (4) the area for naming objects located in the anterior part of the occipital lobe.
- *Prefrontal association area* functions in close relation with motor areas of the frontal lobe to plan complex patterns and sequences of movement. Much of its input comes from the parieto-occipito-temporal association cortex, whereas its principal output is sent to the caudate nucleus for additional processing. It is also involved in nonmotor functions, which include memory-related transformations related to problem solving and other internally guided behavior. It contains one specialized region, *Broca's area*, that is involved in the motor aspects of speech and receives its input from *Wernicke's area* in the temporal lobe. Broca's area provides output to the nearby motor cortex that controls the muscles required for speech production.
- *Limbic association cortex* includes the anterior pole of the temporal lobe, the ventral aspect of the frontal lobe, and a portion of the cingulate cortex. It is involved with the complex processes of emotional and motivational behavior and is connected with limbic system structures, such as the hypothalamus, amygdala, and hippocampus.
- *Facial recognition area* is located on the ventromedial surfaces of the occipital and temporal lobes.

Concept of the Dominant Hemisphere (p. 737)

The interpretive functions of Wernicke's area, the angular gyri and frontal motor speech areas are more highly developed in one hemisphere—the dominant hemisphere. In approximately 95 per cent of all individuals, the left hemisphere is dominant regardless of handedness. How one hemisphere comes to be dominant is not yet understood.

Wernicke's area is often assigned a *general interpretive function* because damage to this area results in the inability to understand spoken or written language even though the individual has no hearing deficit and may be able to read the words on a page. Likewise, damage to the angular gyrus (with Wernicke's area intact) may leave undamaged the ability to understand spoken language, but the ability to comprehend *written* words is lost; this is called *word blindness*.

Interestingly, the area in the nondominant hemisphere that corresponds to *Wernicke's area* is also involved in language function. It is responsible for understanding the emotional content or intonation of spoken language. In a way, it also is "dominant" for a particular function.

Higher Intellectual Functions of the Prefrontal Association Cortex (p. 738)

The function of the prefrontal cortex is complex and multifactorial and is typically explained by describing the deficits seen in individuals with large lesions in this cortex.

- *Decreased aggressiveness and inappropriate social responses*—This is most apparent when lesions involve the ventral aspect of the prefrontal cortex, the limbic association area.
- *Inability to progress toward goals or carry through sequential thoughts*—Prefrontal cortex gathers information from widespread areas of the brain to develop solutions to problems, whether they require motor or nonmotor responses. Without this function, thoughts lose their logical progression, and the individual loses the ability to focus attention and becomes highly distractible.
- *The prefrontal cortex is the site of "working memory"*—The ability to hold and sort bits of information to be used in a problem-solving function is described

as "working memory." By combining these stored bits of information, we can prognosticate, plan for the future, delay a response while further information is gathered, consider the consequences of actions before they are performed, correlate information from many different sources, and control actions in accordance with societal or moral laws. All of these are considered intellectual functions of the highest order and seem to be definitive for the human experience.

Function of the Brain in Communication—Language Input and Output (p. 739)

There are two aspects to communication: language input (the sensory aspect) and language output (the motor aspect). Some individuals are capable of hearing or identifying written or spoken words but do not comprehend the meaning of the words. This is the result of a lesion in Wernicke's area; the condition is known as *receptive* or *sensory aphasia* and may simply be called *Wernicke's aphasia*. If the lesion extends beyond the confines of Wernicke's area, a total inability to use language communication ensues; this is termed *global aphasia*.

If an individual is able to formulate verbal language in his or her mind but cannot vocalize the response, the condition is called *motor aphasia*. This indicates a lesion involving Broca's area in the frontal lobe; the condition can also be referred to as *Broca's aphasia*. The defect is not in the control of the musculature needed for speech but rather in the elaboration of the complex patterns of neural and muscle activation that in effect define the motor aspects of language.

Function of the Corpus Callosum and Anterior Commissure in the Transfer of Information Between Hemispheres (p. 741)

The corpus callosum provides abundant interconnections for most areas of the cerebral hemispheres except for the anterior portions of the temporal lobes, which are connected via the anterior commissure. Some of the more important functional connections mediated by these two fiber bundles are listed as follows:

- The corpus callosum allows Wernicke's area in the left hemisphere to communicate with the motor cortex in the right hemisphere. In the absence of this connection, voluntary movement of the left side of the body to a communicated command is not possible.
- Visual and somatosensory information from the left side of the body reaches the right hemisphere. Without a corpus callosum, this sensory information cannot extend to Wernicke's area in the left hemisphere; as a result, such information cannot be used for processing by Wernicke's area, and the left body and left visual field are ignored.
- Without a corpus callosum, only the left half of the brain can understand both the written and spoken word. The right side of the brain can comprehend only the written word—not verbal language. Emotional responses, however, can involve both sides of the brain (and body) if the anterior commissure is intact.

THOUGHTS, CONSCIOUSNESS, AND MEMORY (p. 742)

The neural substrates for these three processes are very poorly understood at present. The *holistic theory* suggests that a thought results from patterned stimulation of the cerebral cortex, thalamus, and limbic system; each of these areas contributes its particular character or quality to the process.

Memory—The Role of Synaptic Facilitation and Inhibition

Memories derive from the changes in synaptic transmission between neurons that occur as a result of previous neural activity. These changes cause new, facilitated, or inhibited pathways to develop in the appropriate neural circuitry. The new or altered pathways are called *memory traces*. Although we think of memories as positive collections of previous experiences, probably many are in a sense negative memories. Our minds are inundated with sensory information, and an important brain function is the capability to ignore irrelevant or extraneous information. This process is called *habituation*. Conversely, the brain also has the capacity to enhance or store certain memory traces through *facilitation* of synaptic circuits; this mechanism is referred to as *memory sensitization*.

It is obvious that some memories last for only a few seconds, whereas others last for hours, days, months, or years. Consequently, three categories of memories have been described: (1) short-term memory, which lasts for only seconds or minutes unless converted into long-term memory; (2) intermediate long-term memory, which lasts for days to weeks but is eventually lost; and (3) long-term memory, which once stored, can be recalled years later or for a lifetime.

Short-Term Memory (p. 743). Short-term memory is typified by the memory of a new telephone number that is recalled for a few seconds or minutes as one continues to think about the number. Several theories concerning the substrate for this mechanism are under investigation: (1) this type of memory is due to continuous neural activity in a reverberating circuit; (2) it occurs as a result of activation of synapses on presynaptic terminals that typically result in prolonged facilitation or inhibition; and (3) the accumulation of calcium in axon terminals may eventually lead to enhanced synaptic output from that terminal.

Intermediate Long-Term Memory. This type of memory can result from temporary chemical or physical changes in either the presynaptic or postsynaptic membrane that can persist for a few minutes to several weeks. Some experimental observations on such mechanisms have come from studies in the snail *Aplysia*, as shown in Figure 57–1. Stimulation of a facilitator terminal at the same time as activation of another

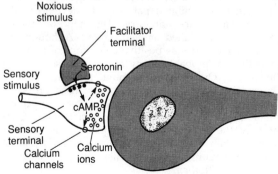

Figure 57–1 Memory system that has been discovered in the snail *Aplysia*.

sensory input causes serotonin to be released at synaptic sites on the sensory terminal. Stimulation of serotonin receptors activates adenyl cyclase in the main sensory terminal, resulting in the formation of cyclic AMP, which causes the release of a protein kinase and leads to phosphorylation of a protein that blocks potassium channels in the sensory terminal. Decreased potassium conductance causes a prolongation of action potentials that reach the sensory terminal; this in turn allows increased calcium to enter the sensory terminal, resulting in increased neurotransmitter release from the sensory terminal and thereby facilitating transmission at this synapse.

Long-Term Memory. Long-term memory is thought to result from *structural changes* at the synapse that enhance or suppress signal conduction. These structural changes include (1) an increase in the number of synaptic vesicle release sites, (2) an increase in the number of available synaptic vesicles, (3) an increase in the number of synaptic terminals, and (4) changes in the shape or number of postsynaptic spines.

Consolidation of Memory (p. 745)

For memories to be converted to long-term memory, they must be *consolidated*; that is, they must initiate the chemical or structural changes that underlie the formation of a long-term memory. In general, 5 to 10 minutes is required for minimal consolidation, whereas 1 hour or longer may be needed for strong consolidation. The mechanism of *rehearsal* is thought to represent the consolidation process.

Rehearsal of the same information again and again in the mind potentiates the transfer from short- to long-term memory. Over time, the important features of sensory experience become progressively more fixed in memory stores. During consolidation, memories are codified into different classes of information. For example, new and old experiences relative to a topic are compared for similarities and differences, and it is the latter information that is stored.

Roles of Specific Brain Parts in the Memory Process
(p. 746)

Lesions of the hippocampus lead to *anterograde amnesia*, or the inability to form or store *new* memories.

Memories formed before the onset of the lesion are not affected; this appears to be because the hippocampus (and the dorsomedial thalamic nucleus) is connected to the *punishment and reward* centers. That is, our experiences may be associated in the hippocampus with pleasure or punishment, and that is the substrate for initiation of the memory process. The loss of long-term memory occurs with thalamic lesions and, in some instances, with damage to the hippocampus. The hypothesis is that the thalamus may be a part of the mechanism that searches the memory stores and "reads" them. Interestingly, individuals with hippocampal lesions do not have difficulty in learning physical skills that only require manual repetition and do not involve verbalization or other types of symbolic higher-order intelligence. This suggests that memory mechanisms for different functions are distributed in more than one brain location.

58

Behavioral and Motivational Mechanisms of the Brain — Limbic System and Hypothalamus

ACTIVATING-DRIVING SYSTEMS OF THE BRAIN (p. 749)

Signals from the brain stem activate the cerebrum in two ways: (1) by stimulating the background level of activity throughout wide areas of the brain and (2) by activating neurohumoral systems that release specific facilitatory or inhibitory hormone–like neurotransmitters into selected areas of the brain.

Control of Cerebral Activity by Continuous Excitatory Signals from the Brain Stem

A *reticular excitatory area* is located in the reticular formation of the pons and midbrain. It has descending spinal projections to the spinal cord that provide an excitatory influence on motor neurons that innervate antigravity musculature. This same reticular area also sends fibers rostrally to a variety of locations, including the thalamus, where in turn neurons distribute to all regions of the cerebral cortex.

The signals that reach the thalamus are of two types. One type arises from the large cholinergic reticular neurons, is rapidly transmitted, and excites the cerebrum for only a few milliseconds. The second type of signal originates from the small reticular neurons, which generate relatively slow action potentials that terminate mainly in the intralaminar and reticular nuclei of the thalamus. The excitatory signal from the latter input builds up slowly and produces a widespread effect that presumably controls the background level of excitability of neurons in the cortex.

The level of activity in the reticular excitatory area is determined largely by input from the various ascending somatosensory pathways—the pain pathway, in particular. This was deduced from animal experiments in which the brain stem was transected just rostral to the entry of the trigeminal nerve. This effectively eliminated all ascending somatosensory input, and the excitatory reticular area went silent as the animal entered a coma-like state. Curiously, the cortex also provides descending excitatory input to the excitatory reticular area, which serves as positive feedback and allows cerebral activity to reinforce the action of the ascending reticular system. The thalamus and cortex are linked by reciprocal connections. Part of the "thinking" process may involve memory formation resulting from the back-and-forth signal transfer between the thalamus and cortex.

The lower brain stem in the ventromedial medulla contains a *reticular inhibitory area.* Like the more rostral excitatory reticular area, this region provides descending spinal projections, and these fibers inhibit the activity of antigravity muscles. Similarly, the inhibitory reticular projects rostrally to decrease the excitatory levels of the cerebrum through serotonergic systems (discussed later).

NEUROHORMONAL CONTROL OF BRAIN ACTIVITY
(p. 751)

A second method for altering the background level of activity in the cerebrum involves projections from cell groups that use excitatory or inhibitory neurotransmitter agents that function similar to hormones; these three agents are *norepinephrine, dopamine,* and *serotonin.*

- The *norepinephrine system* originates from the neurons of the locus ceruleus, a cell group located in the rostral pons and caudal midbrain. These cells have unusually long and diffusely projecting axons that reach throughout many areas of the brain, including the thalamus and cerebral cortex. At most of its synaptic targets, norepinephrine exerts excitatory effects, although in some regions, the receptor to which it binds produces inhibition. Often, the effects of norepinephrine are modulatory. That is, they might not cause the target neuron to generate an action potential but rather raise the excitability level of the cell and make it more likely to fire action potentials in response to subsequent input.

- Neurons in the compact portion (pars compacta) of the substantia nigra, a large cell group in the midbrain, represent the important source of *dopamine fibers* that project rostrally into the cerebrum, where they provide dense input to the caudate and putamen, the *nigrostriatal system*. Dopamine projections produce both excitation and inhibition. Neurons in some basal ganglia circuits exhibit receptors that cause excitatory postsynaptic potentials when they bind dopamine, whereas other receptors in other circuits cause just the opposite effect (inhibition).
- The raphe nuclei are relatively small, thin, discontinuous groups of cells located adjacent to the midline at various levels in the brain stem extending from the midbrain to the medulla. Most (but certainly not all) neurons use *serotonin* as a neurotransmitter, and a large number of the serotonin-producing cells project to the thalamus and cortex. When released in the cortex, serotonin nearly always produces inhibitory effects.

A number of other neurotransmitter systems play important functional roles in the thalamus and cerebral cortex, including the enkephalins and endorphins, GABA, glutamate, vasopressin, adrenocorticotropic hormone, angiotensin II, vasoactive intestinal peptide, and neurotensin.

FUNCTIONAL ANATOMY OF THE LIMBIC SYSTEM
(p. 752)

The limbic system is the combined neuronal circuitry that controls emotional behavior and motivational drives. This large complex of brain structures is composed of subcortical and cortical components. *The subcortical group includes the hypothalamus, septum, parolfactory area, epithalamus, anterior thalamic nucleus, portions of the basal ganglia, hippocampus, and amygdala.* Surrounding the subcortical structures is the limbic cortex composed of the orbitofrontal cortex, subcallosal gyrus, cingulate gyrus, and parahippocampal gyrus. Among the subcortical structures, the hypothalamus is the most important output source; it communicates with brain stem nuclei through the medial forebrain bundle, which conducts signals in two directions: toward the brain stem and back to the forebrain.

Hypothalamus: Major Control Point for the Limbic System
(p. 753)

The influence of the hypothalamus is extended via projections down to the brain stem, up to other portions of the diencephalon and limbic cortex, and to the pituitary gland. The hypothalamus controls (1) vegetative and endocrine functions and (2) behavior and motivation.

Vegetative and Endocrine Control Functions. Various studies have parceled the hypothalamus into a number of different cell groups responsible for certain functions; however, the localization of function is less precise than is suggested by these studies.

- *Cardiovascular regulation* involves control of arterial pressure and heart rate and is focused in general in the posterior and lateral hypothalamic areas, which increase blood pressure and heart rate, or in the preoptic area, which decreases blood pressure and heart rate. These effects are mediated by cardiovascular centers in the pontine and medullary reticular formation.
- *Body temperature regulation* is controlled by neurons in the preoptic area that are able to sense changes in the temperature of blood flowing through the area. Increases or decreases in temperature signal the appropriate cells that activate body temperature-lowering or -elevating mechanisms.
- *Regulation of body water intake* is controlled by mechanisms that create thirst or govern the excretion of water into urine. The thirst center is located in the lateral hypothalamus; when the concentration of electrolyte levels here is elevated, a desire to "drink" is initiated. The supraoptic nucleus is involved with mechanisms that control urinary excretion of water, and neurons in this location release antidiuretic hormone (ADH, or vasopressin) into the posterior pituitary gland, which then enters the circulation and acts on the collecting ducts in the kidney to effect reabsorption of water, causing the urine to become more concentrated.
- *Uterine contraction and milk ejection* are stimulated by *oxytocin*, which is released by neurons of the paraventricular nucleus.

- *Gastrointestinal and feeding regulation* are controlled by several different hypothalamic areas. The lateral hypothalamus causes the desire to seek out food, whereas damage to this area may result in starvation. In comparison, the ventromedial nucleus is called the satiety center because its activity produces a stop-eating signal. The mamillary nuclei are involved in certain reflexes related to food intake, such as lip licking and swallowing.
- *Anterior pituitary gland regulation* is achieved by the elaboration of releasing and inhibitory factors from the hypothalamus that are carried by a portal system to the anterior lobe of the pituitary, where they act on glandular cells that produce the anterior pituitary hormones. The hypothalamic neurons that produce these factors are found in the periventricular zone, arcuate nucleus, and ventromedial nucleus.

Behavioral Control Functions of the Hypothalamus and Associated Limbic Structures. Emotional behavior is affected by stimulation of the hypothalamus or by lesions in the hypothalamus. Stimulation effects include (1) increased general level of activity, leading to rage and aggression; (2) sense of tranquillity, pleasure, and reward; (3) fear and feelings of punishment and aversion; and (4) sexual arousal. Effects caused by hypothalamic lesions include (1) extreme passivity and loss of drives and (2) excessive eating and drinking, rage, and violent behavior.

Reward and Punishment Centers (p. 756)

The major locations that evoke a pleasurable feeling or sense of reward when stimulated are found along the course of the medial forebrain bundle, especially in the lateral and ventromedial hypothalamus. Conversely, areas that evoke aversive behavior when stimulated include the midbrain periaqueductal gray, periventricular zones of the thalamus and hypothalamus, amygdala, and hippocampus.

Rage Behavior (p. 757)

In animals, intense stimulation of aversive centers in the lateral hypothalamus and periventricular zone evokes a rage response. This is characterized by a defense posture, extended claws, elevated tail, hissing and spitting, growling, and piloerection. Normally, the rage reaction is held in check by activity in the ventromedial hypothalamus.

Importance of Reward and Punishment in Shaping Behavior

Much of our daily behavior is controlled by punishment and reward. The administration of tranquilizers inhibits both punishment and reward centers and thereby decreases behavioral affect in general. These drugs are not selective, however, and other hypothalamic functions may be depressed as well, creating potentially harmful side effects. In addition, stimulation that effects either the reward or punishment centers tends to build strong memory traces, and the responses to such stimulation are said to be reinforced. Other kinds of stimulation that are essentially indifferent tend to become habituated.

SPECIFIC FUNCTIONS OF OTHER PARTS OF THE LIMBIC SYSTEM (p. 757)

The Hippocampus. Stimulation of the hippocampus can evoke rage, passivity, and excessive sexual drive. It also is hyperexcitable, and weak stimuli can produce epileptic seizures. Lesions of the hippocampus lead to a profound inability to form new memories based on any type of verbal symbolism (language); this is called *anterograde amnesia*. It is suggested that the hippocampus provides the signal for memory consolidation (e.g., the transformation from short- to long-term memory).

The Amygdala. This large aggregate of cells is located in the medial, anterior pole of the temporal lobe and consists of two subdivisions: a corticomedial nuclear group and a basolateral group of nuclei. The amygdala output is varied and extensive, reaching the cortex, hippocampus, septum, thalamus, and hypothalamus. Stimulation of the amygdala produces changes in heart rate, arterial pressure, gastrointestinal motility, defecation and urination, pupillary dilation, piloerection, and secretion of anterior pituitary hormones. In addition, involuntary movements can be elicited; these include tonic posture, circling movements, clonus, and movements associated with olfaction and eating. Behavior such as rage, escape, and sexual activity can be evoked. Bilateral destruction of the temporal poles leads to the *Klüver-Bucy syndrome*; this includes extreme orality, loss of fear, decreased aggressiveness, tameness, changes in eating behavior, psychic blindness, and excessive sexual drive.

The Limbic Cortex. The discrete contributions of various portions of limbic cortex are currently unknown. Knowledge of their function can be derived from lesions that interrupt portions of the cortex. Bilateral destruction of the anterior temporal cortex leads to the Klüver-Bucy syndrome. Bilateral lesions in the posterior orbitofrontal cortex lead to insomnia and restlessness. Bilateral destruction of the anterior cingulate and subcallosal gyri evokes an extreme rage reaction.

59

States of Brain Activity—Sleep; Brain Waves; Epilepsy; Psychoses

SLEEP (p. 761)

Sleep is defined as a state of unconsciousness from which one can be aroused by sensory stimulation. Investigators believe that there are two entirely different types of sleep: slow-wave sleep and rapid-eye-movement (REM) sleep.

Slow-Wave Sleep. This is the deep, restful type of sleep characterized by decreases in peripheral vascular tone, blood pressure, respiratory rate, and metabolic rate. Dreams can occur in slow-wave sleep, but they are not remembered.

REM Sleep. This is called *paradoxical sleep* because the brain is quite active, and skeletal muscle contractions occur. Typically, REM sleep lasts for 5 to 30 minutes and will repeat at approximately 90-minute intervals. When an individual is extremely tired, REM may be absent but will eventually return as the individual becomes more rested. There are several important features of REM sleep: (1) dreaming occurs, and the dream can be recalled, at least in part; (2) waking a person in REM is more difficult, yet in the morning, we typically awaken during an REM period; (3) muscle tone is substantially depressed; (4) heart rate and respiration become irregular; (5) despite decreased tone, muscle contractions do occur, especially REMs; and (6) brain metabolism is increased by as much as 20 per cent, and on electroencephalography (EEG), brain waves are characteristic of the waking state.

Basic Theories of Sleep (p. 762)

Initially, the *passive theory* of sleep was in favor; it suggested that sleep occurred when the reticular activating system simply fatigued. This concept was called into question by animal experiments that involved transection of the brain stem at midpontine levels that resulted in an animal that never slept. It is now believed that sleep is caused by an *active* mechanism that inhibits other parts of the brain.

Mechanisms that Cause Sleep

Sleep can occur by stimulating any one of three brain locations. The most potent site is the *raphe of the caudal pons and medulla*. Many of the neurons in the raphe nuclei use *serotonin* as a transmitter, and it is known that drugs that block the formation of serotonin prevent sleep. In addition, stimulation in the nucleus of the solitary tract promotes sleep, but this occurs only if the raphe nuclei are also functional. Activation of the *suprachiasmatic level of the hypothalamus* or the *midline nuclei of the thalamus* produces sleep. Some studies, however, have shown that blood levels of serotonin are actually lower during sleep than during wakefulness; this suggests that some other substance is responsible for sleep production. One possibility is *muramyl peptide*, which accumulates in cerebrospinal fluid and urine. When microgram amounts of this substance are injected into the third ventricle, sleep is induced within minutes.

REM sleep is enhanced by cholinergic agonists. It is postulated that certain of the projections of cholinergic neurons of the midbrain reticular formation are responsible for the initiation of REM sleep. These projections would activate only neurons that lead to REM sleep and avoid systems that contribute to the waking state and the reticular activating system.

The Cycle Between Sleep and Wakefulness (p. 763)

At present, the mechanism that accounts for the cycling from wakefulness to sleep and back to wakefulness is poorly understood. One explanation is that the systems that maintain the waking state gradually become fatigued, the positive feedback that helps maintain wakefulness fades, and eventually the neurons of the sleep center are able to overcome the waking center; then, sleep ensues. The cycle is repeated when the sleep centers fatigue. Although not attractive, no other plausible mechanisms have been advanced.

Physiological Effects of Sleep

Prolonged wakefulness (absence of sleep) is associated with sluggishness of thought, irritability, and even psychotic behavior. Sleep in some way must restore or renew the normal balance of activity in many parts of the brain, from the higher intellectual centers of the cortex to the vegetative and behavioral functions of the hypothalamus and limbic system. The specifics of this process are unknown. Sleep deprivation affects other systems in the body that regulate blood pressure, heart rate, peripheral vascular tone, muscle activity, and basal metabolic rate. Again, the mechanisms are not yet defined.

BRAIN WAVES (p. 763)

Electrical potentials that originate near the surface of the brain and are recorded from outside the head are called brain waves, and the recording process is the EEG. The recorded potentials range from 0 to 200 microvolts, and their frequency ranges from once every few seconds to 50 or more per second. Distinct wave patterns can appear; some are characteristic for specific brain abnormalities. Four major brain wave patterns have been described: *alpha, beta, theta,* and *delta waves.*

Origin of Brain Waves

Many neurons (thousands to millions) must fire synchronously to be recorded through the skull. In fact, strong, nonsynchronous signals often nullify themselves because they are of opposite polarity. This makes the EEG a relatively crude electrical signal.

- *Alpha waves*—These are rhythmical waves with a frequency of 8 to 12 Hz at about 50 microvolts; they are found in normal, awake, but resting (eyes closed) individuals.
- *Beta waves*—When the eyes are opened in the light, slightly higher frequency (14 to 80 Hz) *beta waves* appear, with voltages below 50 microvolts. For these waves to be recorded, thalamocortical projections must be intact; presumably, the ascending reticular input to the thalamus also must be functional.
- *Theta waves*—These waves have frequencies in the range of 4 to 7 Hz and occur mainly in the parietal and temporal areas in children, but they can

appear in adults during a period of emotional stress. They also appear in association with brain disorders and degenerative brain states.

- *Delta waves*—These are all the waves below 3.5 Hz and occur during deep sleep, in serious organic brain disease, and in infants. It appears that they persist in the absence of cortical input from the thalamus and lower brain centers. Because they can be seen during slow-wave sleep, this sleep state is probably due to release of the cortex from the influence of lower centers.

Electroencephalographic Changes in the Different Stages of Wakefulness and Sleep

As one progresses from alert wakefulness to deep sleep, there is a gradual change in brain wave pattern from low-voltage, high-frequency waves (alpha) to high-voltage, low-frequency waves (delta). These changes can also be described as a progression from *desynchronized* activity (alert) to *synchronous* patterns (deep sleep). REM sleep is again paradoxical because it is a sleep state, yet the brain exhibits asynchronous activity characteristic of the waking state.

EPILEPSY (p. 765)

Epilepsy is characterized by uncontrolled, excessive activity in the nervous system, which is termed a *seizure*. Three types of epilepsy are typically described: *grand mal*, *petit mal*, and focal.

- *Grand mal epilepsy*—This is the most severe variety and seems to be the result of intense discharges in many parts of the brain, including the cortex, thalamus, and brain stem. Initially, generalized tonic seizures affect much of the body; these are followed by alternating tonic-clonic seizures. This activity may persist for 3 to 4 minutes and is followed by a postseizure depression of the nervous system, which can leave the individual stuporous, sleepy, and fatigued for several hours. EEG activity during a seizure of this type shows very characteristic high-voltage, high-frequency patterns. Grand mal seizures can be precipitated in susceptible individuals by (1) strong emotional stimuli, (2) alkalosis caused by hyperventilation, (3) drugs, (4) fever, and (5) loud noise or flashing light. In addition, brain trauma and tumors can lead to seizure activity. It is said that grand mal seizures occur in individuals predisposed to abnormal electrogenic circuitry in the brain, which may be inherited.

- *Petit mal epilepsy*—This is less severe seizure activity in which the individual loses consciousness for 3 to 30 seconds and exhibits small twitching of muscles around the head or face, especially blinking of the eyes. This is also called an *absence seizure* or syndrome, and such activity is thought to be limited to abnormal function in the thalamocortical system. On occasion, a petit mal attack will progress to a grand mal seizure.

- *Focal epilepsy*—This type of seizure activity can involve almost any part of the brain and nearly always is caused by some local abnormality such as scar tissue formation, tumor, ischemia, or congenital abnormality. The typical presentation is a focal muscle twitching that progresses to adjacent body parts. The EEG can often be used to locate the initial focus of abnormal brain activity so it can be surgically removed.

PSYCHOTIC BEHAVIOR AND DEMENTIA (p. 767)

Depression and Manic-Depressive Psychoses

These disorders might be the result of decreased production of norepinephrine, serotonin, or both. Drugs that increase the excitatory effects of norepinephrine are effective in treating depression; these include monoamine oxidase inhibitors, tricyclic antidepressants, and drugs that enhance the action of serotonin. Manic-depressive conditions (bipolar disorder) can be effectively treated with lithium compounds that diminish the actions of norepinephrine and serotonin.

Schizophrenia

There are three possible explanations for this disorder, which is manifest by individuals who hear voices and have delusions of grandeur, intense fear, or paranoia. These are (1) abnormal circuitry in the prefrontal cortex, (2) excessive activity by dopamine systems that project to the cortex, and (3) abnormal function of limbic circuitry related to the hippocampus. The excessive dopamine output theory involves midbrain dopamine neurons (mesolimbic dopamine system) that are separate from those in the substantia nigra, which are related to Parkinson's disease. Evidence supporting this theory derives from the fact that schizophrenic symptoms are reduced by drugs such as chlorpromazine and haloperidol that diminish dopamine release at axon terminals.

Alzheimer's Disease

This disease of the elderly is characterized by the accumulation of *amyloid plaques* in widespread areas of the brain, including the cerebral cortex, hippocampus, and basal ganglia. The severe dementia that ensues may be related to the widespread loss of cholinergic input to the cerebral cortex resulting from the loss of neurons in the basal nucleus of Meynert. Many patients also exhibit a genetic abnormality involving apolipoprotein E, a protein that transports cholesterol.

60

Autonomic Nervous System; Adrenal Medulla

The portion of the nervous system that controls the visceral functions of the body is called the *autonomic nervous system*. This system acts rapidly to control arterial pressure, gastrointestinal motility and secretion, urinary bladder emptying, sweating, body temperature, and many other activities.

GENERAL ORGANIZATION OF THE AUTONOMIC NERVOUS SYSTEM (p. 769)

The central portions of the autonomic nervous system are located in the *hypothalamus, brain stem,* and *spinal cord.* Higher brain centers, such as the *limbic cortex* and portions of the *cerebral cortex,* can influence the activity of the autonomic nervous system by sending signals to the hypothalamus and lower brain areas.

The autonomic nervous system is a motor system for the visceral organs, blood vessels, and secretory glands. The cell body of the *preganglionic neuron* is located in either the brain stem or spinal cord. The axon of this visceral motor neuron projects as a thinly myelinated preganglionic fiber to an *autonomic ganglion.* The postganglionic neuron has its cell body in the ganglia and sends an unmyelinated axon, the *postganglionic fiber,* to visceral effector cells.

In general, sympathetic ganglia are located close to the central nervous system, whereas parasympathetic ganglia are located close to the effector tissues. Sympathetic pathways have short preganglionic fibers and long postganglionic fibers, whereas parasympathetic pathways have long preganglionic fibers and short postganglionic fibers.

523

Physiologic Anatomy of the Sympathetic Nervous System

In the sympathetic division of the autonomic nervous system, visceral motor neurons are located in the *intermediolateral horn* of the spinal cord from level T-1 to L-2. The axons of these motor neurons leave the spinal cord via the *ventral root*. From here, the axon can take one of three paths:

1. It can enter the *sympathetic chain* via the *white ramus* and terminate at its level of origin.
2. It can enter the sympathetic chain via the white ramus and ascend or descend before terminating in the sympathetic chain at a different level.
3. It can enter the sympathetic chain through the white ramus and exit without synapsing via a *splanchnic nerve* and terminate in a *prevertebral ganglion*.

The postganglionic neuron originates in one of the sympathetic chain ganglia or prevertebral ganglia. From either source, the postganglionic fibers travel to their destinations.

Preganglionic sympathetic nerve fibers pass all the way to the adrenal medulla without synapsing. Preganglionic sympathetic nerve fibers that innervate the adrenal medulla originate in the intermediolateral horn of the spinal cord and pass through the sympathetic chains and splanchnic nerves to reach the adrenal medulla, where they end directly on modified neuronal cells that secrete epinephrine and norepinephrine into the blood stream. The secretory cells of the adrenal medulla are derived embryologically from nervous tissue and are analogous to postganglionic neurons.

Physiologic Anatomy of the Parasympathetic Nervous System

In the parasympathetic division of the autonomic nervous system, visceral motor neurons are located in discrete brain stem nuclei or in sacral spinal cord segments 2 to 4. The axons of these motor neurons leave the brain stem via *cranial nerves III, VII, IX,* and *X* or leave the sacral spinal cord via the *pelvic nerves*.

Parasympathetic fibers in the *third cranial nerve* travel to the pupillary sphincters and ciliary muscles of the eye. Fibers from the *seventh cranial nerve* travel to the lacrimal, nasal, and submandibular glands, and fibers from the *ninth cranial nerve* travel to the parotid gland. About 75 per cent of all parasympathetic nerve fibers are located in the *tenth cranial nerve,* the *vagus nerve.* The vagus nerve supplies parasympathetic input to the heart, lungs, esophagus, stomach, small intestine, proximal half of the colon, liver, gallbladder, pancreas, and upper portions of the ureters.

The *sacral parasympathetic fibers* distribute their fibers to the descending colon, rectum, and bladder, and the lower portions of the ureters and external genitalia.

BASIC CHARACTERISTICS OF SYMPATHETIC AND PARASYMPATHETIC FUNCTION (p. 771)

The two primary neurotransmitter substances of the autonomic nervous system are *acetylcholine* and *norepinephrine.* Autonomic neurons that secrete acetylcholine are said to be *cholinergic;* those that secrete norepinephrine are said to be *adrenergic.* All preganglionic neurons in both the sympathetic and parasympathetic divisions of the autonomic nervous system are cholinergic. Acetylcholine and acetylcholine-like substances will therefore excite both the sympathetic and parasympathetic postganglionic neurons.

Virtually all postganglionic neurons of the parasympathetic nervous system secrete acetylcholine and are cholinergic. Most postganglionic sympathetic neurons secrete norepinephrine and are adrenergic. A few postganglionic sympathetic nerve fibers, however, are cholinergic. These fibers innervate sweat glands, piloerector muscles, and some blood vessels.

Synthesis and Secretion of Acetylcholine and Norepinephrine by Postganglionic Nerve Endings

Acetylcholine is synthesized in the terminal endings of cholinergic nerve fibers through the combination of *acetyl-CoA* with *choline.* Once released by the cholinergic nerve endings, acetylcholine is rapidly degraded by the enzyme *acetylcholinesterase.*

Norepinephrine and epinephrine are synthesized from the amino acid *tyrosine.* Tyrosine is converted to *DOPA,* which is then converted to *dopamine;* dopamine is subsequently converted to norepinephrine. In the adrenal medulla, this reaction proceeds one step further to transform 80 per cent of the norepinephrine into *epinephrine*. The action of norepinephrine is terminated by reuptake into the adrenergic nerve endings or by diffusion from the nerve endings into the surrounding fluids.

Cholinergic receptors are subdivided into *muscarinic* and *nicotinic* receptors. Muscarinic receptors are found on all effector cells stimulated by the postganglionic neurons of the parasympathetic nervous system as well as those stimulated by the postganglionic cholinergic neurons of the sympathetic nervous system. Nicotinic receptors are found in the synapses betwen the preganglionic and postganglionic neurons of both the sympathetic and parasympathetic nervous systems as well as in the skeletal muscle neuromuscular junction.

Adrenergic receptors are subdivided into *alpha* and *beta* receptors. Norepinephrine and epinephrine have somewhat different affinities for the alpha and beta receptors. Norepinephrine excites mainly alpha receptors, but it excites beta receptors to a lesser extent. Epinephrine excites both types of receptors approximately equally. The relative effects of norepinephrine and epinephrine on different effector organs are thereby determined by the types of receptors located on these organs.

The stimulation of alpha receptors results in vasoconstriction, dilation of the iris, contraction of the intestinal and bladder sphincters, and contraction of the pilomotor muscles.

The beta receptor is subdivided into $beta_1$ and $beta_2$ receptor subtypes. The stimulation of $beta_1$ receptors causes an increase in heart rate and strength of contraction. The stimulation of $beta_2$ receptors causes skeletal muscle vasodilation, bronchodilation, uterine relaxation, calorigenesis, and glycogenolysis.

Excitatory and Inhibitory Actions of Sympathetic and Parasympathetic Stimulation (p. 774)

Sympathetic stimulation causes excitatory effects in some organs but inhibitory effects in others. Likewise, parasympathetic stimulation causes excitation in some organs but inhibition in others. Occasionally, the two divisions of the autonomic nervous system act reciprocally in an organ, with one system causing an increase in activity and the other system causing a decrease in activity. Most organs, however, are dominantly controlled by one of the two systems.

Effects of Sympathetic and Parasympathetic Stimulation on Specific Organs (p. 774)

Eyes. Two functions of the eyes are controlled by the autonomic nervous system: *pupillary opening* and *focusing of the lens.* Sympathetic stimulation contracts the *radial dilator muscle* of the iris, resulting in pupillary dilation, whereas parasympathetic stimulation contracts the *sphincter muscle* of the iris, resulting in pupillary constriction. Focusing of the lens is controlled almost entirely by the parasympathetic nervous system. Parasympathetic excitation contracts the ciliary muscle, which releases the tension on the suspensory ligament of the lens and allows it to become more convex. This change allows the eye to focus on close objects.

Glands of the Body. The *nasal, lacrimal, salivary,* and *gastrointestinal glands* are strongly stimulated by the parasympathetic nervous system, resulting in copious quantities of watery secretion. Sympathetic stimulation causes vasoconstriction of blood vessels that supply the glands and in this way often reduces the rate of secretion from these glands. Sympathetic stimulation has a direct effect on glandular cells in causing formation of a concentrated secretion that contains extra enzymes and mucus.

The *sweat glands* secrete large quantities of sweat when the sympathetic nerves are stimulated. Parasympathetic stimulation has no effect on sweat gland secretion. The sympathetic fibers to most sweat glands are cholinergic; almost all other sympathetic fibers are adrenergic.

The *apocrine glands* in the axillae secrete a thick odoriferous secretion as a result of a sympathetic stimulation. These glands do not respond to parasympathetic stimulation. The apocrine glands are controlled by adrenergic fibers rather than by cholinergic fibers.

Gastrointestinal System. Sympathetic and parasympathetic stimulation can affect gastrointestinal activity mainly by increasing or decreasing activity on the *enteric nervous system.* In general, parasympathetic stimulation increases the overall degree of activity of the gastrointestinal tract. Normal function of the gastrointestinal tract is not very dependent on sympathetic stimulation. Strong sympathetic stimulation, however, inhibits peristalsis and increases the tone of the various sphincters in the gastrointestinal tract.

Heart. Sympathetic stimulation increases the rate and strength of heart contractions. Parasympathetic stimulation causes the opposite effect.

Systemic Blood Vessels. Sympathetic stimulation causes vasoconstriction of many of the blood vessels of the body, especially the abdominal viscera and the skin of the limbs.

Arterial Pressure. The arterial pressure is determined by two factors: the propulsion of blood by the heart and the resistance to flow of this blood through the blood vessels. Sympathetic stimulation increases both propulsion by the heart and resistance to flow, which results in an increase in arterial pressure. Parasympathetic stimulation decreases the pumping ability of the heart but has no effect on peripheral vascular resistance. This change results in a slight fall in arterial pressure.

Function of the Adrenal Medulla (p. 776)

Stimulation of the sympathetic nerves to the adrenal medulla causes large quantities of epinephrine and norepinephrine to be released into the circulating blood. About 80 per cent of the secretion from the adrenal medulla is epinephrine, and about 20 per cent is norepinephrine. The effect of epinephrine and norepinephrine that is released from the adrenal medulla lasts 5 to 10 times longer than when they are released by sympathetic neurons because these hormones are slowly removed from the blood.

The circulating norepinephrine causes vasoconstriction, increase in the heart rate and contractility, inhibition of the gastrointestinal tract, and dilation of the pupils. The circulating epinephrine, because of its ability to strongly stimulate the beta receptors, has a greater effect on cardiac performance than does norepinephrine. Epinephrine causes only a weak constriction of the blood vessels in muscles, resulting in a slight rise in arterial pressure but a dramatic rise in cardiac output.

Epinephrine and norepinephrine are always released by the adrenal medulla at the same time that the different organs are directly stimulated by generalized sympathetic activation. This dual mechanism of sympathetic stimulation provides a safety factor to ensure optimal performance when it is needed.

Sympathetic and Parasympathetic "Tone" (p. 777)

The basal rate of activity of the autonomic nervous system is known as *sympathetic* and *parasympathetic tone*. Sympathetic tone and parasympathetic tone allow a single division of the autonomic nervous system to increase or decrease the activity of a visceral organ or to constrict or dilate a vascular bed. Normally, sympathetic tone constricts systemic arterioles to about one half of their maximum diameter, whereas parasympathetic tone maintains normal gastrointestinal motility.

Discrete or Mass Discharges of the Autonomic Nervous System

In some instances, the sympathetic nervous system becomes very active and causes a widespread reaction throughout the body called the *alarm* or *stress response*. At other times, sympathetic activation occurs in isolated areas of the body; for example, local vasodilation and sweating occur in response to a local increase in temperature. The parasympathetic nervous system is usually responsible for highly specific changes in visceral function, such as changes in salivary and gastric secretion or in bladder and rectal emptying.

Widespread activation of the sympathetic nervous system can be brought about by fear, rage, or severe pain. The alarm or stress response that results is often called the *fight or flight reaction*. Widespread sympathetic activation causes increases in arterial pressure, muscle blood flow, metabolic rate, blood glucose concentration, glycogenolysis, and mental alertness and decreases in blood flow to the gastrointestinal tract and kidneys and in coagulation time. These effects allow an individual to perform far more strenuous activity than would otherwise be possible.

PHARMACOLOGY OF THE AUTONOMIC NERVOUS SYSTEM (p. 780)

Drugs That Act on the Adrenergic Effector Organs— Sympathomimetic Drugs

Drugs that act like norepinephrine and epinephrine at the sympathetic nerve terminal are called *sympathomimetic* or *adrenergic drugs.* There are many drugs in this category. These drugs differ from one another in the degree to which they stimulate the various adrenergic receptors and in the duration of action. Most sympathomimetic drugs have a duration of action of 30 minutes to 2 hours, whereas the action of norepinephrine and epinephrine is only 1 to 2 minutes.

The drug *phenylephrine* specifically stimulates alpha receptors. The drug *isoproterenol* stimulates both beta$_1$ and beta$_2$ receptors, and the drug *albuterol* stimulates only beta$_2$ receptors.

Drugs that stimulate the release of norepinephrine from nerve terminals. Certain drugs have an indirect sympathomimetic action by inducing the release of norepinephrine from storage vesicles in sympathetic nerve endings instead of by directly activating adrenergic receptors. The drugs *ephedrine, amphetamine,* and *tyramine* belong to this class of compounds.

Drugs that block adrenergic activity. Adrenergic activity can be blocked at several points in the stimulatory process: (1) the synthesis and storage of norepinephrine in sympathetic nerve endings can be blocked by *reserpine,* (2) the release of norepinephrine from sympathetic terminals can be blocked by *guanethidine,* and (3) the adrenergic receptors can be blocked by *phenoxybenzamine* and *phentolamine,* which block alpha receptors, or by *propranolol,* which blocks both beta$_1$ and beta$_2$ receptors.

Drugs That Act on Cholinergic Effector Organs

Acetylcholine receptors located on the postganglionic nerve cells of both the sympathetic and parasympathetic nervous systems are the *nicotinic type* of acetylcholine receptor, whereas the acetylcholine receptors located on the parasympathetic effector organs are the *muscarinic type* of acetylcholine receptor. Drugs that act like acetylcholine at the effector organs are therefore called *parasympathomimetic* or *mus-*

carinic drugs. *Pilocarpine* acts directly on the muscarinic type of cholinergic receptor. The muscarinic action of the drug also stimulates the cholinergic sympathetic fibers that innervate sweat glands, resulting in profuse sweating.

Drugs that prolong the activity of acetylcholine. Some drugs do not have a direct effect on the cholinergic receptors but rather prolong the action of acetylcholine by blocking *acetylcholinesterase;* examples of these drugs include *neostigmine, pyridostigmine,* and *ambenonium.*

Drugs that block cholinergic activity. Drugs that block the effect of acetylcholine on the muscarinic type of cholinergic receptors are called *antimuscarinic drugs.* These drugs include *atropine, homatropine,* and *scopolamine.* These drugs do not affect the nicotinic action of acetylcholine on the postganglionic neurons or skeletal muscle.

Drugs That Stimulate or Block Sympathetic and Parasympathetic Postganglionic Neurons

All postganglionic sympathetic and parasympathetic neurons contain the nicotinic type of acetylcholine receptor. Drugs that stimulate the postganglionic neurons in the same manner as acetylcholine are called *nicotinic drugs.* Nicotine excites both sympathetic and parasympathetic postganglionic neurons at the same time, which results in a strong sympathetic vasoconstriction and an increase in gastrointestinal activity.

Drugs That Block Impulse Transmission From Preganglionic to Postganglionic Neurons

Drugs in this category block the effect of acetylcholine to stimulate the postganglionic neurons in both the sympathetic and parasympathetic systems simultaneously and are called *ganglionic blocking drugs.* The drugs *tetraethyl ammonium, hexamethonium,* and *pentolinium* are used to block sympathetic activity but are rarely used to block parasympathetic activity. The effect of sympathetic blockade far overshadows the effect of parasympathetic blockade in many tissues. The ganglionic blocking drugs can be given to reduce arterial pressure in patients with hypertension. These drugs have several side effects, however, and are difficult to control, which limits their use.

61

Cerebral Blood Flow, Cerebrospinal Fluid, and Brain Metabolism

Functioning of the brain is closely tied to the level of cerebral blood flow. Total cessation of blood flow to the brain causes unconsciousness within 5 to 10 seconds because of the decrease in oxygen delivery and the resultant cessation of metabolic activity.

CEREBRAL BLOOD FLOW (p. 783)

The normal cerebral blood flow in an adult averages 50 to 65 milliliters per 100 grams, or about 750 to 900 ml/min; therefore, the brain receives approximately 15 per cent of the total resting cardiac output.

Cerebral blood flow is related to the level of metabolism. Three metabolic factors—*carbon dioxide, hydrogen ions,* and *oxygen*—have potent effects on cerebral blood flow. Carbon dioxide combines with water to form carbonic acid, which partially dissociates to form hydrogen ions. The hydrogen ions induce cerebral vasodilation in proportion to their concentration in the cerebral blood. Any substance that increases the acidity of the brain and, therefore, the hydrogen ion concentration will increase cerebral blood flow; such substances include lactic acid, pyruvic acid, or other acidic compounds that are formed during the course of metabolism. A decrease in cerebral tissue PO_2 will cause an immediate increase in cerebral blood flow due to local vasodilation of the cerebral blood vessels.

Measurements of local cerebral blood flow have revealed that blood flow in individual segments of the brain changes within seconds in response to local neuronal activity. The act of making a fist with the hand causes an immediate increase in blood flow in the motor cortex of the opposite cerebral hemisphere. The act of reading elevates the blood flow in the occipital cortex and in the language perception area of the temporal cortex.

Cerebral blood flow is autoregulated. Cerebral blood flow is nearly constant between the limits of 60 and 140 mm Hg mean arterial pressure. Arterial pressure, therefore, can fall to as low as 60 mm Hg or rise to as high as 140 mm Hg without significant changes occurring in cerebral blood flow. When arterial pressure falls below 60 mm Hg, cerebral blood flow becomes extremely compromised. If the arterial pressure rises above the limit of autoregulation, blood flow rises rapidly, and overstretching or rupture of the cerebral blood vessels can result in brain edema or cerebral hemorrhage.

The sympathetic nervous system has a role in regulation of cerebral blood flow. The cerebral circulation has a dense sympathetic innervation; under certain conditions, sympathetic stimulation can cause marked constriction of the cerebral arteries. During strenuous exercise or states of enhanced circulatory activity, sympathetic impulses can constrict the large- and intermediate-sized arteries and prevent the high pressure from reaching the small-sized blood vessels. This mechanism is important in prevention of vascular hemorrhage. Under many conditions in which the sympathetic nervous system is moderately activated, however, cerebral blood flow is maintained relatively constant by autoregulatory mechanisms.

A cerebral "stroke" occurs when cerebral blood vessels are blocked or ruptured. Most strokes are caused by arteriosclerotic plaques that occur in one or more of the large arteries of the brain. The plaque material can initiate the clot mechanism, which results in clot formation, artery blockage, and subsequent brain function loss in the areas supplied by the vessel. In about one fourth of persons who develop strokes, the cerebral blood vessels rupture as a result of high blood pressure. The resulting hemorrhage compresses the brain tissue, leading to local ischemia and edema.

The neurological effects of a stroke are determined by the brain area that is affected. If the middle cerebral artery in the dominant hemisphere is involved, the person is likely to become almost totally debilitated due to loss of Wernicke's area, which is involved in speech comprehension. In addition, these individuals often become unable to speak because of damage to Broca's motor area for word formation, and a loss in the function of other motor control areas of the dominant hemisphere can create spastic paralysis of the muscles of the opposite side of the body.

Cerebral Microcirculation

The density of capillaries is four times greater in the gray matter of the brain than that in the white matter. The level of blood flow to the gray matter is, therefore, four times as great as that to the white matter. The brain capillaries are much less "leaky" than capillaries in other portions of the body. Capillaries in the brain are surrounded by "glial feet," which provide physical support to prevent overstretching of the capillaries in the event of exposure to high pressure.

CEREBROSPINAL FLUID SYSTEM (p. 785)

The entire cavity enclosing the brain and spinal cord has a volume of about 1650 milliliters; about 150 milliliters of this volume is occupied by cerebrospinal fluid, and the remainder is occupied by the brain and spinal cord. This fluid, as shown in Figure 61–1, is found in the *ventricles of the brain*, the *cisterns around the brain*, and the *subarachnoid space*

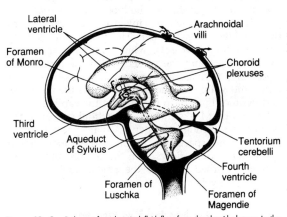

Figure 61–1 Pathway of cerebrospinal fluid flow from the choroid plexuses in the lateral ventricles to the arachnoidal villi protruding into the dural sinuses.

around both the brain and the spinal cord. These chambers are interconnected, and the pressure of the cerebrospinal fluid is regulated at a constant level.

A major function of the cerebrospinal fluid is to cushion the brain. The brain and the cerebrospinal fluid have about the same specific gravity. The brain, therefore, essentially floats in the cerebrospinal fluid. A blow to the head moves the entire brain simultaneously with the skull, causing no single portion of the brain to be momentarily contorted.

Formation and Absorption of Cerebrospinal Fluid

About 500 milliliters of cerebrospinal fluid is formed each day. Most of this fluid originates from the choroid plexuses of the four ventricles. Additional amounts of fluid are secreted by the ependymal surfaces of the ventricles and the arachnoidal membranes. The choroid plexus is a cauliflower-like growth of blood vessels covered by a thin layer of epithelial cells. This structure projects into the temporal horn of each lateral ventricle, the posterior portion of the third ventricle, and the roof of the fourth ventricle.

The cerebrospinal fluid is absorbed by multiple arachnoidal villi that project into the large sagittal venous sinus as well as into other venous sinuses of the cerebrum. The cerebrospinal fluid empties into the venous blood through the surface of these villi.

The perivascular space functions as a lymphatic system for the brain. As the blood vessels that supply the brain penetrate inward, they carry with them a layer of *pia matter*. The pia is only loosely adherent to the vessels, which creates a space between the pia and the vessels called the *perivascular space*. The perivascular space follows both the arteries and veins into the brain as far as the arterioles and venules but not to the level of the capillaries.

Protein that leaks into the interstitial spaces of the brain flows through the perivascular spaces into the subarachnoid space. On reaching the subarachnoid space, the protein flows with the cerebrospinal fluid and is absorbed through the arachnoidal villi into the cerebral veins.

Cerebrospinal Fluid Pressure

Cerebrospinal fluid is formed at a constant rate; therefore, the rate of absorption of this fluid by the arachnoidal villi determines both the quantity of fluid present in the ventricular system and the level of cerebrospinal fluid pressure.

The arachnoidal villi function like one-way valves that allow the cerebrospinal fluid to flow into the blood of the venous sinuses but prevent the flow of blood into the cerebrospinal fluid. Normally, the valve-like action of the villi allows cerebrospinal fluid to flow into the venous sinuses when the pressure in the fluid is approximately 1.5 mm Hg greater than the pressure of the blood in the venous sinuses. When the villi become blocked by large particulate matter or fibrosis, cerebrospinal fluid pressure can rise dramatically.

The normal level of cerebrospinal fluid pressure is 10 mm Hg. Brain tumors, hemorrhage, or infective processes can disrupt the absorptive capacity of the arachnoidal villi and cause the cerebrospinal fluid pressure to increase to levels three or four times that of normal.

Obstruction to flow of the cerebrospinal fluid causes *hydrocephalus.* This condition is frequently defined as *communicating hydrocephalus* or *noncommunicating hydrocephalus.* In communicating hydrocephalus, fluid flows readily from the ventricular system into the subarachnoid space, whereas in noncommunicating hydrocephalus, fluid flow out of one or more of the ventricles is blocked.

The communicating type of hydrocephalus is usually caused by blockage of fluid flow into the subarachnoid space around the basal regions of the brain or blockage of the arachnoidal villi themselves. The noncommunicating type of hydrocephalus is usually caused by blockade of the *aqueduct of Sylvius* as a result of a congenital defect or brain tumor. The continual formation of cerebrospinal fluid by the choroid plexuses in the two lateral ventricles and the third ventricle causes the volume of these ventricles to increase greatly. This flattens the brain into a thin shell against the skull. In neonates, the increased pressure also causes the entire head to swell because the skull bones have not yet fused.

Blood–Cerebrospinal Fluid and Blood-Brain Barriers

The constituents of the cerebrospinal fluid (CSF) are not exactly the same as those of the extracellular fluid elsewhere in the body. Furthermore, many large molecular substances do not pass from the blood into the cerebrospinal fluid or into the interstitial fluids of the brain. Barriers, called the *blood-cerebrospinal fluid barrier* and the *blood-brain barrier* thereby exist between the blood and the cerebrospinal fluid and the brain fluid. These barriers are highly permeable to water, carbon dioxide, oxygen, most lipid-soluble substances such as alcohol, and most anesthetics; slightly permeable to electrolytes such as sodium, chloride, and potassium; and almost totally impermeable to plasma proteins and most non–lipid-soluble large organic molecules.

The cause of the low permeability of these barriers is the manner in which the endothelial cells of the capillaries are joined to one another. The membranes of the adjacent endothelial cells are tightly fused with one another rather than having extensive slit pores between them, as is the case with most other capillaries of the body. These barriers often make it impossible to achieve effective concentrations of therapeutic drugs, such as protein antibodies and non–lipid-soluble compounds in the cerebrospinal fluid or parenchyma of the brain.

Brain Edema

One of the most serious complications of abnormal cerebral hemodynamics and fluid dynamics is the development of brain edema. Because the brain is encased in a solid vault, the accumulation of the fluid of edema compresses the blood vessels, resulting in a decrease in blood flow and destruction of brain tissue. Brain edema can be caused by greatly increased capillary pressure or by a concussion in which the brain's tissues and capillaries are traumatized and capillary fluid leaks into this tissue.

Once brain edema begins, it often initiates a vicious circle. The edema fluid compresses the vasculature, which, in turn, decreases the blood flow and causes brain ischemia. The ischemia causes arteriolar dilation with further increases in capillary pressure. The greater capillary pressure causes more edema fluid, and the edema becomes progressively worse. The reduced blood flow also decreases the oxygen delivery, which increases the permeability of the capillaries, allowing more fluid leakage. The decreased oxygen delivery depresses brain metabolism, which turns off the sodium pumps of the brain cell, causing them to swell.

Once this process has begun, heroic measures must be taken to prevent total destruction of the brain. One measure is to infuse intravenously a concentrated osmotic substance such as mannitol. This pulls fluid from the brain tissue through osmosis and breaks the vicious circle. Another procedure is to remove fluid quickly from the lateral ventricles of the brain through ventricular puncture, thereby relieving intracerebral pressure.

BRAIN METABOLISM (p. 789)

Under resting conditions, the metabolism of the brain accounts for 15 per cent of the total metabolism of the body even though the mass of the brain is only 2 per cent of the total body mass. Under resting conditions, therefore, brain metabolism is about 7.5 times the average metabolism of the remainder of the body.

The brain has a limited anaerobic capability. Most tissues of the body can go without oxygen for several minutes. During this time, the cells obtain their energy through anaerobic metabolism. Because of the high metabolic rate of the brain, anaerobic breakdown of glycogen cannot supply the energy needed to sustain neuronal activity. Most neuronal activity, therefore, depends on the second-by-second delivery of glucose and oxygen from the blood.

Under normal conditions, most brain energy is supplied by glucose derived from the blood. A special feature of glucose delivery to the neurons is that its transport through the cell membranes of the neurons does not depend on insulin. Even in patients who have serious diabetes, therefore, glucose diffuses readily into the neurons. When a diabetic patient is overtreated with insulin, the blood glucose concentration can fall to an extremely low level because the excess insulin causes almost all of the glucose in the blood to be transported rapidly into the insulin-sensitive, non-neural cells throughout the body. When this happens, insufficient glucose is left in the blood to supply the neurons, and mental function can become seriously impaired, leading to mental imbalances, psychotic disturbances, and, sometimes, coma.

UNIT

XII

Gastrointestinal Physiology

62

General Principles of Gastrointestinal Function—Motility, Nervous Control, and Blood Circulation

The alimentary tract provides the body with a continual supply of water, electrolytes, and nutrients. This requires (1) movement of food through the alimentary tract; (2) secretion of digestive juices and digestion of food; (3) absorption of digestive products, water, and various electrolytes, (4) circulation of blood to carry away absorbed substances; and (5) nervous and hormonal control of all these functions. The basic principles of function in the entire alimentary tract are discussed in this chapter.

GENERAL PRINCIPLES OF GASTROINTESTINAL MOTILITY
(p. 793)

Characteristics of the Gastrointestinal Wall

The motor functions of the gut are performed by the different layers of smooth muscle. The intestinal wall is composed of the following layers (from the outer surface inward): (1) *serosa*, (2) *longitudinal muscle layer*, (3) *circular muscle layer*, (4) *submucosa*, and (5) *mucosa*. In addition, a sparse layer of smooth muscle fibers, the muscularis mucosae, lies in the deeper layers of the mucosa.

The gastrointestinal smooth muscle functions as a syncytium. The smooth muscle fibers in the longitudinal and circular muscle layers are electrically connected through gap junctions that allow ions to move from one cell to the next. Each muscle layer functions as a syncytium; when an action potential is elicited within the muscle mass, it generally travels in all directions in the muscle. The distance that it travels depends on the excitability of the muscle; sometimes, it stops after only a few millimeters, and at other times, it travels many centimeters or even the entire length of the intestinal tract.

Electrical Activity of Gastrointestinal Smooth Muscle
(p. 794)

The smooth muscle of the gastrointestinal tract is subject to almost continual but slow electrical activity. This activity tends to have two basic types of electrical waves: *slow waves* and s*pikes*. In addition, the voltage of the resting membrane potential of the gastrointestinal smooth muscle can change to different levels, which can also have important effects in control of motor activity of the gastrointestinal tract.

The rhythm of most gastrointestinal contractions is determined by the frequency of slow waves in the smooth muscle membrane potential. These waves are not action potentials; instead, they are slow, undulating changes in the resting membrane potential. Their intensity usually varies from 5 to 15 millivolts, and their frequency varies in different parts of the human gastrointestinal tract from 3 to 12 per minute. The cause of slow waves is not known, but they could result from a slow undulation of the activity of the sodium-potassium pump. The slow waves mainly control intermittent spike potentials.

The spike potentials are true action potentials that cause muscle contraction. They occur when the resting membrane potential becomes more positive than about -40 millivolts (the normal resting membrane potential is between -50 and -60 millivolts). The channels responsible for the action potentials allow particularly large numbers of calcium ions to enter along with smaller numbers of sodium ions and therefore are called *calcium-sodium channels*. The movement of large amounts of calcium ions to the interior of the muscle fiber during the action potential plays a special role in causing the intestinal smooth muscle to contract.

The basic level of resting membrane potential of gastrointestinal smooth muscle can be increased or decreased. The resting membrane potential normally averages about -56 millivolts.

- *Factors that depolarize the membrane*—that is, make it more positive and more excitable—include (1) stretching the muscle, (2) stimulation by acetylcholine, (3) stimulation by parasympathetic nerves that secrete acetylcholine at their endings, and (4) stimulation by gastrointestinal hormones.

- *Factors that hyperpolarize the membrane*—that is, make it more negative and less excitable—include (1) the effect of norepinephrine or epinephrine on the muscle membrane and (2) the stimulation of the sympathetic nerves that secrete norepinephrine at their endings.

NEURAL CONTROL OF GASTROINTESTINAL FUNCTION
(p. 795)

The gastrointestinal tract has its own nervous system called the enteric nervous system. It lies entirely in the wall of the gut, beginning in the esophagus and extending all the way to the anus. The enteric system is composed mainly of two plexuses:

- The *myenteric plexus,* or *Auerbach's plexus,* is an outer plexus located between the longitudinal and circular muscle layers. Stimulation causes (1) increased "tone" of the gut wall, (2) increased intensity of rhythmical contractions, (3) increased rate of contraction, and (4) increased velocity of conduction, creating more rapid movement of the peristaltic waves. The myenteric plexus is also useful for inhibiting the pyloric sphincter, which controls the emptying of the stomach, and the sphincter of the ileocecal valve, which controls the emptying of the small intestine into the cecum.
- The *submucosal plexus,* or *Meissner's plexus,* is an inner plexus that lies in the submucosa. In contrast to the myenteric plexus, it is mainly concerned with controlling function within the inner wall of each minute segment of the intestine. For instance, many sensory signals originate from the gastrointestinal epithelium and are integrated in the submucosal plexus to help control local intestinal secretion, local absorption, and local contraction of the submucosal muscle, which causes various degrees of infolding of the stomach mucosa.

Autonomic Control of the Gastrointestinal Tract (p. 797)

The parasympathetic nerves increase the activity of the enteric nervous system. This in turn enhances the activity of most gastrointestinal functions. The parasympathetic supply to the gut is made up of cranial and sacral divisions.

- The *cranial parasympathetics* innervate, by way of the vagus nerves, the esophagus, stomach, pancreas, and first half of the large intestine.
- The *sacral parasympathetics* innervate, by way of the pelvic nerves, the distal half of the large intestine. The sigmoidal, rectal, and anal regions have an especially rich supply of parasympathetic fibers that function in the defecation reflexes.

The sympathetic nervous system usually inhibits activity in the gastrointestinal tract, causing many effects opposite to those of the parasympathetic system. The sympathetics innervate all portions of the gastrointestinal tract rather than being more extensively supplied to the portions nearest the oral cavity and anus, as is true of the parasympathetics. The sympathetic nerve endings secrete norepinephrine, which exerts its effects in two ways: (1) to a slight extent by a direct action that inhibits smooth muscle (except the muscularis mucosae, which it excites), and (2) to a major extent by an inhibitory effect on the neurons of the enteric nervous system. Strong stimulation of the sympathetic system can block movement of food through the gastrointestinal tract.

Gastrointestinal Reflexes (p. 797)

Three types of reflexes are essential for gastrointestinal control.

- *Reflexes that occur entirely within the enteric nervous system* control gastrointestinal secretion, peristalsis, mixing contractions, local inhibitory effects, and so forth.
- *Reflexes from the gut to the sympathetic ganglia and then back to the gut* transmit signals for long distances: signals from the stomach cause evacuation of the colon (*gastrocolic reflex*), signals from the colon and small intestine inhibit stomach motility and stomach secretion (*enterogastric reflexes*), and reflexes from the colon inhibit emptying of ileal contents into the colon (*coloileal reflex*).
- *Reflexes from the gut to the spinal cord or brain stem and then back to the gut* include in particular (1) reflexes from the stomach and duodenum to the brain stem and back to the stomach—by way of the vagus nerves—that control gastric motor and secretory activity; (2) pain reflexes that cause general inhibition of the entire gastrointestinal tract; and (3) defecation reflexes that travel to the spinal

cord and back again to produce the powerful colonic, rectal, and abdominal contractions required for defecation.

Gastrointestinal Hormones

The four major gastrointestinal hormones are secretin, gastrin, cholecystokinin, and gastric inhibitory peptide. The gastrointestinal hormones are released into the portal circulation and have physiological actions on target cells with specific receptors for the hormone; the effects of the hormones persist even after all nervous connections between the site of release and the site of action have been severed. Table 62–1 outlines the actions of each gastrointestinal hormone as well as the stimulus for secretion and site of secretion. There are a number of gastrointestinal peptides, called *candidate hormones*, that for one reason or another have failed to satisfy all the criteria required to become an "official" hormone.

FUNCTIONAL TYPES OF MOVEMENTS IN THE GASTROINTESTINAL TRACT (p. 798)

Two types of movements occur in the gastrointestinal tract: (1) *propulsive movements*, which cause food to move forward along the tract at an appropriate rate for digestion and absorption, and (2) *mixing movements*, which keep the intestinal contents thoroughly mixed at all times.

Peristalsis is the basic propulsive movement of the gastrointestinal tract. Distention of the intestinal tract causes a contractile ring to appear around the gut, which moves forward a few centimeters before ending. At the same time, the gut sometimes relaxes several centimeters down toward the anus, which is called *receptive relaxation*, allowing the food to be propelled more easily toward the anus. This complex pattern does not occur in the absence of the myenteric plexus; therefore, the complex is called the *myenteric reflex*, or *peristaltic reflex*. The peristaltic reflex plus the direction of movement toward the anus is called the *law of the gut*. In addition to distention, irritation of the epithelium lining the gut and extrinsic nervous signals, particularly parasympathetic, can initiate peristalsis.

Peristalsis and local constrictive contractions cause mixing in the alimentary tract. In some areas, the peristaltic contractions themselves cause most of

TABLE 62-1 GASTROINTESTINAL HORMONE ACTIONS, STIMULI FOR SECRETION, AND SITE OF SECRETION

Hormone	Stimuli for Secretion	Site of Secretion	Actions
Secretin	Acid Fatty acids	S cells of the duodenum	Stimulates: Pepsin secretion Pancreatic bicarbonate secretion Biliary bicarbonate secretion Growth of exocrine pancreas Inhibits: Gastric acid secretion Effect of gastrin on growth of gastric mucosa
Gastrin	Small peptides Amino acids Gastric distention Vagal stimulation	G cells of the antrum	Stimulates: Gastric acid secretion Growth of gastric mucosa

Cholecystokinin	Small peptides
	Amino acids
	Fatty acids
		I cells of the duodenum and jejunum	Stimulates:
			  Pancreatic enzyme secretion
			  Pancreatic bicarbonate secretion
			  Gallbladder contraction
			  Growth of exocrine pancreas
			Inhibits:
			  Gastric emptying

Gastric inhibitory peptide	Fatty acids
	Amino acids
	Oral glucose
		Duodenum and jejunum	Stimulates:
			  Insulin release
			Inhibits:
			  Gastric acid secretion

the mixing. This is especially true when forward progression of the intestinal contents is blocked by a sphincter, so that a peristaltic wave can only churn the intestinal contents, rather than propel them forward. At other times, local constrictive contractions occur every few centimeters in the gut wall. These constrictions usually last for only a few seconds; then, new constrictions occur at other points in the gut, "chopping" the contents first here and then there. These peristaltic and constrictive movements are modified in different parts of the gastrointestinal tract for proper propulsion and mixing.

GASTROINTESTINAL BLOOD FLOW (p. 799)

The blood vessels of the gastrointestinal tract are part of the splanchnic circulation. The splanchnic circulation includes the blood flow through the gut itself plus the blood flow through the spleen, pancreas, and liver. The blood that courses through the splanchnic circulation flows immediately into the liver by way of the portal vein. In the liver, the blood passes through liver sinusoids and finally leaves the liver by way of hepatic veins.

Gastrointestinal blood flow usually is proportional to the level of local activity. For instance, during active absorption of nutrients, blood flow in the villi and adjacent regions of the submucosa is greatly increased. Likewise, blood flow in the muscle layers of the intestinal wall is greater with increased motor activity in the gut. Although the precise cause or causes of increased blood flow during increased gastrointestinal activity are still unclear, some facts are known:

- Vasodilator substances are released from the mucosa during the digestive process. Most of these are peptide hormones, including cholecystokinin, gastrin, and secretin. These same hormones are also important in controlling certain motor and secretory activities of the gut.
- Some of the gastrointestinal glands also release into the gut wall two kinins, kallidin and bradykinin. These kinins are powerful vasodilators that are believed to cause much of the increased mucosal vasodilation that occurs with secretion.
- Decreased oxygenation of the gut wall can increase intestinal blood flow by at least 50 per cent;

therefore, tissue hypoxia due to greater gut activity probably causes much of the vasodilation. The lack of oxygen can also lead to the release of adenosine, a well known vasodilator that could be responsible for much of the increased flow.

Nervous Control of Gastrointestinal Blood Flow (p. 801)

Parasympathetic stimulation increases blood flow. Stimulation of the parasympathetic nerves to the stomach and lower colon increases local blood flow while it increases glandular secretion. This greater flow probably results secondarily from the greater glandular activity and is not a direct effect of nervous stimulation.

Sympathetic stimulation decreases blood flow. After a few minutes of sympathetic induced vasoconstriction, the flow often returns almost to normal by means of a mechanism called *autoregulatory escape*: the local metabolic vasodilator mechanisms that are elicited by ischemia become prepotent over the sympathetic vasoconstriction and therefore redilate the arterioles, causing return of necessary nutrient blood flow to the gastrointestinal glands and muscle.

Sympathetic vasoconstriction is important when other parts of the body need extra blood flow. A major value of sympathetic vasoconstriction in the gut is that it allows the shutting off of gastrointestinal and other splanchnic blood flow for short periods during heavy exercise when increased flow is needed by the skeletal muscle and heart. In addition, in circulatory shock, when all the vital tissues are in danger of cellular death for lack of blood flow—especially the brain and the heart—sympathetic stimulation can block splanchnic blood flow almost entirely for as long as 1 hour.

63

Transport and Mixing of Food in the Alimentary Tract

For food to be processed optimally in the alimentary tract, the time that it remains in each part of the tract is critical, and appropriate mixing must occur. The purpose of this chapter is to discuss these movements and the mechanisms that control them.

INGESTION OF FOOD (p. 803)

The pharyngeal stage of swallowing is involuntary and constitutes the passage of food through the pharynx into the esophagus. When the food is ready for swallowing, it is "voluntarily" squeezed, or rolled posteriorly, into the pharynx by the tongue, which constitutes the voluntary stage of swallowing. The bolus of food stimulates swallowing receptors all around the opening of the pharynx, and impulses from these pass to the brain stem to initiate a series of automatic pharyngeal muscle contractions as follows:

- The soft palate is pulled upward to close the posterior nares, preventing reflux of food into the nasal cavities.
- The palatopharyngeal folds on either side of the pharynx are pulled medially, forming a sagittal slit through which the food must pass. This slit impedes the passage of large objects into the posterior pharynx.
- The vocal cords are strongly approximated, the larynx is pulled upward and anteriorly by the neck muscles, and the epiglottis swings backward over the opening of the larynx. These effects prevent passage of food into the trachea.
- The upper esophageal sphincter relaxes, allowing food to move easily from the posterior pharynx into the upper esophagus. This sphincter, between

swallows, remains strongly contracted, thereby preventing air from going into the esophagus during respiration.
• A fast peristaltic wave originating in the pharynx forces the bolus of food into the upper esophagus.

The pharyngeal stage of swallowing is controlled by the nervous system. Impulses from around the pharyngeal opening are transmitted through the sensory portions of the trigeminal and glossopharyngeal nerves into the medulla oblongata. The areas in the medulla and lower pons that control swallowing are collectively called the *deglutition* or *swallowing center*. The motor impulses from the swallowing center to the pharynx and upper esophagus that cause swallowing are transmitted by the 5th, 9th, 10th, and 12th cranial nerves and a few of the superior cervical nerves.

The esophagus exhibits two types of peristaltic movements: primary peristalsis and secondary peristalsis.

• *Primary peristalsis* is a continuation of the peristaltic wave that begins in the pharynx and spreads into the esophagus during the pharyngeal stage of swallowing. This wave passes all the way from the pharynx to the stomach in about 8 to 10 seconds.
• *Secondary peristalsis* results from distention of the esophagus when the primary peristaltic wave fails to move the food into the stomach. These secondary waves are initiated by (1) intrinsic neural circuits in the esophageal myenteric nervous system and (2) reflexes transmitted from the esophagus to the medulla and then back to the esophagus by way of vagal fibers.

Receptive relaxation of the stomach prepares it to receive the food. As the esophageal wave passes toward the stomach, a wave of relaxation, transmitted through myenteric inhibitory neurons, precedes the peristalsis. Furthermore, the entire stomach and, to a lesser extent, the duodenum become relaxed.

The lower esophageal sphincter relaxes ahead of the peristaltic wave. At the lower end of the esophagus, the esophageal circular muscle functions as a lower esophageal sphincter. It remains tonically constricted until a peristaltic swallowing wave passes down the esophagus. The sphincter then relaxes ahead of the peristaltic wave, allowing easy propulsion of the swallowed food into the stomach.

MOTOR FUNCTIONS OF THE STOMACH (p. 806)

The motor functions of the stomach are threefold:

- Storage of food until the food can be processed in the duodenum.
- Mixing of food with gastric secretions until it forms a semifluid mixture called *chyme*.
- Emptying of food into the small intestine at a rate suitable for proper digestion and absorption by the small intestine.

The stomach relaxes when food enters it. Normally, when food enters the stomach, a "vagovagal reflex" from the stomach to the brain stem and then back to the stomach reduces the tone in the muscular wall of the stomach. The wall can bulge progressively outward, accommodating about 1.5 liters in the completely relaxed stomach. The pressure in the stomach remains low until this limit is approached.

"Retropulsion" is an important mixing mechanism of the stomach. Each time a peristaltic wave passes over the antrum toward the pylorus, the pyloric muscle contracts, which further impedes emptying through the pylorus. Most of the antral contents are squirted backward through the peristaltic ring toward the body of the stomach. The moving peristaltic constrictive ring, combined with this squirting action called *retropulsion*, is an exceedingly important mixing mechanism of the stomach.

The pyloric sphincter is important for controlling stomach emptying. The pyloric sphincter, which is located at the distal opening of the stomach, remains slightly contracted most of the time. It usually opens sufficiently for fluids to empty from the stomach. The constriction, however, normally prevents passage of food particles until they have become mixed in the chyme to an almost fluid consistency.

Gastric emptying is inhibited by enterogastric reflexes from the duodenum. When food enters the duodenum, multiple nervous reflexes are initiated from its wall that pass back to the stomach and slow or even stop stomach emptying as the volume of chyme in the duodenum becomes too much. Factors that can excite the enterogastric reflexes include the following:

- Degree of distention of the duodenum
- Presence of any degree of irritation of the duodenal mucosa
- Degree of acidity of the duodenal chyme
- Degree of osmolality of the chyme
- Presence of certain breakdown products in the chyme, especially breakdown products of proteins.

Cholecystokinin inhibits gastric emptying. Cholecystokinin is released from the mucosa of the jejunum in response to fatty substances in the chyme; the contents of the stomach are released very slowly after the ingestion of a fatty meal.

MOVEMENTS OF THE SMALL INTESTINE (p. 808)

Distention of the small intestine elicits mixing contractions called *segmentation contractions*. These are concentric contractions that have the appearance of a chain of sausages. As one set of contractions relaxes, a new set begins, mainly at new points between the previous contractions. These segmentation contractions usually "chop" the chyme about two or three times a minute, promoting progressive mixing of the solid food particles with the secretions of the small intestine.

Maximum frequency of segmentation contractions is determined by the slow-wave frequency. The frequency of slow waves is about 12 per minute in the duodenum and proximal jejunum, which is the maximum frequency of segmentation contractions in these areas. In the terminal ileum, the maximum frequency is usually 8 or 9 contractions per minute. The segmentation contractions become exceedingly weak after the administration of atropine; therefore, even though slow waves control the segmentation contractions, these are not effective without background excitation by the enteric nervous system.

Chyme is propelled through the small intestine by peristaltic waves. These move toward the anus at a velocity of 0.5 to 2.0 cm/sec. Movement of chyme along the small intestine averages only 1 cm/min. About 3 to 5 hours are required for passage of chyme from the pylorus to the ileocecal valve.

Peristalsis is controlled by nervous and hormonal signals. Peristaltic activity of the small intestine is greatly increased after a meal for the following reasons:

- *Nervous signals.* These are caused in part by the entry of chyme into the duodenum but also by a so-called *gastroenteric reflex* that is initiated by distention of the stomach and conducted principally through the myenteric plexus down along the wall of the small intestine.
- *Hormonal signals.* Gastrin, cholecystokinin, and insulin are released after a meal and can enhance intestinal motility. Secretin and glucagon inhibit small intestinal motility. The physiological significance of these factors in controlling motility is uncertain.

Peristaltic rush is caused by intense irritation of the intestinal mucosa. A peristaltic rush is a very powerful and rapid peristalsis; it is initiated in part by extrinsic nervous reflexes that travel to the autonomic nervous system ganglia and brain stem and then back to the gut and in part by direct enhancement of the myenteric plexus reflexes. The powerful peristaltic contractions then travel long distances in the small intestine, sweeping the contents into the colon and thereby relieving the small intestine of either irritative chyme or excessive distention.

The ileocecal valve prevents backflow of fecal contents from the colon into the small intestine. The lips of the ileocecal valve protrude into the lumen of the cecum and are forcefully closed when excess pressure builds up in the cecum and the cecal contents push backward against the lips. The wall of the ileum near the ileocecal valve has a thickened muscular coat called the *ileocecal sphincter*. This sphincter normally remains mildly constricted and slows the emptying of ileal contents into the cecum, except immediately after a meal.

The ileocecal sphincter and the intensity of peristalsis in the terminal ileum are controlled by reflexes from the cecum. Whenever the cecum is distended, the contraction of the ileocecal sphincter is intensified, and ileal peristalsis is inhibited, which greatly delays emptying of additional chyme from the ileum. Any irritant in the cecum delays emptying. These reflexes from the cecum to the ileocecal sphincter and ileum are mediated by way of the myenteric plexus in the gut wall itself and through extrinsic nerves, especially reflexes by way of the prevertebral sympathetic ganglia.

MOVEMENTS OF THE COLON (p. 810)

The principal functions of the colon are (1) absorption of water and electrolytes from the chyme and (2) storage of fecal matter until it can be expelled. The proximal half of the colon is concerned principally with absorption, and the distal half is concerned with storage.

Contraction of circular and longitudinal muscles in the large intestine causes haustrations to develop. These combined contractions cause the unstimulated portion of the large intestine to bulge outward into bag-like sacs called *haustrations*. The haustral contractions perform two main functions:

- *Propulsion*—Haustral contractions at times move slowly toward the anus during their period of contraction, especially in the cecum and ascending colon, and thereby provide a minor amount of forward propulsion of the colonic contents.
- *Mixing*—Haustral contractions dig into and roll over the fecal material in the large intestine. In this way, all the fecal material is gradually exposed to the surface of the large intestine, and fluid and dissolved substances are progressively absorbed.

Mass movements are important for propelling the fecal contents through the large intestine. A mass movement is a modified type of peristalsis characterized by the following sequence of events: A constrictive ring occurs at a distended or irritated point in the colon, usually in the transverse colon, and then the colon distal to the constriction contracts as a unit, forcing the fecal material in this segment en masse down the colon. The entire series of mass movements usually persists for only 10 to 30 minutes. When they have forced a mass of feces into the rectum, the desire for defecation is felt.

The appearance of mass movements after meals is facilitated by gastrocolic and duodenocolic reflexes. These reflexes result from distention of the stomach and duodenum. They occur either not at all or hardly at all when the extrinsic nerves are removed; therefore, these reflexes almost certainly are conducted through the extrinsic nerves of the autonomic nervous system. Irritation in the colon can also initiate intense mass movements. Mass movements can also be initiated by intense stimulation of the parasympathetic nervous system or by overdistention of a segment of the colon.

Defecation can be initiated by an intrinsic reflex mediated by the local enteric nervous system. When feces enter the rectum, distention of the rectal wall initiates afferent signals that spread through the myenteric plexus to initiate peristaltic waves in the descending colon, sigmoid, and rectum, forcing feces toward the anus. As the peristaltic wave approaches the anus, the internal anal sphincter is relaxed by inhibitory signals from the myenteric plexus; if the external anal sphincter is consciously, voluntarily relaxed at the same time, defecation will occur.

The intrinsic defecation reflex functioning by itself is relatively weak. To be effective in causing defecation, the reflex usually must be fortified by a parasympathetic defecation reflex that involves the sacral segments of the spinal cord. When the nerve endings in the rectum are stimulated, signals are transmitted first into the spinal cord and then reflexly back to the descending colon, sigmoid, rectum, and anus by way of parasympathetic nerve fibers in the pelvic nerves. These parasympathetic signals greatly intensify the peristaltic waves and relax the internal anal sphincter and thus convert the intrinsic defecation reflex from a weak movement into a powerful process of defecation.

64

Secretory Functions of the Alimentary Tract

Secretory glands subserve two primary functions in the alimentary tract: (1) digestive enzymes are secreted in most areas, and (2) mucous glands provide mucus for lubrication and protection of all parts of the alimentary tract. The purpose of this chapter is to describe the different alimentary secretions and their functions and the regulation of their production.

GENERAL PRINCIPLES OF ALIMENTARY TRACT SECRETION (p. 815)

Contact of food with the epithelium stimulates secretion. Direct mechanical stimulation of glandular cells by food causes the local glands to secrete digestive juices. In addition, epithelial stimulation activates the enteric nervous system of the gut wall. The types of stimuli that do this are (1) tactile stimulation, (2) chemical irritation, and (3) gut wall distention.

Parasympathetic stimulation increases the rate of glandular secretion. This is especially true of salivary glands, esophageal glands, gastric glands, the pancreas, Brunner's glands in the duodenum, and the glands of the distal portion of the large intestine. Secretion in the remainder of the small intestine and in the first two thirds of the large intestine occurs mainly in response to local neural and hormonal stimuli.

Sympathetic stimulation can have a dual effect on glandular secretion. Sympathetic stimulation may increase or decrease glandular secretion, depending on the existing secretory activity of the gland. This dual effect can be explained as follows:

- Sympathetic stimulation alone usually slightly increases secretion.

- If secretion has already increased, superimposed sympathetic stimulation usually reduces the secretion because it reduces blood flow to the gland.

SECRETION OF SALIVA (p. 817)

Saliva contains a serous secretion and a mucus secretion.

- The *serous secretion* contains ptyalin (an α-amylase), which is an enzyme for digesting starches.
- The *mucus secretion* contains mucin for lubrication and for surface protection.

Saliva contains high concentrations of potassium and bicarbonate ions and low concentrations of sodium and chloride ions. Salivary secretion is a two-stage operation: a primary secretion contains ptyalin and/or mucin in a solution with an ionic composition similar to extracellular fluid. The primary secretion is then modified in the ducts, as follows:

- Sodium ions are actively reabsorbed and potassium ions are actively secreted into the ducts. An excess of sodium reabsorption creates a negative charge in the salivary ducts; this in turn causes chloride ions to be reabsorbed passively.
- Bicarbonate ions are secreted into the duct, caused in part by exchange of bicarbonate for chloride ions but also by an active secretory process.

Salivation is controlled mainly by parasympathetic nervous signals. The salivatory nuclei in the brain stem are excited by taste and tactile stimuli from the tongue, mouth, and pharynx. Salivation can also be affected by higher centers of the brain (e.g., salivation increases when a person smells favorite foods). Salivation also occurs in response to reflexes originating in the stomach and upper intestines—particularly when irritating foods are swallowed or when a person is nauseated.

GASTRIC SECRETION (p. 819)

The stomach mucosa has two important types of tubular glands.

- The *oxyntic (acid-forming) glands* are located in the body and fundus. The three types of cells are (1) the *mucous neck cells*, which secrete mainly mucus but also some pepsinogen; (2) the *peptic (or chief) cells,* which secrete pepsinogen; and (3) the *parietal (or oxyntic) cells,* which secrete hydrochloric acid and intrinsic factor.

- The *pyloric glands,* which are located in the antrum, secrete mainly mucus for protection of the pyloric mucosa but also some pepsinogen and, very importantly, the hormone gastrin.

Gastric acid is secreted by parietal cells. The parietal cell contains many large branching intracellular canaliculi. When these cells secrete their acid juice, the membranes of the canaliculi empty their secretion directly into the lumen of the oxyntic gland. The final secretion entering the canaliculus contains concentrated hydrochloric acid (155 mEq/liter), potassium chloride (15 mEq/liter), and small amounts of sodium chloride. The concentrated hydrochloric acid is then secreted into the lumen of the gland.

Hydrochloric acid is as necessary as pepsin for protein digestion in the stomach. The pepsinogens have no digestive activity when they are first secreted; however, as soon as they come into contact with hydrochloric acid and especially when they come into contact with previously formed pepsin plus the hydrochloric acid, they are changed to form active pepsin.

Parietal cells also secrete "intrinsic factor." Intrinsic factor is essential to absorption of vitamin B_{12} in the ileum. When the acid-producing cells of the stomach are destroyed, which frequently occurs in chronic gastritis, the person develops not only achlorhydria but often also pernicious anemia due to failure of the red blood cells to mature.

Basic factors that stimulate gastric secretion are acetylcholine, gastrin, and histamine. Acetylcholine excites secretion of pepsinogen by peptic cells, hydrochloric acid by parietal cells, and mucus by mucous cells. In comparison, both gastrin and histamine stimulate strongly the secretion of acid by parietal cells but have little effect on the other cells.

Gastric secretion can be initiated by long vagovagal reflexes and short reflexes.

- *Long vagovagal reflexes* are transmitted from the stomach mucosa all the way to the brain stem and then back to the stomach through the vagus nerves.
- *Short reflexes* originate locally and are transmitted entirely through the local enteric nervous system.

Acid secretion is stimulated by gastrin. Nerve signals from the vagus nerves and local enteric reflexes cause gastrin cells (G cells) in the antral mucosa to secrete gastrin. Gastrin is carried by blood to the oxyntic glands, where it stimulates parietal cells strongly and peptic cells to a lesser extent.

Histamine stimulates acid secretion by parietal cells. Small amounts of histamine are formed continually in the gastric mucosa, but this causes little acid secretion. Whenever acetylcholine and gastrin stimulate the parietal cells at the same time, the histamine enhances acid secretion. Thus, histamine is a cofactor for stimulating acid secretion.

Pepsinogen secretion is stimulated by acetylcholine and gastric acid. Acetylcholine is released from vagus nerves or other enteric nerves. Gastric acid probably does not stimulate peptic cells directly but elicits additional enteric reflexes. When the ability to secrete normal amounts of acid is lost, pepsinogen level is low even though the peptic cells are normal.

Gastric secretion is inhibited by excess acid in the stomach. When the pH of gastric juice falls below 3.0, gastrin secretion is decreased for two reasons: (1) the high acidity directly depresses gastrin secretion by the G cells, and (2) the acid causes an inhibitory nervous reflex that inhibits gastric secretion. This mechanism protects the stomach against excessive acidity.

There are three phases of gastric secretion.

- The *cephalic phase* accounts for 30 per cent of the response to a meal and is initiated by the anticipation of eating and the odor and taste of food. It is mediated by the vagus nerve.
- The *gastric phase* accounts for 60 per cent of the acid response to a meal. It is initiated by distention of the stomach, which leads to nervous stimulation of gastric secretion. In addition, partial digestion products of proteins in the stomach cause *gastrin* to be released from the antral mucosa. The gastrin then causes secretion of a highly acidic gastric juice.
- The *intestinal phase* (10 per cent of the response) is initiated by nervous stimuli associated with distention of the small intestine. The presence of digestion products of proteins in the small intestine can also stimulate gastric secretion via a humoral mechanism.

Chyme in the small intestine inhibits secretion during the gastric phase. This inhibition results from at least two influences:

- *Enterogastric reflex*—The presence of food in the small intestine initiates this reflex, which is transmitted through the enteric nervous system and through the extrinsic sympathetic and vagus nerves, and inhibits stomach secretion. The reflex can be initiated by the distention of the small bowel, the presence of acid in the upper intestine, the presence of protein breakdown products, or the irritation of the mucosa.
- *Hormones*—The presence of chyme in the upper small intestine causes the release of several intestinal hormones. Secretin and gastric inhibitory peptide are especially important in the inhibition of gastric secretion.

PANCREATIC SECRETION (p. 824)

Digestive enzymes are secreted by the pancreatic acini.

- The more important enzymes for digestion of proteins are trypsin, chymotrypsin, and carboxypolypeptidase, which are secreted in the inactive forms trypsinogen, chymotrypsinogen, and procarboxypolypeptidase.
- The pancreatic digestive enzyme for carbohydrates is pancreatic amylase, which hydrolyzes starches, glycogen, and most other carbohydrates (except cellulose) to form disaccharides and a few trisaccharides.
- The main enzyme for fat digestion is pancreatic lipase, which hydrolyzes neutral fat into fatty acids and monoglycerides; cholesterol esterase, which causes hydrolysis of cholesterol esters; and phospholipase, which splits fatty acids from phospholipids.

Bicarbonate ions and water are secreted by epithelial cells of the ductules and ducts. Bicarbonate ion in the pancreatic juice serves to neutralize acid emptied into the duodenum from the stomach.

Pancreatic secretion is stimulated by acetylcholine, cholecystokinin, and secretin.

- *Acetylcholine*, which is released from nerve endings, mainly stimulates secretion of digestive enzymes.

- *Cholecystokinin*, which is secreted by the duodenal and upper jejunal mucosa, mainly stimulates the secretion of digestive enzymes.
- *Secretin*, which is secreted by the duodenal and jejunal mucosa when highly acid food enters the small intestine, mainly stimulates the secretion of sodium bicarbonate.

Pancreatic secretion occurs in three phases.

- *Cephalic phase*—The nervous signals that cause gastric secretion also cause acetylcholine release by vagal nerve endings in the pancreas; this accounts for about 20 per cent of the pancreatic enzymes after a meal.
- *Gastric phase*—The nervous stimulation of enzyme secretion continues, accounting for another 5 to 10 per cent of the enzymes secreted after a meal.
- *Intestinal phase*—After chyme enters the small intestine, pancreatic secretion becomes copious, mainly in response to the hormone secretion. In addition, cholecystokinin causes still much more increase in the secretion of enzymes.

Secretin stimulates secretion of bicarbonate, which neutralizes acidic chyme. When acid chyme enters the duodenum from the stomach, the hydrochloric acid causes the release and activation of secretin, which is subsequently absorbed into the blood. Secretin in turn causes the pancreas to secrete large quantities of fluid that contain a high concentration of bicarbonate ion. The secretin mechanism is especially important for the following two reasons:

- The acid contents emptied into the duodenum from the stomach become neutralized, and the peptic activity of the gastric juices is immediately blocked.
- Bicarbonate secretion provides an appropriate pH for action of pancreatic enzymes, which function optimally in a slightly alkaline or neutral medium.

Cholecystokinin stimulates enzyme secretion by the pancreas. The presence of food in the upper small intestine also causes cholecystokinin to be released from cells called I cells in the mucosa of the duodenum and upper jejunum. This effect results in particular from the presence of proteases and peptones (which are products of partial protein digestion) and of long-chain fatty acids; hydrochloric acid from the stomach juices also causes cholecystokinin release in smaller quantities.

SECRETION OF BILE BY THE LIVER; FUNCTIONS OF THE BILIARY TREE (p. 827)

Bile is important for (1) fat digestion and absorption and (2) waste product removal from the blood.

- *Fat digestion and absorption*—Bile salts help to emulsify the large fat particles of the food into many minute particles that can be attacked by the lipase enzyme secreted in pancreatic juice. They also aid in the transport and absorption of the digested fat end products to and through the intestinal mucosal membrane.
- *Waste product removal*—Bile serves as a means for excretion of several important waste products from the blood; these include especially bilirubin, an end product of hemoglobin destruction, and excess cholesterol synthesized by the liver cells.

Bile is secreted in two stages by the liver.

- The initial portion, which is secreted by liver hepatocytes, contains large amounts of bile acids, cholesterols, and other organic constituents. It is secreted into the minute bile canaliculi that lie between the hepatic cells in the hepatic plates.
- A watery solution of sodium and bicarbonate ions is added to the bile as it flows through the bile ducts. This second secretion is stimulated by secretin, thus causing increased quantities of bicarbonate ions that supplement pancreatic secretions for neutralizing acid.

The most abundant substance secreted in bile is the bile salts. They account for about one half of the total solutes of bile; also secreted or excreted in large concentrations are bilirubin, cholesterol, and lecithin, and the usual electrolytes of plasma.

Bile is concentrated in the gallbladder. Active transport of sodium through the gallbladder epithelium is followed by secondary absorption of chloride ions, water, and most other soluble constituents. Bile is normally concentrated in this way about fivefold.

Cholecystokinin stimulates contraction of the gallbladder. Fatty foods that enter the duodenum cause cholecystokinin to be released from the local glands. Cholecystokinin causes rhythmical contractions of the gallbladder, but effective emptying also requires simultaneous relaxation of the *sphincter of Oddi,* which guards exit of the common bile duct into the duodenum. Cholecystokinin has a relaxing effect on the sphincter of Oddi, but this effect alone is usually not sufficient to allow significant emptying. A more important effect is the following: during the relaxation phase of duodenal peristalsis, the sphincter of Oddi relaxes along with the muscle of the gut wall. As a result, bile usually enters the duodenum in the form of squirts that are synchronized with the relaxation phase of the duodenal peristaltic waves.

SECRETIONS OF THE SMALL INTESTINE (p. 830)

Brunner's glands secrete an alkaline mucus in the small intestine. The secretion of mucus is stimulated by the following:

- Tactile stimuli or irritating stimuli of the overlying mucosa
- Vagal stimulation, which causes secretion concurrently with increase in stomach secretion
- Gastrointestinal hormones, especially secretin.

Mucus protects the duodenal wall from digestion by gastric juice. Brunner's glands respond rapidly and intensely to irritating stimuli. In addition, the secretin-stimulated secretion by the glands contains a large excess of bicarbonate ions, which add to the bicarbonate ions from pancreatic secretion and liver bile in neutralizing acid that enters the duodenum from the stomach.

Intestinal digestive juices are secreted by the crypts of Lieberkuhn. The crypts of Lieberkuhn lie between the intestinal villi, and the intestinal surfaces of both the crypts and villi are covered by an epithelium composed of two types of cells:

- *Goblet cells,* which secrete mucus that provides its usual functions of lubrication and protection of the intestinal mucosa
- *Enterocytes,* which, in the crypts, secrete large quantities of water and electrolytes and, over the surfaces of the villi, reabsorb the water and electrolytes along with the end products of digestion. This circulation of fluid from the crypts to the villi supplies a watery vehicle for absorption of substances from the chyme as it comes in contact

with the villi, which is the primary function of the small intestine.

Enterocytes that cover the villi contain enzymes that digest specific food substances while they are being absorbed through the epithelium. These enzymes are the following:

- Several peptidases for splitting small peptides into amino acids
- Four enzymes for splitting disaccharides into monosaccharides—sucrase, maltase, isomaltase, and lactase
- Intestinal lipase for splitting neutral fats into glycerol and fatty acids; most if not all of these enzymes are mainly in the brush border of the enterocytes.

Secretion in the small intestine is regulated by local stimuli and hormones.

- *Local stimuli*—Secretion is mainly regulated by reflexes initiated by tactile or irritative stimuli and by enteric nervous activity associated with gastrointestinal movements; therefore, secretion in the small intestine occurs simply in response to the presence of chyme in the intestine.
- *Hormonal regulation*—Some of the same hormones that promote secretion elsewhere in the gastrointestinal tract also increase small intestinal secretion, especially secretin and cholecystokinin.

SECRETIONS OF THE LARGE INTESTINE (p. 831)

Most of the secretion in the large intestine is mucus. Mucus in the large intestine protects the wall against excoriation, but it also provides the adherent medium for fecal matter. Furthermore, it protects the intestinal wall from the great amount of bacterial activity that takes place inside the feces. Combined with the alkalinity of the secretion, it provides a barrier to keep acids formed deep in the feces from attacking the intestinal wall.

Water and electrolytes are secreted in response to irritation. This secretion acts to dilute the irritating factors and to cause rapid movement of the feces toward the anus; the usual result is diarrhea.

65

Digestion and Absorption in the Gastrointestinal Tract

The primary foods on which the body lives can be classified as *carbohydrates*, *fats*, and *proteins*. In general, these are useless as nutrients without the preliminary process of digestion. This chapter discusses (1) the digestion of carbohydrates, fats, and proteins and (2) the mechanisms by which the end products of digestion as well as water, electrolytes, and other substances are absorbed.

DIGESTION OF VARIOUS FOODS (p. 833)

Hydrolysis is the basic chemical process of digestion for carbohydrates, fats, and proteins.

- *Carbohydrates*—Most dietary carbohydrates are combinations of monosaccharides that are bound together by condensation. A hydrogen ion has been removed from one of the monosaccharides while a hydroxyl ion has been removed from the next one, and the hydrogen and hydroxyl ions combine to form water. When carbohydrates are digested, specific enzymes return the hydrogen and hydroxyl ions to the polysaccharides and thereby separate the monosaccharides from each other; this process is called *hydrolysis*.
- *Fats*—Most dietary fat consists of triglycerides (neutral fats), which are combinations of three fatty acid molecules condensed with a single glycerol molecule. In condensation, three molecules of water have been removed. Digestion of the triglycerides consists of returning water molecules to the triglyceride molecule and thereby splitting the fatty acid molecules away from the glycerol.
- *Proteins*—These are formed from amino acids bound together by peptide linkages. In this linkage, a hydroxyl ion is removed from one amino acid and a hydrogen ion is removed from the

571

succeeding one; the amino acids in the protein chain are bound together by condensation, and digestion occurs via the reverse effect of hydrolysis.

Digestion of Carbohydrates (p. 834)

The digestion of carbohydrates begins in the mouth and stomach. Saliva contains the enzyme ptyalin (an α-amylase), which hydrolyzes starch into maltose and other small polymers of glucose. Food spends so little time in the mouth that less than 5 per cent of the starch content of a meal is hydrolyzed before swallowing. Digestion continues in the stomach for about 1 hour before the activity of salivary amylase is blocked by gastric acid. Nevertheless, α-amylase hydrolyzes as much as 30 to 40 per cent of the starches to maltose.

Pancreatic secretion, like saliva, contains a large quantity of α-amylase. The pancreatic α-amylase is almost identical in function to the α-amylase of saliva but is several times as powerful; therefore, soon after chyme empties into the duodenum and mixes with pancreatic juice, virtually all the starches are digested. In general, the starches are almost totally converted into maltose and other very small glucose polymers before they have passed beyond the duodenum or upper jejunum.

Disaccharides and small glucose polymers are hydrolyzed into monosaccharides by intestinal epithelial enzymes. The microvilli brush border contains enzymes that split the disaccharides lactose, sucrose, and maltose as well as small glucose polymers into their constituent monosaccharides. Glucose usually represents more than 80 per cent of the final products of carbohydrate digestion.

- *Lactose* splits into a molecule of *galactose* and a molecule of glucose.
- *Sucrose* splits into a molecule of *fructose* and a molecule of glucose.
- *Maltose* and the other small glucose polymers all split into molecules of *glucose*.

Digestion of Proteins (p. 834)

Protein digestion begins in the stomach. The ability of pepsin to digest collagen is especially important because collagen is a major constituent of the inter-cellular connective tissue of meats. For enzymes to penetrate meats and digest cellular proteins, the collagen fibers must be digested. Consequently, in persons who lack peptic activity in the stomach, the ingested meats are less well penetrated by these enzymes and may be poorly digested.

Most protein digestion results from the actions of pancreatic proteolytic enzymes. Proteins leaving the stomach in the form of proteoses, peptones, and large polypeptides are digested into dipeptides, tripeptides, and some larger peptides by proteolytic pancreatic enzymes; only a small percentage of proteins is digested to amino acids by pancreatic juices.

- *Trypsin and chymotrypsin* split protein molecules into small polypeptides.
- *Carboxypolypeptidase* cleaves amino acids from the carboxyl ends of the polypeptides.
- *Proelastase* gives rise to elastase, which in turn digests the elastin fibers that hold together meat.

Amino acids represent more than 99 per cent of protein digestive products. The last digestion of proteins in the intestinal lumen is achieved by enterocytes that line the villi.

- *Digestion at the brush border*—Aminopolypeptidase and several dipeptidases, located at the brush border, succeed in splitting larger polypeptides into tripeptides, dipeptides, and some amino acids. These are transported into the enterocyte.
- *Digestion inside the enterocyte*—The enterocyte contains multiple peptidases that are specific for linkages between the various amino acids. Within minutes, virtually all the last dipeptides and tripeptides are digested to amino acids, which then enter the blood.

Digestion of Fats (p. 835)

The first step in fat digestion is emulsification by bile acids and lecithin. Emulsification is the process by which fat globules are broken into smaller pieces by the detergent actions of bile salt and especially lecithin. The emulsification process increases the total surface area of the fats by as much as 1000-fold. The lipases are water-soluble enzymes and can attack fat globules only on their surfaces. Consequently, it can be readily understood how important this detergent action of bile salts is for the digestion of fats.

Triglycerides are digested by pancreatic lipase. The most important enzyme for digestion of triglycerides is pancreatic lipase. This is present in such enormous quantities in pancreatic juice that all triglycerides are digested into free fatty acids and 2-monoglycerides within a few minutes.

Bile salts form micelles that accelerate fat digestion. The hydrolysis of triglycerides is highly reversible; therefore, accumulation of monoglycerides and free fatty acids in the vicinity of digesting fats quickly blocks further digestion. Bile salts form micelles that remove monoglycerides and free fatty acids from the vicinity of the digesting fat globules. Micelles are composed of a central fat globule (containing monoglycerides and free fatty acids) with molecules of bile salt projecting outward to cover the surface of the micelle. The bile salt micelles also carry monoglycerides and free fatty acids to the brush borders of the intestinal epithelial cells.

BASIC PRINCIPLES OF GASTROINTESTINAL ABSORPTION
(p. 837)

The folds of Kerckring, villi, and microvilli increase the mucosal absorptive area by nearly 1000-fold. The total area of the small intestinal mucosa is 250 or more square meters—about the surface area of a tennis court.

- *Folds of Kerckring* increase the surface area of the absorptive mucosa by about threefold.
- *Villi* project about 1 millimeter from the surface of the mucosa, increasing the absorptive area by an additional 10-fold.
- *Microvilli* covering the villar surface (brush border) increase the surface area exposed to the intestinal materials by at least an additional 20-fold.

ABSORPTION IN THE SMALL INTESTINE (p. 839)
Absorption of Water

Water is transported through the intestinal membrane entirely by diffusion. Water is absorbed from the gut when the chyme is dilute, and water moves into the intestine when hyperosmotic solutions enter the duodenum. As dissolved substances are absorbed from the gut, the osmotic pressure of the chyme tends to decrease, but water diffuses so readily through the intestinal membrane that it almost instantaneously "follows" the absorbed substances into the blood.

Absorption of Ions (p. 839)

Sodium is actively transported through the intestinal membrane. Sodium is actively transported from inside the intestinal epithelial cells through the basal and side walls (basolateral membrane) of these cells into the paracellular spaces, which decreases the intracellular sodium concentration. This low concentration of sodium provides a steep electrochemical gradient for sodium movement from the chyme through the brush border into the epithelial cell cytoplasm. The osmotic gradient created by the high concentration of ions in the paracellular space causes water to move by osmosis through the tight junctions between the apical borders of the epithelial cells and, finally, into the circulating blood of the villus.

The absorption of sodium ions creates slight electronegativity in the chyme and electropositivity on the basal side of the epithelial cells. Chloride ions move along this electrical gradient to "follow" the sodium ions.

Aldosterone greatly enhances sodium absorption. Dehydration leads to aldosterone secretion by the adrenal glands, which greatly enhances sodium absorption by the intestinal epithelial cells. The increased sodium absorption then causes secondary increased absorption of chloride ions, water, and some other substances. This effect of aldosterone is especially important in the colon because it allows virtually no loss of sodium chloride in the feces and little water loss.

Cholera causes extreme secretion of chloride ions, sodium ions, and water from the crypts of Lieberkuhn. The toxins of cholera and some other types of diarrheal bacteria can stimulate the secretion of sodium chloride and water so greatly that as much as 5 to 10 liters of water and salt can be lost each day as diarrhea. Within 1 to 5 days, many severely affected patients die of this loss of fluid alone. In most instances, the life of the cholera victim can be saved by the simple administration of tremendous amounts of sodium chloride solution to make up for losses.

Calcium, iron, potassium, magnesium, and phosphate ions are actively absorbed.

- *Calcium ions* are actively absorbed in relation to the need of the body for calcium. Calcium absorption is controlled by parathyroid hormone and vitamin D; the parathyroid hormone activates vitamin D in the kidneys. The activated vitamin D in turn greatly enhances calcium absorption.
- *Iron ions* are also actively absorbed from the small intestine. The principles of iron absorption and the regulation of its absorption are discussed in Chapter 32.
- *Potassium, magnesium, phosphate,* and probably *other ions* can also be actively absorbed through the mucosa.

Absorption of Carbohydrates (p. 841)

Essentially all carbohydrates are absorbed in the form of monosaccharides. Only a small fraction of carbohydrates are absorbed as disaccharides, and almost none are absorbed as larger carbohydrate compounds. The most abundant of the absorbed monosaccharides is glucose, usually accounting for more than 80 per cent of the absorbed carbohydrate calories. Glucose is the final digestion product of our most abundant carbohydrate food, the starches.

Glucose is transported by a sodium co-transport mechanism. Active transport of sodium through the basolateral membranes into the paracellular spaces depletes the sodium inside the cells. This decrease causes sodium to move through the brush border of the enterocyte to its interior by facilitated diffusion. The sodium combines with a transport protein that requires another substance, such as glucose, to bind simultaneously. When intestinal glucose combines with the transport protein, sodium and glucose are transported into the cell at the same time.

Other monosaccharides are transported. *Galactose* is transported by almost exactly the same mechanism as glucose. In contrast, *fructose* is transported by facilitated diffusion all the way through the enterocyte but is not coupled with sodium transport. In addition, much of the fructose is converted into glucose on its way through the enterocyte and finally is transported in the form of glucose the remainder of the way into the paracellular space.

Absorption of Proteins (p. 842)

Most proteins are absorbed through the luminal membranes of the intestinal epithelial cells in the form of dipeptides, tripeptides, and a few free amino acids. The energy for most of this transport is supplied by a sodium co-transport mechanism in the same way that sodium co-transport of glucose occurs. A few amino acids do not require this sodium co-transport mechanism but instead are transported by special membrane transport proteins in the same way that fructose is transported—via facilitated diffusion.

Absorption of Fats (p. 842)

Monoglycerides and fatty acids diffuse passively through the enterocyte cell membrane to the interior of the enterocyte. Lipids are soluble in the enterocyte membrane. After entering the enterocyte, the fatty acids and monoglycerides are mainly recombined to form new triglycerides. A few of the monoglycerides are further digested into glycerol and fatty acids by an intracellular lipase. These free fatty acids also are reconstituted by the smooth endoplasmic reticulum into triglycerides.

Chylomicrons are excreted from the enterocytes by exocytosis. The reconstituted triglycerides aggregate within the Golgi apparatus into globules that contain cholesterol and phospholipids. The phospholipids arrange themselves with the fatty portions toward the center and the polar portions on the surface, providing an electrically charged surface that makes the globules miscible with water. The globules are released from the Golgi apparatus and excreted by exocytosis into the basolateral spaces; from there, they pass into the lymph in the central lacteal of the villus. These globules are then called *chylomicrons.*

Chylomicrons are transported in the lymph. From the basolateral surfaces of the enterocytes, the chylomicrons wend their way into the central lacteals of the villi and are then propelled, along with the lymph, by the lymphatic pump upward through the thoracic duct to be emptied into the great veins of the neck. Between 80 and 90 per cent of all fat absorbed from the gut is absorbed in this manner and transported to the blood by way of the thoracic lymph in the form of chylomicrons.

ABSORPTION IN THE LARGE INTESTINE: FORMATION OF THE FECES (p. 843)

The proximal half of the colon is important for absorption of electrolytes and water. The mucosa of the large intestine has a high capability for active absorption of sodium, and the electrical potential created by absorption of sodium causes chloride absorption as well. The tight junctions between the epithelial cells are tighter than those of the small intestine, which decreases back-diffusion of ions through these junctions. This allows the large intestinal mucosa to absorb sodium ions against a higher concentration gradient than can occur in the small intestine. The absorption of sodium and chloride ions creates an osmotic gradient across the large intestinal mucosa, which in turn causes absorption of water.

The large intestine can absorb a maximum of about 5 to 7 liters of fluid and electrolytes each day. When the total quantity entering the large intestine through the ileocecal valve or by way of large intestinal secretion exceeds this maximum absorptive capacity, the excess appears in the feces as diarrhea.

The feces normally are about three-fourths water and one-fourth solid matter. The solid matter is composed of about 30 per cent dead bacteria, 10 to 20 per cent fat, 10 to 20 per cent inorganic matter, 2 to 3 per cent protein, and 30 per cent undigested roughage of the food and dried constituents of digestive juices, such as bile pigment and sloughed epithelial cells. The brown color of feces is caused by stercobilin and urobilin, which are derivatives of bilirubin. The odor is caused principally by indole, skatole, mercaptan, and hydrogen sulfide.

66

Physiology of Gastrointestinal Disorders

The logical therapy of most gastrointestinal disorders depends on a basic knowledge of gastrointestinal physiology. In this chapter, we discuss a few representative types of malfunction that have special physiological bases or consequences.

DISORDERS OF SWALLOWING AND THE ESOPHAGUS
(p. 845)

Paralysis of the swallowing mechanism can result from nerve damage, brain damage, or muscle dysfunction.

- *Nerve damage*—Damage to the 5th, 9th, or 10th nerve can cause paralysis of significant portions of the swallowing mechanism.
- *Brain damage*—Diseases such as poliomyelitis and encephalitis can prevent normal swallowing by damaging the swallowing center in the brain stem.
- *Muscle dysfunction*—Paralysis of the swallowing muscles, as occurs in muscular dystrophy or in failure of neuromuscular transmission in myasthenia gravis or botulism, can also prevent normal swallowing.

When the swallowing mechanism is partially or totally paralyzed, the abnormalities that can occur include the following:

- Complete abrogation of the swallowing act so that swallowing cannot occur.
- Failure of the glottis to close so that food passes into the lungs instead of the esophagus.
- Failure of the soft palate and uvula to close the posterior nares so that food refluxes into the nose during swallowing.

Achalasia is a condition in which the lower esophageal sphincter fails to relax during the swallowing mechanism. As a result, the movement of food into the stomach is impeded or prevented. Pathological studies have shown damage in the myenteric plexus in the lower two thirds of the esophagus. The musculature remains spastically contracted, and *receptive relaxation* of the gastroesophageal sphincter does not occur. Swallowed material builds up, stretching the esophagus; over months and years, the esophagus becomes tremendously enlarged.

DISORDERS OF THE STOMACH (p. 845)

Gastritis **means inflammation of the gastric mucosa.** The inflammation can penetrate into the gastric mucosa, causing atrophy. Gastritis can be acute and severe, with ulcerative excoriation of the stomach mucosa. It may be caused by chronic bacterial infection of the gastric mucosa; in addition, irritant substances, such as alcohol and aspirin, can damage the protective gastric mucosal barrier.

The stomach is protected by the gastric mucosal barrier. Absorption from the stomach is normally low for two reasons: (1) the gastric mucosa is lined with mucous cells that secrete a viscid and adherent mucus, and (2) the mucosa has tight junctions between adjacent epithelial cells. These impediments to gastric absorption are called the *gastric mucosal barrier*. In gastritis, this barrier becomes leaky, allowing hydrogen ions to diffuse into the stomach epithelium. A vicious circle of progressive mucosal damage and atrophy can develop, making the mucosa susceptible to peptic digestion, frequently resulting in gastric ulcer.

Chronic gastritis can lead to hypochlorhydria or achlorhydria. Chronic gastritis can cause atrophy of the gastric mucosal glandular function.

- *Achlorhydria* means simply that the stomach fails to secrete hydrochloric acid.
- *Hypochlorhydria* means diminished acid secretion.

Pernicious anemia is a common accompaniment of achlorhydria and gastric atrophy. Intrinsic factor is a glycoprotein secreted by parietal cells. It combines with vitamin B_{12} in the stomach to protect it from being destroyed in the gut. When the intrinsic factor–vitamin B_{12} complex reaches the terminal ileum, the intrinsic factor binds with receptors on the ileal epithelial surface, making it possible for the vitamin B_{12} to be absorbed.

Peptic ulcer is an excoriated area of the mucosa caused by the digestive action of gastric juice. The most frequent site of ulcer formation is the first few centimeters of the duodenum. A peptic ulcer can be caused by either of two ways:

- Excess secretion of acid and pepsin by the gastric mucosa
- Diminished ability of the gastroduodenal mucosal barrier to protect against the digestive properties of the acid-pepsin complex.

Bacterial infection by *Helicobacter pylori* breaks down the gastroduodenal mucosal barrier. At least 75 per cent of peptic ulcer patients have recently been found to have chronic infection of the gastric and duodenal mucosa by the bacterium *H. pylori*. The bacterium is capable of penetrating the mucosal barrier by burrowing through the barrier and releasing digestive enzymes that liquefy the barrier. As a result, gastric secretions can penetrate the mucosa and digest the epithelial cells; this leads to peptic ulceration.

Almost all patients with peptic ulceration can be treated effectively. The following treatments are commonly used:

- *Antibiotics*, such as tetracycline, plus other agents kill the infectious bacteria.
- *Antihistamines* block the stimulatory effect of histamine on gastric gland histamine receptors, reducing gastric acid secretion by 70 to 80 per cent.

DISORDERS OF THE SMALL INTESTINE (p. 847)

Abnormal digestion results from failure of the pancreas to secrete its juice. The loss of pancreatic juice means the loss of many digestive enzymes. As a result, large portions of the ingested food are not used for nutrition, and copious, fatty feces are excreted. The lack of pancreatic secretion frequently occurs in the following instances:

- Pancreatitis (which is discussed later)
- When the pancreatic duct is blocked by a gallstone at the papilla of Vater
- After the head of the pancreas has been removed because of malignancy.

Pancreatitis **means inflammation of the pancreas.** Ninety per cent of all cases are caused by excess alcohol ingestion or blockage of the papilla of Vater by a gallstone. When the main secretory duct is blocked by a gallstone, the pancreatic enzymes are dammed up in the pancreas. These enzymes rapidly digest large portions of the pancreas, sometimes completely and permanently destroying the ability of the pancreas to secrete digestive enzymes.

Nutrients may not be adequately absorbed from the small intestine even though the food is well digested. Several diseases can cause decreased absorption by the mucosa; they are often classified together under the general heading of *sprue.*

- *Nontropical sprue*—One type of sprue results from the toxic effects of gluten present in certain types of grains, especially wheat and rye. In susceptible persons, gluten has a destructive effect on the intestinal enterocytes.
- *Tropical sprue*—This type frequently occurs in the tropics and often can be treated with antibacterial agents.

DISORDERS OF THE LARGE INTESTINE (p. 848)

Constipation **means slow movement of feces though the large intestine.** It is often associated with large quantities of dry, hard feces in the descending colon that accumulate because of the long time available for absorption of fluid.

Severe constipation can lead to megacolon. When large quantities of fecal matter accumulate in the colon for an extended time, the colon can distend to a diameter of 3 to 4 inches; this condition is called *megacolon*, or *Herchsprung's disease*. The most frequent cause is a lack or deficiency of ganglion cells in the myenteric plexus in a segment of the sigmoid colon. As a consequence, neither defecation reflexes nor strong peristaltic motility can occur through this area of the large intestine.

Diarrhea often results from the rapid movement of fecal matter through the large intestine. The following are several causes of diarrhea:

- *Enteritis*—This is infection in the intestinal tract, most often occurring in the large intestine. The result is increased motility and increased rate of secretion by the irritated mucosa, both of which contribute to diarrhea.

- *Psychogenic diarrhea*—Many students are familiar with the diarrhea that occurs during examination time. This type of diarrhea is caused by parasympathetic stimulation, which excites both motility and secretion of mucus in the distal colon.
- *Ulcerative colitis*—This is a disease in which the walls of the large intestine become inflamed and ulcerated. The motility of the ulcerated colon is often so great that mass movements occur most of the time. In addition, the secretions of the colon are greatly enhanced.

GENERAL DISORDERS OF THE GASTROINTESTINAL TRACT
(p. 849)

Antiperistalsis is a prelude to vomiting. The antiperistalsis may begin as far down in the intestinal tract as the ileum and push the intestinal contents back to the duodenum and stomach within a few minutes. As the duodenum becomes excessively distended, the actual vomiting act can be initiated.

The vomiting act results from a squeezing action of abdominal muscles with sudden opening of the esophageal sphincters. Once the vomiting center has been stimulated and the vomiting act is instituted, the first effects are a (1) deep breath, (2) raising of the hyoid bone and larynx to pull open the upper esophageal sphincter, (3) closing of the glottis, and (4) lifting of the soft palate to close the posterior nares. Next, the diaphragm and abdominal muscles contract simultaneously, building the intragastric pressure to a high level. Finally, the lower esophageal sphincter relaxes, allowing expulsion of gastric contents.

The abnormal consequences of obstruction depend on the point in the gastrointestinal tract that becomes obstructed.

- *If the obstruction occurs at the pylorus*, which often results from fibrotic constriction after peptic ulceration, persistent vomiting of stomach contents occurs. This depresses bodily nutrition; it also causes excessive loss of hydrogen ions and can result in alkalosis.
- *If the obstruction is beyond the stomach*, antiperistaltic reflux from the small intestine causes intestinal juices to flow into the stomach, and these juices are vomited along with the stomach secretions. The person becomes severely dehydrated, but the

loss of acids and bases may be approximately equal, so that little change occurs in acid-base balance.

- *If the obstruction is near the lower end of the small intestine*, it is possible to vomit more basic than acidic substances; in this case, acidosis may result. In addition, after a few days of obstruction, the vomitus becomes fecal in character.
- *If the obstruction is near the distal end of the large intestine*, feces can accumulate in the colon for several weeks. The patient develops an intense feeling of constipation, but finally it becomes impossible for additional chyme to move from the small intestine into the large intestine; then severe vomiting begins.

UNIT
XIII

Metabolism and Temperature Regulation

67

Metabolism of Carbohydrates and Formation of Adenosine Triphosphate

The next few chapters deal with the chemical processes that make it possible for the cells to continue living. These chapters are devoted to a review of the principal chemical processes of the cell and an analysis of their physiological implications, especially in relation to the manner in which they fit into the overall concept of homeostasis.

Adenosine triphosphate (ATP) plays a role in metabolism. A great proportion of the chemical reactions in cells are concerned with making the energy in foods available to the various physiological systems of the cell. The substance *adenosine triphosphate (ATP)* plays a key role in making the energy of the foods available for this purpose. ATP is a labile chemical compound containing two high-energy phosphate bonds. The amount of free energy in each of these phosphate bonds is approximately 12,000 calories under conditions found in the body.

ATP is present in the cytoplasm and nucleoplasm of all cells. Essentially all of the physiological mechanisms that require energy for operation obtain this energy directly from ATP (or other, similar high-energy compounds, such as *guanosine triphosphate [GTP]*). In turn, the food in the cells is gradually oxidized and the released energy is used to re-form the ATP, so that a supply of this substance is continually maintained.

TRANSPORT OF MONOSACCHARIDES THROUGH THE CELL MEMBRANE (p. 857)

The final products of carbohydrate digestion in the alimentary tract are almost entirely *glucose, fructose,* and *galactose.* These monosaccharides cannot diffuse through the usual pores in the cell membrane. To enter the cell, these monosaccharides combine with protein carriers in the membrane that allow them to pass through the membrane via facilitated diffusion into the cell. After passing through the membrane, the monosaccharides become dissociated from the carriers.

Insulin increases facilitated diffusion of glucose. The rate of glucose transport through the cell membranes is greatly increased by *insulin*. The amount of glucose that can diffuse into the cells of the body in the absence of insulin, with the exception of the liver and brain, is too little to supply the amount of glucose normally required for energy metabolism; therefore, the rate of carbohydrate utilization by the body is controlled mainly by the rate of insulin secretion from the pancreas.

Glucose is phosphorylated in the cell by the enzyme *glucokinase*. The phosphorylation of glucose is almost completely irreversible, except in the liver cells, renal tubular epithelium, and intestinal epithelial cells, in which *glucose phosphatase* is available for reversing the reaction. In most tissues of the body, phosphorylation serves to capture the glucose in the cell. Once in the cell, the glucose will not diffuse out except from those special cells that have the necessary phosphatase.

STORAGE OF GLYCOGEN IN LIVER AND MUSCLE
(p. 857)

After absorption into cells, glucose can be used immediately for energy or stored in the form of *glycogen*, a large polymer of glucose. All cells of the body are capable of storing some glycogen, but liver and muscle cells can store large quantities of glycogen. The glycogen molecule can be polymerized to form very large molecules with an average molecular weight of 5 million. These large glycogen molecules precipitate to form solid granules.

Glycogenesis is the process of glycogen formation. *Glycogenolysis* is the process of glycogen breakdown to re-form glucose; it is not the reverse process of glycogenesis. In glycogenolysis, the glucose molecule on each branch of the glycogen polymer is split away by the process of phosphorylation catalyzed by the enzyme *phosphorylase*.

Under resting conditions, the phosphorylase enzyme is in the inactive form. When it is required to reform glucose from glycogen, phosphorylase can be activated by the hormones *epinephrine* and *glucagon*. The initial effect of each of these hormones is to increase the formation of *cyclic adenosine monophosphate (cAMP)*. The cAMP initiates a cascade of chemical reactions that activates the phosphorylase.

RELEASE OF ENERGY FROM THE GLUCOSE MOLECULE BY THE GLYCOLYTIC PATHWAY (p. 858)

The complete oxidation of 1 mole of glucose releases 686,000 calories of energy, but only 12,000 calories of energy are required to form 1 mole of ATP. It would be an extreme waste of energy if glucose decomposed to water and carbon dioxide while forming only a single molecule of ATP. Fortunately, cells contain an extensive series of different protein enzymes that cause the glucose molecule to split a little at a time in many successive steps. The energy in glucose is released in small packets to form one molecule of ATP at a time. A total of 38 moles of ATP is formed for each mole of glucose utilized by the cells.

Glycolysis involves the formation of pyruvic acid. In glycolysis the glucose molecule is split to form two molecules of pyruvic acid. This process occurs in 10 successive steps, with each step being catalyzed by at least one specific protein enzyme.

Despite the many chemical reactions in the glycolytic series, only 2 moles of ATP are formed for each mole of glucose that is utilized; this amounts to 24,000 calories of energy stored in the form of ATP. The total amount of energy lost from the original glucose molecule is 56,000 calories, so that the overall efficiency for ATP formation during glycolysis is 43 per cent. The remaining 57 per cent of the energy is lost in the form of heat.

Pyruvic acid is converted to *acetyl-coenzyme A (acetyl-CoA)*. The next stage in the degradation of glucose is the conversion of the two molecules of pyruvic acid into two molecules of acetyl-CoA. During this reaction, two carbon dioxide molecules and four hydrogen atoms are released. No ATP is formed. Six molecules of ATP are produced, however, when the four hydrogen atoms are later oxidized in the process of *oxidative phosphorylation*.

Continued degradation of the glucose molecule occurs in the citric acid cycle. This is a sequence of chemical reactions in which the acetyl portion of acetyl-CoA is degraded to carbon dioxide and hydrogen atoms. These reactions occur *in the matrix of the mitochondrion.* The hydrogen atoms released are subsequently oxidized, liberating tremendous amounts of energy to form ATP. No large amount of energy is released during the citric acid cycle; however, for each molecule of glucose metabolized, two molecules of ATP are formed.

FORMATION OF ADENOSINE TRIPHOSPHATE BY OXIDATIVE PHOSPHORYLATION OF HYDROGEN ATOMS
(p. 860)

Despite the complexities of glycolysis and the citric acid cycle, only small amounts of ATP are formed during these processes. Two ATP molecules are formed in the glycolytic scheme, and another two molecules are formed in the citric acid cycle. Almost 95 per cent of the total amount of ATP is formed during subsequent oxidation of the hydrogen atoms that are released during these early stages of glucose degradation. The principal function of these earlier stages is to make the hydrogen of the glucose molecule available in a form that can be utilized for oxidation.

Oxidative phosphorylation is accomplished through a series of enzymatically catalyzed reactions. During this process, the hydrogen atoms are converted into hydrogen ions and electrons. The electrons eventually combine with the dissolved oxygen of the fluids to form hydroxyl ions. The hydrogen and hydroxyl ions combine with each other to form water. During this sequence of oxidative reactions, tremendous quantities of energy are released to form ATP; this is called *oxidative phosphorylation.* This process occurs entirely in the mitochondria by a highly specialized process called the *chemiosmotic mechanism.*

The electrons removed from the hydrogen atoms enter an electron transport chain that is an integral component of the inner membrane of the mitochondria. This transport chain consists of a series of electron acceptors that can be reversibly reduced or oxidized by accepting or giving up electrons. The important members of the electron transport chain include *flavoprotein,* several *iron sulfide proteins, ubi-*

quinone, and *cytochromes B, C_1, C, A,* and *A_3.* Each electron is shuttled from one of the acceptors to the next until it reaches cytochrome A_3. Cytochrome A_3 is called *cytochrome oxidase* because by giving up two electrons, it is capable of causing elemental oxygen to combine with hydrogen ions to form water. During the transport of these electrons through the electron transport chain, energy is released and used to synthesize ATP.

ATP is formed. The energy released as the electrons pass through the electron transport chain is used to create a gradient of hydrogen ions across the inner membrane of the mitochondria. The high concentration of hydrogen ions across this space creates a large electrical potential difference across the membrane, which causes hydrogen ions to flow into the mitochondrial matrix through a molecule called *ATP synthetase.* The energy derived from the hydrogen ions is utilized by the ATP synthetase to convert ADP into ATP. For each two hydrogen atoms ionized by the electron transport chain, up to three molecules of ATP are synthesized.

SUMMARY OF ADENOSINE TRIPHOSPHATE FORMATION DURING BREAKDOWN OF GLUCOSE

- The total number of ATP molecules formed from one molecule of glucose is 38.
- Two molecules of ATP are formed during glycolysis.
- Two molecules of ATP are formed during the citric acid cycle.
- Thirty-four molecules of ATP are formed during oxidative phosphorylation.

Thus, 456,000 calories of energy are stored in the form of ATP, whereas 686,000 calories are released during the complete oxidation of each mole of glucose; this represents an overall efficiency of 66 per cent. The remaining 34 per cent of the energy becomes heat.

Glycolysis and glucose oxidation are regulated. Continuous release of energy from glucose when the energy is not needed by the cells would be an extremely wasteful process. Glycolysis and the subsequent oxidation of hydrogen atoms are continually controlled in accordance with the needs of the cell for ATP. This control is accomplished by a feedback mechanism related to the concentrations of both ADP and ATP.

One important way in which ATP helps control energy metabolism is an allosteric inhibition of the enzyme *phosphofructokinase*. This enzyme promotes the formation of fructose-1,6-diphosphate in the initial steps of the glycolytic series. The net effect of excess cellular ATP is to stop glycolysis, which in turn stops most carbohydrate metabolism. Conversely, ADP causes the opposite allosteric change in this enzyme, greatly increasing its activity. Whenever ATP is used by the tissues for energy, the ATP inhibition of the enzyme is reduced, but at the same time its activity is increased as a result of the ADP that is formed; the glycolytic process is set in motion. When cellular stores of ATP are replenished, the enzyme is again inhibited.

RELEASE OF ENERGY IN THE ABSENCE OF OXYGEN— ANAEROBIC GLYCOLYSIS (p. 862)

Occasionally, oxygen becomes either unavailable or insufficient so that cellular oxidation of glucose cannot take place. Under these conditions, a small amount of energy can still be released to the cells through glycolysis because the chemical reactions in the glycolytic breakdown of glucose to pyruvic acid do not require oxygen. The process of *anaerobic glycolysis* is extremely wasteful of glucose because only 24,000 calories of energy are used to form ATP for each mole of glucose. This represents a little over 3 per cent of the total energy in the glucose molecule; however, this release of glycolytic energy to the cells can be a lifesaving measure for a few minutes when oxygen is unavailable.

The formation of *lactic acid* during anaerobic glycolysis allows the release of extra anaerobic energy. The end products of the glycolytic reactions, pyruvic acid and NADH, combine under the influence of the enzyme *lactic dehydrogenase* to form lactic acid and NAD^+. This prevents the buildup of pyruvic acid and NADH, which would inhibit the glycolytic reactions. The lactic acid formed readily diffuses out of the cells into the extracellular fluids. Lactic acid represents a type of "sinkhole" into which the glycolytic end products can disappear, allowing glycolysis to proceed far longer than would otherwise be possible.

PENTOSE PHOSPHATE PATHWAY—AN ALTERNATIVE MEANS OF RELEASING ENERGY FROM GLUCOSE
(p. 862)

As much as 30 per cent of the glucose breakdown in the liver and fat cells is accomplished independent of glycolysis and the citric acid cycle. The pentose phosphate pathway is a cyclic process that removes one carbon atom from a glucose molecule to produce carbon dioxide and hydrogen during each turn of the cycle. The hydrogen produced eventually enters the oxidative phosphorylation pathway to form ATP. This pathway provides the cell with another mechanism of glucose utilization in the event of enzymatic abnormalities.

GLUCONEOGENESIS—FORMATION OF CARBOHYDRATES FROM PROTEINS AND FATS (p. 863)

When the body stores of carbohydrates decrease below normal levels, moderate quantities of glucose can be formed from amino acids and the glycerol portion of fat through the process of gluconeogenesis. Approximately 60 per cent of the amino acids in the body proteins can be easily converted into carbohydrates; each amino acid is converted into glucose through a slightly different chemical process. A low level of carbohydrates in the cells and a decrease in blood glucose are the basic stimuli that increase the rate of gluconeogenesis.

68

Lipid Metabolism

Several different chemical compounds in food and in the body are classified as lipids, including (1) *neutral fat*, or *triglycerides*; (2) *phospholipids*; and (3) *cholesterol*. Chemically, the basic lipid moiety of the triglycerides and phospholipids is fatty acids, which are simply long-chain hydrocarbon organic acids. Although cholesterol does not contain fatty acids, its sterol nucleus is synthesized from degradation products of fatty acid molecules, giving it many of the physical and chemical properties of other lipid substances.

The triglycerides are mainly used in the body to provide energy for the different metabolic processes; this function is shared almost equally with the carbohydrates. Some lipids, especially cholesterol, phospholipids, and derivatives of these compounds, are utilized throughout the body to perform other intracellular functions.

TRANSPORT OF LIPIDS IN THE BODY FLUIDS (p. 865)

Chylomicrons transport lipids from the gastrointestinal tract to the blood via lymph. Essentially all fats in the diet are absorbed into the lymph in the form of chylomicrons. The chylomicrons are transported in the thoracic duct and emptied into the venous blood. Chylomicrons are removed from the plasma as they pass through the capillaries of adipose and liver tissue. The membranes of the adipose and liver cells contain large quantities of an enzyme called *lipoprotein lipase*; this enzyme hydrolyzes the triglycerides of the chylomicrons into fatty acids and glycerol. The fatty acids immediately diffuse into the cells; once inside, they are resynthesized into triglycerides.

Albumin transports fatty acids released from adipose tissue. When the fat that has been stored in the fat cells is to be used elsewhere in the body, it must be transported to the other tissues; it is mainly transported in the form of *free fatty acids*. On leaving the fat cells, the fatty acids ionize strongly in the plasma and immediately combine loosely with albumin of the plasma proteins. The fatty acid bound with proteins in this manner is called *free fatty acid* to distinguish it from other fatty acids in the plasma that exist in the form of esters of glycerol, cholesterol, or other substances.

Lipoproteins transport cholesterol, phospholipids, and triglycerides. Lipoproteins are particles that are much smaller than chylomicrons but similar in composition; they contain mixtures of triglycerides, phospholipids, cholesterols, and proteins. The three major classes of lipoproteins are (1) *very low density lipoproteins*, which contain high concentrations of triglycerides and moderate concentrations of both phospholipids, and cholesterol; (2) *low density lipoproteins*, which contain relatively few triglycerides but a very high concentration of cholesterol; and (3) *high density lipoproteins*, which contain about 50 per cent protein with smaller concentrations of lipids.

Lipoproteins are formed almost entirely in the liver. The principal function of the various types of lipoproteins in the plasma is to transport a specific type of lipid throughout the body. Triglycerides are synthesized in the liver mainly from carbohydrates and transported to adipose and other peripheral tissue in the very low density lipoproteins. The low density lipoproteins are the residuals of the very low density lipoproteins after they have delivered most of their triglycerides to the adipose tissue and left behind large concentrations of cholesterol and phospholipids in the low density lipoproteins. The *high density lipoproteins* transport cholesterol away from the peripheral tissues to the liver; this type of lipoprotein plays a very important role in the prevention of the development of atherosclerosis.

FAT DEPOSITS (p. 867)

Large quantities of fat are stored in the adipose tissue. The major function of the adipose tissue is to store triglycerides until they are needed to provide energy elsewhere in the body. A secondary function of the adipose tissue is to provide heat insulation for the body.

Fat cells of the adipose tissue are modified fibroblasts that are capable of storing almost pure triglycerides in quantities equal to 80 to 95 per cent of their volume. Large quantities of lipases are present in adipose tissue. Some of these enzymes catalyze the deposition of triglycerides derived from the chylomicrons and other lipoproteins. Others, when activated by hormones, cause splitting of the triglycerides in the fat cells to release free fatty acids. Because of the rapid exchange of the fatty acids, the triglycerides in the fat cells are renewed approximately once every 2 to 3 weeks, making fat a very dynamic tissue.

The liver contains large quantities of triglycerides, phospholipids, and cholesterol. The liver has multiple functions in lipid metabolism: (1) to degrade fatty acids into smaller compounds that can be used for energy; (2) to synthesize triglycerides, mainly from carbohydrates and proteins; and (3) to synthesize other lipids from fatty acids, especially cholesterol and phospholipids.

When large quantities of triglycerides are mobilized from the adipose tissue, which occurs during starvation or in diabetes mellitus, the triglycerides are redeposited in the liver, where the initial stages of fat degradation begin. Under normal physiological conditions, the amount of triglycerides present in the liver is determined by the rate at which lipids are being used for energy.

USE OF TRIGLYCERIDES FOR ENERGY (p. 867)

The first stage in the use of triglycerides for energy is hydrolysis of the triglycerides into fatty acids and glycerol. The fatty acids and glycerol are transported to active tissues, where they are oxidized to release energy. Almost all cells, with some degree of exception for brain tissue, can use fatty acids almost interchangeably with glucose for energy.

The degradation and oxidation of fatty acids occur only in the mitochondria; the first step in the metabolism of fatty acids is their transport into the mitochondria. This is a carrier-mediated process that employs carnitine as a carrier substance. Once inside the mitochondria, the fatty acids split away from carnitine and are degraded and oxidized.

Fatty acids are degraded in the mitochondria by *beta oxidation*, which releases two-carbon segments to form acetyl-coenzyme A (acetyl-CoA). The acetyl-CoA molecule formed by the beta oxidation of fatty acids enters the citric acid cycle and is degraded into carbon dioxide and hydrogen atoms. The hydrogen is subsequently oxidized by the oxidative enzymes of the mitochondria to form ATP.

Acetoacetic acid is formed in the liver. A large share of the degradation of fatty acids into acetyl-CoA occurs in the liver, but the liver uses only a small portion of the acetyl-CoA for its own intrinsic metabolic processes. Instead, pairs of acetyl-CoA condense to form molecules of *acetoacetic acid*. A large part of the acetoacetic acid is converted into *β-hydroxybutyric acid* and minute quantities of *acetone*. The acetoacetic acid and β-hydroxybutyric acid freely diffuse through the liver cell membranes and are transported by the blood to peripheral tissues. In the peripheral tissue, these compounds diffuse into the cells, in which reverse reactions occur and acetyl-CoA molecules are re-formed. These molecules enter the citric acid cycle of the cells and are oxidized for energy.

Synthesis of Triglycerides from Carbohydrates

Whenever quantities of carbohydrates enter the body that are greater than can be used immediately for energy or stored in the form of glycogen, the excess is rapidly converted into triglycerides and stored in this form in adipose tissue. Most triglyceride synthesis occurs in the liver, but a small amount occurs in fat cells. The triglycerides formed in the liver are mainly transported by the lipoproteins to fat cells of the adipose tissue and stored until needed for energy.

Carbohydrates are converted into fatty acids. The first step in the synthesis of triglycerides from carbohydrates is conversion of the carbohydrates into acetyl-CoA; this occurs during the normal degradation of glucose by the glycolytic system. Fatty acids are actually large polymers of the acetyl portion of acetyl-CoA, so it is not difficult to understand how acetyl-CoA can be converted into fatty acids.

Fatty acids are combined with α-glycerophosphate to form triglycerides. Once the synthesized fatty acid chains have grown to contain 14 to 18 carbon atoms, they automatically bind with glycerol to form triglycerides. The glycerol portion of the triglyceride is furnished by α-glycerophosphate, which is also a product of the glycolytic breakdown of glucose. The real importance of this mechanism in the formation of triglycerides is that the final combination of fatty acids with glycerol is controlled mainly by the concentration of the α-glycerophosphate, which in turn is determined by the availability of carbohydrates. When carbohydrates form large quantities of α-glycerophosphate, the equilibrium shifts to promote formation and storage of triglycerides. When carbohydrates are not available, the process shifts in the opposite direction; the excess fatty acids become available to substitute for the lack of carbohydrate metabolism.

Fat storage and synthesis are important. Fat synthesis from carbohydrates is especially important because the different cells of the body have limited capacities for storing carbohydrates in the form of glycogen. The average person has about 200 times as much energy stored in the form of fat as in the form of carbohydrate. Storage of energy in the form of fat is also important because each gram of fat contains approximately 2¼ times as many calories of usable energy as each gram of glycogen. For a given weight gain, a person can store far more energy in the form of fat than in the form of carbohydrate.

Synthesis of Triglycerides from Proteins

Many amino acids can be converted into acetyl-CoA, which subsequently can be converted into triglycerides. When more protein is available in the diet than can be utilized as protein or directly for energy, a large share of the excess energy is stored as fat.

Hormonal regulation of fat utilization. At least seven hormones secreted by the endocrine system have marked effects on fat utilization:

- *Epinephrine* and *norepinephrine* released by the adrenal medulla dramatically increase fat utilization during heavy exercise. These two hormones directly activate *hormone-sensitive triglyceride lipase*, which is present in abundance in the fat cells. The activated hormone causes rapid breakdown of triglycerides and mobilization of fatty acids. Other stressors that activate the sympathetic nervous system increase fatty acid mobilization and utilization in a similar manner.
- *Corticotropin* is released by the anterior pituitary gland in response to stress and causes the adrenal cortex to secrete *glucocorticoids* (*cortisol*). Both corti-

cotropin and the glucocorticoids activate hormone-sensitive triglyceride lipase, which increases the release of fatty acids from the fat tissue.

- *Growth hormone* has an effect similar to but less effective than that of corticotropin and glucocorticoids in activating the hormone-sensitive lipases. Growth hormone can also have a mild fat-mobilizing effect. A lack of *insulin* activates hormone-sensitive lipase and causes rapid mobilization of fatty acids. When carbohydrates are not available in the diet, insulin secretion diminishes; this promotes fatty acid metabolism.

- *Thyroid hormone* causes rapid mobilization of fat. This process is believed to result indirectly from an increased rate of energy metabolism in all the cells of the body under the influence of this hormone.

PHOSPHOLIPIDS AND CHOLESTEROL (p. 871)

Phospholipids. The three major types of phospholipids in the body are *lecithins, cephalins,* and *sphingomyelins.* Phospholipids are used throughout the body for various structural purposes; they are an important constituent of lipoproteins in the blood and are essential to the formation and function of these compounds. The absence of phospholipids can cause serious abnormalities in the transport of cholesterol and other lipids. Thromboplastin, which is needed to initiate the clotting process, is composed mainly of one of the cephalins. Large quantities of sphingomyelins are present in the nervous system. This substance acts as an insulator in the myelin sheath around the nerve fibers. Perhaps the most important function of the phospholipids is participation in the formation of the structural elements, mainly membranes, within the cells throughout the body.

Cholesterol. Cholesterol is present in all diets; it is absorbed from the gastrointestinal tract into the intestinal lymph. In addition to the cholesterol that is absorbed each day from the gastrointestinal tract *(exogenous cholesterol),* a large quantity *(endogenous cholesterol)* is formed in the cells of the body. Essentially all the endogenous cholesterol that circulates in the lipoproteins of the plasma is formed by the liver. Cholesterol is a structural component in cell membranes.

By far the most abundant nonmembranous use of cholesterol in the body is in the formation of *cholic acid* in the liver; about 80 per cent of the cholesterol is converted into cholic acid. Cholic acid is conjugated with other substances to form bile salts, which promote digestion and absorption of fats.

A small quantity of cholesterol is used by the (1) adrenal glands to form *adrenal cortical hormones*, (2) ovaries to form *progesterone* and *estrogen*, and (3) testes to form *testosterone*.

ATHEROSCLEROSIS (p. 873)

Atherosclerosis is principally a disease of large and intermediate arteries in which lipid deposits called *atheromatous plaques* appear in the intimal and subintimal layers of the arteries. These plaques contain a large amount of cholesterol and are associated with degenerative changes in the arterial wall. In the later stages of the disease, fibroblasts infiltrate the degenerative areas and cause a progressive sclerosis of the arteries. In addition, calcium often precipitates with the lipids to develop calcified plaques. When these two processes occur, the arteries become extremely hard; the resulting disease is called *arteriosclerosis*, or simply "hardening of the arteries."

Arteriosclerotic arteries lose most of their distensibility and, because of the degenerative areas, are easily ruptured. In addition, the atheromatous plaques often break through the intima into the flowing blood. The protrusion of the plaques promotes the development of blood clots with resultant thrombus or embolus formation. Almost half of all humans die of some complication of arteriosclerosis.

Role of Low Density Lipoproteins in Causing Atherosclerosis

Essentially all the lipoproteins are formed in the liver. Most of the lipoproteins formed are the *very low density lipoproteins*, which contain very large quantities of triglycerides and cholesterol. Much of the triglyceride and cholesterol portions of the lipoproteins are released into the tissues, however, and the lipoproteins change from very low density to simply *low density lipoproteins*. At this stage, many of these are recaptured by the liver, and their constituents are reused to transport more triglycerides and cholesterols. The recapture process requires the presence of receptors on the liver cell membranes that attach to the protein portion of the lipoprotein. Many individuals have a hereditary deficiency of these receptors, so that the low density lipoproteins are not captured and build up in the blood, causing more cholesterol deposition in the tissues and arterial walls. In fact, when recapture fails to occur, the liver synthesizes more lipoproteins, making the situation worse.

The *high density lipoproteins* are an entirely separate entity from the very low density and low density lipoproteins. The high density lipoproteins are formed mainly in the liver; they have the capability of removing cholesterol from the tissues rather than causing additional deposition. Individuals with high blood levels of high density lipoproteins have a diminished likelihood of developing atherosclerosis.

69

Protein Metabolism

About three fourths of the solids of the body are proteins, including *structural proteins, enzymes, proteins that transport oxygen, proteins of the muscle that causes contraction,* and many other types that perform specific functions both intracellularly and extracellularly.

The principal constituents of proteins are amino acids, of which 20 are present in the body in significant quantities. The amino acids are aggregated into long chains by means of *peptide linkages*. A complicated protein molecule may have as many as 100,000 amino acids. Some protein molecules are composed of several peptide chains rather than a single chain; these chains may be linked by hydrogen bonding; electrostatic forces; or sulfhydryl, phenolic, or salt entities.

TRANSPORT AND STORAGE OF AMINO ACIDS (p. 878)

The normal concentration of amino acids in the blood is between 35 and 65 mg/dl. Recall that the end products of protein digestion in the gastrointestinal tract are almost entirely amino acids and that polypeptides or protein molecules are only rarely absorbed into the blood. After a meal, the amino acids entering the blood are absorbed within 5 to 10 minutes by cells throughout the entire body.

The molecules of essentially all the amino acids are much too large to diffuse through the pores of the cell membranes; therefore, the amino acids are transported through the membrane only by active transport or facilitated diffusion using a carrier mechanism.

Amino acids are stored as proteins in the cells. Almost immediately after entry into the cells, amino acids are conjugated under the influence of intracellular enzymes into cellular proteins, so that the concentration of free amino acids inside the cells almost always remains low. The amino acids are stored mainly in the form of actual proteins. Many intracellular proteins can be rapidly decomposed into amino acids under the influence of intracellular lysosomal digestive enzymes; these amino acids can be transported back into the blood. Special exceptions to this are the proteins in the chromosomes of the nucleus and structural proteins such as collagen and muscle contractile proteins; these proteins do not participate significantly in this reversible storage of amino acids.

Whenever the plasma amino acid concentration falls below the normal level, amino acids are transported out of the cell to replenish the supply in the plasma. Simultaneously, intracellular proteins are degraded into amino acids.

Each particular type of cell has an upper limit to the amount of proteins it can store. After the cells have reached their limits, the excess amino acids in the circulation are degraded into other products and utilized for energy or converted to fat or glycogen and stored.

FUNCTIONAL ROLES OF THE PLASMA PROTEINS (p. 879)

The major types of proteins present in the plasma are *albumin*, *globulin*, and *fibrinogen*. The principal function of albumin is to provide colloid osmotic pressure in the plasma. The globulins are mainly responsible for immunity against invading organisms. Fibrinogen polymerizes into long, branching fibrin threads during blood coagulation, thereby forming blood clots that help repair leaks in the circulatory system.

Plasma proteins form in the liver. Essentially all the albumin and fibrinogen and 50 to 80 per cent of the globulins are formed in the liver. The remaining globulins (mainly gamma globulins in antibodies) are formed in the lymphoid tissue. The rate of plasma protein formation by the liver can be as much as 30 grams per day. The rapid production of plasma proteins by the liver is valuable in preventing death from conditions such as those found in severe burns, which cause the loss of many liters of plasma through the denuded areas of the skin, and in severe renal disease, in which as much as 20 grams of plasma protein per day can be lost in the urine.

When the tissues become depleted of proteins, the plasma proteins can act as a source for rapid replacement. Whole plasma proteins can be absorbed by the liver, split into amino acids, transported back into the blood, and used throughout the body to build cellular proteins. In this way, the plasma proteins function as a labile storage medium and represent a rapidly available source of amino acids.

There are essential and nonessential amino acids. Ten of the 20 amino acids normally present in animal proteins can be synthesized in the cells; the other 10 amino acids either cannot be synthesized or are synthesized in quantities too small to supply the needs of the body. This second group of amino acids are called *essential amino acids* because they must be supplied in the diet. Synthesis of the nonessential amino acids depends on the formation of the appropriate *α-keto acid* precursor of the respective amino acid. Pyruvic acid, which is formed in large quantities during the glycolytic breakdown of glucose, is the α-keto acid precursor of the amino acid *alanine*. During the process of transamination, an amino radical is transferred to the α-keto acid to form the alanine molecule, and the keto oxygen is transferred to the donor of the amino radical.

Proteins are used for energy. Once the protein stores of the cell are full, additional amino acids in the body fluids are degraded and used for energy or stored mainly as fat or glycogen. This degradation occurs almost entirely in the liver. The first step in the degradation process is the removal of amino groups through the process of deamination. This generates the specific α-keto acid that can enter into the citric acid cycle. The amount of adenosine triphosphate formed from each gram of protein oxidized is slightly less than that formed from each gram of glucose. The ammonia released during the deamination process is removed from the blood almost entirely through conversion into *urea* by the liver. In the absence of the liver or in severe liver disease, ammonia accumulates in the blood. The ammonia is very toxic, especially to the brain, and often leads to the state of *hepatic coma*.

Obligatory degradation of proteins can occur. When the diet contains no proteins, a certain proportion of the proteins of the body continue to be degraded into amino acids. These amino acids are deaminated and oxidized; this process involves 20 to

30 grams of protein per day and is called the *obligatory loss of proteins*. To prevent a net loss of proteins from the body, one must ingest at least 20 to 30 grams of protein per day. The minimum recommended amount of protein in the diet is 60 to 75 grams per day.

HORMONAL REGULATION OF PROTEIN METABOLISM
(p. 881)

Growth hormone increases the rate of synthesis of cellular proteins, causing the tissue proteins to increase. The mechanism of action of growth hormone on protein synthesis is not known, but growth hormone is believed to enhance the transport of amino acid through the cell membrane and accelerate the DNA and RNA transcription and translation processes for protein synthesis. Part of the action might also result from the effect of growth hormone on fat metabolism. Growth hormone causes an increased rate of fat liberation from the fat depots; this reduces the rate of oxidation of amino acids and subsequently increases the quantity of amino acids that are available for synthesis into proteins.

Insulin **accelerates the transport of amino acids into the cells.** Insulin deficiency reduces protein synthesis to almost zero; insulin also increases the availability of glucose to the cells, so that the use of amino acids for energy is correspondingly reduced.

Glucocorticoids **decrease the quantity of proteins in most tissues and increase the amino acid concentration in the plasma.** It is believed that glucocorticoids act by increasing the rate of breakdown of extrahepatic proteins, making greater quantities of amino acid available in the body fluids. The effects of glucocorticoids on protein metabolism are especially important in promoting ketogenesis and gluconeogenesis.

Testosterone **increases the deposition of protein in tissues throughout the body, especially muscles.** The mechanism of this effect is not known, but it is different from the effect of growth hormone. Growth hormone causes tissues to continue growing almost indefinitely, whereas testosterone causes the muscles and other protein tissues to enlarge for only several months. Beyond this time, despite the continued administration of testosterone, further protein deposition ceases.

Estrogen **causes slight deposition of protein.** The effect of estrogen is relatively insignificant in comparison with that of testosterone.

Thyroxin **increases the rate of metabolism in all cells and indirectly affects protein metabolism.** If insufficient carbohydrates and fats are available for energy, thyroxin causes rapid degradation of proteins for energy. If adequate quantities of carbohydrates and fats are available, the excess amino acids are utilized to increase the rate of protein synthesis.

A deficiency of thyroxin causes growth to be greatly inhibited because of a lack of protein synthesis. It is believed that thyroxin has little specific direct effect on protein metabolism but does have an important general effect on increasing the rates of both normal anabolic and normal catabolic protein reactions.

70

The Liver as an Organ

The basic functions of the liver include the following:

- Filtration and storage of blood
- Metabolism of carbohydrates, fats, proteins, hormones, and xenobiotics
- Formation and excretion of bile
- Storage of vitamins and iron
- Formation of coagulation factors.

HEPATIC BLOOD AND LYMPH FLOW (p. 884)

The rate of blood flow to the liver is high, and the resistance to flow is low. The rate of blood flow from the portal vein to the liver is approximately 1100 ml/min. An additional 350 ml/min enters the liver through the hepatic artery, so that the rate of total blood flow to the liver is 1450 ml/min, or about 29 per cent of the cardiac output. Under normal conditions, resistance to blood flow through the liver is low, as demonstrated by a 9-mm Hg pressure drop from the portal vein (average pressure, 9 mm Hg) to the vena cava (average pressure, 0 mm Hg). Under certain pathological conditions, such as cirrhosis (the development of fibrous tissue in the liver) or blood clots in the portal vein, blood flow through the liver can be greatly impeded. The rise in vascular resistance in the liver can lead to a rise in capillary pressure throughout the splanchnic circulation, causing significant fluid loss from the capillaries of the intestinal tract, ascites, and, possibly, death.

Rate of lymph flow from the liver is very high. The pores of the hepatic sinusoids are very permeable, which readily allows the passage of both proteins and fluids into the lymphatic system. The protein concentration in the lymph from the liver is approximately 6 gm/dl (slightly less than the plasma protein concentration). The relatively high

609

permeability of the liver sinusoidal epithelium allows leakage of large amounts of protein, causing large quantities of lymph to form. About half of all lymph formed in the body under normal conditions arises from the liver.

A rise in hepatic pressure (due to cirrhosis or congestive heart failure) causes a corresponding rise in liver lymph flow. A rise in vena cava pressure from 0 to 15 mm Hg can increase liver lymph flow to as much as 20 times the normal rate. Under certain pathological conditions, the excess amount of lymph formed can begin to transude through the outer surface of the liver directly into the abdominal cavity, resulting in ascites.

METABOLIC FUNCTIONS OF THE LIVER (p. 885)

Taken together, the hepatic cells are a large chemically reactant pool of cells that share substrates and energy from a myriad of metabolic systems. The liver processes and synthesizes multiple substances that are transported to and from other areas of the body.

Carbohydrate Metabolism. In carbohydrate metabolism, the liver performs the following functions:

- Stores large quantities of glycogen
- Converts galactose and fructose to glucose
- Acts as the primary site for gluconeogenesis
- Produces intermediate products of carbohydrate metabolism.

One of the major functions of the liver in carbohydrate metabolism is the maintenance of a normal blood glucose concentration. The liver can remove excess glucose from the blood and store it in the form of glycogen. When blood glucose level begins to fall, the liver can convert the glycogen back to glucose; this is called the *glucose buffer function* of the liver. When the blood glucose concentration falls below normal, the liver begins to convert amino acids and glycerol into glucose through the process of gluconeogenesis in an effort to maintain a normal blood glucose concentration.

Fat Metabolism. Although almost all cells of the body metabolize fat, certain aspects of fat metabolism occur mainly in the liver.

- *Beta oxidation of fats to acetyl-coenzyme A (acetyl-CoA) occurs very rapidly in the liver.* The excess acetyl-CoA that is formed is converted to acetoacetic acid, a highly soluble molecule that can be transported to other tissues, where it can be reconverted to acetyl-CoA and used for energy.
- *The liver synthesizes large quantities of cholesterol, phospholipids, and most lipoproteins.* About 80 per cent of the cholesterol synthesized in the liver is converted to bile salts; the remainder is transported by lipoproteins to the tissues of the body. Phospholipids are also transported in the blood by lipoproteins. Both cholesterol and phospholipids are utilized by various cells of the body to form membranes and intracellular structures.
- *Almost all fat synthesis from carbohydrates and proteins occurs in the liver.* The fat synthesized in this way is transported by the lipoproteins to adipose tissue for storage.

Protein Metabolism. The body cannot dispense with the services of the liver in protein metabolism for more than a few days without death ensuing. The most important functions of the liver in protein metabolism are as follows:

- *Deamination of amino acids,* which is required before they can be used for energy or converted into carbohydrates or fats. Almost all deamination of amino acids takes place in the liver.
- *Formation of urea,* which removes ammonia from the body fluids. Large amounts of ammonia are formed by the deamination process and produced by the action of the gut bacteria. In the absence of this function in the liver, the plasma ammonia concentration can rise rapidly.
- *Formation of plasma proteins;* essentially all plasma proteins are formed in the liver (with the exception of the gamma globulins, which are formed in the lymphoid tissues).
- *Interconversion of the various amino acids and synthesis of metabolic compounds from amino acids;* an important function of the liver is its ability to synthesize the nonessential amino acids and to convert amino acids into other metabolically important compounds.

Miscellaneous Metabolic Functions of the Liver (p. 886)

Vitamins and iron are stored. The liver has a propensity for storing vitamins and iron; it stores a sufficient quantity of vitamin D to prevent vitamin D deficiency for about 4 months, sufficient vitamin A to prevent vitamin A deficiency for approximately 10 months, and sufficient vitamin B_{12} to prevent vitamin B_{12} deficiency for 1 year.

When iron is available in extra quantities in body fluids, it combines with the protein apoferritin to form ferritin and is stored in this form in the hepatic cells.

Clotting factors are formed. The liver forms the following substances needed in the coagulation process: *fibrinogen*, *prothrombin*, *accelerator globulin*, and *Factor VII*; therefore, liver dysfunction can lead to blood coagulation abnormalities.

Hormones and xenobiotics are metabolized. The liver is well known for its ability to detoxify and excrete many drugs and hormones, such as estrogen, cortisol, and aldosterone. Liver damage can lead to the accumulation of drugs and hormones in the body.

FORMATION AND EXCRETION OF BILIRUBIN BY THE LIVER (p. 886)

Bilirubin is a toxic end product of hemoglobin metabolism that is excreted in the bile. When the heme portion of hemoglobin is metabolized, a substance called *biliverdin* is formed; this substance is rapidly reduced to *bilirubin*, which immediately combines with plasma albumin. This combination of plasma albumin and bilirubin is called *free bilirubin*.

The free bilirubin is absorbed by the hepatic cells, where it is released from the plasma albumin and conjugated with either glucuronide to form *bilirubin glucuronide* or sulfate to form *bilirubin sulfate*. The conjugated forms of bilirubin are excreted in the bile into the intestine, where they are converted through bacterial action to *urobilinogen*. Urobilinogen is very soluble; some of the urobilinogen is reabsorbed by the intestinal mucosa into the blood. About 5 per cent of the urobilinogen absorbed in this way is excreted into the urine by the kidneys; the remaining urobilinogen is re-excreted by the liver.

Jaundice is an excess of either free bilirubin or conjugated bilirubin in the extracellular fluids.

Jaundice can be caused by an (1) increased destruction of red blood cells (i.e., hemolytic jaundice) and (2) obstruction of the bile ducts or damage to the liver cells so that bilirubin cannot be excreted into the gastrointestinal tract (i.e., obstructive jaundice).

In *hemolytic jaundice*, excretory function of the liver is not impaired, but red blood cells are hemolyzed so rapidly that the hepatic cells cannot excrete the bilirubin as rapidly as it is formed. The plasma concentration of free bilirubin rises to levels much above normal. In *obstructive jaundice*, the bile ducts may be obstructed by gallstones or cancer, or the hepatic cells may be damaged, as in hepatitis. The rate of bilirubin formation and the conjugation of bilirubin by the liver are near normal, but the conjugated bilirubin cannot pass into the intestines. In obstructive jaundice, the level of conjugated bilirubin in the blood rises, so that most of the bilirubin in the plasma is in the conjugated form rather than in the free form.

71

Dietary Balances; Regulation of Feeding; Obesity and Starvation; Vitamins and Minerals

The intake of food must be sufficient to supply the metabolic needs of the body; too little food intake leads to starvation, and too much intake leads to obesity.

Energy is available in foods. The approximate amount of energy that is liberated from each gram of carbohydrate as it is oxidized to carbon dioxide and water is 4.0 Calories. For each gram, the amount of energy liberated from fat is 9.0 Calories and the amount liberated from protein is 4.0 Calories.

Average Americans receive about 15 per cent of their energy from protein, 40 per cent from fat, and 45 per cent from carbohydrates. In non-Western diets most of the energy comes from carbohydrates; proteins and fats compose only 15 to 20 per cent of the total energy consumed.

There is a daily protein requirement. About 20 to 30 grams of protein per day is degraded by the body to manufacture other compounds, so that all cells must continue to form new proteins to take the place of those being destroyed. *The average person can maintain normal protein stores when consuming 30 to 40 grams of protein per day.*

Some proteins have inadequate amounts of certain *essential amino acids* and cannot replace the degraded proteins. Proteins that lack the essential amino acids are called *partial proteins*. For example, corn meal lacks the amino acid tryptophan. An individual consuming corn meal as the only source of protein will develop a protein-deficient syndrome called *kwashiorkor*, which consists of failure to grow, depressed mentality, and low plasma protein that in turn leads to severe edema.

METHODS FOR DETERMINING UTILIZATION OF NUTRIENTS BY THE BODY (p. 890)

Nitrogen balance is an index of the quantity of protein breakdown each day. The average protein contains about 16 per cent nitrogen. During metabolism, about 90 per cent of this nitrogen is excreted in the urine in the form of *urea, uric acid*, and *creatinine*. The remaining 10 per cent is excreted in the feces. By measuring the amount of nitrogen in the urine, adding 10 per cent for fecal excretion, and multiplying by 6.25 (100/16), the amount of protein breakdown (in grams) can be determined.

If the daily intake of protein is less than the daily breakdown of protein, the person is said to have a *negative nitrogen balance*; this indicates that the body stores of protein are decreasing.

Respiratory quotient is the ratio of carbon dioxide output to oxygen utilization. When carbohydrates are metabolized with oxygen, 1 carbon dioxide molecule is formed for every molecule of oxygen consumed. For carbohydrates, the respiratory quotient is 1.0. When fat is metabolized with oxygen, 7 carbon dioxide molecules are formed for every 10 molecules of oxygen consumed, so that the respiratory quotient for fat metabolism is 0.70. For proteins, the respiratory quotient is 0.80.

The respiratory quotient can be an index of the relative utilization of different foods by the body. A person consuming mostly fat would have a respiratory quotient close to 0.70, whereas a person consuming mostly carbohydrates would have a respiratory quotient close to 1.0.

REGULATION OF FOOD INTAKE (p. 891)

Hunger is the intrinsic desire for food; it is associated with a number of objective sensations, such as rhythmic contractions of the stomach and restlessness.

Appetite is the desire for a particular type of food; it is useful in helping a person choose the quality of food to be eaten.

Satiety is the opposite of hunger; it is the feeling of fullness in the quest for food.

Neural centers regulate food intake. Stimulation of the *lateral hypothalamic nuclei* induces eating behaviors; this area is referred to as the *feeding center*. Stimulation of the *ventromedial hypothalamic nuclei* induces satiety, making this area of the hypothalamus the *satiety center*. Lesions of these area produce the opposite effect of stimulation. Other areas of the brain also have been implicated as sensors of the nutritional status of the body or neural centers that drive an animal to search for and devour food. Lesions in the *paraventricular nuclei* often cause excess eating, specifically of carbohydrates. Conversely, lesions in the *dorsomedial nuclei* of the hypothalamus usually depress eating. In addition, other areas of the lower brain stem, such as the *area postrema, caudal nucleus of the solitary tract,* or *vagus nerve,* can affect the degree of eating. How all these areas of the brain regulate the overall pattern of feeding is unclear.

Another aspect of feeding is the mechanical act of the feeding process itself. The actual mechanics of feeding, such as chewing, swallowing, and salivating, are controlled by centers in the brain steam. The function of the higher centers in feeding is to control the quantity of food intake and to stimulate the lower feeding-mechanics centers to activity.

The *prefrontal cortex* and *amygdala* are thought to play important roles in the control of appetite. The activities of these centers are closely coupled with the hypothalamus. Bilateral destruction of the amygdala produces a "psychic blindness" in the choice of foods and an inability to control the type or quality of the food that is consumed.

Factors that Regulate Food Intake

The regulation of the quantity of food intake can be divided into *long-term regulation,* which is concerned primarily with the long-term maintenance of normal quantities of energy stores in the body, and *short-term regulation,* which is concerned with the prevention of overeating at each meal.

Long-term regulation of food intake may be related to the concentration of glucose, lipids, and amino acids in the blood. An increase or a decrease in the blood concentration of these nutrients causes a corresponding decrease or increase in food intake. Our knowledge of the long-term regulation of food intake is imprecise, as demonstrated by the prevalence of obesity in modern society. In general, when energy stores of the body fall below normal, the feeding centers become active. When energy stores are adequate (mainly the fat store), the satiety centers become active, and a person loses the desire for food.

Short-term regulation of food intake is accomplished through several feedback signals from the alimentary tract. Distention of the stomach and duodenum causes inhibitory signals to be transmitted to the feeding center by way of the *vagi*, reducing the desire for food. The gastrointestinal hormone *cholecystokinin*, which is released in response to fat entering the duodenum, has a strong direct effect on the feeding center to decrease further eating. There also is a food-metering effect that decreases hunger sensation after a certain amount of food has passed through the mouth; the metering mechanism is not as strong as distention of the stomach to decrease food intake.

ABNORMALITIES OF FOOD INTAKE (p. 893)

Obesity results from a greater energy input than energy expenditure. The excess input results in an increase in fat stores and a corresponding increase in body weight. For each 9.3 Calories of excess energy that enters the body, 1 gram of fat is stored. The excess input occurs only during the development of obesity. Once a person has become obese, that person can eat the same amount of food as an individual of normal weight and remain obese. To reduce weight, a person must decrease energy input and increase energy output.

Obesity usually results from an abnormality in the regulation of food intake; this abnormality may be *psychogenic*. People are known to gain weight during or after stressful life situations.

Neurogenic abnormalities may also contribute to obesity. The functional organization of the hypothalamic centers that regulate food intake may be different in the obese individual.

Genetic factors are known to contribute to obesity. Obesity is known to run in certain families. Genetic factors may contribute to obesity by causing an abnormality in the neural feeding centers or may be related to the chemistry of fat storage.

Inanition and starvation can occur. Inanition is the exact opposite of obesity. In addition to being the result of an inadequate food supply, inanition can result from psychogenic factors, such as *anorexia nervosa*.

When food intake is chronically insufficient, starvation results. During starvation, the energy stores of the body are depleted at different rates. Carbohydrate stores (glycogen) are depleted within 12 to 24 hours. Fat is the main source of energy during starvation, and it is depleted at a constant rate. Proteins are utilized rapidly at first as they are converted to glucose through the process of gluconeogenesis. As starvation continues and the readily available stores of protein are exhausted, the rate of gluconeogenesis is reduced to about one fourth of its previous rate, and the rate of protein depletion is greatly reduced.

When almost all the available fat stores become depleted, the rate of protein utilization increases again as the proteins become the only remaining energy source. Because proteins are essential to the maintenance of cellular function, death ordinarily occurs when the body's proteins are depleted to about one half their normal level.

VITAMINS (p. 895)

Vitamins are organic compounds that are needed in small quantities for normal metabolism. Vitamins cannot be synthesized in the cells of the body and therefore must be supplied in the diet. Vitamin deficiency causes specific metabolic deficits.

Vitamin A occurs in animal tissues as retinol. Vitamin A does not occur in foods of vegetable origin, but provitamins for the formation of vitamin A occur in abundance in many vegetable foods. These provitamins can be converted to vitamin A in the liver. The basic function of vitamin A in metabolism is not clear except in relation to its use in the formation of retinal pigments in the eye. Vitamin A deficiency causes (1) night blindness, (2) scaliness of the skin and acne, (3) failure of skeletal growth in young animals, and (4) failure of reproduction.

Thiamine (vitamin B_1) is needed for the final metabolism of carbohydrates and many amino acids. Thiamine operates in metabolic systems of the body as a cocarboxylase in conjunction with a protein decarboxylase for decarboxylation of pyruvic acid and other α-keto acids. Thiamine deficiency *(beriberi)* causes a decreased utilization of pyruvate and some amino acids by the tissues; it can affect the central nervous system, cardiovascular system, and gastrointestinal tract.

Niacin (nicotinic acid) functions in the body as a hydrogen acceptor. Niacin in the form of *nicotinamide adenine dinucleotide (NAD)* and *nicotinamide adenine dinucleotide phosphate (NADP)* functions as a coenzyme in the metabolic cascades. When a niacin deficiency exists, the normal rate of dehydrogenation cannot be maintained. Oxidative delivery of energy from food to the functional elements of the cells cannot occur at normal rates. Niacin deficiency *(pellagra)* causes lesions of the central nervous system, irritation and inflammation of the mucous membranes, muscle weakness, poor glandular secretion, and gastrointestinal hemorrhage.

Riboflavin (vitamin B$_2$) functions as a hydrogen carrier. Riboflavin combines with phosphoric acid to form *flavin adenine dinucleotide (FAD)*, which operates as a hydrogen carrier of the important oxidative systems of the body. Riboflavin deficiencies can cause many of the same effects as lack of niacin in the diet. These debilities result from a generalized depression of the oxidative process within the cells.

Vitamin B$_{12}$ functions as a hydrogen acceptor coenzyme. Perhaps the most important function of vitamin B$_{12}$ is its ability to act as a coenzyme for reducing ribonucleotides to deoxyribonucleotides, a step necessary for the replication of genes. Vitamin B$_{12}$ is important for red blood cell formation, growth, and maturation. Vitamin B$_{12}$ deficiency leads to poor growth and *pernicious anemia,* a type of anemia caused by failure of red blood cell maturation.

Vitamin B$_{12}$ deficiency is not caused by lack of this substance in foods but rather by a deficiency of *intrinsic factor.* Intrinsic factor is normally secreted by the parietal cells of the gastric glands and is essential to the absorption of vitamin B$_{12}$ by the ileal mucosa.

Folic acid (pteroylglutamic acid) is a potent promoter of growth and maturation of red blood cells. One of the significant effects of folic acid deficiency is the development of *macrocytic anemia,* an anemia that is almost identical to pernicious anemia (which results from vitamin B$_{12}$ deficiency).

Pyridoxine (vitamin B$_6$) is a coenzyme for many chemical reactions related to amino acid and protein metabolism. The most important role of pyridoxine is that of a coenzyme in the transamination process for the synthesis of amino acids. Pyridoxine deficiency can cause dermatitis, decreased rate of growth, development of a fatty liver, anemia, and evidence of mental deterioration.

Pantothenic acid is incorporated in the body into *coenzyme A (CoA)*. A lack of pantothenic acid can lead to depressed metabolism of both carbohydrates and fats.

Ascorbic acid (vitamin C) is essential for collagen formation. Ascorbic acid activates the enzyme *prolyl hydroxylase*, which promotes the hydroxylation step in the formation of *hydroxyproline*, an integral component of collagen. Without ascorbic acid, the collagen fibers are defective and weak. This vitamin is essential for the growth and strength of collagen fibers in subcutaneous tissue, cartilage, bone, and teeth. Deficiency of ascorbic acid *(scurvy)* results in failure of wounds to heal, inhibition of bone growth, and petechial hemorrhages throughout the body.

Vitamin D increases calcium absorption from the gastrointestinal tract and helps control calcium deposition in the bone. Vitamin D promotes the active transport of calcium through the epithelium of the ileum. Vitamin D deficiency *(rickets)* causes abnormalities of calcium metabolism, which can affect the strength and growth of bones.

Vitamin E prevents the oxidation of unsaturated fats. In the absence of vitamin E, the quantity of unsaturated fats in the cells becomes diminished, causing abnormal structure and function of the mitochondria, lysosomes, and cell membranes.

Vitamin K is necessary for the formation of clotting factors. The synthesis in the liver of *prothrombin, Factor VII, Factor IX*, and *Factor X* requires vitamin K. Deficiency of vitamin K causes a retardation of blood clotting. Vitamin K is normally synthesized by bacteria in the colon and absorbed by the colonic epithelium.

MINERAL METABOLISM (p. 899)

Magnesium, calcium, phosphorus, and iron are present in almost all tissues of the body.

Magnesium is required as a catalyst for many intracellular enzymatic reactions, particularly those related to carbohydrate metabolism.

Calcium is present in the body mainly in the form of calcium phosphate in the bone.

Phosphorus is the major anion of extracellular fluids.

Phosphates have the ability to combine reversibly with many coenzyme systems that are necessary for operation of the metabolic processes.

Iron functions in the body as a carrier of oxygen and as an electron acceptor; it is absolutely essential to both the transport of oxygen to the tissues and operation of oxidative systems within the tissue cells.

Trace Elements. *Iodine, zinc,* and *fluorine* are present in the body in such small quantities that they are called *trace elements.* Iodine is important for the formation and function of thyroid hormone. Zinc is an important component of *carbonic anhydrase,* the enzyme responsible for the rapid combination of carbon dioxide and water in the blood, gastrointestinal mucosa, and kidney tubules. Zinc also is a component of *lactic dehydrogenase,* which is important for the interconversions of pyruvic acid and lactic acid. Fluorine does not seem to be necessary for metabolism, but does function to prevent tooth decay.

72

Energetics and Metabolic Rate

The intracellular substance used to energize almost all cellular functions is *adenosine triphosphate (ATP)*. ATP is often referred to as an "energy currency" in metabolism. It energizes the synthesis of cellular components, muscle contraction, active transport across membranes, glandular secretions, and nerve conduction.

Phosphocreatine serves as an "ATP buffer." Phosphocreatine, which is another substance that contains high-energy phosphate bonds, is present in cells in quantities several times as great as ATP. Phosphocreatine cannot act in the same manner as ATP as a direct coupling agent for the transfer of energy between food substances and functional cellular systems; however, phosphocreatine can transfer energy interchangeably with ATP. When extra amounts of ATP are available, phosphocreatine is synthesized; this builds a storehouse of energy in the form of phosphocreatine. When ATP utilization increases, the energy in phosphocreatine is transferred rapidly back to ATP. This effect keeps the concentration of ATP at an almost constant level as long as any phosphocreatine remains.

ANAEROBIC VERSUS AEROBIC ENERGY (p. 904)

Anaerobic energy is energy derived from food without utilization of oxygen. Aerobic energy is energy derived from food by oxidative metabolism. Under anaerobic conditions, carbohydrates are the only significant source of energy. In fact, glycogen is the best source of energy under anaerobic conditions because it is already phosphorylated, whereas glucose must be phosphorylated (a step requiring the expenditure of energy) before it can be utilized.

Anaerobic energy is utilized during strenuous bursts of activity. Oxidative processes are too slow to provide the needed energy required for a strenuous burst of activity; this energy must be supplied

623

from (1) the ATP already present in the muscle cells, (2) the phosphocreatine, and (3) the glycolytic breakdown of glycogen to lactic acid.

Oxygen debt is the excess consumption of oxygen after the completion of strenuous activity. After a period of strenuous exercise, a person continues to breathe hard and consume extra amounts of oxygen for a few minutes. This excess oxygen is used to (1) reconvert the accumulated lactic acid back to glucose, (2) reconvert AMP and ADP to ATP, (3) re-establish phosphocreatine levels, (4) re-establish normal concentrations of oxygen bound to hemoglobin and myoglobin, and (5) raise the oxygen concentration in the lungs back to normal levels.

METABOLIC RATE (p. 906)

The metabolic rate is normally expressed in terms of the rate of heat liberation during the chemical reactions in all the cells of the body. Heat is the end product of almost all of the energy released in the body. On the average, 35 per cent of the energy in foods becomes heat during ATP formation. More energy becomes heat as it is transferred from ATP to the functional systems of the body. Under the best conditions, approximately 27 per cent of all the energy from food is used by the functional systems; almost all this energy eventually becomes heat. The only significant exception occurs when muscles are used to perform some form of work outside the body, such as elevating an object or walking up steps. In these cases, potential energy is created by raising a mass against gravity. When external expenditure of energy is not taking place, it is safe to consider that all the energy released by the metabolic processes eventually becomes body heat.

The *calorie* is the unit used for expressing the quantity of energy released from different foods or expended by the different functional processes of the body. The *gram calorie* is the quantity of heat required to raise the temperature of 1 gram of water 1°C. The gram calorie is much too small a unit for ease of expression when speaking of energy in the body, so that the "large calorie" (sometimes spelled with a capital "C" and often called the *kilocalorie*, which is equivalent to 1000 calories) is the unit ordinarily used in discussing energy metabolism.

Metabolic rate can be measured. Because a person is ordinarily not performing any external work, the whole body metabolic rate can be determined by measuring the total quantity of heat liberated from the body in a given time. *Direct calorimetry* is physically difficult to perform, however, so that other indirect methods are employed to determine the metabolic rate. One of the most accurate indirect methods is to determine the rate of oxygen utilization. For the average diet, the quantity of energy liberated per liter of oxygen consumed in the body is about 4.825 Calories. This is called the *energy equivalent* of oxygen. With this equivalent, one can calculate with a high degree of precision the rate of heat liberated in the body from the quantity of oxygen used in a given period of time.

Basal metabolic rate can be determined. The basal metabolic rate is a measure of the inherent metabolic rate of the tissues independent of exercise or other extraneous factors; it is the rate of energy utilization in the body during absolute rest while the person is awake. The usual method for determining the basal metabolic rate is to measure the rate of oxygen utilization in a given period of time. The basal metabolic rate is then calculated in terms of Calories/hr. The basal metabolic rate normally averages about 60 Calories/hr in young men and about 53 Calories/hr in young women. To correct for body size, the basal metabolic rate is normally expressed in proportion to the body surface area; this allows a comparison of basal metabolic rates between individuals of different sizes.

Factors that Affect the Metabolic Rate (p. 907)

When an average 70-kilogram man lies in bed all day, he uses approximately 1650 Calories of energy. The performance of other basic functions, such as sitting in a chair and eating, increases the amount of energy used. The daily energy requirement for simply existing (i.e., performing essential functions only) is about 2000 Calories/day.

Several factors can raise or lower the metabolic rate. The metabolic rate increases after a meal is ingested; this is mainly the result of the stimulatory effect of amino acids derived from the proteins of the ingested food on the chemical processes in the cell. Thyroid hormone, male sex hormone, growth hormone, sympathetic stimulation, and fever all increase the metabolic rate. Sleep, malnutrition, and age all decrease the metabolic rate.

73

Body Temperature, Temperature Regulation, and Fever

NORMAL BODY TEMPERATURES (p. 911)

The temperature of the deep tissues of the body (core temperature) remains constant within $\pm 1°F$ ($\pm 0.6°C$) despite large fluctuations in environmental temperature. The average normal body temperature is generally considered to be between 98.0°F and 98.6°F when measured orally and about 1°F higher rectally.

Temperature is controlled by the balance between heat production and heat loss. Heat production is a by-product of metabolism. Extra heat can be generated by muscle contraction (shivering) in the short term or by an increase in thyroxin in the long term. Most of the heat produced in the body is generated in the deep tissues. The rate of heat loss is determined by the rate of heat conduction to the skin and the rate of heat conduction from the skin to the surroundings.

Blood flow to the skin from the body core provides heat transfer. Blood vessels are distributed profusely immediately beneath the skin. An increase in blood flow to these vessels can cause more heat loss, and a decrease in blood flow to these vessels can cause less heat loss. The rate of flow to these vessels can vary from 0 to 30 per cent of the cardiac output. The skin is a very effective "heat radiator" system for the transfer of heat from the body core to the skin.

HEAT LOSS (p. 912)

Heat loss from the skin to the surroundings occurs by *radiation*, *conduction*, *convection*, and *evaporation* as follows.

Radiation causes loss of heat in the form of infrared rays. All objects above absolute zero radiate infrared waves in all directions. If the body temperature is greater than the surroundings, the body will radiate heat to the surroundings. Conversely, if the body temperature is lower than the surroundings, the surroundings will radiate heat to the body. About 60 per cent of the body heat is lost through radiation.

Conductive heat loss occurs by direct contact with an object. The body loses about 3 per cent of its heat by conduction to objects. An additional 15 per cent of the body heat is lost by conduction to air; the air in contact with the surface of the skin warms to near body temperature. This warm air has a tendency to rise away from the skin.

Convective heat loss results from air movement. The air next to the skin surface is warmed by conduction. When this warm air is removed, the skin will conduct heat to the "new" layer of unwarmed air.

Convective heat loss is the mechanism for the cooling effect of wind. The cooling effect of water is similar. Because water has such a high specific heat, however, the skin cannot warm a thin layer of water next to the body. As a consequence, heat is continuously removed from the body if the water is below body temperature.

Evaporation is a necessary mechanism of heat loss at very high temperatures. As water evaporates, 0.58 Calorie of heat is lost for each gram of water that is converted into the gaseous state. The energy to change water from a liquid to a gas is derived from the body temperature.

Evaporation accounts for 22 per cent of the heat lost by the body; evaporation of water through the skin (insensible water loss) accounts for about 12 to 16 Calories of heat loss per hour.

Evaporative heat loss is very important when the environmental temperatures are at or near the body temperature. Under these conditions, heat loss by radiation diminishes greatly. Evaporative heat loss becomes the only way to cool the body when environmental temperatures are high.

Air movement across the skin increases the rate of evaporation and as a result increases the effectiveness of evaporative heat loss (e.g., the cooling effect of a fan).

Sweating and Its Regulation by the Autonomic Nervous System (p. 914)

Sweat glands contain a deep coiled glandular portion and a straight ductal portion that exits on the surface of the skin. A *primary secretion* similar to plasma but without plasma proteins is formed by the glandular portion of the sweat gland. As the solution moves up the duct toward the surface of the skin, most of the electrolytes are reabsorbed, leaving a dilute watery secretion.

Sweat glands are innervated by *sympathetic cholinergic fibers*. When sweat glands are stimulated, the rate of precursor solution secretion is increased. The reabsorption of electrolytes occurs at a constant rate. If large volumes of precursor solution are secreted and at the same time electrolyte reabsorption remains constant, more electrolytes (primarily sodium chloride) will be lost in the sweat.

The sweating mechanism can adapt to meet environmental needs. Exposure to a hot climate causes an increase in the maximum rate of sweat production from about 1 liter/hr in the nonacclimatized person to as much as 2 to 3 liters/hr in the acclimatized individual. This higher amount of sweat increases the rate of evaporative heat loss and helps to maintain normal body temperature. Associated with an increase in the rate of sweat production is a decrease in the sodium chloride content of the sweat; this allows better conservation of body salt. The decline in the sodium chloride content of the sweat is primarily the result of an increased secretion of *aldosterone*, which enhances sodium reabsorption from the ductal portion of the sweat gland.

REGULATION OF BODY TEMPERATURE—ROLE OF THE HYPOTHALAMUS (p. 916)

The *anterior hypothalamic-preoptic area* contains large numbers of heat-sensitive neurons; the septum and reticular substance of the midbrain contain large numbers of cold-sensitive neurons. When the temperature centers detect that the body is either too hot or too cold, these areas institute appropriate and familiar temperature-increasing or -decreasing procedures.

Temperature-Decreasing Mechanisms. Three important mechanisms are used to cool the body, as follows:

- *Vasodilatation* of the blood vessels of the skin can increase the amount of heat transfer to the skin by as much as eightfold.
- *Sweating* increases the rate of evaporative heat loss. A 1°C increase in body temperature induces sufficient sweating to remove 10 times the basal rate of heat production.
- *Strong inhibition* takes place of mechanisms that increase heat production, such as shivering and chemical thermogenesis.

Temperature-Increasing Mechanisms. When the body is too cold, the temperature control systems initiate mechanisms to reduce heat loss and increase heat production through the following:

- *Vasoconstriction* of the blood vessels of the skin, which decreases the transfer of heat from the core of the body
- *Piloerection*, which raises the hair to trap air next to the skin and create a layer of warm air to act as an insulator. This mechanism works best in animals that have a complete layer of fur. The vestiges of this system are present in humans (goosebumps), but the effectiveness of this mechanism in humans is limited due to the relative sparseness of body hair.
- *Greater heat production by metabolic systems* such as sympathetic excitation of heat production, increase of thyroxine secretion, and shivering. Shivering can increase the rate of heat production by four-fold to fivefold. The *motor center* for shivering is located in the dorsomedial posterior hypothalamus; this area is inhibited by an increase in body temperature and stimulated by a decrease in body temperature. The output signals from this area are not rhythmic and do not cause the actual muscle shaking; instead, the output signals from this area cause a generalized increase in muscle tone. The greater muscle tone sets up an oscillation in the muscle spindle reflex, which leads to muscle shaking.

***Set-point* for temperature control.** The body maintains a critical core temperature of 37.1°C. When body temperature increases above this level, heat-losing mechanisms are initiated. When body temperature falls below this level, heat-generating mechanisms are initiated. This critical temperature is called the *set-point* of the temperature control system. All temperature control mechanisms continually attempt to bring the body temperature back to this level.

ABNORMALITIES OF BODY TEMPERATURE REGULATION
(p. 920)

Fever is a body temperature above normal. The elevated body temperature can be caused by an abnormality in the brain itself or toxic substances that affect the temperature-regulating centers. Fever results from a resetting of the set-point for temperature control; this resetting can be the result of proteins, protein breakdown products, or bacterial toxins (lipopolysaccharides) collectively called *pyrogens*. Some pyrogens act directly on the temperature control center; most pyrogens act indirectly.

When bacterial or viral particles are present in the body, they are phagocytized by *leukocytes, tissue macrophages,* and *large granular killer lymphocytes.* These cells release *interleukin-1* in response to the phagocytized particles. Interleukin-1 induces the formation of prostaglandin E_2, which acts on the hypothalamus to elicit the fever reaction. When prostaglandin formation is blocked by drugs, the fever is completely abrogated or at least reduced. This is the proposed mechanism of action for *aspirin* to reduce the level of fever and explains why aspirin does not lower the body temperature in a normal person (a normal person does not have elevated levels of interleukin-1).

When the interleukin-1 mechanism resets the set-point for temperature control, body temperature is maintained at a higher level. Raising the set-point of body temperature induces the subjective sensations of being cold, and nervous mechanisms initiate shivering and piloerection. Once the body temperature has reached the new set-point, the individual no longer has the subjective sensation of being cold, and body temperature is elevated above normal. When the pyrogens have been cleared from the body, the set-point for temperature control returns to normal levels. At this point, the body temperature is too warm; this induces the subjective sensations of being too hot, and nervous mechanisms initiate vasodilatation of the skin blood vessels and sweating.

UNIT
XIV

Endocrinology and Reproduction

74

Introduction to Endocrinology

CELL COMMUNICATION VIA CHEMICAL MESSENGERS

The types of intercellular communication by chemical messengers in the extracellular fluid include the following:

- *Neural,* in which neurotransmitters are released at synaptic junctions and act locally
- *Paracrine,* in which cell secretion products diffuse into the extracellular fluid and affect neighboring cells
- *Autocrine,* in which cell secretion products affect the function of the same cell by binding to cell surface receptors
- *Endocrine,* in which hormones reach the circulating blood and influence the function of cells some distance away
- *Neuroendocrine (neurocrine),* in which secretion products from neurons *(neurohormones)* reach the circulating blood and influence the function of cells some distance away.

MAINTENANCE OF HOMEOSTASIS AND REGULATION OF BODY PROCESSES

In many instances, neural and endocrine control of body processes is achieved through interactions between these two systems. These systems are linked by *neuroendocrine cells* located in the hypothalamus, whose axons terminate in the posterior pituitary gland and median eminence. The *neurohormones* secreted from these neuroendocrine cells include *ADH,* *oxytocin,* and *hypophysiotropic hormones* (which control the secretion of the anterior pituitary hormones). Hormones and neurohormones play a critical role in the regulation of almost all aspects of body function, including metabolism, growth and development, water and electrolyte balance, reproduction, and behavior.

CHEMISTRY, SYNTHESIS, STORAGE, AND SECRETION OF HORMONES (p. 927)

Hormones Classified According to Chemical Structure

Chemically, hormones and neurohormones are of the following three types:

- *Proteins and peptides*—Included in this group are peptides ranging from as small as three amino acids (thyrotropin-releasing hormone) to proteins almost 200 amino acids long (growth hormone and prolactin).
- *Steroids*—These are derivatives of cholesterol and include the adrenocortical (cortisol and aldosterone) and gonadal (testosterone, estrogen, and progesterone) hormones.
- *Derivatives of the amino acid tyrosine*—Included in this group are the hormones from the thyroid gland (thyroxine and triiodothyronine) and adrenal medulla (epinephrine and norepinephrine).

Synthesis, Storage, and Secretion of Hormones

Protein/peptide hormones are synthesized like most proteins. Protein/peptide hormones are synthesized on the rough endoplasmic reticulum in the same fashion as most other proteins. Typically, the initial protein formed by the endoplasmic reticulum is a larger protein than the active hormone and is called a *preprohormone*. This large protein is cleaved and modified at the endoplasmic reticulum. Subsequently, within the Golgi apparatus, the protein is packaged as a smaller *prohormone* for storage in *secretion granules*. The secretion granules may contain enzymes for further modification of prohormone, and when the endocrine cell is stimulated, the secretion granules migrate from the cytoplasm to the cell membrane. Free hormone and co-peptides are then released into the extracellular fluid by exocytosis.

Steroid hormones are synthesized from cholesterol. In contrast to protein/peptide hormones, there is little storage of hormone in steroid-producing endocrine cells. Typically, there are large stores of cholesterol esters in cytoplasmic vacuoles that can be rapidly mobilized for synthesis of steroid hormones after stimulation of the steroid-producing cell. Once the steroid hormone appears in the cytoplasm, storage does not take place, and the hormone diffuses through the cell membrane into the extracellular fluid. Much of the cholesterol in steroid-producing cells is removed from the plasma, but there is also de novo synthesis of cholesterol from acetate.

Thyroid hormones and catecholamines are synthesized from tyrosine. As with steroid hormones, there is no storage of thyroid hormones in discrete granules, and once thyroid hormones appear in the cytoplasm of the cell, they leave the cell via diffusion through the cell membrane. In contrast to steroid hormones, there are large stores of thyroxine and triiodothyronine as part of a large iodinated protein *(thyroglobulin)* that is stored in large follicles within the thyroid gland.

In comparison, the other group of hormones derived from tyrosine, the adrenal medullary hormones *epinephrine* and *norepinephrine*, are taken up into preformed vesicles and stored until secreted. As with protein hormones stored in secretion granules, catecholamines are released from adrenal medullary cells through exocytosis.

Control of Hormonal Secretion and Negative Feedback

In most instances, the rate of hormonal secretion is controlled by negative feedback. In general, endocrine glands tend to oversecrete hormone, which in turn drives target cell function. When too much function of the target cell occurs, some factor about the function feeds back to the endocrine gland and causes a negative effect on the gland to decrease its secretory rate.

MECHANISMS OF HORMONAL ACTION (p. 928)

Hormonal Receptors and Role in Hormonal Action

Hormones control cellular processes by interacting with receptors on target cells; these receptors are (1) either on or within the cell membrane, as in the case of peptide/protein and catecholamine hormones, and (2) within the cell, either in the cytoplasm or nucleus, as is the case for steroid and thyroid hormones. Receptors are specific for a single hormone. The hormone-receptor interaction is coupled to a signal-generating mechanism that then causes a change in intracellular processes by altering the activity or concentration of enzymes, carrier proteins, and so forth.

Mediating Hormonal Responses

Cell responses to protein/peptide and catecholamine hormones are mediated by second messengers. In the case of peptide/protein and catecholamine hormones that do not readily pass the cell membrane, interaction with the receptor on or within the cell membrane often results in the generation of a second messenger, which in turn mediates the hormonal response. Often, coupling *G-proteins* in the cell membrane link hormone receptors to the second messenger mechanisms. Second messenger mechanisms include the following:

- *Adenylyl cyclase–cyclic AMP (cAMP)*—Hormone receptor interaction may stimulate (or inhibit) the membrane-bound enzyme adenylyl cyclase. Stimulation of this enzyme results in the synthesis of the second messenger cAMP. The cAMP activates protein kinase A, leading to phosphorylations that either activate or inactivate target enzymes.
- *Calcium-calmodulin*—Hormone receptor interaction activates calcium channels in the plasma membrane, permitting calcium to enter cells. Calcium may also be mobilized from intercellular stores such as the endoplasmic reticulum. The calcium ions bind with the protein calmodulin, and this complex alters the activity of calcium-dependent enzymes and, thus, intercellular reactions.
- *Plasma membrane phospholipids*—Hormone receptor interaction activates the membrane-bound enzyme *phospholipase C*, which in turn causes phospholipids in the cell membrane (especially those derived from *phosphatidylinositol*) to split into the second messengers *diacylglycerol* and *inositol triphosphate*. Inositol triphosphate mobilizes calcium from internal stores, such as the endoplasmic reticulum, and the calcium in turn activates *protein kinase C*. Phosphorylation of enzymes by protein kinase C activates and deactivates enzymes mediating the hormone responses. In addition, the activity of protein kinase C is further enhanced by the second messenger diacylglycerol. Finally, diacylglycerol is hydrolyzed to *arachidonic acid*, which is the precursor for prostaglandins, which also influence hormonal responses.

Other second messenger mechanisms may transduce hormonal responses such as *cyclic GMP*, which mediates the effects of atrial natriuretic peptide. Protein/peptide hormones may exert actions independent of G-protein linked second messenger events. For example, in the case of the peptide hormone insulin, hormone binding to the cell surface receptor results in phosphorylation of an intracellular site of the receptor, which in turn alters enzymatic activity by phosphorylating (or dephosphorylating) other proteins within the cell.

Cell responses to steroid and thyroid hormones are mediated by stimulating protein synthesis. In contrast to protein/peptide hormones and catecholamines, steroid and thyroid hormones enter the cell and bind to intracellular receptors located in the cytoplasm or nucleus of the cell. The hormone-receptor interaction results in a conformational change in the receptor. This permits binding of the hormone-receptor complex to specific points on DNA strands in the chromosomes, which results in activation of specific genes, transcription, and translation of proteins that are essential to mediating the hormonal response. Because the transcription mechanism is involved in mediating the hormonal response, hours are usually required for biological effects to become evident.

MEASUREMENT OF HORMONE CONCENTRATIONS IN THE BLOOD (p. 931)

Most hormones are present in the blood in very minute concentrations (often in nanograms per liter or even picograms per liter). These low concentrations of hormones can be measured by *radioimmunoassay*. The principle of radioimmunoassay is based on the combined incubation of the following substances:

- A fixed amount of antibody specific for the hormone
- A fixed amount of radioactive-labeled hormone
- The plasma sample.

Because the amount of antibody present is limiting, the radioactive and unlabeled native hormones compete for the binding sites on the antibody. High concentrations of native hormone will displace more of the labeled hormone from the antibody. At the end of the incubation period, bound and free hormones are separated, and the amount of radioactivity is determined. The greater the amount of native hormone in the plasma sample, the lower is the amount of radioactivity in the bound fraction. The amount of native hormone in the sample is calculated by comparison with a standard curve generated by incubation of different amounts of unlabeled hormone (rather than the plasma sample), with antibody and radioactively labeled hormone as described.

Other competitive binding procedures can be used to measure hormone levels in the plasma. For example, tissue receptor or plasma binding proteins can be used instead of antibody as the binding protein.

75

Pituitary Hormones and Their Control by the Hypothalamus

PITUITARY GLAND AND ITS RELATION TO THE HYPOTHALAMUS (p. 933)

The hypothalamus and pituitary gland have intimate anatomical and functional relationships; in turn, these structures regulate the function of a number of endocrine glands, including the thyroid, adrenal, and gonads, and play an important role in the regulation of growth, metabolism, lactation, and water balance.

The pituitary gland is composed of two distinct components: (1) the *anterior pituitary gland*, or *adenohypophysis*, which is derived embryologically from an upward invagination of cells from the oral cavity (Rathke's pouch), and (2) the *posterior pituitary gland*, or *neurohypophysis*, which is derived from a downgrowth of cells from the third ventricle of the brain. The pituitary gland is connected to the hypothalamus by the hypothalamic or pituitary stock.

Neurohypophysis—Axons and Nerve Terminals for Storage of Neurohypophyseal Hormones

Magnocellular neurons whose cell bodies are located in the *supraoptic* and *paraventricular nuclei* of the hypothalamus synthesize the *neurohypophyseal hormones antidiuretic hormone (ADH)* and *oxytocin*. Secretion granules containing these neurohormones are transported from the cell bodies in the hypothalamus down axons in the pituitary stalk to storage sites in nerve terminals located in the posterior pituitary gland. ADH and oxytocin are released from secretion granules into the capillary plexus of the inferior hypophyseal artery, the primary blood supply to the neurohypophysis.

Adenohypophysis—Cells that Both Synthesize and Secrete Adenohypophysial Hormones

There are five cell types in the anterior pituitary gland that synthesize, store, and secrete six different polypeptide or peptide *adenohypophysial* hormones. One hormone, prolactin, acts on the breast; the other five are *tropic hormones* that stimulate secretion of hormones by other endocrine glands or, in the case of growth hormone (GH), the liver and other tissues. One cell type, the gonadotrope, secretes two hormones, follicle-stimulating hormone (FSH) and luteinizing hormone (LH). The cells that secrete the anterior pituitary hormones and the chemical structure and physiological actions of the adenohypophysial hormones are listed in Table 75–1.

There is considerable similarity in the chemical structures of the *glycoprotein hormones thyroid-stimulating hormone (TSH)*, *FSH*, and *LH*, all of which are secreted from *basophilic cells*. Similarly, there is structural homology between *prolactin* and *GH*, both of which are secreted from *acidophilic cells*. The corticotropes synthesize a preprohormone containing the amino acid sequences for *adrenocorticotropic hormone (ACTH)* and *melanocyte-stimulating hormones (MSHs)*. In humans, ACTH is generated in the anterior pituitary, but no appreciable amounts of MSHs are secreted. Although the administration of MSHs in humans causes darkening of the skin by increasing synthesis of the black pigment *melanin*, it is likely that the pigmentary changes in endocrinologic diseases are due to changes in circulating ACTH because ACTH has MSH activity.

CONTROL OF PITUITARY SECRETION BY THE HYPOTHALAMUS (p. 935)

Blood Supply to the Anterior Pituitary Gland—the Hypothalamic-Hypophyseal Portal Vessels

There is an extensive network of capillary sinuses surrounding the anterior pituitary cells; most of the blood entering these sinuses has first passed through another capillary plexus in the lower hypothalamus or *median eminence*. The blood from this latter capillary plexus comes from the superior hypophyseal artery and flows through the *hypothalamic-hypophyseal portal vessels* of the pituitary stalk to bathe the adenohypophysial cells.

TABLE 75–1 ADENOHYPOPHYSEAL CELLS AND HORMONES

Cell	Hormone	Chemistry	Physiological Actions
Corticotropes	Adrenocorticotropic hormone (corticotropin; ACTH)	Single chain of 39 amino acids	Stimulates production of glucocorticoids and androgens by the adrenal cortex; maintains size of zona fasciculata and zona reticularis of cortex
Thyrotropes	Thyroid-stimulating hormone (thyrotropin; TSH)	Glycoprotein of two subunits, α (89 amino acids) and β (112 amino acids)	Stimulates production of thyroid hormones by thyroid follicular cells; maintains size of follicular cells
Gonadotropes	Follicle-stimulating hormones (FSH)	Glycoprotein of two subunits, α (89 amino acids) and β (112 amino acids)	Stimulates development of ovarian follicles; regulates spermatogenesis in the testis
Gonadotropes	Luteinizing hormone (LH)	Glycoprotein of two subunits, α (89 amino acids) and β (115 amino acids)	Causes ovulation and formation of the corpus luteum in the ovary; stimulates production of estrogen and progesterone by the ovary; stimulates testosterone production by the testis
Mammotropes, lactotropes	Prolactin (PRL)	Single chain of 198 amino acids	Stimulates secretion of milk
Somatotropes	Growth hormone (somatotropin; GH)	Single chain of 191 amino acids	Stimulates body growth; stimulates secretion of IGF-1; stimulates lipolysis; inhibits actions of insulin on carbohydrate and lipid metabolism

Hypophysiotropic Hormones (Releasing and Inhibiting Hormones) — Secretion of Anterior Pituitary Hormones

In addition to the hypothalamic neuroendocrine cells, which synthesize neurohypophyseal hormones, other neurons in discrete areas of the hypothalamus synthesize the *hypophysiotropic neurohormones (releasing and inhibiting hormones)*, which control the secretion of the anterior pituitary hormones. Although the axons from the magnocellular neurons of the supraoptic and paraventricular nuclei terminate in the posterior pituitary gland, the nerve fibers from the hypothalamic cell bodies that synthesize the hypophysiotropic hormones lead to the *median eminence*. Here, the releasing and inhibiting hormones are stored in secretion granules in the nerve terminals. On stimulation of these hypothalamic neuroendocrine cells, their neurohormones are released into the capillary plexus of the median eminence, flow through the hypothalamic-hypophyseal portal vessels, and reach the sinusoids around the adenohypophyseal cells. The anterior pituitary cells respond to the hypophysiotropic hormones by either increasing or decreasing the synthesis and secretion of adenohypophyseal hormones.

The six established hypophysiotropic hormones are listed subsequently. For secretion of most adenohypophyseal hormones, releasing hormones are most important, but for prolactin, an inhibitory hormone has the most control. Note that the secretion of GH is influenced by both a releasing and inhibiting hormone, and a single hypophysiotropic hormone, gonadotropin-releasing hormone (GnRN), stimulates the gonadotropes to secrete both FSH and LH. All hypophysiotropic hormones are peptides, polypeptides, or derivatives of the amino acid tyrosine (Table 75–2).

The hypothalamus receives neural inputs from many areas of the brain. This information related to the well-being of the body is integrated in the hypothalamus and has an impact on endocrine function in large part by the influence of the hypophysiotropic hormones on the secretion of the anterior pituitary hormones. In turn, the tropic hormones from the anterior pituitary gland stimulate target endocrine glands and tissues. The resultant changes in target gland hormones and metabolic substrates in the peripheral blood exert negative feedback control on the secretion of anterior pituitary hormones

TABLE 75-2 HYPOPHYSIOTROPIC HORMONES

Hormone	Structure	Primary Action on Anterior Pituitary
Thyrotropin-releasing hormone (TRH)	Peptide of 3 amino acids	Stimulates secretion of TSH by thyrotropes
Gonadotropin-releasing hormone (GnRH)	Single chain of 10 amino acids	Stimulates secretion of FSH and LH by gonadotropes
Corticotropin-releasing hormone (CRH)	Single chain of 41 amino acids	Stimulates secretion of ACTH by corticotropes
Growth hormone–releasing hormone (GHRH)	Single chain of 44 amino acids	Stimulates secretion of GH by somatotropes
Growth hormone–inhibiting hormone (somatostatin)	Single chain of 14 amino acids	Inhibits secretion of GH by somatotropes
Prolactin-inhibiting factor (PIF)	Dopamine	Inhibits secretion of PRL by lactotropes

through a direct effect on the adenohypophyseal cells and through an indirect action at the level of the hypothalamus to alter the release of hypophysiotropic hormones.

PHYSIOLOGICAL FUNCTIONS OF GROWTH HORMONE
(p. 936)

Growth Hormone—Multiple Physiological Effects

In contrast to the other pituitary hormones, which stimulate specific target glands, GH has multiple effects throughout the body:

- *Promotion of linear growth*—GH stimulates the *epiphyseal cartilage* or growth plates of the long bones. Under the influence of GH, the chondrocytes in the growth plate are stimulated, leading to proliferation of these cells and deposition of new cartilage, followed by conversion of this cartilage into bone. This process elongates the shaft of the long bones. By late adolescence, when there is no remaining epiphyseal cartilage and the shafts have fused with the epiphyses (epiphyseal closure), GH can no longer cause lengthening of the long bones. Because GH also increases osteoblastic activity, total bone mass is increased by GH even after epiphyseal closure.

- *Promotion of protein deposition in tissues*—GH is a protein *anabolic* hormone and produces a positive nitrogen balance. GH increases amino acid uptake in most cells and the synthesis of amino acids into proteins.

- *Promotion of fat utilization for energy*—GH causes the mobilization of fatty acids from adipose tissue and the preferential utilization of free fatty acids for energy. This action of GH, together with its protein anabolic effects, produces an increase in lean body mass. The lipolytic effects of GH require several hours to occur. At least part of this effect is due to the actions of GH to impair glucose uptake into adipose cells. Because GH increases plasma levels of free fatty acids and keto acids, it is *ketogenic*.

- *Impairment of carbohydrate utilization for energy*—GH decreases the uptake and utilization of glucose by many insulin-sensitive cells, such as muscle and adipose tissue. As a result, blood glucose concentration tends to rise and insulin secretion increases to compensate for the GH-induced insulin resistance; thus, GH is *diabetogenic*.

Somatomedins and the Anabolic Effects of Growth Hormone

The effects of GH on linear growth and protein metabolism are not direct, but they are indirectly mediated via the generation of polypeptides called *somatomedins* or *insulin-like growth factors (IGFs)*. Somatomedins are secreted by the liver and other tissues. *Somatomedin C*, or *IGF-1*, is a circulating 70-amino acid peptide produced by the liver that reflects plasma GH levels. The growth-promoting effects of GH, however, are due to locally produced as well as circulating somatomedins; in cartilage and muscle, locally produced somatomedins act in an autocrine or a paracrine fashion to stimulate growth.

Growth Hormone Secretion—Metabolic Stimuli

GH secretion is under the influence of both a hypothalamic-releasing (GHRH) and -inhibiting hormone (somatostatin). Feedback regulation of GH secretion is mediated by circulating somatomedin C via actions at both the hypothalamus and pituitary. High plasma levels of somatomedin C decrease GH release by increasing the secretion of somatostatin from the hypothalamus and acting directly on the pituitary to decrease responsiveness to GHRH.

GH secretion is highest during puberty and decreases in adult life. This may be partially responsible for the decline in lean body mass and increase in adipose mass characteristic of senescence. There are three general categories of stimuli that increase GH secretion:

- *Fasting, chronic protein deprivation*, or other conditions in which there is an acute fall in plasma levels of metabolic substrates such as glucose and free fatty acids
- *Increased plasma levels of amino acids*, such as occur after a protein meal
- *Exercise and stressful stimuli*, such as pain and fever.

Clearly, the rise in GH during fasting would be beneficial because GH enhances lipolysis and decreases peripheral utilization of glucose. After a protein meal, increased plasma levels of GH would favor the utilization of amino acids for protein synthesis.

Abnormalities in the Impact of Growth Hormone Secretion on the Skeletal System

The importance of GH in linear growth is reflected by the clinical states associated with a deficiency or excess secretion of GH before epiphyseal closure. Short stature (*dwarfism*) occurs when the pituitary secretion of GH is deficient. In comparison, children grow tall (*gigantism*) when tumors of the somatotropes of the anterior pituitary secrete large amounts of GH. If a pituitary tumor secreting GH occurs after epiphyseal closure, the adult form of the disease occurs. In *acromegaly*, linear growth is normal, but there is enlargement of the hands and feet, protrusion of the lower jaw (prognathism), and overgrowth of facial bones. In addition, virtually all internal organs are increased in size. The anti-insulin effects of GH may ultimately lead to diabetes mellitus in states of chronic GH excess.

POSTERIOR PITUITARY GLAND AND ITS RELATION TO THE HYPOTHALAMUS (p. 942)

The *neurohypophyseal hormones ADH* and *oxytocin* are synthesized as preprohormones in the cell bodies of *magnocellular neurons* located in the *supraoptic* and *paraventricular nuclei* and transported in secretion granules down axons to nerve terminals in the posterior pituitary gland. ADH is synthesized largely in the supraoptic nucleus, and oxytocin is synthesized largely in the paraventricular nucleus, although each hormone is synthesized in the alternate site. The secretion granules containing either ADH or oxytocin also contain an additional protein, or *neurophysin*, that is part of the preprohormone. When a nerve impulse travels from the cell body of the magnocellular neurons down the axon to the nerve terminal, both the neurohormone and the corresponding neurophysin are released from secretion granules into the capillary blood as separate polypeptides. ADH and oxytocin are nonapeptides with a similar chemical structure; only the amino acids in positions 3 and 8 differ.

Physiological Functions of Antidiuretic Hormone

ADH regulates the osmolality of body fluids by altering the renal excretion of water. ADH plays an important role in the regulation of plasma osmolality. As discussed in Chapter 28, in the absence of ADH, the collecting tubules and collecting ducts are largely impermeable to water, which prevents significant reabsorption of water in this portion of the nephron. This results in a large volume of dilute urine and net loss of water; consequently, the osmolality of body fluids rises. In comparison, when increased ADH activates V_2-receptors on the basolateral side of the tubules via a *cyclic AMP* second messenger system, cytoplasmic vesicles containing water channels are inserted in the apical membrane. This increases the permeability of the tubules to water; therefore, water moves by osmosis from the tubular to the peritubular capillary fluid. In the collecting ducts, the urine becomes concentrated, and its volume decreases. As a result, there is retention of water in excess of solute, and the osmolality of body fluids decreases.

In accordance with its role in the regulation of the osmotic pressure of plasma, ADH secretion is sensitive to small changes in plasma osmolality (approximately 1 per cent). When plasma osmolality increases above normal, the rate of discharge of ADH-secreting neurons in the supraoptic and paraventricular nuclei increases, and ADH is secreted from the posterior pituitary gland into the systemic circulation. Circulating ADH increases the permeability of the collecting ducts to water, which ultimately decreases plasma osmolality to normal levels. The opposite changes in neuronal discharge and ADH secretion occur when plasma osmolality declines. ADH secretion is regulated by *osmoreceptors* in the anterior hypothalamus that send nervous signals to the supraoptic and paraventricular nuclei. Osmoreceptors are outside the blood-brain barrier and appear to be located in the *circumventricular organs*, primarily the organum vasculosum of the lateral terminalis. These same osmoreceptors may also mediate the thirst response to increased plasma osmolality.

ADH secretion is influenced by multiple factors. Other than increased plasma osmolality, stimuli that increase ADH secretion include hypovolemia, hypotension, nausea, pain, and stress, and a number of drugs, including morphine, nicotine, and barbiturates. Factors that decrease ADH secretion include hypervolemia, hypertension, and alcohol. The influence of these factors on the neurons in the supraoptic and paraventricular nuclei that secrete ADH may have an impact on the regulation of body fluid osmolality. For example, in hypovolemic states, elevated plasma levels of ADH may decrease plasma osmolality.

ADH contributes to the maintenance of blood pressure during hypovolemia. Stimulation of ADH secretion by hypovolemia and/or hypotension is achieved by reflexes initiated from receptors in both the high- and low-pressure regions of the circulation. The high-pressure receptors are those in the carotid sinus and aortic arch; the low-pressure receptors are those in the cardiopulmonary circulation, especially in the atria. Although at least a 5 per cent decrease in blood volume is necessary to increase ADH secretion by this reflex mechanism, greater degrees of hypovolemia and hypotension can result in very large increases in plasma ADH concentration to levels much higher than those required to achieve maximal antidiuresis. When these unusually high plasma levels of ADH occur, such as during hypotensive hemorrhage, ADH constricts vascular smooth muscle and helps to restore blood pressure to normal levels. This action of ADH is a result of the peptide binding to vascular V_1-receptors on arteriolar smooth muscle. The vasoconstriction induced by ADH is mediated by *calcium-* and *phospholipase C*–generated second messengers.

Physiological Functions of Oxytocin Hormone

Oxytocin plays an important role in lactation by causing milk ejection. Oxytocin causes contraction of the *myoepithelial cells* of the alveoli of the mammary glands; this forces milk from the alveoli into the ducts so that the baby can obtain it by suckling. The *milk ejection* reflex is initiated by receptors on the nipples of the breast. Suckling causes reflex stimulation of oxytocin-containing neuroendocrine cells in the supraoptic and paraventricular nuclei and secretion of oxytocin from the posterior pituitary gland. The circulating oxytocin then causes the myoepithelial cells to contract initiating milk ejection.

Oxytocin contributes to parturition. Oxytocin also causes contraction of the smooth muscle of the uterus; the sensitivity of this response is enhanced by plasma levels of estrogen, which increase during pregnancy. During labor, the descent of the fetus down the birth canal stimulates receptors on the cervix, which send signals to the supraoptic and paraventricular nuclei and cause secretion of oxytocin. Secretion of oxytocin in turn contributes to labor by causing contraction of the uterus.

76

Thyroid Metabolic Hormones

FORMATION AND SECRETION OF THYROID HORMONES
(p. 945)

The thyroid gland is composed of a large number of *follicles*. Each follicle is surrounded by a single layer of cells and filled with a proteinaceous material called *colloid*. The primary constituent of colloid is the large glycoprotein *thyroglobulin*, which contains the thyroid hormones within its molecule. The following steps are required for the synthesis and secretion of thyroid hormones into the blood:

- *Iodide trapping (iodide pump)*—Iodine is essential to thyroid hormone synthesis. Ingested iodine is converted to iodide and absorbed from the gut. Most circulating iodide is excreted by the kidneys; much of the remainder is taken up and concentrated by the thyroid gland. To achieve this, the thyroid follicular cells actively transport iodide from the circulation across their basal membrane into the cell. Normally, a free iodide in thyroid and plasma versus a bound iodide ratio of approximately 30 is maintained. Several anions, such as thiocyanate and perchlorate, decrease iodide transport by competitive inhibition. Because they decrease the synthesis of thyroid hormones, these compounds can be used in the treatment of hyperthyroidism.
- *Synthesis of thyroglobulin*—This glycoprotein is synthesized by the follicular cells and secreted into the colloid through exocytosis of secretion granules that also contain *thyroid peroxidase*. Each thyroglobulin molecule contains approximately 100 tyrosine molecules.
- *Oxidation of iodide*—Once in the thyroid gland, iodide is rapidly oxidized to iodine by *thyroid peroxidase*; this occurs at the apical membrane of the follicular cells.
- *Iodination (organification) and coupling*—Once iodide is oxidized to iodine, it is rapidly attached to

the 3 position of tyrosine molecules of thyroglobulin to generate *monoiodotyrosine (MIT)*. MIT is next iodinated in the 5 position to give *diiodotyrosine (DIT)*. Thereafter, two DIT molecules are coupled to form *thyroxine (or T_4)*, the major product of the coupling reaction, or one MIT and one DIT molecule are coupled to form *triiodothyronine (or T_3)*. A small amount of *reverse T_3* is formed by condensation of DIT with MIT. These reactions are catalyzed by *thyroid peroxidase* and blocked by antithyroid drugs such as propylthiouracil.

Approximately two thirds of the iodinated compounds bound to thyroglobulin are MIT or DIT; most of the remainder are the active hormones T_3 and, especially, T_4. Thyroglobulin is stored in the lumen of the follicle as colloid until the gland is stimulated to secrete thyroid hormones.

- *Proteolysis, deiodination, and secretion*—The release of T_3 and T_4 into the blood requires proteolysis of the thyroglobulin. At the apical surface of the follicular cells, colloid is taken up from the lumen of the follicles through endocytosis. Colloid vesicles then migrate from the apical to the basal cell membrane and fuse with *lysosomes*. Lysosomal *proteases* release free T_3 and T_4, which then leave the cell. Free MIT and DIT are not secreted into the blood but instead deiodinated within the follicular cell by the enzyme *deiodinase*; the free iodine is reused in the gland for hormone synthesis. More than 90% of the thyroid hormone released from the gland is T_4. The remaining secretion products are T_3 and very small amounts of the inactive compound reverse T_3.

Transport and Metabolism of Thyroid Hormones

Thyroid hormones are highly bound to plasma proteins. On entering the blood, both T_4 and T_3 are highly bound to plasma proteins, especially to *thyroxine-binding globulin (TBG)*, but also to other plasma proteins such as *albumin* and *thyroid-binding prealbumin*. Approximately 99.9% of T_4 is bound to plasma proteins, and less than 0.1% is free hormone. The binding of T_3 to plasma proteins is slightly less than that of T_4, but still less than 1% is free hormone. In the case of the thyroid hormones, it is the free hormone that is taken up by tissues, in which it exerts biological effects and is metabolized. As a result of the high degree of binding to plasma proteins, the half-lives of T_4 and T_3 are very long (7 and 1 day, respectively).

Alterations in plasma TBG levels do not influence free thyroid hormone concentration. Reductions (e.g., during liver and kidney disease) and elevations (e.g., during estrogen administration and pregnancy) in plasma TBG levels decrease and increase, respectively, the total amount of thyroid hormones in the plasma, but produce no more than a transient change in the free hormone concentration. This is because of the negative feedback effect of free thyroid hormones on the pituitary secretion of TSH. For example, in pregnancy, a fall in free thyroid hormone concentration induced by increased TBG levels in the plasma causes a compensatory rise in TSH secretion. Consequently, the total thyroid hormone concentration in the plasma is increased, but free plasma levels of the hormones are normal.

Most of the T_4 secreted by the thyroid gland is metabolized to T_3. Although T_4 is the dominant secreted and circulated thyroid hormone, large amounts of the hormone are deiodinated in either the 5' or the 5 position in peripheral tissues to produce T_3 and reverse T_3. In fact, most of the T_3 and reverse T_3 in the plasma come from circulating T_4 that has been deiodinated in peripheral tissues rather than secreted from the thyroid gland. Because most of the T_4 that enters cells is converted to T_3 (and reverse T_3), and because the T_3 within cells has a greater affinity than does T_4 for thyroid hormone receptors in the nucleus, T_4 has been considered to be a prohormone for T_3.

FUNCTIONS OF THE THYROID HORMONES IN THE TISSUES (p. 948)

Thyroid Hormones and Transcription of Many Genes

After thyroid hormones enter the cell and bind to nuclear receptors, the hormone-receptor complex binds to DNA and stimulates (or inhibits) transcription of a large number of genes. This leads to the synthesis of numerous enzymes that alter cell function. The actions of T_3 occur more rapidly and are more potent than are those of T_4 because T_3 is bound less tightly to plasma proteins and has a greater affinity for nuclear receptors. Because thyroid hormones act in large part through influencing transcription, a delay of several hours occurs before most hormonal effects are evident; these effects may last several days.

Physiological Effect of Thyroid Hormones—Cellular Metabolic Rate

In most tissues of the body, thyroid hormones increase oxygen consumption and heat production. Mitochondria increase in size and number, the membrane surface areas of the mitochondria increase, and the activities of key respiratory enzymes increase. A complete accounting of the cellular mechanisms that are responsible for the higher oxygen consumption is not possible at present. Because thyroid hormones increase the activity of membrane-bound *Na, K-ATPase*, the greater ATP consumption associated with the greater sodium transport is believed to contribute to the greater metabolic rate induced by thyroid hormone.

Specific Physiological Effects of Thyroid Hormones

Many of the effects of thyroid hormones are secondary to increased metabolic rate. Thyroid hormones are responsible for the following functions:

- *Increased thermogenesis and sweating*—Skin blood flow increases because of the need for heat elimination.
- *Increased rate and depth of respiration* due to the need for oxygen
- *Increased cardiac output* because increased metabolism and utilization of oxygen in tissues cause local vasodilatation. Increased cardiac output is associated with elevations in both stroke volume and heart rate, in part because thyroid hormones have direct and indirect effects on the heart to raise heart rate and force of contraction.
- *Increased pulse pressure but not mean arterial pressure*—Because of the increased cardiac output (stroke volume) and reduced peripheral vascular resistance, systolic arterial pressure is elevated and diastolic arterial pressure is reduced. This results in an increase in pulse pressure but usually no change in mean arterial pressure.
- *Increased utilization of substrates for energy*—Increased metabolic rate is dependent on oxidation of metabolic substrates. Thyroid hormones increase the utilization of carbohydrates, fats, and

proteins for energy. If food intake is not increased sufficiently, there is depletion of body fats and proteins and weight loss. Although thyroid hormones promote lipolysis of triglycerides and increments in plasma levels of free fatty acids, they also decrease circulating levels of cholesterol; this action is due to increased formation of low-density lipoprotein receptors in the liver, resulting in increased removal of cholesterol from the circulation. Since thyroid hormones increase the rate of metabolic reactions, the need for vitamins is greater, and excess thyroid hormone can lead to vitamin deficiency.

Thyroid hormones are essential to normal growth and development. Thyroid hormones are essential to many aspects of growth and development; they play an important role in the development of the skeletal system, teeth, epidermis, and central nervous system. In hypothyroid children, the rate of growth is greatly reduced. A very important effect of thyroid hormone is to promote growth and development of the central nervous system in utero and for the first few years of postnatal life. If thyroid hormone is deficient at this time, irreversible brain damage occurs.

Thyroid hormones have excitatory effects on the nervous system. Thyroid hormones enhance wakefulness, alertness, and responsiveness to various stimuli and increase the speed and the amplitude of peripheral nerve reflexes; they also improve memory and learning capacity.

REGULATION OF THYROID HORMONE SECRETION
(p. 951)

Thyroid Stimulating Hormone — Primary Controller of Thyroid Hormone Secretion

To maintain normal levels of metabolic activity in the body, the free plasma levels of thyroid hormone must be regulated. Thyroid hormone secretion is primarily regulated by plasma levels of *thyroid-stimulating hormone (thyrotropin, or TSH)*. TSH secretion from the pituitary gland is increased by the hypophysiotropic hormone *thyrotropin-releasing hormone (TRH)* and is inhibited in a negative feedback fashion by circulating T_4 and T_3. Although some feedback occurs at the hypothalamus by influencing the secretion of TRH, the predominant feedback occurs at the level of the pituitary. Because T_4 is deiodinated to T_3 in the pituitary gland, T_3 appears to be the final effector that mediates the negative feedback.

TSH promotes the synthesis and secretion of thyroid hormones. Binding of TSH to its receptors on the cell membrane of the thyroid gland activates *adenylyl cyclase* so that *cyclic AMP* mediates at least some of the actions of TSH. An immediate effect of TSH is to promote endocytosis of colloid, proteolysis of thyroglobulin, and release of T_4 and T_3 into the circulation. In addition, TSH stimulates steps in the synthesis of thyroid hormones, including iodine trapping, iodination, and coupling to form thyroid hormones.

TSH has chronic effects to promote growth of the thyroid gland. The chronic effects of TSH include increased blood flow to the thyroid gland and induction of hypertrophy and hyperplasia of the follicular cells. With prolonged TSH stimulation, the thyroid enlarges, and a *goiter* occurs. In the absence of TSH, there is marked atrophy of the gland.

DISEASES OF THE THYROID (p. 953)

Graves' disease is the most common form of hyperthyroidism. Graves' disease is an autoimmune disease in which antibodies, *thyroid-stimulating immunoglobulins (TSI)*, form against the TSH receptor of the thyroid, bind to it, and mimic the actions of TSH. This leads to goiter and secretion of large amounts of thyroid hormones. As a result, several predictable changes occur: (1) increased metabolic rate, (2) heat intolerance and sweating, (3) increased appetite but weight loss, (4) palpitations and tachycardia, (5) nervousness and emotional lability, (6) muscle weakness, and (7) tiredness but inability to sleep.

Many patients with Graves' disease have protrusion of the eyeballs, or *exophthalmos*. This is due to the degenerative changes in the extraocular muscles as a result of an autoimmune reaction. TSH secretion from the pituitary gland is depressed in Graves' disease because of the feedback exerted by the high plasma levels of thyroid hormones.

Many of the effects of hypothyroidism are opposite to those of hyperthyroidism. Hypothyroidism may have several causes but it often results from autoimmune destruction of the thyroid gland. The symptoms are, in general, opposite to those of hyperthyroidism and include (1) decreased metabolic rate, (2) cold intolerance and decreased sweating, (3) weight gain without increased caloric intake, (4) bradycardia, (5) slowness of movement, speech, and thought, and (6) lethargy and sleepiness. There is accumulation of mucopolysaccharides in interstitial spaces, giving rise to nonpitting edema. The puffiness of the skin is referred to as *myxedema*, a term used synonymously for adult hypothyroidism. If severe hypothyroidism occurs in utero or in infancy, irreversible mental retardation results, and growth is impaired; this condition is referred to as *cretinism*. If the hypothalamic-pituitary axis is normal, hypothyroidism is associated with increased plasma levels of TSH due to feedback.

Hypothyroidism can also be associated with *goiter*. In certain areas of the world, dietary iodine is deficient, so that thyroid hormone secretion is depressed. Many persons in a specific region have enlarged thyroids, or *endemic goiter*, because high plasma levels of TSH stimulate the gland. The practice of adding iodine to table salt has decreased the incidence of endemic goiter in many areas of the world.

77

Adrenocortical Hormones

The adrenal gland is composed of two distinct parts: (1) an inner *adrenal medulla*, which is functionally related to the sympathetic nervous system and secretes mainly *epinephrine* but some *norepinephrine*, and (2) an outer *adrenal cortex*, which forms the bulk of the gland and secretes *corticosteroids*. The corticosteroids secreted by the adrenal cortex are as follows:

- *Mineralocorticoids*—C_{21} steroids that have important effects on sodium and potassium balance
- *Glucocorticoids*—C_{21} steroids that influence carbohydrate, fat, and protein metabolism
- *Sex hormones*—C_{19} steroids that are mostly *weak androgens* and contribute to secondary sex characteristics.

The secretion of mineralocorticoids and glucocorticoids is essential to life. Only small amounts of sex hormones are normally secreted by the adrenal cortex, and they have little effect on reproductive function.

CHEMISTRY OF ADRENOCORTICAL SECRETION (p. 957)

The adrenal cortex is composed of three distinct layers or cell types—the zona glomerulosa, zona fasciculata, and zona reticularis.

- The *zona glomerulosa*, or outer zone, is relatively thin; it is the exclusive site of the enzyme *aldosterone synthase* (Fig. 77–1). Its major secretion product is the principal mineralocorticoid *aldosterone*. The primary controllers of aldosterone secretion are *angiotensin II* and *potassium*. Chronic increases in plasma angiotensin II concentration, such as occur during sodium depletion, cause hypertrophy and hyperplasia of zona glomerulosa cells only. Because the zona glomerulosa lacks the en-

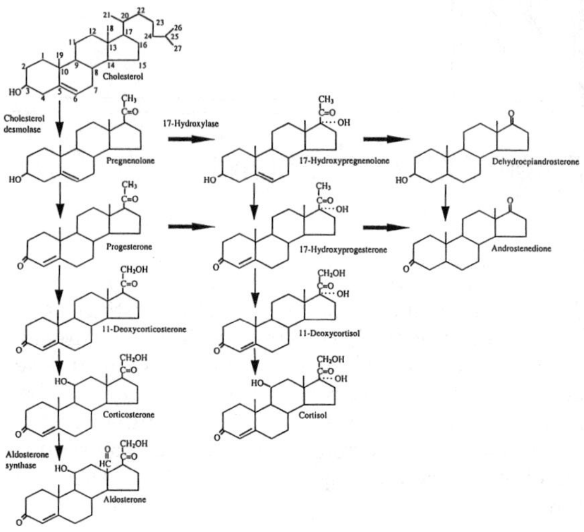

Figure 77–1 Hormone biosynthesis in the adrenal cortex.

zyme *17-hydroxylase* (see Fig. 77–1), it cannot synthesize cortisol or sex hormones.

- The *zona fasciculata*, or middle zone, is the widest zone; it secretes the glucocorticoids *cortisol* (the principal glucocorticoid) and *corticosterone*. This zone also secretes small amounts of sex hormones. The major controller of cortisol secretion is *adrenocorticotropic hormone (corticotropin, or ACTH)*.
- The *zona reticularis*, or inner zone, secretes sex hormones and some glucocorticoids; like the zona fasciculata, it is stimulated by ACTH. Chronic ACTH excess causes hypertrophy and hyperplasia of the inner two zones of the adrenal cortex. The most prevalent adrenal androgens are *dehydroepiandrosterone (DHEA)* and *androstenedione*.

Adrenocortical hormones are synthesized from cholesterol. Most of the cholesterol in adrenocortical cells is taken up from the circulation and then esterified and stored in lipid droplets. The rate-limiting step in the synthesis of adrenocortical hormones is the side-chain cleavage of cholesterol to form *pregnenolone* (see Fig. 77–1). This step includes the delivery of cholesterol to the inner mitochondrial membrane and the enzymatic cleavage (through *cholesterol desmolase*) of a six-carbon unit from cholesterol to yield pregnenolone. In all three zones of the adrenal cortex, this initial step in steroid biosynthesis is stimulated by the controllers of the major hormone products (aldosterone and cortisol). The conversion of cholesterol to pregnenolone and all the subsequent steps in the synthesis of adrenocortical hormones occur in the *endoplasmic reticulum* or *mitochondria*. Not all of the compounds shown in Figure 77–1 are produced in all three zones of the adrenal cortex.

Adrenocortical hormones are bound to plasma proteins. Approximately 90 to 95 per cent of the cortisol in the plasma is bound to plasma proteins, especially *transcortin* or *corticosteroid-binding globulin (CBG)*. As a result of this high degree of binding to plasma proteins, cortisol has a long half-life of about 60 to 90 minutes. Corticosterone is bound to plasma proteins to a lesser degree than cortisol and has a half-life of approximately 50 minutes. Even smaller amounts of aldosterone are bound to plasma proteins; consequently, aldosterone has a half-life of only approximately 20 minutes.

Adrenocortical hormones are metabolized in the liver. Cortisol and aldosterone are metabolized to various compounds in the liver and then conjugated to *glucuronic acid*. These inactive conjugates are freely soluble in plasma and are not bound to plasma proteins. Once released into the circulation, they are readily excreted in urine. The rate of inactivation of adrenocortical hormones is depressed in liver disease.

FUNCTIONS OF THE MINERALOCORTICOIDS— ALDOSTERONE (p. 959)

Aldosterone is the primary mineralocorticoid secreted by the adrenal cortex. Aldosterone accounts for approximately 90 per cent of the mineralocorticoid activity of adrenocortical hormones. Most of the remainder of the mineralocorticoid activity can be attributed to (1) *deoxycorticosterone*, which has approximately 3 per cent of the mineralocorticoid activity of aldosterone and is secreted at a comparable rate, and (2) *cortisol*, a glucocorticoid with weak mineralocorticoid activity that is normally present at plasma concentrations more than 1000 times that of aldosterone. In vitro studies have shown that cortisol binds very well to mineralocorticoid receptors. Because the kidneys have the enzyme *11β-hydroxy-*

steroid dehydrogenase, cortisol is converted to *cortisone,* which does not avidly bind mineralocorticoid receptors. Consequently, cortisol does not normally exert significant mineralocorticoid effects in vivo. Under conditions in which 11β-hydroxysteroid dehydrogenase is either congenitally absent or inhibited (e.g., during excessive licorice ingestion), cortisol will have substantial mineralocorticoid effects.

Aldosterone increases sodium reabsorption and potassium secretion. Aldosterone and other mineralocorticoids act on the distal nephron, especially the *principal cells* of the *collecting duct,* to increase sodium reabsorption and potassium secretion. These effects occur after the binding of aldosterone to intracellular receptors and the subsequent synthesis of proteins, including *Na, K-ATPase* in the *basolateral membrane* and sodium and potassium *channel proteins* in the *apical membrane.* As a result of increased Na, K-ATPase activity, sodium is pumped out of the tubular cells into the blood and exchanged for potassium. Potassium then diffuses into the tubular urine. As sodium is reabsorbed under the influence of aldosterone, there is enhanced tubular secretion of hydrogen and potassium ions. Because protein synthesis is required to mediate the tubular actions of aldosterone, a lag time of at least 60 minutes occurs between exposure to aldosterone and its onset of action.

Aldosterone affects electrolyte transport in organs other than the kidneys. Aldosterone binds to mineralocorticoid receptors in epithelial cells other than those of the kidney. Aldosterone increases sodium reabsorption from the colon and promotes potassium excretion in the feces. Similarly, aldosterone has effects on sweat and salivary glands to decrease the ratio of sodium to potassium in their respective secretions.

Controllers of Aldosterone Secretion—Angiotensin II and Potassium

Angiotensin II stimulates aldosterone secretion. *Angiotensin II* directly stimulates the cells of the *zona glomerulosa* to secrete aldosterone. This effect of angiotensin II is mediated via increments in intracellular levels of *calcium* and *phosphatidylinositol* products. These second messengers activate *protein kinase C,* which in turn stimulates both early *(cholesterol des-*

molase) and late *(aldosterone synthase)* steps in the biosynthesis of aldosterone.

The control of aldosterone secretion by angiotensin II is closely linked to the regulation of extracellular fluid volume and arterial pressure (see Chapter 29). Under the conditions of hypovolemia and hypotension, the renin-angiotensin system is activated, and high plasma levels of angiotensin II stimulate aldosterone secretion. In turn, aldosterone increases sodium reabsorption in the distal nephron; as fluid retention returns body fluid volumes and arterial pressure to normal levels, the stimulus for activation of the renin-angiotensin system wanes, and aldosterone secretion falls to basal levels. Accordingly, the activity of the renin-angiotensin system is inversely related to dietary sodium intake.

Potassium stimulates aldosterone secretion. The cells of the zona glomerulosa are sensitive to small changes in plasma potassium concentration. Increments in plasma potassium concentration increase aldosterone secretion by depolarizing the cell membrane, opening *calcium channels*, and increasing intracellular calcium concentration. In response to these events, aldosterone secretion increases as a result of stimulation of the same early and late biosynthetic steps affected by angiotensin II (see previous discussion).

Aldosterone plays a critical role in the elimination of ingested potassium and feedback regulation of plasma potassium concentration (see Chapter 29). Increments in plasma potassium concentration increase aldosterone secretion, which in turn stimulates tubular secretion of potassium. As plasma potassium concentrations fall to normal levels, the stimulus for aldosterone secretion is removed. The opposite sequence of events occurs when plasma potassium concentration decreases.

ACTH plays a permissive role in the regulation of aldosterone secretion. As long as normal plasma levels of ACTH are present, the responsiveness of the zona glomerulosa to its major controllers, angiotensin II and potassium, is maintained. In comparison, if ACTH is chronically deficient, the aldosterone response to angiotensin II and potassium is diminished. High plasma levels of ACTH, which occur acutely during stress, stimulate aldosterone secretion, but in states of chronic ACTH excess (e.g., in Cushing's syndrome), hyperaldosteronism is not sustained.

FUNCTIONS OF THE GLUCOCORTICOIDS (p. 962)

Cortisol is the primary glucocorticoid secreted by the adrenal cortex. More than 95 per cent of the glucocorticoid activity exerted by the adrenocortical hormones can be attributed to cortisol; most of the remaining glucocorticoid activity is due to corticosterone. Cortisol mediates most of its effects by binding with intracellular receptors in target tissues and inducing (or repressing) gene transcription; this results in the synthesis of enzymes that alter cell function.

Cortisol has widespread effects on metabolism. There are pronounced disturbances in carbohydrate, fat, and protein metabolism in adrenal insufficiency. Some of the metabolic effects of cortisol are permissive in that cortisol does not initiate the changes, but its presence at normal plasma levels permits certain metabolic processes. Cortisol exerts the following effects on metabolism:

- *Decreases protein stores in extrahepatic tissues.* In muscle and in other extrahepatic tissues, cortisol decreases amino acid uptake and inhibits protein synthesis; at the same time, it increases the degradation of proteins. As a result of these *catabolic* and *antianabolic effects* of cortisol, amino acids tend to increase in the blood and are taken up by the liver, in which they are converted into glucose and proteins, including gluconeogenic enzymes.
- *Tends to increase blood glucose concentration in two ways.* First, cortisol increases the hepatic production of glucose by increasing *gluconeogenesis.* The proteins mobilized from peripheral tissues are converted into glucose and glycogen in the liver. By maintaining glycogen reserves, cortisol allows other glycolytic hormones, such as epinephrine and glucagon, to mobilize glucose in times of need, such as between meals. A second way in which cortisol tends to increase blood glucose concentration is by impairing the utilization of glucose by peripheral tissues; cortisol has an *anti-insulin effect* in tissues such as muscle and adipose tissue and impairs the uptake and utilization of glucose for energy. Like growth hormone, cortisol is *diabetogenic* because it tends to raise blood glucose concentration.
- *Plays an important role in the mobilization of fatty acids from adipose tissue.* Although weakly lipolytic itself, normal levels of cortisol exert a permissive

effect on mobilization of fatty acids during fasting. During fasting, cortisol allows other lipolytic hormones, such as epinephrine and growth hormone, to mobilize fatty acids from lipid stores.

Increased cortisol secretion is important in the resistance to stress. Physical or mental stress increases ACTH secretion, which in turn stimulates the adrenal cortex to secrete cortisol. Although it is not clear how hypercortisolism mediates this response, the large rise in cortisol secretion in response to many stressors is essential to survival. Patients with adrenal dysfunction who are administered maintenance doses of steroids require extra glucocorticoid under stressful conditions.

Pharmacological doses of glucocorticoids have anti-inflammatory and antiallergic effects and suppress immune responses. Large doses of glucocorticoids decrease the *inflammatory response* to tissue trauma, foreign proteins, or infections through several effects, including the following:

- *Inhibition of phospholipase*—This decreases the synthesis of *arachidonic acid*, which is the precursor of *leukotrienes*, *prostaglandins*, and *thromboxanes*, mediators of the local inflammatory response that includes dilation of capillaries, increased capillary permeability, and migration of leukocytes into the area of tissue injury.
- *Stabilization of lysosomal membranes*—This decreases the release of proteolytic enzymes by damaged cells.
- *Suppression of the immune system*—Suppression is a result of decreased production of T cells and antibodies that contribute to the inflammatory process.
- *Inhibition of fibroblastic activity*.

Controller of Cortisol Secretion—ACTH

ACTH stimulates cortisol secretion. The secretion of cortisol is under the control of the hypothalamic-pituitary, *corticotropin-releasing hormone (CRH)-ACTH axis*. The release of ACTH (corticotropin) from the pituitary is dependent on the hypophysiotropic hormone CRH. Once ACTH is secreted into the blood, it has a rapid effect on the inner two zones of the adrenal cortex, especially the *zona fasciculata*, to increase the secretion of *cortisol*. This effect of ACTH is achieved by increasing the conversion of cholesterol to pregnenolone and is mediated via the second messenger *cyclic AMP*. Chronic stimulation of the adrenal cortex by ACTH causes hypertrophy and hyperplasia of the zona fasciculata and zona reticularis and increased synthesis of several enzymes that convert cholesterol into the final product cortisol. Under conditions of chronic ACTH excess, such as in some forms of Cushing's syndrome, there are sustained increases in the secretion of cortisol and adrenal androgens.

Blood levels of free (unbound) cortisol are controlled in a negative feedback fashion. Increased plasma levels of cortisol decrease ACTH secretion through a direct effect on the pituitary as well as an indirect inhibition of CRH release from the hypothalamus. The secretion of cortisol is highest in the early morning and reaches its lowest in the late evening because there is a *diurnal* or *circadian rhythm* in ACTH secretion as a result of changes in the frequency and duration of CRH bursts from the hypothalamus. Because of the cyclic changes in cortisol secretion, plasma levels of cortisol are meaningful only when expressed in terms of the time of day when blood sampling occurred.

Stress increases ACTH secretion. Several physical and mental stressors stimulate the neuroendocrine cells of the hypothalamus to secrete CRH; as a result, there is increased ACTH secretion, which stimulates release of cortisol. Under conditions of stress, the inhibitory effect of cortisol on ACTH secretion is insufficient to counteract the extra neural input to the neuroendocrine cells secreting CRH. Consequently, plasma levels of ACTH are increased.

ADRENAL ANDROGENS (p. 967)

The *adrenal androgens DHEA* and *androstenedione* are secreted in appreciable amounts, but they have only weak androgenic effects. Consequently, the normal plasma concentrations of these hormones exert little effect on secondary sex characteristics, especially in males, in whom large amounts of testosterone, the most potent androgen, are secreted by the testes. In females, adrenal androgens are responsible for pubic and axillary hair. Most of the androgenic activity of adrenal hormones may be due to the conversion of adrenal androgens to testosterone in peripheral tissues. In contrast to the normal state, when adrenal androgens are secreted in excessive amounts, as in

Cushing's disease, appreciable masculinization may be produced in both males and females. The secretion of adrenal androgens is stimulated by ACTH.

ABNORMALITIES OF ADRENOCORTICAL SECRETION
(p. 967)

Increased plasma levels of glucocorticoids (cortisol) produce *Cushing's syndrome.* Excess cortisol secretion can be caused by an adrenal tumor, a pituitary tumor that is secreting large amounts of ACTH and causing bilateral adrenal hyperplasia, or a tumor of the lungs or other tissues *(ectopic tumors)* that is secreting large amounts of ACTH and causing bilateral adrenal hyperplasia. Cushing's syndrome may also be produced by the administration of large amounts of exogenous glucocorticoids.

Symptoms of Cushing's syndrome include the following:

- Mobilization of fat from the extremities to the abdomen, face, and supraclavicular areas
- Hypertension and hypokalemia due to high plasma levels of cortisol and 11-deoxycorticosterone (when secreted in excess)
- Protein depletion resulting in muscle weakness, loss of connective tissue and thinning of the skin (leading to purple striae), and impaired growth in children
- Osteoporosis and vertebral fractures due to protein depletion, decreased calcium absorption from the gut (anti–vitamin D action), and increased glomerular filtration rate and renal excretion of calcium
- Impaired response to infections due to a suppressed immune system
- Impaired carbohydrate metabolism, hyperglycemia, and even insulin-resistant diabetes mellitus
- Masculinizing effects when adrenal androgens are secreted in excess.

Conn's syndrome **(primary aldosteronism) is caused by a tumor in the zona glomerulosa.** When a tumor is present in the zona glomerulosa that produces large amounts of aldosterone, the most notable features are hypertension and hypokalemia; usually hypertension is relatively mild because there is only a small increase in extracellular fluid volume due to "sodium escape" (Chapter 29). The hypertension and hypokalemia are exacerbated by increased sodium intake. Because of the expansion of extracellular fluid volume and rise in arterial pressure, plasma renin activity is suppressed. The potassium depletion in Conn's syndrome decreases the concentrating ability of the kidneys, leading to polyuria, and causes muscle weakness and metabolic alkalosis.

Impaired secretion of adrenocortical hormones occurs in *Addison's disease.* Destruction of the adrenal cortex can result from autoimmune disease, tuberculosis, and cancer. These processes usually are gradual, leading to a progressive reduction in glucocorticoid and mineralocorticoid function. As a result of the decrease in cortisol secretion, there is a compensatory increase in ACTH secretion, which produces *hyperpigmentation*. Symptoms of Addison's disease include the following:

Mineralocorticoid Deficiency

- Excessive loss of sodium, hypovolemia, hypertension, and increased plasma renin activity
- Excessive potassium retention and hyperkalemia
- Mild acidosis

Glucocorticoid Deficiency

- Abnormal carbohydrate, fat, and protein metabolism resulting in muscle weakness, fasting hypoglycemia, and impaired utilization of fats for energy
- Loss of appetite and weight loss
- Poor tolerance to stress; the inability to secrete increased amounts of cortisol during stress leads to an *Addisonian crisis* that may culminate in death if supplemental doses of adrenocortical hormones are not administered.

78

Insulin, Glucagon, and Diabetes Mellitus

CHEMISTRY, SYNTHESIS, AND METABOLISM OF PANCREATIC HORMONES

Insulin and glucagon are synthesized in the islets of Langerhans. The pancreas is composed of two types of tissue: (1) *acini*, which secrete digestive juices via the pancreatic duct into the duodenum (exocrine function), and (2) the *islets of Langerhans*, which do not secrete into ducts but instead empty their secretions into the blood (endocrine function). In humans, there are 1 to 2 million islets of Langerhans, which contain at least four distinct cell types:

- *Beta cells* account for approximately 60 per cent of the cells and secrete *insulin*.
- *Alpha cells* make up about 25 per cent of the cells and are the source of *glucagon*.
- *Delta cells* secrete *somatostatin*.
- *PP cells* secrete *pancreatic polypeptide*.

The secretion of pancreatic hormones into the portal vein via the pancreatic vein provides for a much higher concentration of pancreatic hormones in the liver than in the peripheral tissues; this is in keeping with the important metabolic effects of insulin and glucagon in the liver. The physiological functions of *somatostatin* and *pancreatic polypeptide* have not been established.

Insulin and glucagon are synthesized and metabolized like most peptide hormones. Both insulin and glucagon are synthesized as large *preprohormones*. In the Golgi apparatus, the *prohormones* are packaged in granules and then largely cleaved into free hormone plus peptide fragments. In the case of the *beta cells*, insulin and *connecting*, or *C, peptide* (which connects the two peptide chains of insulin) are released into the circulating blood in equimolar amounts. C peptide levels can be measured with radioimmunoassay. Insulin is a polypeptide containing two amino acid chains (21 and 30 amino acids) connected by disulfide bridges. Glucagon is a straight-chain polypeptide of 29 amino acid residues. Both insulin and glucagon circulate unbound to carrier proteins and have short half-lives of 5 to 10 minutes. Approximately 50 per cent of the insulin and glucagon in the portal vein is metabolized on first pass by liver; most of the remaining hormone is metabolized by the kidneys.

INSULIN AND ITS METABOLIC EFFECTS (p. 971)

Insulin is a hormone associated with energy abundance. In response to an influx of nutrients into the blood, insulin is secreted and permits these nutrients to be used by tissues for energy and anabolic processes; it also induces the storage of excess nutrients for later use when energy supplies are deficient. In the presence of insulin, stores of carbohydrates, fats, and proteins increase. Insulin has rapid (e.g., increased glucose and amino acid uptake into cells), intermediate (e.g., stimulation of protein synthesis, inhibition of protein degradation, activation and inactivation of enzymes), and delayed (e.g., increased transcription) actions on carbohydrate, fat, and protein metabolism that occur within seconds, minutes, and hours, respectively.

Most of the actions of insulin are achieved through autophosphorylation of receptors. Insulin does not mediate its physiological effects through generation of second messengers as do most protein hormones; instead, signal transduction is achieved through *autophosphorylation* of the intracellular domains of its own receptor. The insulin receptor is a tetramer made up of *two alpha subunits* that lie outside the cell membrane and *two beta subunits* that penetrate the cell membrane and protrude into the cytoplasm. Binding of insulin to the alpha subunit of the receptor triggers *tyrosine kinase* activity of the beta subunits, producing autophosphorylation of the beta subunits on tyrosine residues. This results in phosphorylation of other intracellular proteins and enzymes, which mediates a multitude of responses.

Effects of Insulin on Carbohydrate Metabolism

In muscle, insulin promotes the uptake and metabolism of glucose. An important effect of insulin in muscle is that it facilitates glucose diffusion down its concentration gradient from the blood into cells. This is achieved by increasing the number of *glucose transporters* in the cell membrane. These transporters are recruited from a cytoplasmic pool of vesicles to the cell membrane. The increased glucose transported into muscle cells undergoes glycolysis and oxidation and is stored as glycogen. Because glucose entry into muscle cells is usually highly dependent on insulin, glucose uptake by these cells is restricted to the postprandial period when insulin is secreted or periods of exercise when glucose transport is non–insulin dependent.

In the liver, insulin promotes glucose uptake and storage, and inhibits glucose production. In the liver, insulin will

- *Increase the flux of glucose into cells*—This is achieved not by increasing the number of glucose transporters in the cell membranes but by inducing *glucokinase*, which increases the phosphorylation of glucose to glucose-6-phosphate.
- *Increase glycogen synthesis by activating glycogen synthase* (as well as by increasing glucose uptake).
- *Direct the flow of glucose through glycolysis by increasing the activity of key glycolytic enzymes* (e.g., phosphofructokinase and pyruvate kinase).
- *Decrease the hepatic output of glucose* in several ways. First, insulin impairs *glycogenolysis* by inhibiting *glycogen phosphorylase*. Second, insulin decreases the exit of glucose from the liver by inhibiting *glucose-6-phosphatase*. Third, insulin inhibits *gluconeogenesis* by decreasing the amino acid uptake into the liver (see discussion on effects on protein metabolism) and by decreasing the activity or levels of key *gluconeogenic enzymes* (e.g., pyruvate carboxylase and fructose-1,6-diphosphatase).
- *Enhance synthesis of fatty acids* in two ways—First, insulin increases the flow of glucose to pyruvate (glycolysis) and the subsequent conversion to *acetyl-coenzyme A (acetyl-CoA)*. Second, insulin stimulates *acetyl-CoA carboxylase*, which converts acetyl-CoA to malonyl-CoA. This is the rate-limiting step in the synthesis of fatty acids.

In adipose tissue, insulin facilitates glucose entry into cells. This is achieved in much the same way that insulin promotes glucose uptake into muscle cells—by increasing glucose transporters in the cell membrane. Subsequently, the metabolism of glucose to *α-glycerol phosphate* provides the glycerol that is needed for esterification of fatty acids for storage as triglycerides (see the following discussion).

Insulin has little effect on glucose uptake and use by the brain. In the brain, insulin has little effect on glucose transport into cells. Because brain cells are quite permeable to glucose and highly dependent on this substrate for energy, it is essential that blood glucose concentration is maintained at normal levels. If blood glucose concentration falls too low, symptoms of *hypoglycemic shock* will occur, including fainting, seizure, and even coma.

Effects of Insulin on Fat Metabolism

In adipose tissue, insulin enhances storage and inhibits mobilization of fatty acids. This effect of insulin is accomplished in several ways:

- *Insulin inhibits hormone-sensitive lipase.* This decreases the rate of lipolysis of triglycerides and the release of stored fatty acids into the circulation.
- *Insulin increases glucose transport.* The subsequent metabolism of glucose to *α-glycerol phosphate* increases the rate of *esterification* of fatty acids for storage as *triglycerides*.
- *Insulin induces lipoprotein lipase.* This enzyme is present in the capillary wall and splits circulating triglycerides into fatty acids, which is necessary for their transport into fat cells.

In the liver, insulin promotes the synthesis and inhibits the oxidation of fatty acids. As discussed previously, insulin promotes the synthesis of fatty acids from glucose in the liver. Because of the increased availability of α-glycerol phosphate from glycolysis, fatty acids are esterified to form triglycerides. Oxidation of fatty acids is impaired because of the increased conversion of acetyl-CoA to malonyl-CoA by acetyl-CoA carboxylase, as discussed. Malonyl-CoA inhibits *carnitine acyltransferase*, which is responsible for shuttling fatty acids from the cytoplasm into the mitochondria for *beta oxidation* and *conversion to keto acids*; insulin is *antiketogenic*.

Effects of Insulin on Protein Metabolism

Insulin is an *anabolic hormone*. It increases the uptake of several amino acids from the blood into cells by stimulating transport across the cell membrane; this limits the rise in plasma levels of certain amino acids after a protein meal. In addition, insulin increases protein synthesis by stimulating both gene transcription and translation of mRNA. Finally, insulin inhibits catabolism of proteins and therefore decreases the release of amino acids from muscle.

Insulin, like growth hormone, is essential to growth. Diabetic animals fail to grow. The anabolic effects of insulin and growth hormone are synergistic.

Control of Insulin Secretion

Glucose is the most important controller of insulin secretion. Although several factors can increase or decrease insulin secretion, the major control of insulin secretion is exerted by a feedback effect of blood glucose on the beta cells of the pancreas. When blood glucose concentration rises above fasting levels, insulin secretion increases. As a result of the subsequent effects of insulin to stimulate glucose uptake by the liver and peripheral tissues, blood glucose concentration returns to normal levels. This provides an important negative feedback mechanism for the control of blood glucose concentration.

Multiple stimuli other than hyperglycemia increase insulin secretion. These stimuli include the following:

- *Amino acids*, especially arginine, lysine, leucine, and alanine—As a result, dietary amino acids are removed from the blood and used by cells to synthesize proteins. Amino acids have a synergistic effect with glucose in stimulating insulin secretion.
- *Gastrointestinal hormones*, especially *gastric inhibitory polypeptide* and *glucagon-like polypeptide 1*—These hormones are released from the gastrointestinal tract after a meal is eaten and account for the greater increase in insulin secretion when glucose is administered orally than when comparable amounts are administered intravenously.
- Other hormones, including *cortisol* and *growth hormone*—These hormones increase insulin secretion in large part because they antagonize the effects of insulin on glucose uptake in peripheral tissues, leading to increased blood glucose concentration. Indeed, chronic increments in cortisol (in Cush-

ing's syndrome) and growth hormone (in acromegaly) lead to hypertrophy and exhaustion of the beta cells of the pancreas and thereby cause *diabetes mellitus.*

GLUCAGON AND ITS FUNCTIONS (p. 978)

Most of the actions of glucagon are achieved by activation of adenylyl cyclase. At physiological doses, the primary effects of glucagon occur at the liver and are opposite those of insulin. The binding of glucagon to hepatic receptors results in activation of *adenylyl cyclase and generation of the second messenger cyclic AMP*, which in turn activates *protein kinase A*, leading to phosphorylations that result in the activation or deactivation of a number of enzymes.

Glucagon promotes hyperglycemia in several ways.

- *Glucagon stimulates glycogenolysis*—Glucagon has immediate and pronounced effects on the liver to increase glycogenolysis and the release of glucose into the blood. This effect is achieved through activation of *glycogen phosphorylase* and simultaneous inhibition of *glycogen synthase.*
- *Glucagon inhibits glycolysis*—Glucagon inhibits several key steps in glycolysis, including phosphofructokinase and pyruvate kinase. Consequently, *glucose-6-phosphate* levels tend to rise, leading to increased glucose release from the liver.
- *Glucagon stimulates gluconeogenesis*—Glucagon increases the hepatic extraction of amino acids from the plasma and increases the activities of key *gluconeogenic enzymes*, including pyruvate carboxylase and fructose-1,6-diphosphatase. Consequently, glucagon has delayed and protracted actions to promote glucose output by the liver.

Glucagon is ketogenic. Because glucagon inhibits *acetyl-CoA carboxylase*, there is decreased production of malonyl-CoA, an inhibitor of *carnitine acyltransferase.* Consequently, fatty acids are directed into the mitochondria for *beta oxidation* and *ketogenesis.*

Control of Glucagon Secretion

Glucose is the most important controller of glucagon secretion. Glucose is the most important controller of both glucagon and insulin secretion; however, glucose has opposing effects on the secretion of these two hormones. Hypoglycemia increases glucagon secretion; as a result of the hyperglycemic actions of glucagon, blood glucose concentration returns toward normal. Conversely, increases in blood glucose concentration decrease glucagon secretion; glucagon and insulin provide important, but opposing, mechanisms for the regulation of blood glucose concentration.

Amino acids, **especially arginine and alanine, stimulate glucagon secretion.** After a protein meal, both insulin and glucagon secretion are stimulated, but the glucagon response is depressed if glucose is ingested simultaneously. The glucagon response to a protein meal is valuable because without the hyperglycemic effects of glucagon, increased insulin secretion would cause hypoglycemia.

Fasting and *exercise* **stimulate glucagon secretion.** Under these conditions, the stimulation of glucagon secretion helps to prevent large decreases in blood glucose concentration. The secretion of glucagon may be mediated in part by stimulation of the sympathetic nervous system.

Insulin/Glucagon Ratios

As discussed above, *insulin* is a hormone of *energy storage*. It promotes the synthesis of glycogen, fatty acids, triglycerides, and proteins. On the other hand, *glucagon* is a hormone of *energy release*. In the liver, glucagon increases glycogenolysis and gluconeogenesis, and the release of glucose into the blood; furthermore, it promotes oxidation of fatty acids and the synthesis of ketone bodies. Because insulin and glucagon have opposite effects on metabolism, one must consider the relative changes in plasma levels of each hormone. This can be done by determining the *insulin/glucagon ratio* in the blood.

The usual molar ratio is approximately 2.0. Under conditions of fasting, the ratio decreases to 0.5 or less because of the fall in insulin secretion and an increase in glucagon release. As a result, stores of carbohydrates, fats, and proteins are metabolized, and blood glucose concentration is preserved, which maintains the glucose supply for the central nervous system. Under these conditions, most tissues oxidize fatty acids and ketone bodies for energy. In contrast, after a normal meal, the insulin/glucagon ratio may increase to more than 10. There is a high rate of glucose and amino acid uptake from the blood and the synthesis of glycogen, fats, and proteins. The high insulin/glucagon ratio minimizes the magnitude and duration of postprandial hyperglycemia.

SOMATOSTATIN—ITS EFFECT TO INHIBIT GLUCAGON AND INSULIN SECRETION (p. 979)

Somatostatin is synthesized by the *delta cells* in the pancreas, as well as in the gut and hypothalamus, where it is a hypophysiotropic hormone (Chapter 75). In the pancreas, the major product of the somatostatin prohormone is a 14-amino acid peptide. Pancreatic somatostatin secretion is stimulated by factors related to the ingestion of food, including increased blood levels of glucose, amino acids, and fatty acids and a number of gastrointestinal hormones. Somatostatin inhibits gastrointestinal motility, secretion, and absorption and is a potent inhibitor of insulin and glucagon secretion; it delays the assimilation of nutrients from the gastrointestinal tract and the utilization of absorbed nutrients by the liver and peripheral tissues.

DIABETES MELLITUS (p. 980)

In diabetes mellitus, carbohydrate, fat, and protein metabolism are impaired because of a deficient response to insulin. There are two forms of *diabetes mellitus*:

- *Insulin-dependent diabetes mellitus (IDDM or type I)*, which is caused by impaired secretion of insulin.
- *Non–insulin-dependent diabetes mellitus (NIDDM or type II)*, which is caused by resistance to the metabolic effects of insulin in target tissues.

IDDM is caused by impaired secretion of insulin by the beta cells of the pancreas. Often, IDDM is a result of autoimmune destruction of beta cells, but it can also arise from the loss of beta cells due to viral infections. Because the usual onset of IDDM occurs in childhood, it is referred to as *juvenile diabetes*.

Most of the pathophysiological features of *IDDM* can be attributed to the following major effects of insulin deficiency:

- *Hyperglycemia* as a result of impaired glucose uptake into tissues and increased glucose production by the liver (increased gluconeogenesis)

- *Depletion of proteins* due to decreased synthesis and increased catabolism
- *Depletion of fats stores and increased ketogenesis.*

As a result of these fundamental derangements, the following occur:

- Glucosuria, osmotic diuresis, hypovolemia, and hypotension
- Hyperosmolality of the blood, dehydration, and polydipsia
- Hyperphagia but weight loss; lack of energy
- Acidosis progressing to diabetic coma; rapid and deep breathing
- Hypercholesterolemia and atherosclerotic vascular disease.

Insulin resistance is the hallmark of NIDDM. NIDDM is far more common than IDDM and is usually associated with obesity. This form of diabetes is characterized by an impaired ability of target tissues to respond to the metabolic effects of insulin, which is referred to as *insulin resistance*. In contrast to IDDM, pancreatic beta cell morphology is normal throughout much of the disease, and there is an elevated rate of insulin secretion. NIDDM usually develops in adults and therefore is also called *adult-onset diabetes*.

Although hyperglycemia is a prominent feature of NIDDM, accelerated lipolysis and ketogenesis usually do not occur. Caloric restriction and weight reduction usually improve insulin resistance in target tissues, but in the late stages of the disease when insulin secretion is impaired, insulin administration is required.

79

Parathyroid Hormone, Calcitonin, Calcium and Phosphate Metabolism, Vitamin D, Bone, and Teeth

The physiology of calcium and phosphate metabolism, the function of vitamin D, and the formation of bone and teeth are all tied together in a common system with the two regulatory hormones, parathyroid hormone (PTH) and calcitonin.

CALCIUM AND PHOSPHATE IN THE EXTRACELLULAR FLUID AND PLASMA—FUNCTION OF VITAMIN D
(p. 985)

Control of Vitamin D Formation

The active form of vitamin D, *1,25-dihydroxycholecalciferol*, is carefully regulated via the following steps:

- *In the skin*, 7-dehydrocholesterol is converted by ultraviolet light to vitamin D_3.
- *In the liver*, vitamin D_3 is converted to 25-hydroxycholecalciferol.
- *In the cortex of the kidney*, 25-hydroxycholecalciferol is converted to 1,25-dihydroxycholecalciferol in a reaction *stimulated by and tightly controlled by PTH.*

Because PTH formation is stimulated by reduction in the extracellular fluid (ECF) concentration of calcium, formation of 1,25-dihydroxycholecalciferol will also increase when calcium concentration in the ECF falls.

Gastrointestinal Calcium Absorption— 1,25-Dihydroxycholecalciferol

In the epithelial cells of the small intestine, 1,25-dihydroxycholecalciferol stimulates formation of *calcium binding protein*, *calcium-stimulated ATPase*, and *alkaline phosphatase*, all of which promote absorption of calcium ions out of the lumen of the intestine.

Being a divalent cation, Ca^{2+} cannot cross the cell membrane of the epithelial cells without the mechanisms activated by 1,25-dihydroxycholecalciferol; *therefore, calcium absorption will occur at a rate determined specifically by the activity of the mechanisms regulated by 1,25-dihydroxycholecalciferol.*

Phosphate ions are absorbed in a relatively unregulated manner, although the rate of absorption is elevated by 1,25-dihydroxycholecalciferol.

Calcium and Phosphate in the Extracellular Fluid (ECF) and Plasma

Accurate regulation of calcium ion concentration is essential to normal function of the neuromuscular system and the skeletal system. If the concentration of calcium in the ECF falls to less than 50 per cent of normal for even brief periods, neuromuscular dysfunction of the skeletal muscles will result, initially in the form of hyperreflexivity and finally as tetanic contractions. If calcium ion concentration increases to 50 per cent greater than normal, central nervous system depression will occur, along with slowing of the contractions of the smooth muscle of the gastrointestinal tract.

Calcium is found normally in the ECF in a total concentration of 2.4 mmol/liter, or 9.4 mg/dl. In the ECF, 50 per cent of the calcium is in the free divalent cation form, 40 per cent is loosely bound to proteins, and 10 per cent is in the nonionized form.

Phosphate ion concentration in the ECF can vary rather widely without physiological impact. Phosphate in the ECF can be either monobasic ($H_2PO_4^-$) or dibasic ($HPO_4^=$). The normal concentration of $H_2PO_4^-$ is 0.26 mmol/liter, whereas $HPO_4^=$ is found at a concentration of 1.05 mmol/liter. The relative concentrations of the two are affected by the pH of the ECF, with reduction in pH increasing the amount of $H_2PO_4^-$ and decreasing the concentration of $HPO_4^=$. Clinically, total phosphate concentration is usually expressed in milligrams per deciliter (mg/dl) and is normally 3 to 4 mg/dl.

BONE AND ITS RELATION TO EXTRACELLULAR CALCIUM AND PHOSPHATES (p. 989)

Bone is composed mostly of calcium and phosphate salts along with organic matrix. Approximately 70 per cent of bone is calcium salts; most is in the form of large crystals of *hydroxyapatite*, $Ca_{10}(PO4)_6(OH)_2$. Bone is about 30 per cent organic matrix, made up of collagen fibers and cells. Some calcium in bone is not in crystalline form and therefore is rapidly exchangeable with the calcium in the ECF.

Bone calcification occurs. Bone formation begins with secretion of *collagen fibers* by *osteoblast cells*; the uncalcified collagen structure is referred to as *osteoid*. Calcification of the osteoid takes place over a period of weeks.

Bone is continually deposited by *osteoblasts* and absorbed by *osteoclasts*, a dynamic process referred to as *remodeling*. Bone has the capacity to undergo remodeling throughout life, although the process takes place much more rapidly in children and young adults than in the elderly. Osteoclast cells digest bone, after which *osteoblasts* deposit new bone. The balance between the two processes is affected by the following:

- *Mechanical stress* on the bone, which stimulates remodeling and formation of stronger bone at points of stress
- *PTH* and *1,25-dihydroxycholecalciferol*, which stimulate osteoclast activity and formation of new osteoclasts
- *Calcitonin*, which decreases the absorptive capacity of osteoclasts and decreases the rate of formation of new osteoclasts.

The calcium and phosphate present in bone serve as reservoirs for the ions in the ECF. About 99 per cent of the total body calcium is in the bone, whereas less than 1 per cent is in the ECF. If the calcium ion concentration in the ECF falls below normal, calcium ions will move from the bone into the ECF. The distribution between bone and ECF is affected by PTH and 1,25-dihydroxycholecalciferol, which stimulate movement of calcium and phosphate from bone to ECF, and by calcitonin, which has the opposite effect. Conversely, when calcium concentration in the ECF increases above normal, calcium can be deposited in the bone.

PARATHYROID HORMONE (p. 992)

PTH secretion increases in response to a reduction in extracellular calcium concentration. The hormone is formed in the *chief cells* of the parathyroid glands located immediately behind the thyroid gland. The rate of formation of PTH is strongly regulated by the ECF calcium ion concentration; small decreases in the concentration of the ion will result in large increases in the rate of PTH formation. If the reduction below the normal level of calcium concentration persists, the parathyroid glands will hypertrophy, as occurs in pregnancy and disease states such as rickets that are characterized by inadequate calcium absorption from the gastrointestinal tract.

Increases in PTH concentration decrease renal calcium excretion. Normally, greater than 99 per cent of calcium filtered at the glomerulus is reabsorbed along the tubule. Approximately 5 per cent of the filtered calcium is reabsorbed in the collecting duct, and it is calcium transport in this segment that is stimulated by PTH. Other factors that affect calcium excretion include the following:

Increase Calcium Excretion	Decrease Calcium Excretion
Decreased [PTH]	Increased [PTH]
Increased ECFV	Decreased ECFV
Decreased [$HPO_4^=$]	Increased [$HPO_4^=$]
Metabolic acidosis	Metabolic alkalosis

Increases in PTH concentration elevate phosphate excretion. Phosphate excretion is regulated as a tubular maximum (T_m) system (see Chapter 29). Approximately 80 per cent is reabsorbed in the proximal tubule, with additional absorption taking place at more distal sites in the nephron. PTH inhibits phosphate reabsorption in the proximal tubule; other factors that affect phosphate excretion include the following:

Increase $HPO_4^=$ Excretion	Decrease $HPO_4^=$ Excretion
Increased [PTH]	Decreased [PTH]
Increased ECFV	Decreased ECFV
Increased [$HPO_4^=$]	Decreased [$HPO_4^=$]
Metabolic acidosis	Metabolic alkalosis

CALCITONIN (p. 995)

Calcitonin secretion increases in response to elevation of extracellular calcium concentration. The hormone is a polypeptide with 32 amino acids secreted from the *parafollicular cells* found in the interstitial tissue of the thyroid gland. In general, its effects are opposite those of PTH in the bone and renal tubule, and the magnitude of its effects is much less than that of PTH.

OVERALL CONTROL OF CALCIUM ION CONCENTRATION
(p. 996)

Calcium concentration in the ECF is controlled by a system that affects the distribution among the calcium stored in bone and the ECF, rate of intake from the gastrointestinal tract, and rate of excretion by the kidneys.

Regulation of Calcium Distribution between Bone and Extracellular Fluid

When ECF calcium concentration falls:

- Readily exchangeable calcium ions diffuse into the ECF.
- PTH formation increases, stimulating the activity of osteoclasts and causing movement of calcium from bone to ECF.

Regulation of Absorption from the Gastrointestinal Tract

When calcium concentration in the ECF falls:

- PTH formation increases, causing a higher rate of formation of 1,25-dihydroxycholecalciferol.
- Elevated concentration of 1,25-dihydroxycholecalciferol stimulates the formation of calcium-binding protein and other factors in the epithelium of the small intestine, which increase the rate of absorption of calcium from the lumen of the gut.

Regulation of Renal Calcium and Phosphate Excretion

When calcium concentration in the ECF falls:

- PTH formation increases; as a result:
  1. Calcium absorption from the collecting duct increases, and excretion of calcium decreases.
  2. Phosphate reabsorption from the proximal tubule decreases, and phosphate excretion increases.

In humans, the most important feedback control mechanism is the effect of a reduction in ECF calcium concentration to elevate PTH formation. The involvement of calcitonin in the control system is of minor importance in adults.

PHYSIOLOGY OF PARATHYROID AND BONE DISEASES
(p. 997)

Hypoparathyroidism decreases extracellular calcium concentration. The condition results from inadequate formation of PTH. Osteoclasts become inactive, and the formation of 1,25-dihydroxychole-calciferol declines to low levels. Transfer of calcium from the bone to the ECF decreases, calcium absorption from the gut decreases to low levels, and calcium excretion by the kidneys is greater than the rate of absorption from the gut. As a result, calcium concentration in the ECF falls below normal levels, and phosphate concentration remains normal or is elevated. The condition can be treated with very large doses of vitamin D, which has some effect to stimulate gastrointestinal calcium absorption, or by the administration of 1,25-dihydroxycholecalciferol.

Excessive formation of PTH by the parathyroid gland (hyperparathyroidism) causes loss of calcium from bone and increased extracellular calcium concentration. The excessive PTH levels stimulate osteoclastic activity, renal retention of calcium and excretion of phosphate, and increased formation of 1,25-dihydroxycholecalciferol. Calcium concentration in the ECF is greater than normal, and phosphate levels are below normal. The most serious consequences are related to the damage done by excessive osteoclastic absorption of bone, which results in weakening of the bone.

Rickets is caused by inadequate absorption of calcium from the gastrointestinal tract. This can be due to inadequate calcium in the diet or failure to form adequate amounts of 1,25-dihydroxycholecalciferol. If the kidneys are damaged or absent, 1,25-dihydroxycholecalciferol cannot be formed. Because of inadequate absorption of calcium, PTH levels are elevated, which stimulates osteoclastic resorption of bone and release of calcium to the ECF. In addition, the elevated PTH levels exert renal effects, causing retention of calcium and excretion of phosphate. The net results of these effects are weakening of the bones, a below-normal phosphate concentration, and for periods of months, only slightly below normal calcium concentration due to the transfer of calcium from the bone to the ECF.

Osteoporosis is caused by depressed deposition of new bone by the osteoblasts. As a result, the rate of osteoclastic resorption of bone exceeds the rate of deposition of new bone.

The most common cause of the condition is loss of anabolic sex steroids, estrogen and testosterone, which strongly stimulate the activity of the osteoblasts. In men, testosterone levels decline gradually and provide a significant anabolic effect into the seventh and eighth decades of life. In women, estrogen formation falls to near zero at menopause, usually at about 50 years of age. The decline in estrogen concentration shifts the balance between deposition and resorption of bone, although no symptoms are apparent for many years. Starting even before menopause, calcium is continually lost from the skeleton. After years of the gradual wasting of calcium, the bones become weakened to the point that symptoms appear, such as vertebral compression and brittleness of the long bones and pelvis. The condition can be prevented with estrogen replacement therapy beginning at menopause. Calcium supplements after menopause are not effective because the condition is not characterized by inadequate calcium in the ECF.

PHYSIOLOGY OF THE TEETH (p. 998)

Teeth are composed of four parts: *enamel, dentine, cementum,* and *pulp.*

Enamel makes up the outer layer of the crown of the tooth. It is composed of very large and dense crystals of hydroxyapatite embedded in a tight meshwork of protein fibers similar to keratin in hair. The crystalline structure makes the enamel extremely hard, whereas the protein, which is completely insoluble, provides resistance to enzymes, acids, and other corrosive substances.

Dentine makes up the main body of the tooth. It is composed of hydroxyapatite crystals embedded in a strong meshwork of collagen fibers, a structure similar to bone. Dentine has no cellular components; all of the nourishment of the structure is provided from *odontoblast cells,* which line the inner surface of the dentine along the wall of the pulp cavity.

Cementum is a bony substance that lines the tooth socket. It is secreted by the cells of the periodontal membrane. Collagen fibers pass from the bone of the jaw, through the periodontal membrane, and into the cementum. This arrangement provides the firm attachment between the teeth and jaw.

Pulp is the tissue that fills the pulp cavity of the tooth. It is composed of odontoblasts, nerves, blood vessels, and lymphatic vessels. During formation of the tooth, the odontoblasts lay down new dentine along the lining of the pulp cavity, making it progressively smaller.

80

Reproductive and Hormonal Functions of the Male (and the Pineal Gland)

SPERMATOGENESIS (p. 1003)

Spermatogenesis is the process of formation of spermatocytes from spermatogonia. It is initiated at puberty and continues throughout the remainder of a man's life. It take place within the walls of the *seminiferous tubules*.

The walls of the tubules are composed of two compartments separated by tight junctions between the *Sertoli cells*:

- The *basal layer*, which consists of the *Leydig cells* and the *spermatogonia*
- The *ad luminal layer*, which is made up of Sertoli cells and *spermatocytes*.

The initial step in the process is transformation of *type A spermatogonia*, which are epithelioid in nature, to *type B spermatogonia*, a process involving four divisions. The type B cells imbed in the Sertoli cells. In association with the Sertoli cells, the type B cells are transformed to *primary spermatocytes* and, in a step involving the first meiotic division, *secondary spermatocytes*. The secondary spermatocytes undergo a second meiotic division, yielding *spermatids*, of which each has 23 unpaired chromosomes. *The steps as described are stimulated by testosterone and follicle-stimulating hormone (FSH).*

Spermiogenesis is the process of transformation of the spermatids, which are still epithelioid, to sperm cells. The process takes place with the cells imbedded in the Sertoli cells and *requires estrogen and FSH.*

Once the sperm cells are formed, they are extruded into the lumen of the tubule in a process stimulated by luteinizing hormone (LH). From the first division of the type A spermatogonia to extrusion of the sperm cells requires a period of approximately 64 days.

The newly formed sperm cells are not functional and require a *maturation process*, which takes place in the *epididymis* over a period of 12 days. Maturation *requires both testosterone and estrogen*. The mature sperm are *stored in the vas deferens*.

THE MALE SEXUAL ACT (p. 1008)

The male sexual act is the process that culminates in deposition of several hundred million viable sperm cells at the cervix of the man's sexual partner. The sperm cells are contained in a mixture of fluids produced by the male reproductive organs that is called *semen* and includes the following:

- *Seminal vesicle fluid*, which makes up 60 per cent of the total volume of the semen. It contains mucoid, prostaglandin E_2, fructose, and fibrinogen.
- *Prostatic fluid*, which makes up 20 per cent of the semen volume and contains $NaHCO_3$ (pH 7.5), clotting enzyme, calcium, and profibrinolysin
- *Sperm cells*.

The average volume of semen ejaculated at each coitus is 3.5 milliliters, and each milliliter of semen contains approximately 120 million sperm cells. For normal fertility, the sperm count per milliliter must be greater than 20 million.

The *sexual act* takes place in three stages:

- *Erection and lubrication*—Erection is the process of filling the erectile tissue of the penis with blood at a pressure level near that of the arterial blood. The arteries leading to the erectile tissue dilate in response to parasympathetic impulses, which stimulate release of nitric oxide at the nerve endings on the arterial smooth muscle. Parasympathetic reflexes also stimulate secretion of mucus by the urethral glands and bulbourethral glands. The mucus aids in vaginal lubrication during coitus.
- *Emission*—This is the process of stimulating the smooth muscle surrounding the seminal vesicles, vas deferens, and prostate gland, causing the organs to empty their contents into the internal urethra, a process elicited by sympathetic reflexes from L-1 and L-2.

- *Ejaculation*—This is a skeletal muscle reflex that is elicited in response to distention of the internal urethra. It results in contraction of the ischiocavernosus and bulbocavernosus muscles and the muscles of the pelvis, causing compression in the internal urethra and propelling the semen out of the urethra.

MALE SEX HORMONES (p. 1009)

Testosterone is an anabolic steroid hormone secreted by the Leydig cells of the testes. The hormone is formed from cholesterol in amounts ranging from 2 to 10 mg/day. In the blood, testosterone is carried in association with albumin or tightly bound to *sex hormone binding globulin*. The hormone is removed from the blood within 30 to 60 minutes of secretion by fixation to target tissue cells or degradation to inactive compounds. It is metabolized to dihydrotestosterone (the biologically active androgen) in target tissues and to estrogen in adipose tissue.

Testosterone has effects on reproductive and nonreproductive organs. The hormone is required for stimulation of prenatal differentiation and pubertal development of the testes, penis, epididymis, seminal vesicles, and prostate. Testosterone is required in adult males for maintenance and normal function of the primary sex organs. Testosterone also has effects on bone, stimulating growth and proliferation of bone cells, resulting in increased density of the bones. It also has effects on hair distribution and causes the skin to thicken. Testosterone affects the liver, causing synthesis of clotting factors and hepatic lipases. Under the influence of testosterone, blood high-density lipoprotein levels decrease, and low-density lipoprotein levels increase. Hematocrit and hemoglobin concentrations are elevated because of the effect of testosterone to stimulate production of erythropoietin. The hormone has a generalized effect in many tissues to enhance the rate of protein synthesis.

Being a steroid hormone, testosterone readily enters the cytoplasm of target tissue cells through diffusion through the cell membrane. The enzyme *5-α-keto reductase* converts it to *dihydrotestosterone*, which then binds with a cytoplasmic receptor protein. This combination migrates to the nucleus, where it binds with a nuclear protein that induces DNA-RNA transcription.

Gonadotropin-releasing hormone increases release of LH and FSH from the anterior pituitary gland. The polypeptide hormone, which is also referred to as gonadotropin-releasing hormone (GnRH), is secreted from the hypothalamus into the hypothalamic-hypophyseal portal system. Its formation is inhibited by testosterone and estrogen.

LH stimulates testosterone formation by the Leydig cells, and FSH stimulates spermatogenesis and spermiogenesis. They are secreted from the basophilic cells of the anterior pituitary. Their release is stimulated by GnRH.

Inhibin is formed by Sertoli cells and inhibits FSH secretion. Inhibin formation increases as the rate of sperm cell production increases.

MALE INFERTILITY (p. 1014)

Approximately 15 per cent of couples in the United States are infertile, and approximately 50 per cent of the dysfunction is in the male partner; 5 per cent of males in the United States are believed to be infertile. The causes of male infertility include the following:

- *Androgen dysfunction with normal sperm cell production*, caused by hypothalamic-pituitary defects, Leydig cell defects, or androgen resistance
- *Isolated dysfunction of sperm cell production with normal androgen levels*, resulting from infection or trauma, congenital deformation of passages, or formation of nonmotile or otherwise abnormal sperm
- *Combined androgen and sperm cell production defects* due to developmental defects, such as Klinefelter's syndrome, or abnormal testicular descent or to acquired testicular defects, such as infections, autoimmune reactions, or systemic diseases such as chronic liver and kidney diseases

In 50 per cent of infertile males, no cause can be identified.

81

Female Physiology Before Pregnancy; and the Female Hormones

FEMALE HORMONAL SYSTEM (p. 1017)

Reproductive function in the female is regulated by interactions of hormones from the hypothalamus, anterior pituitary, and ovaries. Several of the hormones are found in males as well as females.

- *Gonadotropin-releasing hormone (GnRH)* is the releasing factor from the hypothalamus that stimulates secretion of follicle-stimulating hormone (FSH) and luteinizing hormone (LH) from the anterior pituitary. The release of GnRH is inhibited by estrogen and progesterone.
- *LH* is secreted from the basophilic cells of the anterior pituitary gland and stimulates development of the corpus luteum in the ovaries.
- *FSH* is secreted from the basophilic cells of the anterior pituitary gland in response to GnRH and stimulates development of the follicles in the ovaries.
- *Estrogen* and *progesterone* are the steroid hormones secreted by the follicle and corpus luteum of the ovary.

The 28-day female sexual cycle is determined by the time required for development of the follicles and corpus lutea after menstruation and the feedback effect of the hormones they secrete on the hypothalamus.

MONTHLY OVARIAN CYCLE (p. 1018)

One mature ovum is released from the ovary during each monthly cycle, and the endometrium of the uterus is prepared for implantation of the fertilized ovum at the appropriate time. To achieve these results, all of the hormones of the female reproductive system must interact. The changes of the concentrations in the blood of the most important hormones

693

of the system over the course of the 28-day cycle are illustrated in Figure 81–1.

Follicular Development

At the beginning of the monthly cycle, no mature follicles or corpus lutea are present. Estrogen and progesterone concentrations in the blood are at their lowest levels (see Fig. 81–1). As a result, the hypothalamus receives no inhibitory signals to block secretion of GnRH. The GnRH that is secreted stimulates FSH and LH secretion from the pituitary, and the FSH stimulates development of 12 to 14 primary ovarian follicles. The follicles are surrounded by *granulosa cells*, which begin to secrete fluid into the center of the structure; this in turn expands to form a fluid-filled antrum that surrounds the oocyte. At this stage, the structure is referred to as an *antral follicle*. The fluid is rich in estrogen, which diffuses into the blood and results in a progressive rise in its concentration. The follicles continue to develop, stimulated by FSH, LH, and the estrogen secreted by the follicles. Proliferation of the granulosa cells proceeds, accompanied by the growth of surround

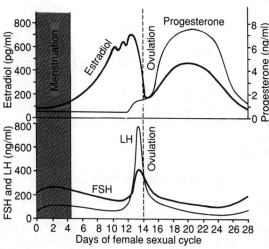

Figure 81–1 Approximate plasma concentrations of the gonadotropins and ovarian hormones during the normal female sexual cycle.

ing layers of *thecal cells* derived from the stroma of the ovary. With accumulation of additional fluid and continued development, the follicle is referred to as a *vesicular follicle*.

After approximately 1 week of development, one follicle begins to outgrow the others. The remaining follicles, which developed to the follicular stage, undergo *atresia* and degenerate; the cause of this process is unknown. The remaining dominant follicle continues to develop rapidly, with proliferation of granulosa and thecal cells stimulated by FSH and estrogen. The estrogen promotes development of additional FSH and LH receptors on the granulosa and thecal cells, which provides a positive feedback cycle for rapid development of the maturing follicle.

Because of the rapidly rising concentration of estrogen in the blood (see Fig. 81–1), the hypothalamus receives an inhibitory signal to depress GnRH secretion. This results in suppression of FSH and LH secretion from the pituitary; the reduction in the secretion of FSH prevents the development of additional follicles. The dominant follicle continues to develop because of its intrinsic positive feedback cycle while the other vesicular follicles involute, and no additional primary follicles begin to develop.

Ovulation

At approximately 12 days after the initiation of follicular development, the pituitary produces a *surge of LH secretion* for reasons that are unknown. The rate of secretion rises 6- to 10-fold.

Associated with the LH surge, the thecal cells begin to secrete progesterone for the first time; they secrete a fluid containing proteolytic enzymes into the vesicle. The blood flow in the thecal layers increases at this time, as does the rate of transudation of fluid into the vesicle.

At a point of weakness in the wall of the follicle on the surface of the ovary, a protrusion, or *stigma*, develops. The wall ruptures at the stigma within 30 minutes of its formation; within minutes of the rupture, the follicle evaginates, and the *oocyte* and surrounding layers of granulosa cells, referred to as the *corona radiata*, leave the vesicle and enter the abdominal cavity at the opening to the fallopian tube. Completion of ovulation occurs an average of 14 days after the initiation of follicular development and 16 hours after the peak of LH concentration in the blood.

Corpus Luteum

The structure of the follicle remaining on the surface of the ovary after ovulation contains layers of granulosa and thecal cells. The high concentration of LH before ovulation converts these cells to *lutein cells*, which enlarge after ovulation and become yellowish; this structure is referred to as the *corpus luteum*. The granulosa cells secrete large amounts of progesterone and smaller amounts of estrogen, and the thecal cells produce androgenic hormones, testosterone, and androstenedione, most of which are converted by the granulosa cells to the female hormones.

The cells of the corpus luteum require stimulation by the preovulatory surge of LH to undergo transformation and proliferation. The corpus luteum secretes large amounts of progesterone and estrogen for approximately 12 days under the continuing stimulatory influence of the declining concentration of LH. After 12 days, when LH levels are minimal due to feedback inhibition of the hypothalamus by estrogen and progesterone (see Fig. 81–1), the corpus luteum degenerates and ceases to secrete hormones. Within 2 days of failure of the corpus luteum, menstruation begins (see subsequent discussion). At the same time, FSH and LH secretion from the pituitary begins to rise owing to the absence of inhibition of the hypothalamus by estrogen and progesterone. As the concentration rises in the blood of the stimulatory hormones from the pituitary, a new group of primary follicles begins to develop, initiating another cycle.

FUNCTIONS OF THE OVARIAN HORMONES—ESTRADIOL AND PROGESTERONE (p. 1022)

The ovaries secrete two classes of hormones: *estrogens* and *progestins*; the most important of the estrogens is *estradiol*, and *progesterone* is the dominant progestin. In the nonpregnant female, essentially all of the estrogen compounds are secreted from the ovaries, with only minute amounts being synthesized in the adrenal cortex. Nearly all of the progesterone in nonpregnant females is produced in the corpus luteum; only small amounts are formed in the mature follicle during the day immediately before ovulation.

Functions of Estrogen

Estrogens cause growth and proliferation of the cells of the female sex organs and other tissues associated with reproduction.

Estrogen stimulates the growth and development of the uterus and external female sex organs. At puberty, the levels of estrogen rise rapidly, causing rapid growth in the ovaries, fallopian tubes, uterus, vagina, and external genitalia. The lining of the uterus, the *endometrium*, becomes thickened under the effect of estrogen, as discussed later.

Estrogens stimulate development of the stroma tissue of the breasts, growth of an extensive ductile system, and deposition of fat in the breasts.

Estrogen causes growth of the skeleton by stimulating osteoblastic activity. At puberty, the effect on the osteoblast causes a period of rapid growth in the long bones, although this "growth spurt" lasts only a few years because of the effect of estrogen to cause closure of the epiphyses of the bones. Longitudinal growth occurs only at the epiphyses, so that once they are closed, additional lengthening of the bones cannot take place.

Estrogen has a weak effect to increase total body protein and metabolic rate. It promotes deposition of fat in the subcutaneous tissue, particularly in the breasts, hips, and thighs.

Functions of Progesterone

The most important function of progesterone is to promote secretory changes in the uterine endometrium during the latter half of the monthly sexual cycle. This prepares the uterus for implantation of the zygote. Progesterone has a similar effect on the lining of the fallopian tubes, causing the secretion of fluid that provides nutrition for the fertilized ovum during its passage to the uterus. The hormone also reduces the excitability and motility of the uterine smooth muscle.

Progesterone stimulates development of the lobules and alveoli of the breasts. This effect causes the alveolar cells to enlarge, proliferate, and become secretory in nature, although the cells do not produce milk in response to progesterone.

Progesterone causes an upward resetting of the body temperature control system by about 0.5°F. This effect can be used to determine the time of ovulation because progesterone is not produced until the preovulatory LH surge, which takes place a few hours before ovulation.

Monthly Endometrial Cycle and Menstruation (p. 1025)

Driven by the cyclic production of ovarian hormones, the endometrium goes through a monthly cycle characterized by three phases: (1) proliferation, (2) development of secretory changes, and (3) menstruation.

The proliferative phase is initiated by secretion of estrogen from the developing follicles. At the beginning of each cycle, most of the endometrium has been lost during menstruation, and only a thin layer of basal endometrial stroma remains. The only remaining epithelial cells are located within the crypts of the endometrium and in the deep portions of the endometrial glands. Estrogen secreted from the developing follicles during the early portion of the cycle stimulates very rapid proliferation of the stromal and epithelial cells. The entire endometrial surface is re-epithelialized within 4 to 7 days of the beginning of menstruation. During the next 10 days, the stimulatory effects of estrogen cause development and thickening of the endometrium of up to 4 millimeters.

The secretory phase results from changes brought about by progesterone. After ovulation, the corpus luteum secretes large amounts of progesterone and estrogen. The effect of the progesterone is to cause swelling and secretory development of the endometrium. The glands secrete fluid, and the endometrial cells accumulate lipids and glycogen in their cytoplasm. The vascularity of the endometrium continues to develop in response to the requirements of the developing tissue. At the peak of the secretory phase, at 1 week after ovulation, the endometrium is approximately 6 millimeters thick.

Menstruation follows within 2 days of the involution of the corpus luteum. Without the stimulation of the estrogen and progesterone secreted by the corpus luteum, the endometrium rapidly involutes, to about 65 per cent of its previous thickness. Then, starting approximately 24 hours before menstruation, the blood vessels supplying the endometrium become vasospastic, resulting in ischemia and finally necrosis of the tissue. Hemorrhagic areas develop in the necrotic tissue, and gradually the outer layers separate from the uterine wall. At about 48 hours after the start of menstruation, all the superficial layers of the endometrium are desquamified. Distention of the uterine cavity, elevated levels of prostaglandin E_2 released from the ischemic and necrotic tissue, and low levels of progesterone contribute to stimulation of uterine contractions, which expel the shed tissue and blood. The menstrual fluid is normally nonclotting due to the presence of *fibrinolysin* released from the endometrial tissue.

REGULATION OF THE FEMALE MONTHLY RHYTHM— INTERPLAY BETWEEN THE OVARIAN AND HYPOTHALAMIC-PITUITARY HORMONES (p. 1026)

At the beginning of each monthly cycle, a new group of primary follicles begins to develop, secreting increasing levels of estrogen in response to the trophic hormones from the pituitary, FSH, and LH.

Estrogen in small amounts strongly inhibits secretion of LH and FSH through a direct pituitary effect, although estrogen also inhibits the hypothalamic secretion of GnRH. Progesterone acts synergistically with estrogen, but it has only a weak inhibitory effect by itself.

As the level of estrogen rises, the rate of secretion of the pituitary hormones begins to fall; however, for unknown reasons, the pituitary gland secretes a large amount of LH immediately before ovulation, when estrogen levels are elevated. This surge of LH at a time when LH secretion "should" be suppressed by the inhibitory influence of estrogen triggers ovulation and the transformation of the granulosa and thecal cells to luteal cells.

After ovulation, the estrogen and progesterone secreted from the corpus luteum again exert an inhibitory effect on the secretion of LH and FSH.

Inhibin also is secreted from the corpus luteum. As in the male, inhibin in the female inhibits secretion of FSH and, to a lesser extent, LH.

Once the levels of LH fall to minimal values, because of the inhibitory influences of the hormones from the corpus luteum, the corpus luteum involutes, and estrogen and progesterone secretion rates decline toward zero. The formation of LH and FSH rises in the absence of inhibition as menstruation begins, initiating the development of a new group of follicles.

Puberty, Menarche, and Menopause

Puberty is the onset of adult sexual life. It is marked by a gradual increase in the secretion of estrogen from developing follicles driven by increasing concentrations of FSH and LH from the pituitary.

Menarche is the onset of menstruation. It marks the completion of the first cycle of the system, although the first several cycles usually do not include ovulation.

Menopause is the period during which the cycles cease and the ovarian hormone levels fall to minimal values. The cessation of the cycling is the result of the presence of an inadequate number of primary follicles in the ovary to respond to the stimulatory effect of FSH. As a result, estrogen-secretory dynamics during the first portion of the cycle are inappropriate to trigger the LH surge, and ovulation does not occur. After several irregular anovulatory cycles, estrogen production declines to near zero. Without inhibition, the rate of secretion of LH and FSH proceeds at very high levels for many years after menopause.

FEMALE SEXUAL ACT (p. 1030)

Both psychic and local sensory stimulation are important for satisfactory performance of the female sexual act. Sexual desire is affected to some extent by estrogen and testosterone levels in the female; consequently, desire may be greatest a few days before ovulation, when estrogen secretion from the follicle is greatest.

Erectile tissue analogous to that in the penis is located around the introitus and extending into the clitoris. Dilation of the arteries leading into the tissue is mediated by parasympathetic nerves that release nitric oxide from their nerve endings on the vascular smooth muscle of the arteries. Parasympathetic stimulation also causes secretion of mucus from Bartholin's gland, which is located beneath the labia minora.

With appropriate local sensory and psychic stimulation, reflexes are initiated that cause the female orgasm.

FEMALE FERTILITY (p. 1030)

Female fertility depends on properly timed ovulation, ability of sperm to reach the ovum in the fallopian tube within 24 hours of ovulation, and ability of the zygote to implant and survive in the endometrium. Several problems that can make a woman infertile can be categorized as follows:

- *Failure to ovulate* can result from
  1. Mechanical obstruction on the surface of the ovary due to (1) the presence of a thickened capsule, (2) scarring from infection, and (3) overgrowth of the surface by cells of endometrial origin, a condition referred to as *endometriosis*
  2. Absence of an LH surge or other hormonal abnormalities.
- *Obstruction of the fallopian tubes*, often as a result of infection or endometriosis.

82

Pregnancy and Lactation

TRANSPORT, FERTILIZATION, AND IMPLANTATION OF THE DEVELOPING OVUM (p. 1033)

The *primary oocyte* undergoes a meiotic division shortly before ovulation, giving rise to the first polar body, which is expelled from the nucleus. With this division, the oocyte is transformed to a *secondary oocyte* containing 23 unpaired chromosomes. A few hours after a sperm cell enters the oocyte, the nucleus divides again and a second polar body is expelled, forming the *mature ovum*, which still contains 23 unpaired chromosomes.

The ovum enters the fallopian tube (oviduct). At ovulation, the ovum and surrounding layers of granulosa cells, referred to as the *corona radiata*, are expelled from the ovary into the peritoneal cavity at the ostium, or opening, of the fallopian tube. The ciliated epithelium lining the tubes creates a weak current that draws the ovum into the tube.

Fertilization takes place in the fallopian tube. Within 5 to 10 minutes of ejaculation, sperm cells reach the ampullae at the ovarian ends of the fallopian tubes, aided by contractions of the uterus and fallopian tubes. Normally, approximately one-half billion sperm are deposited at the cervix during coitus, but only a few thousand reach the ampullae of the fallopian tubes, where fertilization usually takes place.

Before fertilization can occur, the *corona radiata* must be removed through the successive actions of many sperm cells that release the proteolytic enzymes in the *acrosome* at the head of the sperm cell. Once the way is cleared, one sperm cell can bind to and penetrate the *zona pellucida* surrounding the ovum and enter the ovum. The 23 unpaired chromosomes from the sperm cell rapidly form the *male pronucleus*, which then align themselves with the 23 unpaired chromosomes of the *female pronucleus* to form the 23 pairs of chromosomes of the fertilized ovum or *zygote*.

The zygote is transported in the fallopian tubes. Three to 4 days are required for passage of the zygote through the fallopian tube to the cavity of the uterus. During this time, the survival of the organism is dependent on the secretions of the epithelium of the tube. The first series of cellular divisions take place while the ovum is in the fallopian tube, so that by the time it enters the uterus, the structure is referred to as a *blastocyst*. Shortly after ovulation, the *isthmus* of the fallopian tubes (the last 2 centimeters before the tube enters the uterus) become tonically contracted, blocking movement between the tubes and uterus. The final entry into the uterus does not take place until the smooth muscle at the isthmus relaxes under the influence of rising levels of progesterone from the corpus luteum.

The blastocyst implants in the endometrium. The developing blastocyst remains free in the cavity of the uterus for an additional 3 days before implantation begins. On about the seventh day after ovulation, the *trophoblast cells* on the surface of the blastocyst begin to secrete proteolytic enzymes that digest and liquefy the adjacent endometrium. Within a few days, the blastocyst has invaded the endometrium and is firmly attached to it. The contents of the digested cells, which have large amounts of stored nutrients, are actively transported by the trophoblast cells for use as substrates to enable the rapid growth of the blastocyst.

FUNCTION OF THE PLACENTA (p. 1035)

Development of the Placenta

The trophoblast cells form cords that grow into the endometrium. Blood capillaries grow into the cords from the vascular system of the embryo; by day 16 after fertilization, blood flow begins into the capillaries. Simultaneously, on the maternal side, sinuses develop that are perfused with blood from the uterine vessels, surrounding the trophoblast cords. The cords branch extensively as they continue to grow, forming the *placental villi* into which embryonic capillaries grow. The villi contain capillaries carrying fetal blood, and they are surrounded by sinuses filled with maternal blood. The two blood supplies remain separated by several cell layers, and no mixing occurs of the blood from the mother and fetus.

Blood enters the fetal side of the placenta from two umbilical arteries and returns to the fetus by way of a single umbilical vein. The paired uterine arteries of the mother give rise to branches that supply blood for the maternal sinuses, which are drained by branches of the uterine veins.

Placental Permeability and Transport

Oxygen diffuses from the maternal blood through the placental membranes and into the fetal blood, driven by a pressure gradient. The mean PO_2 for the blood in the maternal sinuses is about 50 mm Hg, whereas in the venous end of the fetal capillaries, the PO_2 averages 30 mm Hg; the 20-mm Hg pressure gradient is the driving force for the diffusion of oxygen from the maternal to the fetal blood.

Several factors assist in the diffusion of oxygen from the mother to the fetus.

- The fetal hemoglobin has a greater affinity for oxygen than adult hemoglobin. At the partial pressures of O_2 present in the placenta, fetal hemoglobin can carry 20 to 50 per cent more oxygen than maternal hemoglobin.
- The concentration of hemoglobin in the fetal blood is 50 per cent greater than that in the maternal blood.
- The *Bohr effect* operates in favor of transfer of oxygen from the maternal blood to that of the fetus. The Bohr effect refers to the action of an increase in PCO_2 to decrease the affinity of hemoglobin for O_2. Fetal blood entering the placenta has a high PCO_2, but it rapidly diffuses into the maternal blood because of a favorable pressure gradient. As a result, the PCO_2 in the fetal blood decreases while that of the maternal blood increases, causing the affinity of the fetal hemoglobin for oxygen to increase and the affinity of the maternal hemoglobin to decrease.

Carbon dioxide diffuses readily through the membranes of the placenta. Even though the pressure gradient driving the diffusion averages only about 2 to 3 mm Hg, the CO_2 molecule is extremely soluble in biological membranes and can move easily across the layers of the placenta.

Movement of metabolic substrates such as glucose and fatty acids across the placenta occurs by the same mechanisms that operate in other parts of the body. Glucose diffusion is aided by a facilitated diffusion processes, and fatty acids cross the membranes by simple diffusion. Electrolytes such as sodium and potassium move by both diffusion and active transport.

Diffusion gradients favor movement of nonprotein nitrogenous compounds, such as urea, uric acid, and creatinine, from the fetal to the maternal blood.

HORMONAL FACTORS IN PREGNANCY (p. 1037)

Human Chorionic Gonadotropin—Persistence of the Corpus Luteum and Prevention of Menstruation

Human chorionic gonadotropin (hCG) is a glycoprotein hormone produced by the trophoblast cells beginning 8 to 9 days after fertilization. It reaches the maternal blood and binds to luteinizing hormone (LH) receptors in the cells of the corpus luteum. At about this time, LH levels begin to decline; if fertilization does not occur, the corpus luteum involutes and menstruation begins within a few days. The hCG effect on the corpus luteum is the same as that of LH: the hCG maintains the function of the corpus luteum and continues to stimulate its secretion of large amounts of progesterone and estrogen, so that the endometrium will continue in a viable state that will support the early development of the embryo. As a result of the hCG secretion, menstruation does not occur.

In addition, hCG binds to LH receptors in the Leydig cells of the testes of male embryos; this stimulates testosterone secretion, which is essential to the differentiation of the male sex organs.

Trophoblast Cells—Estrogen and Progesterone

Late in pregnancy, estrogen secretory rate is approximately 30 times the normal rate. The high concentrations of estrogens cause the following:

- Enlargement of the mother's uterus
- Enlargement of the mother's breasts, with growth of the ductile structure
- Enlargement of the mother's external genitalia.

Progesterone is also necessary for pregnancy. The rate of secretion reaches 10 times the maximum level present during nonpregnant cycle. Its functions include the following:

- Promotion of storage of nutrients in the endometrial cells, transforming them into *decidual cells*
- Reduction of contractility of the uterine smooth muscle, *preventing contractions*
- Promotion of secretion of nutrient-rich fluids from the epithelium of the fallopian tubes that *sustain the zygote* before implantation
- Promotion of development of the *alveoli of the breasts.*

Human Chorionic Somatomammotropin

This is a third placental hormone, and it is secreted by the placenta starting in the fifth week of pregnancy. The specific function of the hormone remains unknown, although it does have metabolic effects similar to those of growth hormone. It reduces insulin sensitivity of tissues and decreases glucose utilization. Human chorionic somatomammotropin also promotes releases of fatty acids from fat stores.

PARTURITION (p. 1041)

Increased Uterine Excitability Near Term

Parturition is the process by which the baby is born. Toward the end of pregnancy, the uterus becomes progressively more excitable until it begins strong rhythmical contractions that expel the baby. Changes in hormonal levels and mechanical properties of the uterus and its contents contribute to the increase in uterine contractility.

Hormones increase uterine contractility. Beginning in the seventh month of pregnancy, the rate of progesterone secretion remains constant, whereas the rate of estrogen secretion continues to rise. Although progesterone reduces the contractility of the uterine smooth muscle, estrogen has the opposite effect. *Because the ratio of estrogen increase to progesterone increase during the final weeks of pregnancy, the excitability of the organ increases.*

Oxytocin, which is secreted from the posterior pituitary, can cause uterine contractions. During the final weeks of pregnancy, the oxytocin receptors increase on the cells of the uterine smooth muscle. This will increase the intensity of response for a given concentration of hormone. At the time of labor, the concentration of oxytocin is elevated considerably above normal. There is reason to believe that oxytocin contributes to the mechanism of parturition.

Stretch of the uterus and cervix increases uterine contractility. Stretch of smooth muscle increases its excitability. The size of the fetus near the end of pregnancy provides continual distention of the uterus, and the vigorous movements of the maturing fetus provide intermittent stretch of portions of the smooth muscle wall of the organ. The cervix becomes greatly distended as the end of pregnancy approaches. Contractions initiated by stretch of this part of the uterus can spread upward through the body of the uterus. In addition, stretch and distention of the cervix elicit reflexes that cause releases of oxytocin from the posterior pituitary gland.

Onset of Labor—A Positive Feedback Theory for Its Initiation

Beginning in the sixth month of pregnancy, the uterus undergoes periodic slow rhythmical contractions called *Braxton-Hicks contractions*. As the duration of pregnancy increases, the frequency and intensity of these contractions increase. At some point, a contraction occurs that is sufficiently powerful, and the uterine muscle is sufficiently excitable that the effect of the contraction elevates the level of excitability still more, so that after several minutes, another contraction is initiated. If the second contraction is more powerful than the first, an even greater elevation of excitability results, followed by an even more powerful contraction. Such a *positive feedback cycle* appears to operate during parturition. The cycles continue to intensify the strength of contractions until final delivery occurs.

LACTATION (p. 1044)

High levels of estrogen and progesterone during the later months of pregnancy promote the final developmental changes in the breasts that prepare them for lactation. These hormones do not stimulate milk production by the alveolar cells. Milk formation is achieved through the effects of prolactin, an anterior pituitary hormone that is secreted in rising concentrations throughout pregnancy. The stimulatory effect of prolactin is blocked by the high concentrations of estrogen and progesterone secreted by the placenta, so that no milk is formed until after delivery of the baby. When the levels of estrogen and progesterone fall, the stimulatory effect of prolactin causes the cells of the alveoli to synthesize milk, which accumulates in the alveoli and ducts of the breast.

The mechanical stimulation associated with suckling elicits a reflex to the hypothalamus, releasing oxytocin from the posterior pituitary gland. Oxytocin travels to the breast in the blood and causes contraction of the *myoepithelial cells* that surround the ducts of the breast. The contraction increases the pressure of the milk filling the ducts, causing milk to flow from the nipple to the baby. Milk is not usually ejected from the breast until the baby suckles the nipple.

After delivery, prolactin levels tend to fall toward nonpregnant levels. Stimulation of the nipples associated with suckling, however, increases the release of prolactin, which in turn stimulates milk production. The greater the duration of suckling, the greater is the response of prolactin and the greater is the amount of milk produced by the breast. This feedback control system regulated by the baby's desire for milk and duration of suckling provides for a well-regulated supply of milk for the baby from the time it is born until as long as 1 year or more after birth, when its requirements for milk have increased greatly. When the baby discontinues breast-feeding, the signal for prolactin secretion stops, and milk production declines rapidly.

Prolactin is regulated by hypothalamic release of *prolactin-inhibitory factor (PIF)*, which is believed to be *dopamine*. Elevated dopamine release from the hypothalamus inhibits prolactin secretion from the pituitary gland.

During the period of breast-feeding, the mother's ovarian cycle is interrupted, so that ovulation and menstruation do not occur for several months after delivery. The precise cause for this effect is not known.

Human milk is composed of 88.5 per cent water, 3.3 per cent fat, 6.8 per cent lactose, 0.9 per cent casein, and other proteins and minerals. When a woman is lactating heavily to supply the needs of a rapidly growing, large baby, she may secrete 2 to 3 grams per day of calcium phosphate into the milk. This can lead to depletion of calcium from the bones if the mother does not carefully choose a diet that is rich in calcium.

83

Fetal and Neonatal Physiology

GROWTH AND FUNCTIONAL DEVELOPMENT OF THE FETUS (p. 1047)

Circulatory System. The heart begins to beat during the fourth week after fertilization, which is about the same time that the first non-nucleated red blood cells form. During the first two thirds of gestation, red blood cells are formed outside the bone marrow; only during the final 3 months of gestation are most of the red blood cells formed in the bone marrow.

Respiratory System. Although some respiratory movements take place during the first and second trimesters, respiratory movements are inhibited during the final 3 months of gestation. This inhibition prevents filling of the lungs with debris from the amniotic fluid.

Nervous System. The organization of the central nervous system is completed in the first months of gestation, but full development and even complete myelination do not take place until after delivery.

Gastrointestinal Tract. By midpregnancy, the fetus ingests amniotic fluid and excretes *meconium* from the gastrointestinal tract. Meconium is composed of residue from amniotic fluid and waste products and debris from the epithelium of the gastrointestinal tract. By the final 2 to 3 months of gestation, gastrointestinal tract function approaches maturity.

Kidneys. The fetal kidneys can form urine beginning in the second trimester, and urination takes place during the latter half of gestation. The ability of the kidneys to accurately regulate the composition of the extracellular fluid is poorly developed until several months after birth.

ADJUSTMENTS OF THE INFANT TO EXTRAUTERINE LIFE
(p. 1049)

Onset of Breathing. Normally, a baby begins to breathe within seconds of delivery. The stimuli for the sudden activation of the respiratory system probably include hypoxia incurred during delivery and the sudden cooling of the face on exposure to the air. Within 1 minute of delivery, a normal pattern of breathing develops. In some cases, the onset of breathing may be delayed; newborn infants can tolerate 8 to 10 minutes without breathing before permanent damage occurs; in adults, death or severe damage will take place if breathing is interrupted for 4 to 5 minutes.

Expansion of the Lungs at Birth. The surface tension of the fluid-filled lungs at birth keeps the alveoli in a collapsed state. Approximately 25 mm Hg of negative inspiratory pressure is required to overcome the surface tension. At birth, the first inspirations are powerful and generate as much as 60 mm Hg negative intrapleural pressure.

Circulatory Readjustments at Birth

Two primary changes occur in the fetal circulation at birth:

- A *doubling of systemic vascular resistance* due to loss of the placenta, which has very low vascular resistance—This increases aortic pressure and left ventricular and left atrial pressures.
- A *fivefold decrease in pulmonary vascular resistance* due to expansion of the lungs following the first inspiration—As a result, pulmonary arterial, right ventricular, and right atrial pressures decrease.

After these primary changes, several other alterations follow:

- The *foramen ovale, which is located between the right and left atria, closes* due to the pressure in the left side being greater than the pressure in the right.
- The *ductus arteriosus between the pulmonary artery and descending aorta closes.*
- The *ductus venosus closes.* In fetal life, it carries blood from the umbilical vein and the fetal portal bed, bypassing the fetal liver, directly to the inferior vena cava.

With these adjustments, the fetal circulation is transformed within a matter of hours to the neonatal configuration.

SPECIAL PROBLEMS IN THE NEONATE (p. 1051)

In the newborn, most of the cardiovascular, hormonal, and neural control systems are poorly developed and are often unstable.

Respiratory System. Because of the relatively small residual capacity (less than one half the volume per kilogram of body weight than that of adults), relatively high metabolic rate of the newborn, and immaturity of the neural components of the respiratory control system, blood gas values fluctuate widely during the first weeks of life.

Circulation. *Blood volume* at birth is normally about 300 milliliters. If the baby is left attached to the placenta for a few minutes after birth, approximately 75 milliliters of additional blood can enter the baby's circulatory system, which is equivalent to a transfusion of 25 per cent of the blood volume. This overload could contribute to an elevation of left atrial pressure and a tendency to develop pulmonary edema.

Liver Function. *Bilirubin* formed from the breakdown of hemoglobin from red blood cells is normally excreted by the liver into the bile, conjugated with glucuronic acid; however, the neonatal liver has an inadequate ability to conjugate bilirubin at the rate it is formed. As a result, blood concentration of bilirubin rises for the first 3 days after birth and then returns to normal as the capability of the liver improves. This condition is referred to as *physiological hyperbilirubinemia* and can be seen in some cases as a slight jaundice or yellowish tint in the skin and sclera of the eyes.

In addition to the potential problems associated with bilirubin conjugation, the limited capability of the liver during the first few days of life can lead to difficulty in synthesizing adequate quantities of protein for maintaining colloid osmotic pressure, adequate amounts of glucose, and necessary amounts of factors required for coagulation. These potential limitations of hepatic function improve rapidly during the first weeks of postnatal life.

Fluid Balance and Renal Function. On a per-kilogram of body weight basis, the neonate takes in seven times as much fluid as an adult. In addition, the metabolic rate per kilogram of body weight of the newborn is twice as great as that of the adult. These and other factors can contribute to problems in the newborn in regulation of fluid balance, electrolyte concentrations, pH, and colloid osmotic pressure.

Digestion and Metabolism. The gastrointestinal absorptive capacity and hepatic digestive function of neonates are limited to some extent in the following ways:

- *Absorption of starches* is limited by a deficient rate of secretion of pancreatic amylase, which breaks down complex carbohydrates such as starches.
- *Absorption of fat* is not as great in neonates as it is in older children.
- *Gluconeogenic capacity* of the liver is not sufficient in many newborns to maintain blood glucose concentration in the normal range for long periods after feeding. It is important to maintain the newborn on a schedule of frequent feedings.

All of these gastrointestinal limitations are exacerbated in preterm infants. The limited capacities for absorption of starches and fats are worsened by feeding cow's milk–based formulas to preterm and newborn infants. The carbohydrates and fats in human milk are digested and absorbed more readily than those in nonhuman milk and formula preparations.

The *basal metabolic rate* of the newborn is twice as great per kilogram of body weight as that of an adult, and the surface area–to–body rate ratio is much greater in the neonate than in the adult. As a result, body temperature control is relatively unstable, especially in preterm infants.

Sports Physiology

84

Sports Physiology

The stress on the body caused by heavy exercise greatly exceeds other normal, day-to-day stresses. For example, metabolic rate increases about 100 per cent in a person with a high fever, but the metabolism of a marathon runner may increase to 2000 per cent of normal during a race.

THE FEMALE AND MALE ATHLETE (p. 1059)

Muscle strength, pulmonary ventilation, and cardiac output, all of which depend on muscle mass, in females are two thirds to three fourths of the values in males. If measured in terms of strength per square centimeter of muscle cross-sectional area, however, a female can achieve the same maximum force of contraction as men: 3 to 4 kg/cm^2. Much of the difference in athletic performance of males and females is due to the smaller amount of muscle mass in females.

The increased amount of muscle mass in males is primarily caused by *testosterone*, which has strong *anabolic effects* on protein deposition, especially in the muscles. Even a nonathletic male may have 40 per cent more muscle mass than his female counterpart. In comparison, *estrogen* in females causes increased fat deposition in the breasts and subcutaneous tissue. The nonathletic female may have about 27 per cent body fat in contrast to 15 per cent body fat in a nonathletic man. In addition, testosterone promotes some aggressiveness, which may play a role in some athletic events.

MUSCLES IN EXERCISE (p. 1060)

The *contractile strength* of a muscle is determined mainly by its size. A person with large muscles is generally stronger than one with small muscles. A quadriceps muscle with a cross-sectional area of up to 150 cm^2 has a maximum contractile strength of

525 kilograms (or 1155 pounds). When an athlete is using the quadriceps muscles for lifting, a tremendous amount of stress is applied to the patellar tendon. This or any other highly strenuous activity places much stress on joints, tendons, muscles, and ligaments. The *holding strength* of a muscle is approximately 40 per cent greater than the maximal contractile strength and is the force required to stretch out a muscle after it has contracted.

The *power* **of a muscle is the amount of work per unit time that can be performed**. The power is determined not only by muscle strength but also by the distance it contracts and number of times it contracts each minute; this is usually measured in kg-m/min. The following table shows that muscle power is very high during the early phases of exercise and then decreases.

A large power surge occurs in a race such as a 100-meter dash (Table 84–1), but in a longer-distance race, much lower power levels are available, about one fourth as much. The velocity achieved in a 100-meter dash, however, is only about 1.75 times as great as that achieved in a 10,000-meter run.

Endurance **depends on maintaining a nutrition supply for the muscle**. As seen in Table 84–2, a person on a high-carbohydrate diet stores more glycogen in the muscles, which increases his or her endurance in races at marathon speeds. This is why marathon runners eat a large amount of carbohydrates, such as pasta, on the day before the race.

Muscle Metabolic Systems in Exercise

The basic sources of energy for muscle contraction are the following:

- *Phosphagen system*, which consists of adenosine triphosphate (ATP) and phosphocreatine

TABLE 84–1 MUSCLE POWER DURING EXERCISE

Time	Muscle Power (kg-m/min)
First 8–10 sec	7000
Next 1 min	4000
Next 30 min	1700

TABLE 84-2 EFFECTS OF GLYCOGEN STORAGE ON EXERCISE
ENDURANCE

Diet	Glycogen Stored in Muscle (gm/kg of muscle)	Endurance Time at Marathon Speed (min)
High-carbohydrate	40	240
Mixed	20	120
High-fat	6	85

- *Glycogen–lactic acid system*
- *Aerobic system.*

Adenosine triphosphate is the basic source of energy for muscle contraction. ATP, which consists of adenosine with three high-energy phosphate bonds attached, supplies the short-term energy needs of the muscle fibers. ATP is converted to adenosine diphosphate (ADP) by the removal of one high-energy phosphate radical; this releases 7300 calories per mole of ATP. This energy is used for muscle contraction as ATP combines with the myosin filaments. The removal of another phosphate radical converts ADP to adenosine monophosphate and supplies an additional 7300 calories per mole of ADP.

The amount of ATP present in muscle sustains maximal muscle contraction for only 3 minutes, but the phosphocreatine system is another energy source. The combination of the cellular ATP and phosphocreatine system is called the *phosphagen energy system.*

Phosphocreatine (or creatine phosphate) is the combination of creatine and a phosphate radical connected with a high-energy phosphate bond, which, when broken, provides 10,300 calories per mole. Adding to the importance of this system is the fact that muscle cells have twofold to fourfold more phosphocreatine than ATP. The phosphocreatine system is an important source of energy for the muscle cell.

Phosphocreatine reversibly combines with ADP to form ATP and creatine in the cell. This phosphagen energy system by itself, however, supplies only enough energy for 8 to 10 seconds of maximal muscle contraction, or nearly enough energy for a 100-meter race.

The glycogen–lactic acid system supplies energy through anaerobic metabolism. The glycogen stored in muscle rapidly splits into glucose molecules that can be used for energy. The initial stage of this process is called *glycolysis*; this occurs without the use of oxygen and is referred to as *anaerobic metabolism*. The glycogen in this process is mostly converted to lactic acid and supplies four ATP molecules for each molecule of glucose. An advantage of this glycogen-lactic acid system is that it forms ATP 2.5 times as fast as oxidative metabolism in the mitochondria. This system supplies enough energy for maximal muscle contraction for 1.3 to 1.6 minutes.

For longer periods of muscle use, energy for muscle contraction must be supplied through the anaerobic system. In this system, glucose, fatty acids, and amino acids are oxidized in the mitochondria to form ATP.

Recovery of energy systems after exercise requires oxygen. After exercise is completed, the energy sources of muscle must be reconstituted. Any lactic acid that is formed during exercise is converted to pyruvic acid and then metabolized oxidatively or reconverted into glucose (mainly in the liver). The extra liver glucose forms glycogen, which replenishes the glycogen stores in the muscles.

The anaerobic system is also replenished after exercise by two means:

- *The increased respiration that occurs after exercise replenishes the oxygen debt.* The oxygen debt is the deficit in the oxygen stored in the body as air in the lungs, dissolved in body fluids, and combined with hemoglobin and myoglobin.
- *The glycogen is replaced in the muscle.* This process can take days to complete after extreme long-lasting exercise, with the recovery time highly dependent on the type of the diet of the person. A person on a high-carbohydrate diet replenishes the muscle glycogen stores much faster than one on either a mixed diet or a high-fat diet.

Resistive Training Significantly Enhances Muscle Strength

If the muscles are exercised under no load, even for hours, little increase in strength will occur. If muscles are contracted with at least a 50 per cent maximal force for a few times each day three times a week, an optimal increase in muscle strength occurs, and muscle mass increases through a process called *muscle hypertrophy*. Most of the hypertrophy is caused by an increase in the size of the muscle fibers, but the number of fibers increases moderately. Other changes occur in the muscle during training, including the following:

- Increase in number of myofibrils
- Up to 120 per cent increase in mitochondrial enzymes
- A 60 to 80 per cent increase in the components of the phosphagen energy system
- A 50 per cent increase in stored glycogen
- A 75 to 100 per cent increase in stored triglycerides.

Fast- and Slow-Twitch Muscle Fibers and Different Types of Exercise

Fast-twitch muscle fibers give a person the ability to rapidly and forcefully contract their muscles. *Slow-twitch fibers* are used for prolonged lower leg muscle activity. The differences between fast-twitch and slow-twitch fibers include the following:

- Fast-twitch fibers are about twice as large in diameter.
- Enzymes that release energy from the phosphagen and glycogen–lactic acid energy systems are twofold to threefold as active in the fast-twitch fibers.
- Slow-twitch fibers are used more for endurance exercise while employing the anaerobic system of energy, and the number of mitochondria are greater than in fast-twitch fibers.
- Slow-twitch fibers contain more myoglobin, which is a hemoglobin-like substance that combines with oxygen in muscle.
- Capillary density in slow-twitch fibers also exceeds that of fast-twitch fibers.

Fast-twitch fibers generate a great amount of power in a short period of time, such as during a sprint. In contrast, slow-twitch fibers are used in endurance exercises, such as marathons.

RESPIRATION IN EXERCISE (p. 1065)

Maximum oxygen consumption increases during athletic training. The oxygen consumption of the average untrained male is 3600 ml/min; this increases to 4000 ml/min in the athletically trained male. The male marathon runner has an even higher oxygen consumption of 5100 ml/min, which is provided during exercise by an increase in pulmonary ventilation. The maximum oxygen consumption increases during training, but the high values in marathon runners may be partly genetically determined by factors such as large lung capacity in relation to body size and strength of respiratory muscles.

At maximal exercise, pulmonary ventilation is 100 to 110 liters/min, but maximum breathing capacity exceeds this by 50 per cent. The lungs have a built-in safety mechanism that can be helpful if exercise is attempted at (1) a high altitude, (2) under hot conditions, or (3) with some abnormality in the respiratory system.

Pulmonary oxygen-diffusing capacity increases in athletes. The oxygen-diffusing capacity is the rate at which oxygen diffuses from the alveoli into the blood per mm Hg oxygen pressure. During exercise, diffusing capacity increases in a nonathlete from a resting value of 23 ml/min/mm Hg to 48 ml/min/mm Hg. A trained oarsman at maximum exercise can have a diffusing capacity of 80 ml/min/mm Hg. Diffusing capacity increases in exercise mainly because of an opening of underperfused pulmonary capillaries; this provides a greater surface area for diffusion of oxygen.

CARDIOVASCULAR SYSTEM IN EXERCISE (p. 1066)

As discussed in Chapter 20, the blood flow through muscle increases up to 25 times that of normal during exercise. Most of the muscle blood flow occurs between contractions because the blood vessels are compressed during the contractile process. An increase in arterial pressure during exercise directly raises flow. The stretching of the arteriolar walls by the increase in pressure decreases vascular resistance and raises the flow much more.

Athletic training increases stroke volume and decreases heart rate. If a normal person starts extensive athletic training of the aerobic type, both the heart size and maximum cardiac output increase. The *stroke volume* thus increases, and the resting heart rate decreases. Table 84–3 shows the results of training. Note that stroke volume increases only 50 per cent during maximum exercise in the mara-

TABLE 84–3 COMPARISON OF CARDIAC OUTPUT BETWEEN MARATHONERS AND NONATHLETES

	Stroke Volume (ml)	Heart Rate (beats/min)
Resting		
Nonathlete	75	75
Marathoner	105	50
Maximum		
Nonathlete	110	195
Marathoner	162	185

thoner, and the heart rate increases 270 per cent. Cardiac output can be calculated from the data in this table with the following formula:

$$\text{Cardiac Output} = \text{Stroke Volume} \times \text{Heart Rate}$$

The increase in heart rate provides a much greater proportion of the increase in cardiac output in the marathoner than does the increase in stroke volume.

The heart limits the amount of exercise one can perform. During maximum exercise, both stroke volume and heart rate are at 95 per cent of their maximum values; with the cardiac output formula, one can calculate that cardiac output is at 90 per cent of its maximum value. In contrast, the pulmonary ventilation is only 65 per cent of its maximum. The cardiovascular system usually limits the amount of exercise that can be performed.

During any type of cardiac disease, the maximum cardiac output decreases, and this limits the amount of exercise that can be performed. Any type of respiratory disease that severely limits pulmonary ventilation or oxygen-diffusing capacity also limits exercise.

BODY HEAT IN EXERCISE (p. 1068)

The body produces a great amount of heat during exercise, and problems with elimination of this heat from the body can limit exercise. Hot, humid conditions limit heat loss and can lead to *heat stroke*; symptoms include nausea, weakness, headache, profuse sweating, confusion, dizziness, collapse, and unconsciousness. The person is treated by decreasing body temperature as quickly as possible.

Dehydration also occurs in hot, humid conditions during exercise and can lead to nausea, muscle cramps, and other effects. Therapy is provided by replacing the fluid, sodium, and potassium losses.

Index

Note: Page numbers in *italics* refer to illustrations; page numbers followed by t refer to tables.

725

Contents